Dawn of the Age of Mammals in the northern part of the Rocky Mountain Interior, North America

Edited by

Thomas M. Bown
Paleontology and Stratigraphy Branch
U.S. Geological Survey
Box 25046, Denver Federal Center
Denver, Colorado 80225

Kenneth D. Rose
Department of Cell Biology and Anatomy
Johns Hopkins University School of Medicine
Baltimore, Maryland 21205

SPECIAL PAPER

243

1990

Published by The Geological Society of America, Inc.
3300 Penrose Place, P.O. Box 9140, Boulder, Colorado 80301

GSA Books Science Editor Richard A. Hoppin

Printed in U.S.A.

Library of Congress Cataloging-in-Publication Data

Dawn of the age of mammals in the northern part of the Rocky Mountain
interior, North America / edited by Thomas M. Bown, Kenneth D. Rose.
p. cm. — (Special paper / Geological Society of America ;
243)
Includes bibliographical references.
ISBN 0-8137-2243-8
1. Mammals, Fossil—Rocky Mountains. 2. Paleontology—West (U.S.)
3. Paleontology—Canada, Western. I. Bown, Thomas M. II. Rose,
Kenneth David, 1949- . III. Series: Special papers (Geological
Society of America) ; 243.
QE881.D275 1990
569'.0978—dc20 89-29558
CIP

Cover illustration: Life restoration of *Chriacus* sp., based on USGS 2353 and 15404. Drawn by Elaine Kasmer, under supervision of K. Rose.

Contents

Preface

The last decade has witnessed the expansion of field work and the recovery of many important new fossil vertebrates from a number of areas in the Rocky Mountain Interior. This has resulted in (among other things) discovery of many previously unknown taxa and of much better preserved specimens of known forms, significant geographic and temporal range extensions, the development of greater temporal resolution and sampling density for many taxa (leading to much more precise biochronology), and the creation and applicaton of models for examining faunal relations and evolutionary patterns. When the opportunity for this symposium arose, we were faced with the dilemma of how to incorporate as much of this new information as possible, yet limit the resulting monograph to manageable proportions. We decided to confine the subject matter geographically and temporally, to a region and a time span with which we are most familiar and within which much of the exciting new research is now being conducted—the latest Cretaceous and early Tertiary of the northern Rocky Mountain region. This volume is derived from papers presented at that symposium on May 2, 1987, at the annual regional meeting of the Rocky Mountain Section of the Geological Society of America, held at the University of Colorado in Boulder.

The contributions are diverse, involving many aspects of vertebrate paleontology: systematics, origin, evolution, and extinction of taxa; functional morphology; paleogeography and biogeography; biostratigraphy; taphonomy; paleoecology; and site studies. Chapter organization is temporal, beginning with latest Cretaceous and ending with Oligocene—the interval encompassing most of what has come to be regarded as the "Dawn of the Age of Mammals" in North America. Although some authors have deviated somewhat from their original platform presentations and abstracts presented at the 1987 symposium, the editors believe that the present volume even more forcefully reflects the intentions of that meeting. Within necessary temporal, geographic, and space constraints, our profession is amply represented here in all its diversity, creativity, and vitality.

The first three chapters and a part of the fourth document the biostratigraphy, origins, and paleogeography of mammals of the latest Cretaceous and earliest Tertiary. The volume opens with an innovative and masterfully written study by **Jay Lillegraven** and **Lawrence Ostresh, Jr.**, which exemplifes Lillegraven's long-standing fascination with Cretaceous and Tertiary paleobiogeography. The authors document the migration of Late Cretaceous shorelines along the western edge of the North American epicontinental sea in relation to localities that yield fossil mammals, and conclude that a slow, regional regression took place over the last 20 m.y. of the Cretaceous. Following this they propose a correlation of mammal-bearing sites and North American Land-Mammal "Ages" with ammonite-based marine zonation.

In the second chapter, **Dave Archibald** and **Don Lofgren** detail mammalian faunal composition and biostratigraphy near the Cretaceous-Tertiary boundary. They offer a novel mammalian biostratigraphy for the systematic contact, defining a new boundary interval zone—Pu_0, represented by the Bug Creek faunas—which contains dinosaur remains in association with mammals of both Lancian and earliest Puercan (sensu lato) aspect. Judging from some palynological evidence, the K/T iridium anomaly may lie within the Pu_0 Interval Zone.

Dick Fox (Chapter 3) describes the fossil mammal succession during the first 8 to 10 m.y. of the Tertiary in western Canada, based on fossils from 41 localities in Alberta and Saskatchewan, largely amassed under his supervision during the last 15 years. This new evidence leads Fox to conclude that mammals of Paleocene aspect began to evolve prior to the end of the Cretaceous, and that the late Paleocene decline in species diversity observed in middle Tiffanian faunas of the United States may have resulted from local factors or sampling error rather than a hypothesized global cooling.

An introduction to the late Paleocene and early Eocene is afforded by **Dave Krause** and **Mary Maas** (Chapter 4), who offer a detailed review of the evidence on the biogeographic sources of mammalian first appearances in the Western Interior at that time (as well as important new data on Tiffanian faunas). Their fascinating exposition takes us on a nearly worldwide excursion into the dark continents of the early Tertiary (and back again), tracing archaic groups and the antecedents of modern mammals to their probable geographic and temporal origins. Kraus and Maas challenge the hypothesis that first appearances of mammalian taxa in the early and late Tiffanian represent immigrations from Central or South America. They conclude that taxa appearing in the early Clarkforkian originated in Asia, and that the first appearance of several mammalian higher taxa at the beginning of the Wasatchian was essentially synchronous in Eurasia and North America, the likely site of origin being Africa or India.

An overview of the postcranial anatomy and locomotion of early Eocene mammals from the lower Eocene Willwood Formation of the Bighorn Basin, Wyoming, is presented by **Ken Rose** in Chapter 5. His report is based on an extensive sample of nearly complete and partial skeletons of about 30 genera in 12 orders, almost all collected over the last decade. They include the first evidence of the postcranial skeleton for several taxa. The new specimens offer a unique glimpse of the varied locomotor capabilities of many putatively primitive "archaic" mammals, as well as of the earliest known representatives of certain living orders, and reveal some surprisingly close convergences to extant mammals.

In Chapter 6, **Tom Bown** and **Chris Beard** examine empirically documented lateral genetic relation of paleosols and sedimentary facies in order to explore small-scale distributions of related sympatric groups of early Eocene mammals. Using a data base of more than 16,000 specimens from the Willwood Formation, they have been able to document that the relative proportions of certain sympatric taxa and of relatively "complete" versus incomplete skeletal remains are directly related to paleosol maturity and lateral position relative to ancient stream systems.

Catherine Badgley uses statistical analysis to investigate the role of sampling artifacts in the faunal composition and mammalian biostratigraphy of the lower part of the Willwood Formation (Chapter 7). Her results suggest that some apparent extinctions, previously interpreted as a faunal turnover event (Biohorizon A), may instead be explained as an artifact of small sample size.

Richard Stucky, Leonard Krishtalka, and **Andrew Redline** describe the geology, paleoecology, and vertebrate fauna of the latest Wasatchian Buck Spring Quarries (Wind River Basin) in Chapter 8. These recently discovered sites, representing ancient swampy or ponded areas, have yielded a particularly rich vertebrate fauna of more than 100 species, including some of the best preserved skulls and skeletal associations of this age from anywhere. Dominated by small mammals, many of which were arboreally adapted, the fauna has characteristics typical of recent tropical assemblages.

In Chapter 9, **Bob Emry** describes a middle Eocene fauna from Nevada, the first Paleogene vertebrate assemblage known from the Great Basin. The Elderberry Canyon Local Fauna contains about 40 species of early Bridgerian vertebrates (several of them new) and, like the Buck Spring fauna, appears to represent a marsh community.

The volume concludes with **Malcolm McKenna**'s description of two new genera and species of plagiomenids of middle Eocene and early Oligocene age (Chapter 10). McKenna transfers the supposed Arikareean primate *Ekgmowechashala* to the Plagiomenidae (a family usually included in the Dermoptera), and assigns all three to the late-surviving plagiomenid subfamily Ekgmowechashalinae. With this shift, the primate fossil record in North America is abridged by about 5 m.y., ending in the Chadronian.

Thomas M. Bown
Kenneth D. Rose

Acknowledgments

Completion of this volume would not have been possible without the assistance of many colleagues. In addition to those cited by the authors of each chapter, the editors would like to acknowledge the efforts of those who reviewed chapter manuscripts for the Geological Society of America: W. A. Cobban (USGS), H. H. Covert (University of Colorado), J. L. Franzen (Senckenberg Institute, Frankfurt), M. Godinot (Laboratoire de Paleontologie, Montpellier), R. H. Hunt, Jr. (University of Nebraska), F. A. Jenkins, Jr. (Harvard University), M. J. Kraus (University of Colorado), D. W. Krause (State University of New York, Stony Brook), J. A. Lillegraven (University of Wyoming), M. C. McKenna (American Museum of Natural History), D. E. Russell (Institut de Paleontologie, Paris), L. S. Russell (Royal Ontario Museum), D. M. Schankler (Princeton, New Jersey), P. L. Shipman (Johns Hopkins University), C. Teichert (University of Rochester), W. D. Turnbull (Field Museum of Natural History), R. M. West (Cranbrook Institute of Science), and M. O. Woodburne (University of California, Riverside).

We are especially grateful to Mary J. Kraus and William Bradley of the University of Colorado, who invited us to develop the symposium that resulted in this volume, and to GSA editors Campbell Craddock (University of Wisconsin) and Richard A. Hoppin (University of Iowa) for their valuable advice and editorial efforts during preparation and editing of the manuscripts.

Thomas M. Bown
Kenneth D. Rose

Geological Society of America
Special Paper 243
1990

Late Cretaceous (earliest Campanian/Maastrichtian) evolution of western shorelines of the North American Western Interior Seaway in relation to known mammalian faunas

Jason A. Lillegraven
Departments of Geology/Geophysics and Zoology/Physiology, The University of Wyoming, Laramie, Wyoming 82071-3006
Lawrence M. Ostresh, Jr.
Department of Geography and Recreation, The University of Wyoming, Laramie, Wyoming 82071-3371

ABSTRACT

A series of 33 Late Cretaceous (earliest Campanian through Maastrichtian) paleoshoreline maps was developed to document the migrational evolution of the western edge of the North American Western Interior Seaway. The maps represent a geologic span of roughly 18 million years, and portray the estimated positions of the strandline for each standard Western Interior ammonite zone, beginning with the *Clioscaphites choteauensis* zone and continuing to the end of the Mesozoic. We attempted correlation of all significant mammal-bearing localities known from the Western Interior with the ammonite-based marine zonation. First approximations of correspondence between ammonite zones and North American Land-Mammal "Ages" (NALMAs) include: Lancian (*Sphenodiscus* through *"Triceratops"* zones); "Edmontonian" (a name not yet faunally defined; *Didymoceras cheyennense* through *Baculites clinolobatus* zones); Judithian (the smooth, late form of *Baculites* sp. through *Exiteloceras jenneyi* zones); and Aquilan (*Scaphites hippocrepis* through *Baculites asperiformis* zones). Correlations emphasize use of provincial biostratigraphic terminology designed specifically for use in the Western Interior. On the basis of temporal constraints suggested herein, known mammalian fossils from the upper Fruitland and/or lower Kirtland Formations of New Mexico probably are of "Edmontonian," not Judithian age. Although considerable latitudinally based taxonomic diversification of Judithian mammals is now recognizable across the Western Interior, comparative data are inadequate to defend a similar statement for the remaining Late Cretaceous NALMAs. Quantitative evaluation of geographic patterns of shoreline change suggests occurrence of a general, regional regression of the sea during the entire geologic interval represented in the study. We favor explanation by a slow sea-level depression resulting from topographic evolution of the world's mid-oceanic system of ridges and volcanic plateaus. Local and subregional asynchronous episodes of shoreline transgressions, stillstands, and regressions are superimposed upon the general regressive trend, and probably represent influences of local tectonism, not eustatic changes in sea level. Strandline evolution of the epeiric sea during the last 20 million years of the Cretaceous in the North American Western Interior is inconsistent with: (1) existence of geologically brief (1 to 10 m.y.) global fluctuations in sea level; and (2) the concept that the late Campanian was represented by an unusually high global sea level.

Lillegraven, J. A., and Ostresh, L. M., Jr., 1990, Late Cretaceous (earliest Campanian/Maastrichtian) evolution of western shorelines of the North American Western Interior Seaway in relation to known mammalian faunas, *in* Bown, T. M., and Rose, K. D., eds., Dawn of the Age of Mammals in the northern part of the Rocky Mountain Interior, North America: Boulder, Colorado, Geological Society of America, Special Paper 243.

INTRODUCTION

The history of the Western Interior Seaway is of great interest to paleontologists who study the evolution of North American Late Cretaceous terrestrial vertebrates. Aside from the seaway's presumed influence on paleoclimatic moderation, it must have served as a potent barrier to dispersal between land-based organisms on its opposite shores. Various transgressions and regressions of the western edge of the sea profoundly influenced availability of habitat for terrestrial organisms by altering the areal distribution of dry land across a broad coastal plain of low relief. Also, influences of the Western Interior Seaway led to deposition of an enormous volume of sediments. Studies of sedimentary variations in the Western Interior have had great impact on development of geological theory, especially as applied to interactions of local and regional tectonism with putative eustatic sea-level changes.

We had three principal goals in developing this chapter. The first was to attempt correlation of the various Upper Cretaceous mammal-bearing fossil localities (Fig. 1 and Table 1) with a standard biostratigraphic zonation of the epicontinental marine sequence. Surprisingly little reciprocal attention has been paid by paleontologists who have terrestrial- versus marine-based interests to correlations with the others' zonations. As a result, quite independent time scales have been developed, and it has been difficult to integrate the Late Cretaceous geological/biological histories of the two realms. We approached the general problem of marine/nonmarine correlations by way of detailed review of stratigraphic interdigitation between terrestrial and shallow, epicontinental marine, paleoenvironmental settings. Paleoshoreline maps were a product of that review.

It became clear as the work progressed that a second goal could be a natural partner to the primary mission of paleontological correlation. Specifically, we wanted to use our synthesized stratigraphic data and resulting maps to test the widely accepted concept that the brief but geographically widespread Late Cretaceous marine transgressive/regressive sequences as seen in the North American Western Interior reflect globally synchronous variations in ancient sea level (Fig. 2). Although our study is restricted to the western shoreline of the Western Interior Seaway, the results appear adequate to provide new considerations applicable to continental, even global, issues of geologic history and theory.

Finally, we wanted to develop new techniques in computer cartography that could be applied to a wide variety of problems in paleogeography.

SCOPE AND METHODS OF STUDY

Our investigation was limited to the latter part of the Late Cretaceous. Using the marine, ammonite-based zonation of Obradovich and Cobban (1975), our story begins with the earliest Campanian (sensu Frerichs, 1980) *Clioscaphites choteauensis* zone and ends with the *"Triceratops"* zone of the very late Mesozoic. Radioisotopically, the sequence involves the interval of

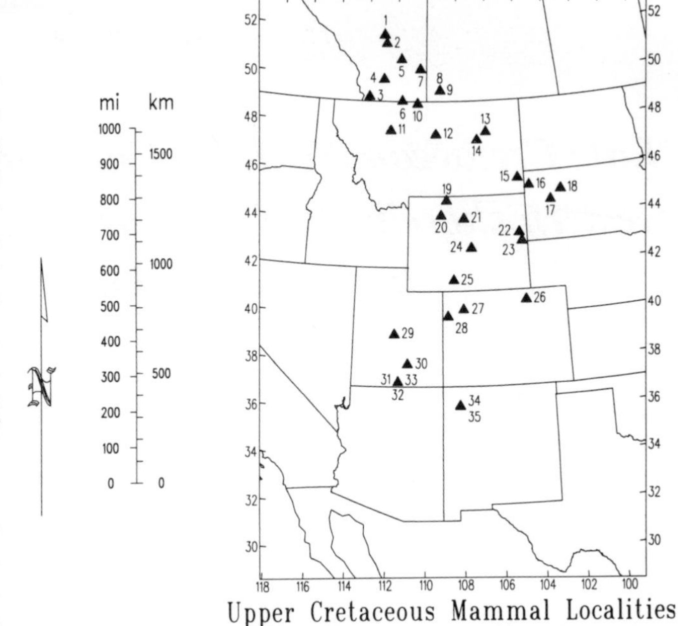

Figure 1. Lambert equal-area reference map to significant Late Cretaceous (earliest Campanian through Maastrichtian) mammal-bearing localities of the Western Interior. Identifying numbers refer to Table 1 and Figure 5. Note the great geographic biases in the Late Cretaceous mammalian record of the Western Interior.

roughly 84 to 66 Ma (with the older age interpolated from Table 1 of Obradovich and Cobban, 1975, using revised constants of Dalrymple, 1979). Geographically, geologic data were summarized from 167 specific localities (Fig. 3 and Appendix 1). The localities extend along deposits of the western shoreline of the ancient seaway from about 55°N latitude (of modern coordinates) in the Slave Lake area of Alberta south to the Big Bend area of Texas, at about 29°N latitude. Stratigraphic data relevant to the chosen interval of study are sparse, however, north of about 53°N and south of about 35°N latitudes. Thus, our attempts at quantification of relative shoreline migrations were limited to areas between these two latitudes.

Geologic data and interpretations used in development of Appendix 1 were gathered and integrated by Lillegraven from primary sources listed in Appendix 2. Outcrop and subsurface data were considered. Needless to say, a major limiting factor in the present study is that much of the Upper Cretaceous stratigraphic record has been removed by erosion, both during the Laramide orogeny and in more recent time. Nevertheless, this geologic interval of the Late Cretaceous is better documented in the Western Interior than in any other comparably sized region of the world.

Mainframe computers (Control Data Corporation CYBER 760 and 840) at The University of Wyoming were used as tools for data management. The series of 33 Lambert equal-area paleoshoreline maps presented as Figure 4 was developed using a cartographic program called CMDP12, written and modified for

TABLE 1. KEY LITERATURE TO AREAS OF SIGNIFICANT LATE CRETACEOUS (LATE SANTONIAN THROUGH MAASTRICHTIAN) MAMMAL-BEARING LOCALITIES OF THE WESTERN INTERIOR, ARRANGED BY NORTH AMERICAN LAND-MAMMAL "AGE" (sensu LILLEGRAVEN AND McKENNA, 1986).
(Numbers on left refer to sites plotted on Figs. 1 and 5)

LANCIAN

Alberta

1 Scollard Formation (vicinities of Griffith and Henry farms)
- Clemens and others, 1979
- Lillegraven, 1969
- Russell, 1987

Saskatchewan

8 Ravenscrag Formation (vicinity of Eastend)
- Johnston and Fox, 1984
- Fox, 1987

9 Frenchman Formation (vicinity of Eastend)
- Johnston, 1980

Montana

13-14 Hell Creek Formation (type area; 13, McCone County; 14, Garfield County)
- Archibald, 1982
- Archibald and Clemens, 1984
- Clemens and others, 1979
- Lillegraven and McKenna, 1986
- Sloan and others, 1986

15 Hell Creek Formation (vicinity of Ekalaka)
- Archibald, in preparation

South Dakota

16 Hell Creek Formation (vicinity of "The Jumpoff")
- Archibald, 1982
- Clemens and others, 1979
- Wilson, 1983

17 Fox Hills Formation (vicinity of Redowl)
- Clemens and others, 1979
- Wilson, 1987

18 Fox Hills Formation (Iron Lightning Member; vicinity of Iron Lightning)
- Waage, 1968

Wyoming

19 Lance Formation (Dumbbell Hill at Polecat Bench)
- Clemens and others, 1979
- Gingerich and others, 1980

20 Lance Formation (Hewitt's Foresight localities at Meeteetse Rim)
- Lillegraven, in preparation

22 Lance Formation (vicinity of Mule Creek Junction)
- Whitmore, 1985

23 Lance Formation (type area in vicinity of Lance Creek)
- Archibald, 1982
- Clemens, 1973a
- Clemens and others, 1979
- Fox, 1974
- Lillegraven and McKenna, 1986

25 Lance Formation (eastern Rock Springs Uplift)
- Breithaupt, 1982

Colorado

26 Laramie Formation (Weld County)
- Carpenter, 1979

Utah

29 North Horn Formation (Dragon Canyon)
- Clemens, 1961
- Clemens and others, 1979

New Mexico

34 Kirtland Formation (Naashoibito Member, San Juan Basin)
- Flynn, 1986
- Lehman, 1984

"EDMONTONIAN"

Alberta

2 Horseshoe Canyon Formation (vicinity of Drumheller)
- Fox and Naylor, 1986

3 St. Mary River Formation (north of Lundbreck)
- Clemens and others, 1979
- Russell, 1975

4 St. Mary River Formation (Scabby Butte)
- Clemens and others, 1979
- Russell, 1975
- Sloan and Russell, 1974

Colorado

27 Williams Fork Formation (Axial Basin)
- Lillegraven, 1987

28 Williams Fork Formation (Piceance Creek Basin)
- Archibald, 1987

New Mexico

35 Fruitland Formation and/or "transition" and lower Kirtland Formation (San Juan Basin)
- Armstrong-Ziegler, 1978
- Clemens, 1973b
- Clemens and Lillegraven, 1986
- Clemens and others, 1979
- Flynn, 1986
- Lillegraven, 1987
- Rigby and Wolberg, 1987

JUDITHIAN

Alberta

5 Judith River Formation (sensu McLean, 1977; upper part; vicinity of Steveville)
- Clemens and others, 1979
- Fox, 1981
- Lillegraven and McKenna, 1986

7 Judith River Formation (sensu McLean, 1977; lower part; north of Medicine Hat)
- Fox, 1976
- Clemens and others, 1979
- Russell, 1952

Montana

10 Judith River Formation (Hill County)
- Montellano, 1986

?11 Two Medicine Formation (vicinity of Choteau)
- Montellano, 1986
- Schultz and others, 1980 (Plate 1)

12 Judith River Formation (type area in breaks of Missouri River)
- Clemens and others, 1979
- Fox, 1981
- Lillegraven and McKenna, 1986
- Sahni, 1972

Wyoming

21 "Mesaverde" Formation (vicinity of Worland)
- Lillegraven and McKenna, 1986

24 "Mesaverde" Formation (Rattlesnake Hills anticline)
- Clemens and Lillegraven, 1986
- Lillegraven and McKenna, 1986

Utah

31 Kaiparowits Formation (Kaiparowits Plateau)
- Eaton, 1987a

AQUILAN

Alberta

6 Milk River Formation (upper part)
- Clemens and others, 1979
- Fox, 1984
- Lillegraven and McKenna, 1986

Montana

?11 Two Medicine Formation (vicinity of Choteau)
- Montellano, 1986
- Schultz and others, 1980 (Plate 1)

Utah

30 Masuk Formation (sensu Eaton, 1987a; Henry Mountains)
- Eaton, 1987a

32 Wahweap Formation (Kaiparowits Plateau)
- Eaton, 1987a

PRE-AQUILAN

Utah

33 Straight Cliffs Formation (John Henry Member; Kaiparowits Plateau)
- Eaton, 1987a

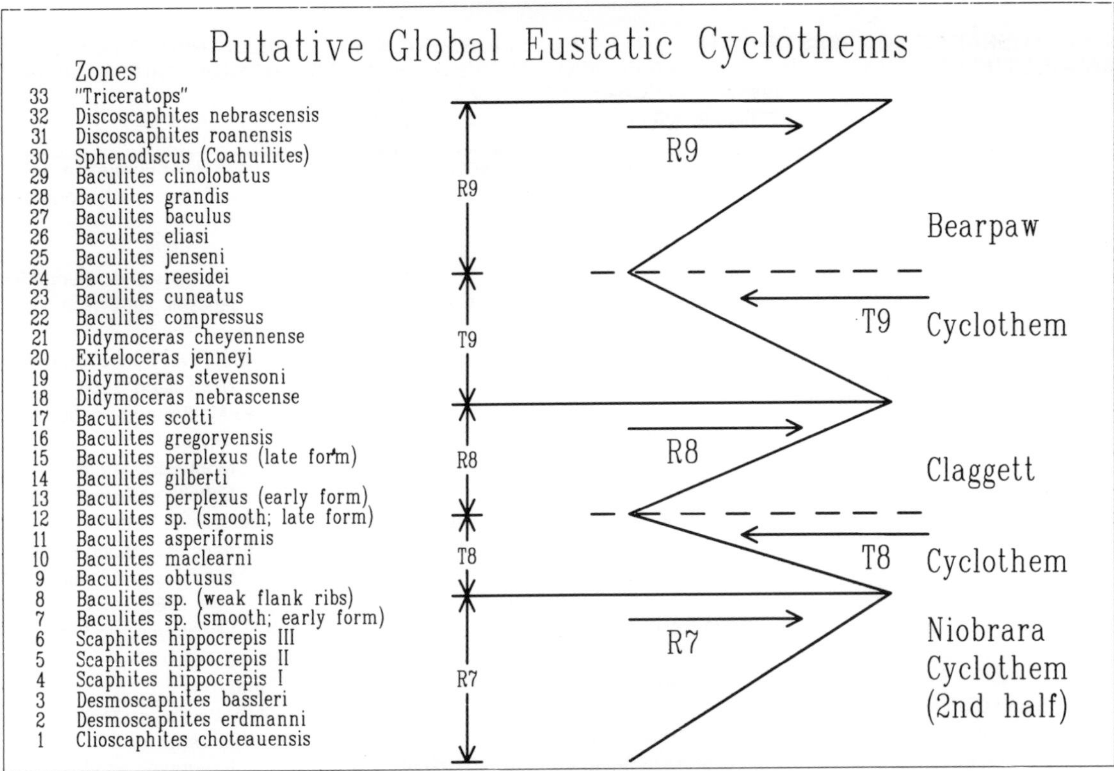

Figure 2. Diagrammatic representation of the stratigraphic distribution (using the ammonite zonation of the Western Interior by Obradovich and Cobban, 1975) of the influence of Late Cretaceous sea-level changes, as interpreted from Kauffman (1977) and Hancock and Kauffman (1979). The transgressive (T) and regressive (R) phases of sea-level changes reputedly were cyclic, globally recognizable, and tectonoeustatic in nature. The zigzag line on the right of the diagram simulates general trends of shoreline motion through time, with west to the left and east to the right. The reputedly global cyclothems would correspond to "third order" cycles (i.e., 1 to 10 m.y. in duration) as propounded by Vail and others (1977). Supposed cyclothemic sea-level changes "T-10" and "R-10" of Kauffman (1977, 1984) are omitted.

present uses by Ostresh. Strandlines were interpolated and drawn in smoothed form using raw input similar to that provided in Appendix 1. It was necessary, however, to add selected dummy data to constrain shoreline plots to geographically reasonable limits for situations in which adequate geological information is unavailable. As an example, western limits of the strandlines are largely unknown in the oldest parts of the study sections; commonly, they have been overridden by younger, eastern edges of the various foreland thrusts. Similarly, margins of the "Sheridan Delta" in the vicinity of eastern Montana were artificially modeled using dummy points in parts 27 to 28 of Figure 4 to simulate interpretations by Gill and Cobban (1973, Fig. 19). A printout of the complete data file is available from Lillegraven upon request.

Because of uncertainties in correlation, effects of postdepositional erosion, and computer-smoothing of lines, details of shoreline positions on the scale of a few tens of kilometers are surely incorrect on many of the maps. We have confidence, however, in the reliability of the general patterns of shoreline migration at greater scales. For the most part, our discussions and interpretations are restricted to geographic and temporal scales appropriate to reliability of available information.

All geographic coordinates cited are those of present Earth. Assuming accuracy of paleogeographic maps 21 and 23 for the world at 80 Ma by Smith and others (1981), the coordinates for our area of interest would need only trivial correction to represent equivalent Late Cretaceous positions. A palinspastic reconstruction was prepared by Hamilton (1978, p. 59) for the Eocene of western North America. Inspection of that map and comparison with modern coordinates suggests that post-Cretaceous deformation of the coastal plain, our area of primary concern, had effects of only minor consequence on geographic shoreline estimates presented in Figure 4.

The principal stratigraphic framework of marine zonations used in development of Appendix 1 (plus Figs. 2, 4, 5, and 6) was derived from Obradovich and Cobban (1975, Table 1), based on ammonites, and was designed principally for use in the contiguous United States. The Canadian picture was completed by correlations to the Obradovich and Cobban scheme, as suggested by various workers (especially by Caldwell and by Jeletzky). Although Cobban (1977) suggested addition of a new zone (*Baculites reduncus*) between the *B. gregoryensis* and *B. scotti* zones, the species on which it is based is limited in known distribution to

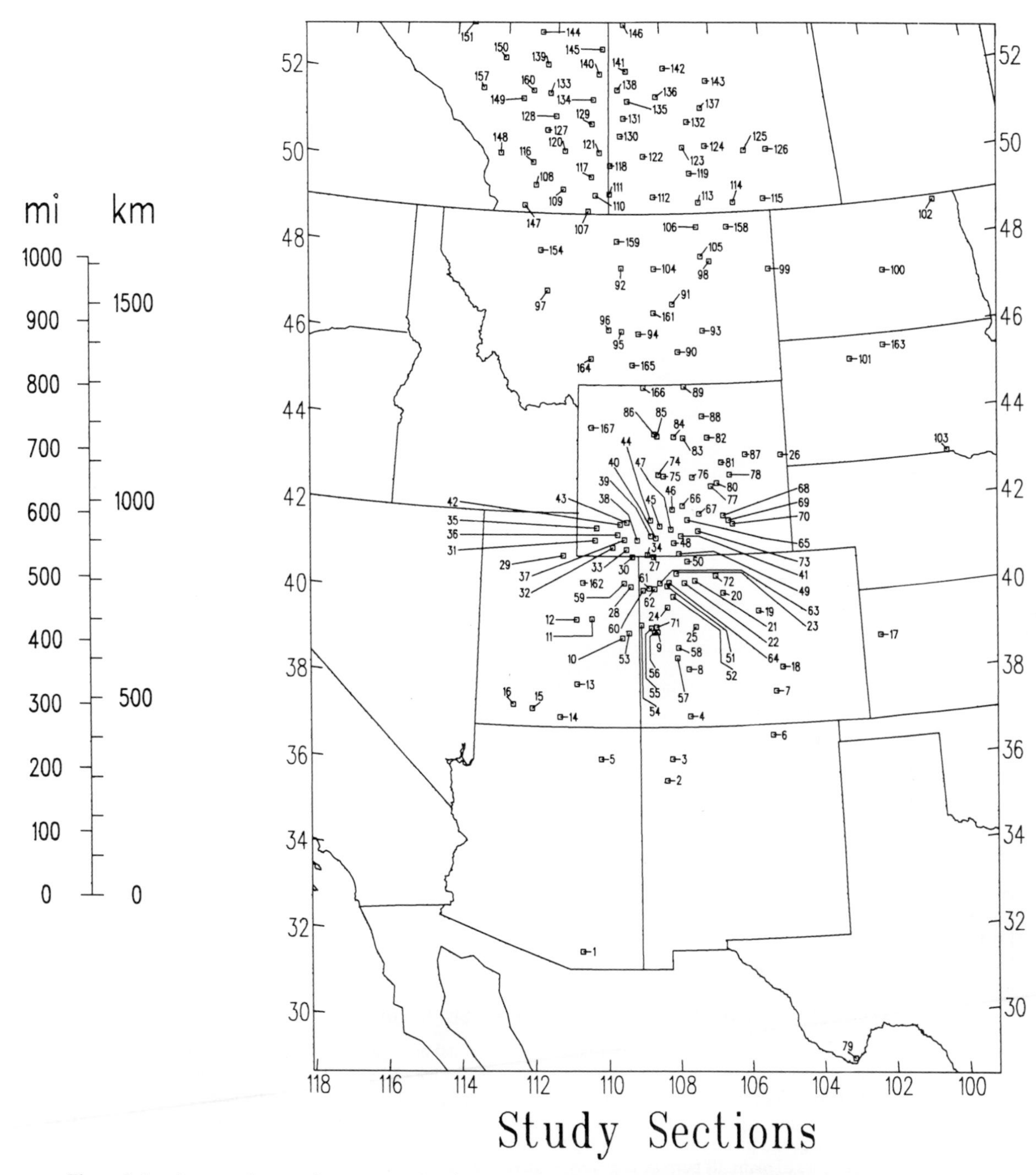

Figure 3. Lambert equal-area reference map showing most of the specific localities from which primary paleoenvironmental data were synthesized. Interpretations of the data are summarized (by locality) in Appendix 1, and are integrated as computer-generated paleoshoreline maps in Figure 4. See Appendix 1 for present-day geographic coordinates of the various localities. Although a few of the localities (152, 153, 155, 156) included in Appendix 1 are outside the map area of Figure 4, all data were used by the computer in estimating positions of strandlines.

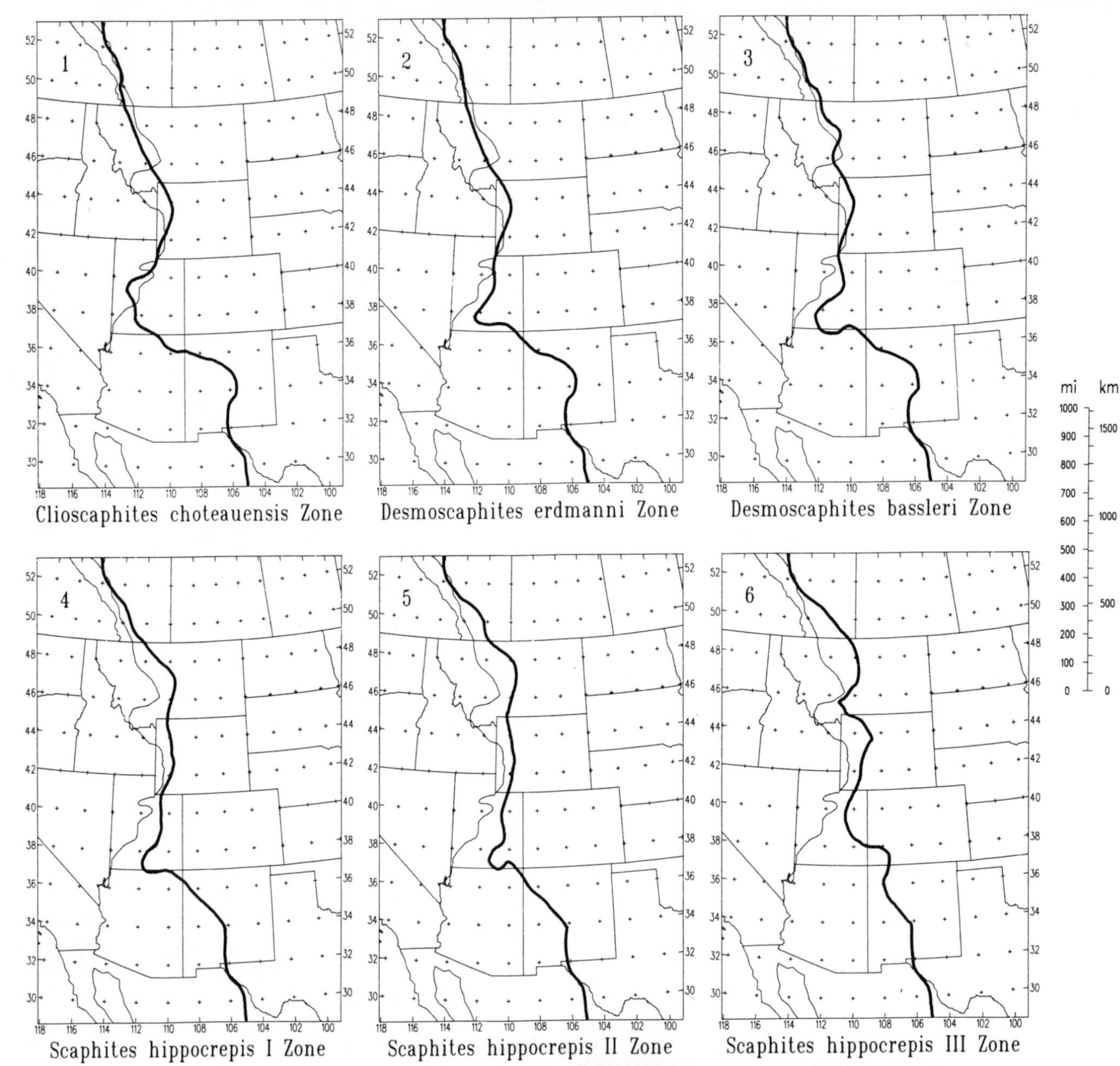

Figure 4 (in 33 parts). Nonpalinspastic, Lambert equal-area paleoshoreline (heavy lines) maps, using modern coordinates, for each of the zones listed on the left of Figure 2. Paleoenvironmental interpretations on which the maps were based are presented in Appendix 1 (sources of data listed in Appendix 2). The irregular lighter line represents the present-day position of the eastern edge of the zone of foreland thrusting, and (in an admittedly geologically unrealistic fashion) remains constant in position throughout the series of maps. Its purpose is merely to serve as a frame of reference for visualizing the relative widths of the evolving coastal plain through the interval of study. Paleoshoreline positions that are presently covered by the eastwardly shifted foreland thrust sheets may never be known with certainty, and as presented here should be considered as minimum westward excursions of the sea.

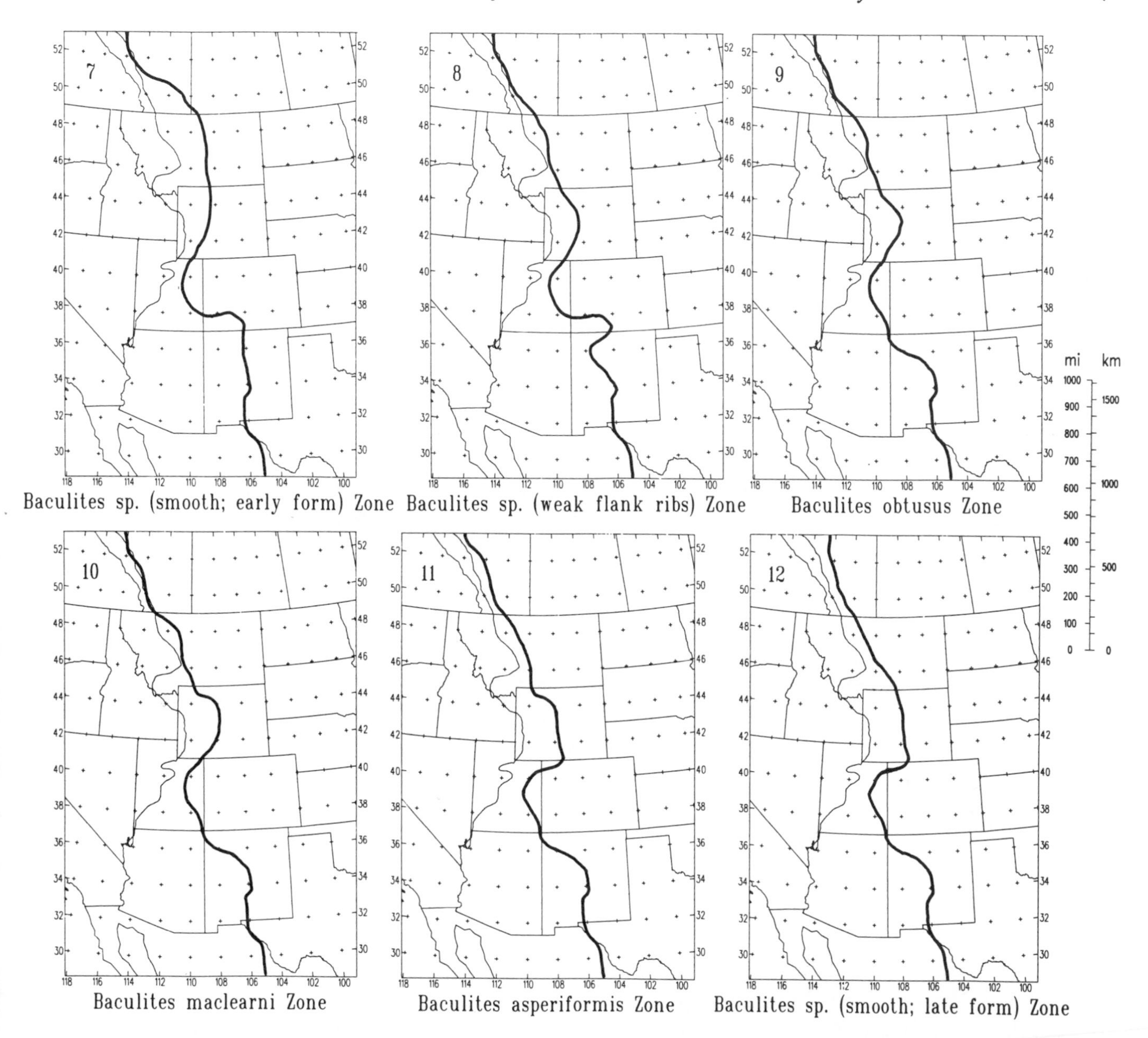

southeastern Wyoming and adjacent parts of Colorado. Thus, the *B. reduncus* zone is not useful as a tool for biostratigraphic correlation at a regional scale and, at the personal recommendation of Cobban (1988), the *B. reduncus* zone was not used in the present study. The actual duration of very few of the ammonite zones is known in terms of radioisotopic time (see Obradovich and Cobban, 1975). Although we have drafted the zones on our various figures to physically fit equal spaces, that procedure is not intended to imply isochrony of the various zones.

Correlations suggested in Figure 5 between stratigraphic positions of the various mammal-bearing localities and the marine zonation are based on two principal constraints. First is the bracketing, where such occurs, of the fossiliferous terrestrial strata by biostratigraphically identified marine rocks. Much detailed work still needs to be done in this regard, and many such correlations, as presented here, perhaps should be considered accurate only within plus or minus two ammonite zones. The second constraint is the relative stratigraphic position of any given mammal-bearing locality within its enclosing nonmarine sequence. Clearly, in the absence of other independent forms of dating, the farther a mammal locality is from a link to the marine zonation, the more equivocal is its age. Thus, in Figure 5 we have provided rather pessimistic estimates of the reliability of the age control for each of the mammal-bearing sites. We made no attempt in this study to introduce data or interpretations from subdisciplines of magnetostratigraphy or radioisotopic dating. Clearly, however, that should be done both as a next logical step and as a means of testing validity of correlations suggested here.

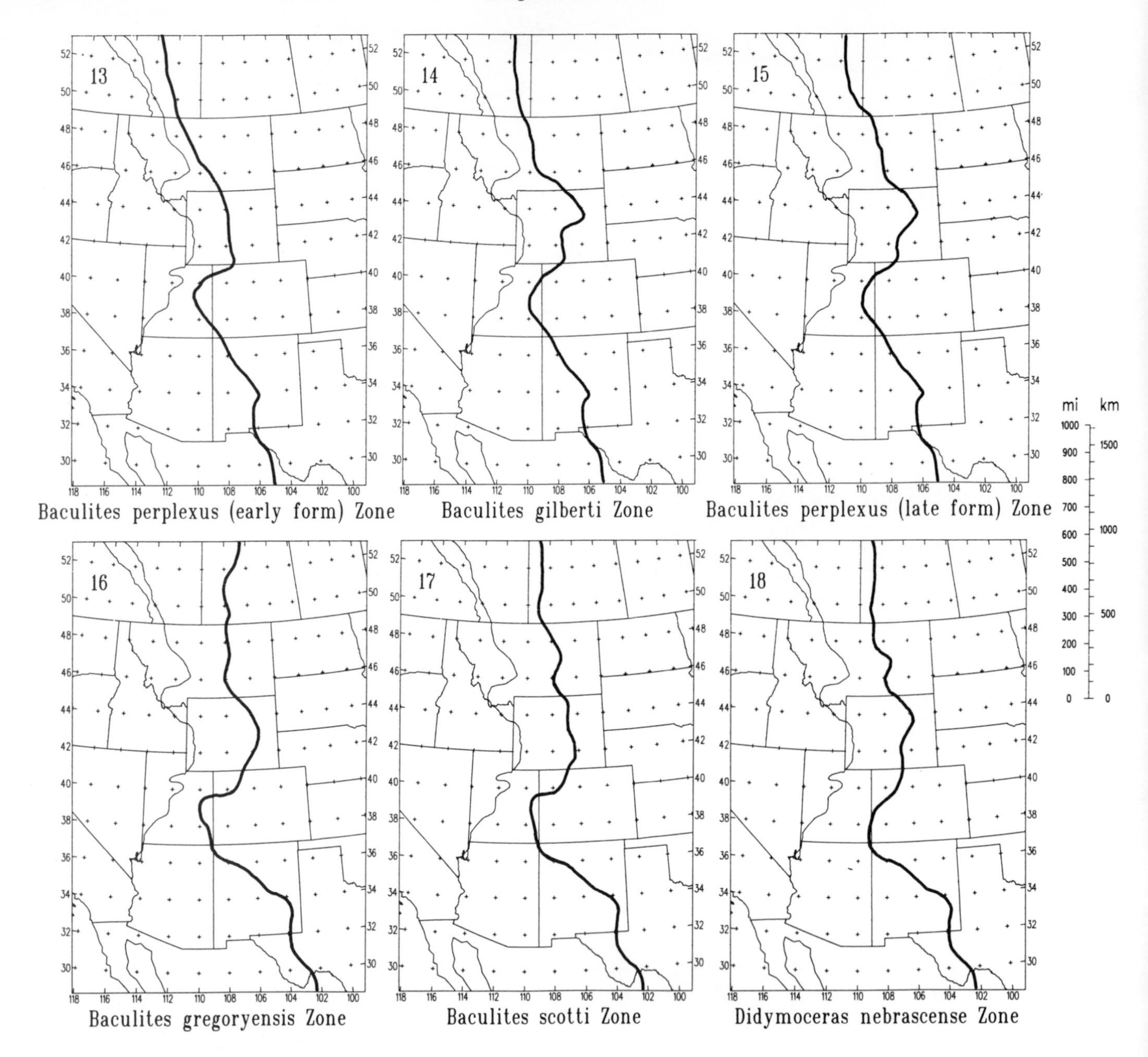

Table 2 (and its graphic equivalent, Fig. 6) presents quantified estimates of kilometers of shoreline migration at each degree (35° to 53°N) of modern latitude through the geologic interval studied. Procedures used in developing those estimates were as follows. Inspection of the maps making up Figure 4 showed that all paleoshorelines are located east of modern longitude 122°W. Thus, that line of longitude provided a convenient reference for measuring to the variously shifting east-west positions of the ancient strandlines. The longitude of each estimated shoreline for each stratigraphic zone was recorded at each degree of latitude between 35°N and 53°N. The longitudinal position for each strandline was then subtracted from 122°W. The subtractions gave angular differences (in tenths of a degree of longitude) of each shoreline's position east of 122°W longitude. The value of each angular difference was then multiplied by a constant (which was unique for each degree of latitude) that converted angular differences into distances (in kilometers) eastward from 122°W as measured along the Earth's latitudinal small circles. A spherical Earth with a radius of about 6,377 km was assumed. The minimum value for shoreline distance east of 122°W longitude at any given latitude was then subtracted from all larger distance values for that same latitude. By this method, the estimates of relative shoreline migration were normalized for each degree of latitude, with the values of zero seen in Table 2 and on Figure 6 representing the shoreline's most westerly excursion through the geologic interval of study.

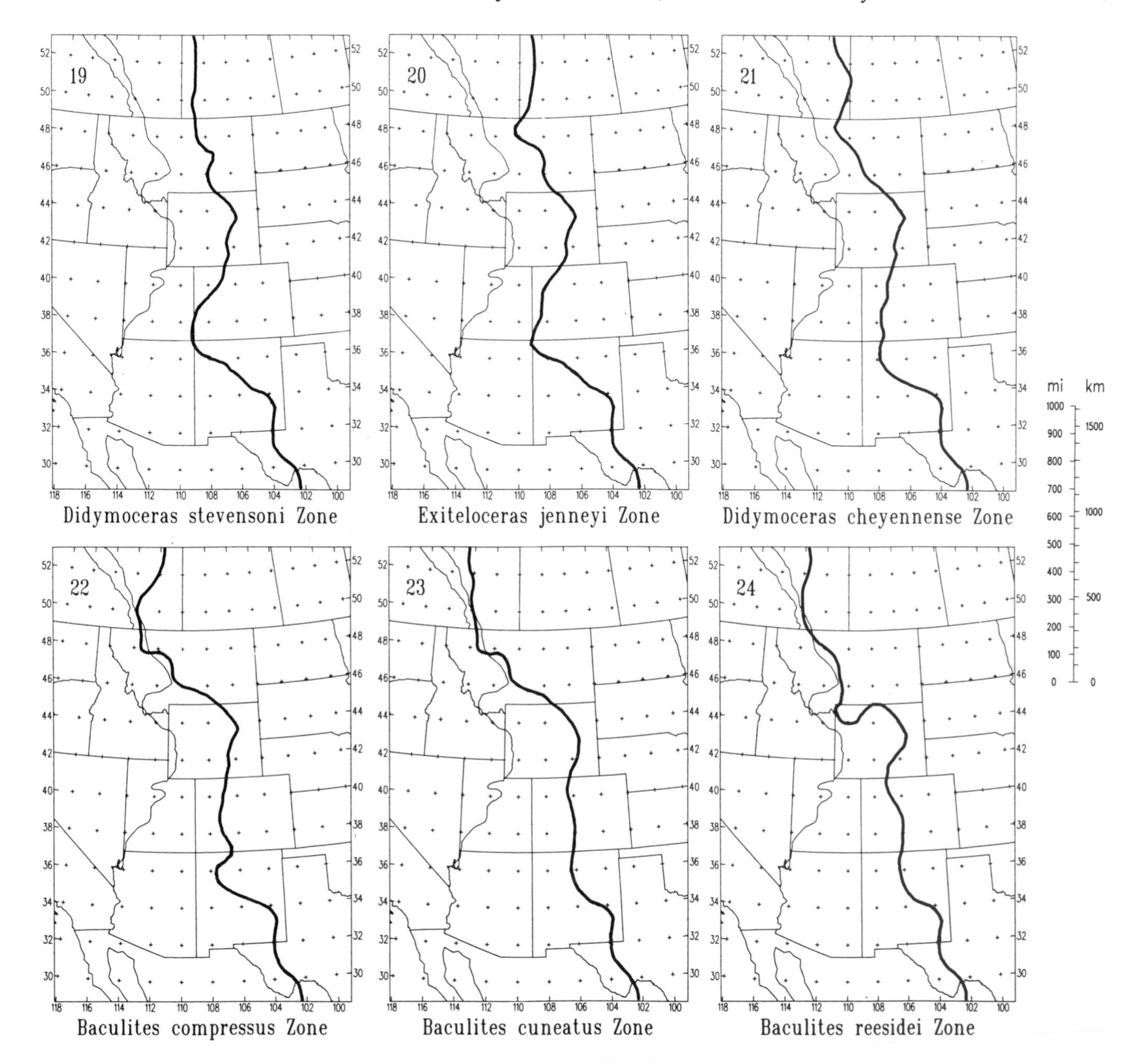

RELEVANCE OF NAMES ASSOCIATED WITH EUROPEAN STAGE BOUNDARIES

The European-based, marine stage terms of Santonian, Campanian, and Maastrichtian traditionally have been applied to Upper Cretaceous sedimentary rocks of the North American Western Interior. The boundaries between them, however, have been controversially assigned, varying widely in stratigraphic position according to different authorities. Few of the largely endemic species of marine macroinvertebrates of the Western Interior are found in common with the European stratotypes (or, for that matter, even with stratigraphic sections of the North American Gulf Coast) of the Campanian or Maastrichtian. Microfossil assemblages have proven to contain more cosmopolitan species ranges, and thus can be linked more directly from the Western Interior to the European stratotypes (by way of the Gulf Coast). One result is that micropaleontologists have defined the marine stage boundaries in the Western Interior quite differently from those paleontologists who use various kinds of marine molluscs. Major nomenclatorial confusion exists, therefore, and the unwary might not realize that the same stage names have been applied in the Western Interior to marine rocks of markedly different ages by paleomalacologists and by micropaleontologists.

Much of the literature on this general subject pertinent to questions of the Campanian/Maastrichtian boundary of the Western Interior was reviewed by Lillegraven and McKenna (1986), with a critical follow-up by Eaton (1987b). Eaton, however, significantly overstated conclusions made by Lillegraven and Mc-

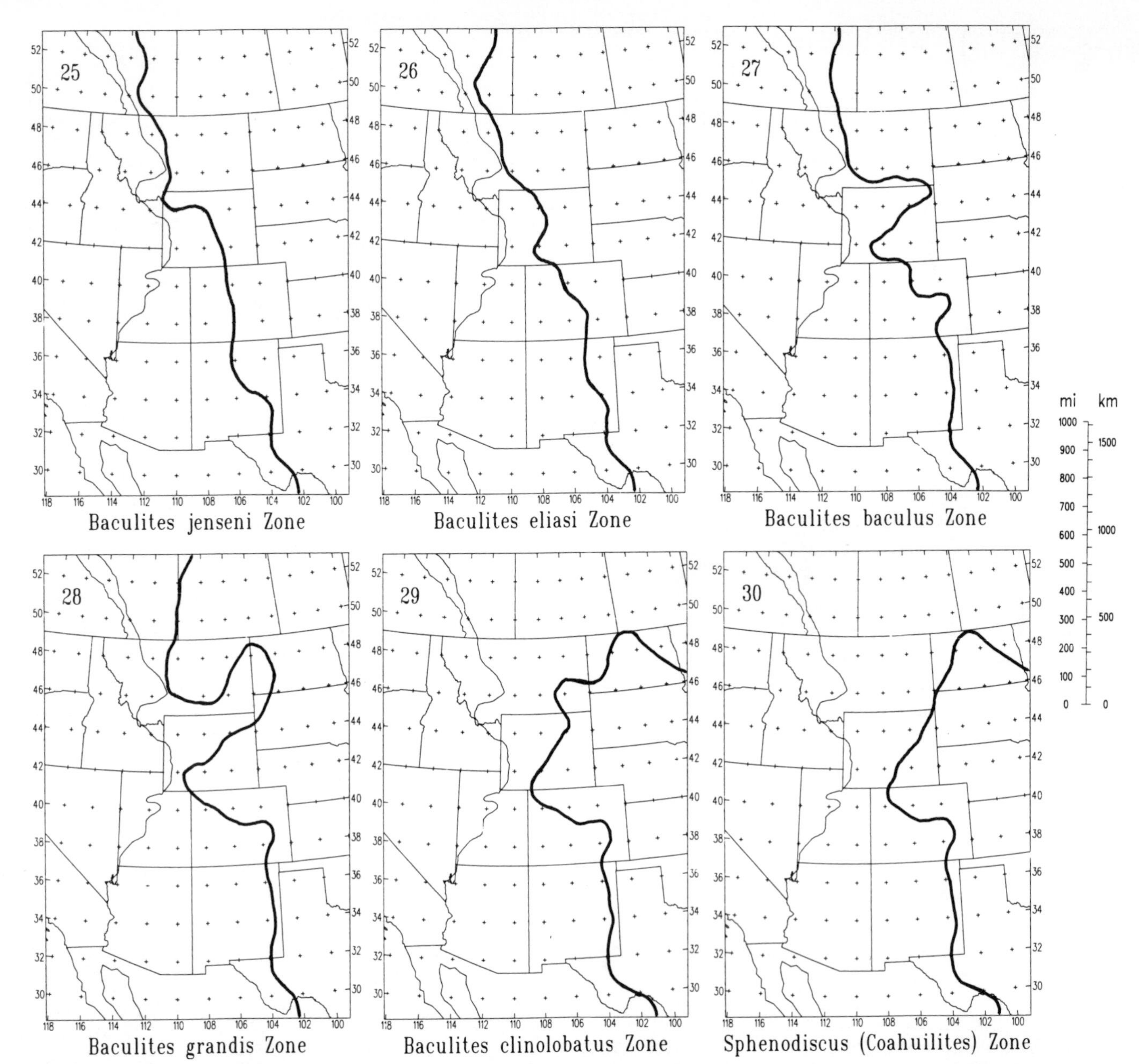

Kenna (1986) concerning placement of the Campanian/Maastrichtian boundary in the Western Interior. Also, important factual errors exist in Eaton's interpretation of Upper Cretaceous foraminiferal stratigraphy of the Western Interior. Nevertheless, his conclusion (p. 38) probably is correct that: "The exact equivalency of boundaries between North American zones and European stages remains to be solved." Paleontological evidence from all available sources concerning placement of the Santonian/Campanian and Campanian/Maastrichtian boundaries in the Western Interior is under review by Lillegraven.

Fortunately, controversial issues associated with use of European stage terminology are mostly irrelevant to the present project. We have restricted our principal use of marine stratigraphic zonal terminology to the provincial systems developed within the Western Interior itself. But when attempting to link the geological story of the Late Cretaceous of the Western Interior to much of the remainder of the world, one immediately becomes entangled in terminological problems. It is important to keep in mind, for example, that much of the Late Cretaceous history of the world's oceanic basins has been interpreted by way of the Deep Sea Drilling Project, applying foraminiferal zonations to deep-sea cores. As alluded to above, the criteria by which paleomalacologists have assigned stage boundaries in the Western Interior differ markedly from criteria for stage definitions used by micropaleontologists across most of the rest of the marine world. Thus, the same stage names are being used in various geographic areas of the world to represent quite different divisions of geologic time. In such situations, of course, nomenclatorial problems translate into true problems of geological interpretation as well.

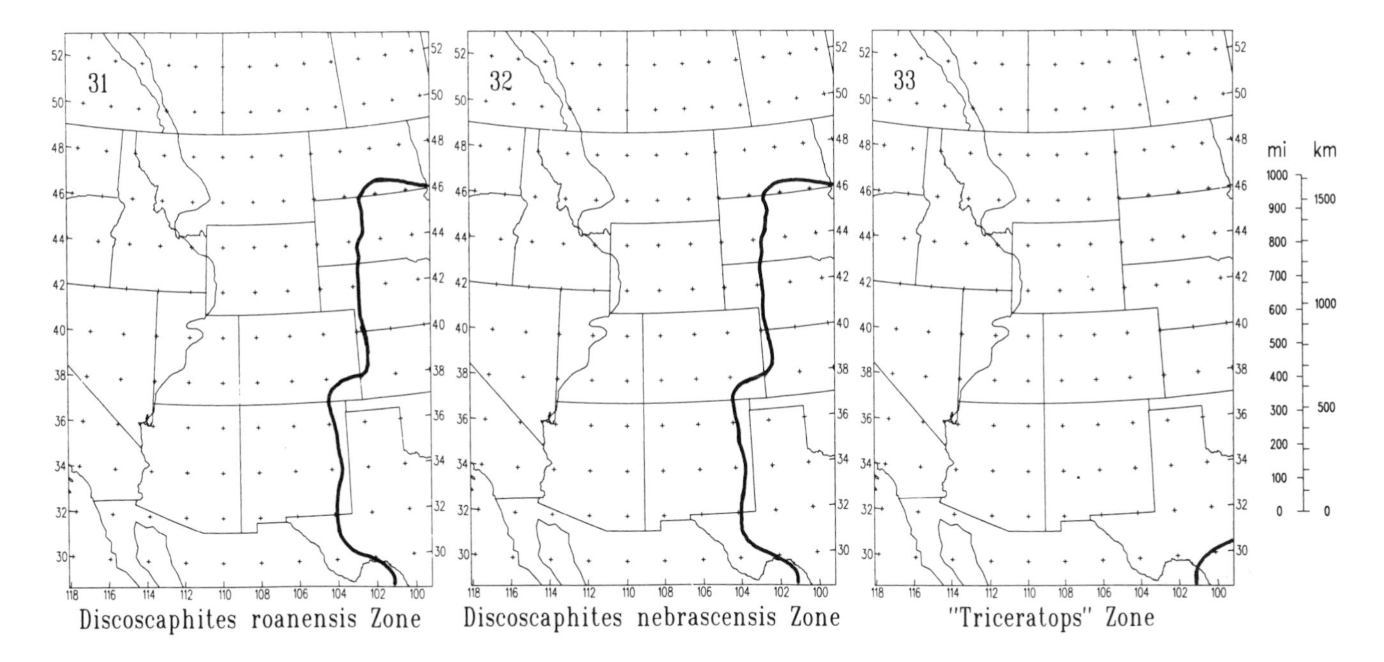

For purposes of general reference, traditional definitions of the Santonian/Campanian boundary in the Western Interior by ammonite specialists (see Obradovich and Cobban, 1975) has been at the top of the *Desmoscaphites bassleri* zone (no. 3 of Figs. 2, 4–6) and by micropaleontologists (Frerichs, 1980; the scheme accepted in this paper) at about the *Clioscaphites choteauensis* zone (no. 1 of Figs. 2, 4–6). Similarly, the Campanian/Maastrichtian boundary of macropaleontologists usually has been put near (or higher than) the *Baculites reesidei* zone (no. 24 of Figs. 2, 4–6); by micropaleontologists (Bergstresser and Frerichs, 1982), placement has been as low as the *Baculites obtusus* zone (no. 9 of Figs. 2, 4–6; no particular boundary is used in this chapter). The European terms of Santonian, Campanian, and Maastrichtian will be applied only in the most general sense in the remainder of the present chapter.

RESULTS AND DISCUSSION

Mammalian correlations

Lillegraven and McKenna (1986) attempted faunal definitions for provincial Late Cretaceous North American Land-Mammal "Ages" (NALMAs). The definitions were based on mammalian species, and used the preexisting terms "Lancian," "Judithian," and "Aquilan." The terms were applied in the temporal sense characteristic of North American land-mammal "ages" (as developed for the Cenozoic by Wood and others, 1941). Another preexisting term, the "Edmontonian," had been used by Russell (1975) to designate terrestrial faunas older than those of the Lancian and younger than those of the Judithian. Although Lillegraven and McKenna (1986) agreed that such an intermediate inteval of geologic time existed, they concluded that, using then-existing paleontological evidence, an Edmontonian "age" could not be defined using purely faunal criteria. Because the "Edmontonian" still has not been defined formally, we bound the word with quotation marks wherever used in the present paper. Yet another mammalian "age," the Bugcreekian, is being defined by Archibald (this volume) for post-Lancian faunas that, in uncertain fashion, approximate the Mesozoic-Cenozoic boundary. We have not attempted to incorporate the Bugcreekian into this chapter.

Our first approximations of correlations between Late Cretaceous North American land-mammal "ages" and standard ammonite zonation for the Western Interior are presented in Figure 5. The Lancian is the best represented of the Late Cretaceous NALMAs in terms of numbers of known localities (Fig. 1, Table 1). Faunally defined Lancian assemblages occur in nonmarine sequences correlated with strata younger than the *Baculites clinolobatus* zone. Most of the Lancian faunas probably represent the closing phases of the interval, within what was referred to by Obradovich and Cobban (1975, Table 1) as the "*Triceratops* zone." Use of the term "*Triceratops* zone," unfortunately, has had a history of confusion. Although shown in Figure 5 to represent nonmarine strata younger than marine correlatives of the *Discoscaphites nebrascensis* zone, the more common use among paleontologists with terrestrially based interests has been for almost any very late Cretaceous nonmarine faunas (roughly comparable to our concept of the Lancian).

Progress is being made toward faunal defensibility of an Edmontonian "age." Although mammalian hypodigms from the St. Mary River Formation (locs. 2 and 3 of Fig. 1 and Table 1) remain largely enigmatic taxonomically, a new, highly specialized, and apparently phylogenetically relict species of *Didelphodon* has been described (Fox and Naylor, 1986) from the correlative Horseshoe Canyon Formation of Alberta. Because of its relict nature, however, the new form of *Didelphodon* may have

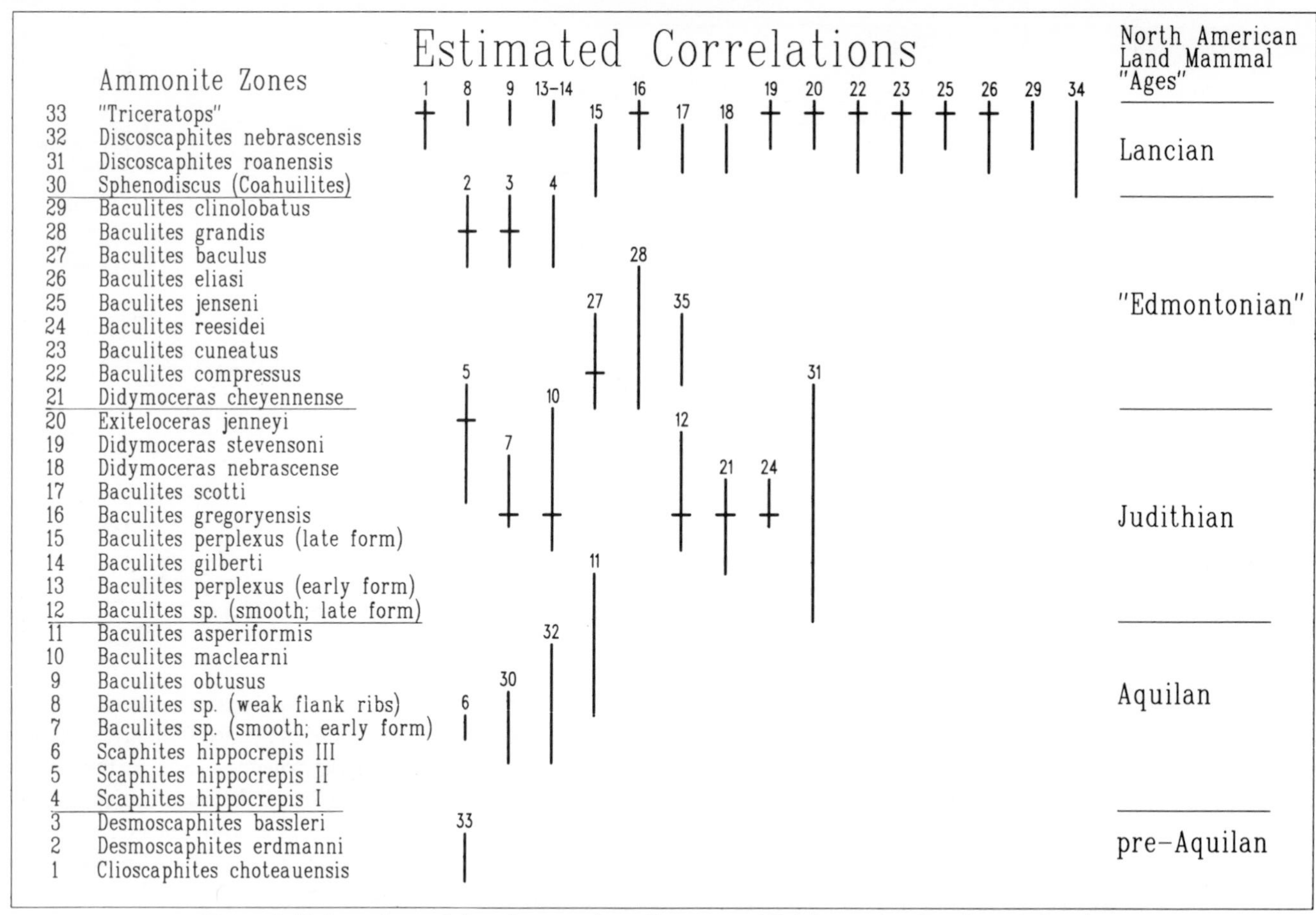

Figure 5. Estimated correlations between Late Cretaceous (slightly pre-Aquilan through Lancian) terrestrial mammal-bearing localities of the Western Interior and the standard ammonite-based zonation (earliest Campanian through Maastrichtian) for marine strata deposited within the Western Interior Seaway. Identifying numbers refer to mammal localities within rock units specified in Table 1; the localities are plotted geographically in Figure 1. For each locality, the vertical bar represents a reasonable range of possible stratigraphic correlation; the horizontal bar, if present, represents our estimate of the most probable correlation. Note the great temporal biases in the Late Cretaceous mammalian record of the Western Interior.

limited use for purposes of biostratigraphy. A new species of *Meniscoessus* from northwestern Colorado (Lillegraven, 1987; Archibald, personal communication, 1987) occurs in rocks shown to be intermediate in age between those yielding Judithian and Lancian mammals. The new form is morphologically intermediate between species of *Meniscoessus* characteristic of the Judithian and Lancian, and it appears that the genus as a whole has considerable utility for purposes of biostratigraphic correlation in nonmarine rocks.

Locality 35, as plotted on Figure 1, represents a series of discrete collecting sites in the San Juan Basin found at approximately the same stratigraphic level near the variously defined boundary of the Fruitland and Kirtland Formations. The age of the mammalian faunas from these localities recently has come into debate. Although previously considered as "Edmontonian" in age (see Clemens and others, 1979), Rigby and Wolberg (1987) considered the assemblage to compare more favorably with northerly faunas of Judithian age. On that basis, Rigby and Wolberg reinterpreted (using older chrons) the magnetostratigraphic sequence of the San Juan Basin from the scheme developed by Lindsay and others (1981) and modified by Butler and Lindsay (1985).

The paleogeographic maps that make up Figure 4 provide help toward solution of this controversy. Correlations in more northerly areas between faunally defined Judithian mammalian assemblages and their contemporary ammonite zones (see Fig. 5) suggest that the Judithian land mammal "age" as presently defined is approximately bracketed stratigraphically between the *Baculties* sp. (smooth; late form) and the *Didymoceras cheyennense* ammonite zones (our nos. 12 and 21, respectively). As can be seen in parts 12 to 21 of Figure 4, the vicinity of the San Juan Basin that yielded the series of mammalian faunas in question experienced dominantly shallow marine or immediate shoreline conditions during that interval. Thus, we suggest in Figure 5 that the local faunas within locality series 35 of Figure 1 were enclosed by younger strata, specifically of early phases of the "Ed-

TABLE 2. CALCULATED RELATIVE DISTANCES (in km) OF EASTERN EXCURSIONS OF THE LATE CRETACEOUS SHORELINE FROM ITS WESTERNMOST EXTENT (DURING THE INTERVAL REPRESENTED BY ZONES 1 to 32) AT EACH DEGREE NORTH LATITUDE (MODERN COORDINATES) FROM 35 TO 53°.*

Zones	35°	36°	37°	38°	39°	40°	41°	42°	43°	44°	45°	46°	47°	48°	49°	50°	51°	52°	53°	Zones
32	346	522	733	886	986	895	792	744	688	725	756	812	......	......	......	......	......	......	......	32
31	346	522	729	886	986	891	767	740	696	725	756	812	......	......	......	......	......	......	......	31
30	346	522	733	798	817	414	275	298	346	440	523	615	740	834	......	......	......	......	......	30
29	351	522	733	802	817	388	192	219	252	356	378	375	740	834	......	......	......	......	......	29
28	351	522	733	807	817	414	158	95	252	388	582	131	121	197	274	304	322	384	476	28
27	351	518	684	724	809	507	375	153	248	380	496	97	118	179	183	172	154	233	301	27
26	219	405	649	684	653	477	321	199	224	216	146	97	121	182	153	100	119	233	305	26
25	150	293	533	579	592	477	379	323	236	0	0	85	121	179	131	86	119	192	174	25
24	150	288	533	570	588	277	329	335	379	4	0	85	118	130	33	25	25	120	174	24
23	150	288	533	574	588	477	396	409	379	356	275	131	121	30	33	25	25	86	100	23
22	59	162	502	478	489	431	354	335	317	356	275	128	118	0	33	25	84	233	301	22
21	59	153	396	417	467	414	346	331	322	356	272	220	235	201	215	236	322	322	305	21
20	123	135	276	333	389	345	342	335	322	356	275	278	330	235	307	379	399	459	496	20
19	123	126	267	303	376	422	354	331	322	352	275	286	379	343	369	379	396	456	496	19
18	123	135	267	303	376	422	359	318	317	356	279	290	383	350	372	383	399	463	482	18
17	123	135	267	281	272	396	346	364	281	292	275	290	398	424	376	379	396	456	496	17
16	119	158	271	263	251	401	354	389	383	360	275	278	361	428	449	476	455	565	643	16
15	50	153	329	263	246	243	275	269	309	356	275	255	307	343	336	265	231	267	305	15
14	46	144	324	263	246	239	275	252	256	352	275	189	201	253	237	229	224	264	298	14
13	68	158	324	267	216	188	300	244	195	208	185	189	159	179	150	154	119	158	174	13
12	123	144	267	263	221	183	300	244	199	212	193	189	156	186	153	100	63	89	104	12
11	119	140	267	263	221	192	300	248	199	196	79	143	163	182	131	82	56	89	17	11
10	132	158	262	254	212	141	150	203	187	184	71	85	118	175	55	21	0	0	3	10
9	114	126	267	254	212	128	117	136	167	136	71	93	118	175	128	39	4	0	3	9
8	46	162	524	482	212	128	117	149	147	128	79	131	118	179	146	50	0	0	7	8
7	132	293	538	491	221	132	100	128	126	160	185	232	285	268	340	268	53	0	0	7
6	0	140	396	381	216	132	83	87	77	128	28	77	201	276	248	175	56	0	7	6
5	9	81	236	140	212	132	71	70	37	36	47	139	216	268	150	154	53	0	10	5
4	14	81	49	88	190	119	33	62	37	36	47	139	220	205	135	82	42	0	7	4
3	141	95	0	9	177	111	0	0	0	36	0	4	99	97	91	39	0	0	7	3
2	141	0	0	9	143	85	0	0	4	36	0	0	0	26	18	25	7	0	7	2
1	146	129	0	0	0	0	0	4	0	36	0	0	0	30	0	0	4	0	7	1

*Missing values above zone 27 indicate areas at which wholly nonmarine conditions had come into being, thus providing no useful quantitative estimates for shoreline positions. Figure 6 is a graphical equivalent of this table.

montonian." Such a suggestion, of course, precludes the simple device of redefinition of the younger end of the range of the Judithian to include these fossils.

Relevant to this general discussion is the observation by Eaton (1987a) that Judithian mammalian faunas of the Kaiparowits Plateau of southern Utah show marked taxonomic differences from correlative assemblages in more northerly parts of the Rocky Mountains. Late Cretaceous climatic zonation of the Western Interior as discussed by Wolfe and Upchurch (1987) would be expected to have had significant latitudinal influence on taxonomic compositions of land vertebrate faunas.

Mammalian faunas of Aquilan age are stratigraphically bracketed by the *Scaphites hippocrepis* and *Baculites asperiformis* zones (our nos. 4 to 11, respectively). The "pre-Aquilan" mammalian occurrence on Figure 5 (loc. 33 of Fig. 1) is set off by itself, because it would be premature on the basis of the scanty mammalian record from the John Henry Member of the Straight Cliffs Formation (see Eaton, 1987a) to (1) lump the occurrence into the Aquilan, or (2) attempt definition of a new, older land mammal "age."

General trends in shoreline transgressions and regressions

Figure 6A, and its numerical source, Table 2, provide a quantitative summary of estimates of east-west components of migration of the western shoreline of the Western Interior Seaway through the interval of interest. The data were derived from maps in Figure 4 as explained above in the section on Scope and Methods of Study. As an aside, the shoreline positions presented in Figure 4 differ markedly from maps of similar geologic intervals developed by Kauffman (1984, Figs. 11 to 13). The discrepancies commonly are dramatic, with differences for comparable intervals on the scale of hundreds of kilometers. We provide Appendices 1 and 2 so that readers can evaluate the data on which our maps are based.

Figure 6A suggests occurrence of a general, regional regression of the sea throughout the entire interval represented by zones 1 to 33. Superimposed upon the general regression, however, were geographically localized perturbations of shoreline transgression and regression, as emphasized by the polygonal patterns in Figure 6B. Presumed causes of the various irregularities to the general regression are identified below. Figure 6C suggests the relative temporal relationships of strandline migrations to the Late Cretaceous North American land mammal "ages."

Polygon 1 of Figure 6B reflects transgression of the Satan Tongue of the Mancos Shale as recognized in the San Juan Basin of New Mexico. The local transgression followed a brief interval of northeastward shoreline progradation, as represented by the Gibson Coal Member of the Crevasse Canyon Formation. The Satan Tongue rests upon a transgressive sandstone, the Hosta Tongue of the Point Lookout Sandstone (see Molenaar, 1983, p. 206). According to the numbering scheme applied to transgressions and regressions of the sea as used by Molenaar (1983a), the Satan Tongue represents "T-4."

Polygon 2 of Figure 6B reflects east-west components of the northeasterly shoreline progradation of lower parts of the nonmarine Menefee Formation. Interface with more northerly marine waters is represented by the classic regressive Point Lookout Sandstone. The shoreline regression is identified as "R-4" by Molenaar (1983a, p. 210), and present data do not suggest its occurrence north of the vicinity of the San Juan Basin.

Polygon 3 of Figure 6B encloses the lengthy interval of local transgression, stillstands, and regression of the seaway as represented by the type Lewis Shale of the northern San Juan Basin. During the transgressive phase ("T-5" of Molenaar, 1983a), the sequence reflects shoreline stillstands that led to deposition of the Cliff House Sandstone plus upper reaches of the Menefee Formation. The regressive phase ("R-5" of Molenaar, 1983a) involves the near-shore Pictured Cliffs Sandstone plus the terrestrial Fruitland Formation. Note on Figure 6 that, within the interval of time represented by polygon 3, the relative phases of shoreline transgression versus regression were diametrically opposed between the southern and northern parts of the region under study. The

Figure 6. A, Migrations of paleoshorelines at each degree of modern latitude from 35° to 53°N, as plotted in kilometers of eastward excursion from the shorelines' most westerly extents during the geologic interval represented by ammonite zones 1 to 32. The vertical line to the left of each column provides a zero reference from which distance measurements can be taken. Calculated distances of eastern excursion of the shoreline are presented in Table 2 (see Scope and Methods of Study section for procedures of calculation). B, Same as A, but with addition of polygonal patterns to suggest geographical distributions of effects of: 1, shoreline transgression following deposition of the Crevasse Canyon Formation in the San Juan Basin (resulting in deposition of the Satan Tongue of the Mancos Shale); 2, shoreline progradation represented in San Juan Basin by the Point Lookout Sandstone/Menefee Formation; 3, transgressive-stillstand-regressive sequence represented by the type Lewis Shale in the northern San Juan Basin; 4, shoreline progradation leading to deposition of the Eagle Sandstone of Montana (and its continuation into Alberta, the Milk River Formation); 5, effect (possibly largely illusionary) of the so-called "Claggett transgression"; 6, effects of the "Bearpaw transgression" (limited to Alberta, Montana, and possibly northwestern Wyoming), following the "Claggett regression" (as evidenced by progradation of the Judith River Formation across Montana and Alberta into western Saskatchewan); 7, effects of regression from Montana and Alberta of the "Bearpaw sea"; 8, transgression of the "Lewis sea," as recognized by deposition in southern Wyoming/northern Colorado of the erroneously named "Lewis" Shale (genetically unrelated to the type Lewis Shale); and 9, effects of regression of the "Lewis sea" from southern Wyoming and northern Colorado. Most remaining parts of the diagram that are not covered by patterns suggest a single, general regression of the Western Interior Seaway. We propose that shoreline perturbations seen most prominently in areas covered by the patterns reflect the primary influences of local to subregional tectonism upon sedimentary transport and deposition. The general earliest Campanian to Paleocene regression, in contrast, reflects larger scale phenomena, such as slow, global depression of sea level and/or continental elevation (perhaps presaging culmination of the Laramide orogeny). C, Same as A, but with estimated correlations of limits of land-mammal ages (of Fig. 5) with standard ammonite zonation of the Western Interior.

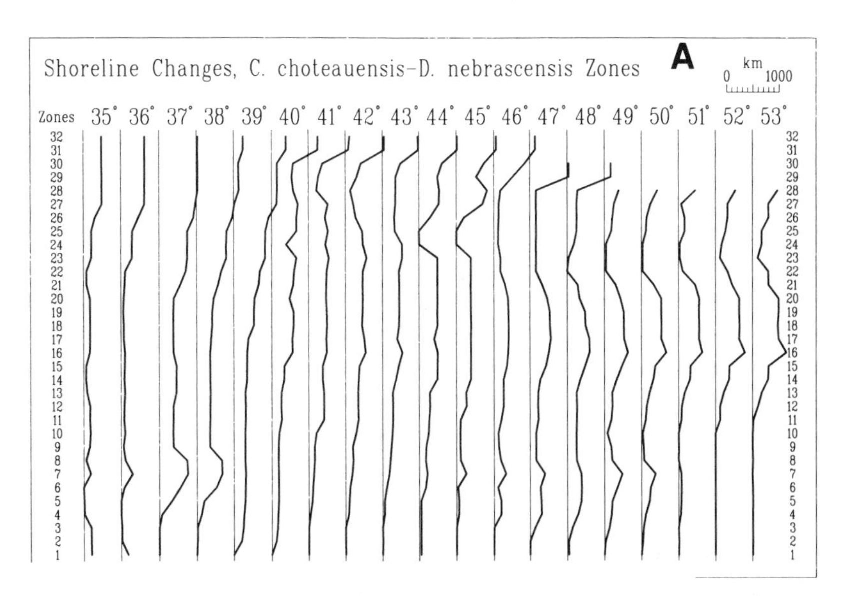
Shoreline Changes, C. choteauensis–D. nebrascensis Zones
A
km
0 1000
Zones 35° 36° 37° 38° 39° 40° 41° 42° 43° 44° 45° 46° 47° 48° 49° 50° 51° 52° 53°
32 31 30 29 28 27 26 25 24 23 22 21 20 19 18 17 16 15 14 13 12 11 10 9 8 7 6 5 4 3 2 1

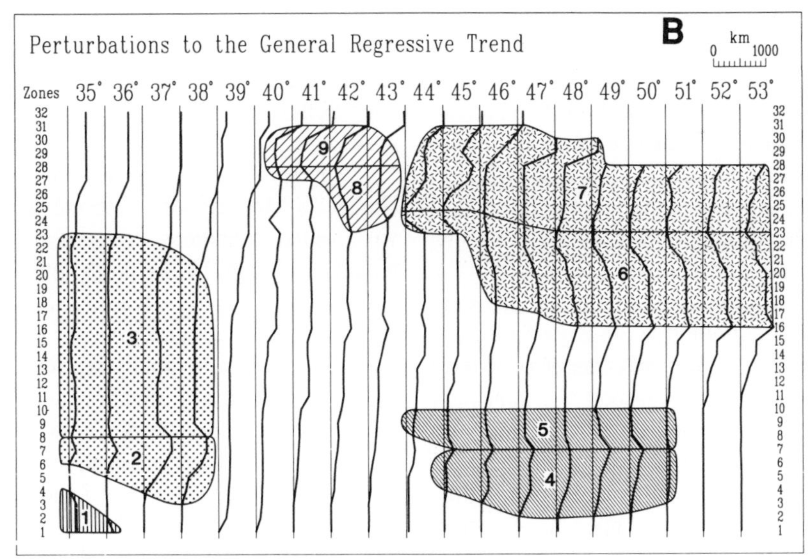
Perturbations to the General Regressive Trend
B
km
0 1000
Zones 35° 36° 37° 38° 39° 40° 41° 42° 43° 44° 45° 46° 47° 48° 49° 50° 51° 52° 53°
32 31 30 29 28 27 26 25 24 23 22 21 20 19 18 17 16 15 14 13 12 11 10 9 8 7 6 5 4 3 2 1
1
2
3
4
5
6
7
8
9

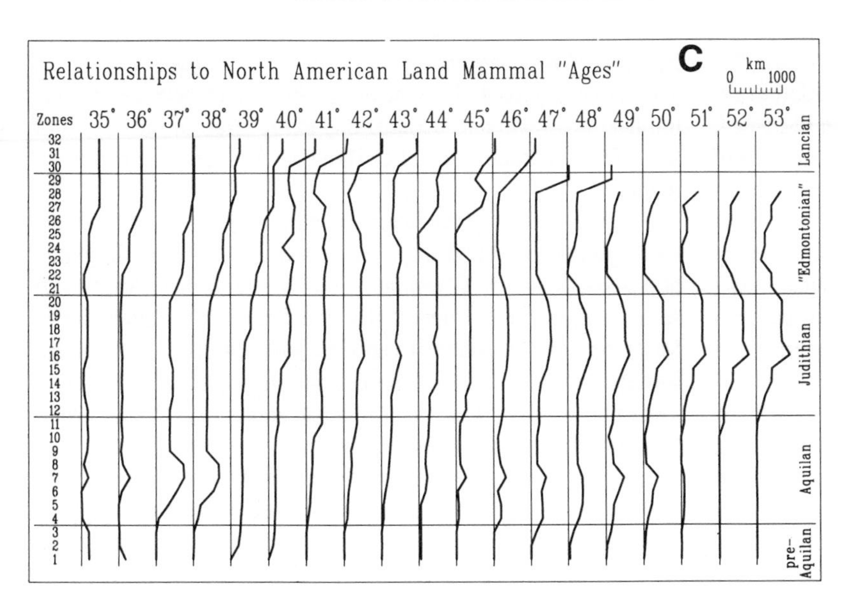
Relationships to North American Land Mammal "Ages"
C
km
0 1000
Zones 35° 36° 37° 38° 39° 40° 41° 42° 43° 44° 45° 46° 47° 48° 49° 50° 51° 52° 53°
32 31 30 29 28 27 26 25 24 23 22 21 20 19 18 17 16 15 14 13 12 11 10 9 8 7 6 5 4 3 2 1
Lancian
"Edmontonian"
Judithian
Aquilan
pre-Aquilan

transition beds (see Flynn, 1986) plus the Kirtland Formation and the McDermott Member of the Animas Formation represent the nonmarine uppermost Cretaceous section of the San Juan Basin above the limits of polygon 3.

As summarized by Cumella (1983), sources for sediments of the Crevasse Canyon, Menefee, Fruitland, and major parts of the Kirtland Formations were in common, and most probably from new orogenic centers in what is now southern Arizona.

Farther to the north, polygon 4 of Figure 6B reflects east- and northeastward progradation of the shoreline in association with deposition of the largely nonmarine Eagle Sandstone of Montana (plus its direct continuation into Alberta, the Milk River Formation). According to the scheme of enumeration of marine transgressions and regressions by Kauffman (1977), polygon 4 could be identified as "R7." Though the Eagle/Milk River progradation was approximately synchronous with the more southerly regression represented by polygon 2, present data do not suggest similarly heightened pulses of eastward shoreline migration in intervening latitudes of Wyoming or Colorado.

Polygon 5 represents a westerly shoreline migration that usually is referred to as the "Claggett transgression." It reflects deposition of the marine Claggett Shale and its Canadian correlative, lower reaches of the Pakowki Formation (Rice and Shurr, 1983, p. 339). The transgressive event was identified by Kauffman (1977) as "T8." Present data suggest restriction of landward migration of the shoreline to the area from northern Wyoming into southern Alberta. It seems worth pursuing the idea that the Claggett transgression was a localized phenomenon related to isostatic subsidence following deposition of the Eagle Sandstone and its lateral marine equivalents. The local transgression may well, in part, reflect reduction of rate of transport of sedimentary debris into the seaway from distant Cordilleran sources and from newly evolving volcanic centers in southwestern Montana. If true, the "Claggett transgression" may represent little more than return of the strand to the less dramatic component of general regional regression.

Polygon 6 represents the so-called "Bearpaw transgression." It reflects generally slow, westward migration of the strand following the prolonged and extensive shoreline progradation related to deposition of the dominantly nonmarine (and ill-named; see Lillegraven and McKenna, 1986) "Mesaverde" Formation of Wyoming, the Two Medicine and Judith River Formations of Montana, and the Belly River/Judith River (the latter being the Foremost/Oldman Formations of older literature) complex of the Prairie Provinces. This major progradational sequence ("R8" of Kauffman, 1977) is not isolated by a polygon on Figure 6B because it merely represents an exaggeration of the general eastward migration of the shoreline that characterizes all but a southern part of the region (where marine transgression was near its maximum); placement of a southern limit to such a polygon on Figure 6B would had to have been purely arbitrary. The easterly shoreline progradation associated with the "Mesaverde," Two Medicine, and Judith River Formations was related to a combination of effects of massive erosion plus transport of resulting sediments from (1) new volcanic centers in southwestern Montana (McGookey and others, 1972; Tysdal and others, 1986); (2) various points farther west in the Cordillera (McLean, 1971; Armstrong and Ward, 1989); and (3) geographically intermediate areas of active foreland thrusting (Lamerson, 1982; Wiltschko and Dorr, 1983).

Figure 6 suggests that effects on shoreline position of the Bearpaw transgression (equivalent to "T9" of Kauffman, 1977) are not recognizable south of northern Wyoming; even there, the reality of effects is uncertain. By explanation, the southerly extreme of polygon 6 of Figure 6B (involving the columns for 44° to 45°N) is intended to show, as a hypothesis, northerly connection of the Bearpaw sea into what is now southeastern Yellowstone National Park. Love (1973) described a series of mollusc-rich strata in that area within the dinosaur-bearing Bobcat Member of the Harebell Formation. The molluscs were identified as juvenile stages of *Mytilus* by Kauffman (1973), and interpreted by him to represent estuarine facies adjacent to more open marine conditions. The presence of "*Veryhachium, Tasmanites,* acritarchs, and dinoflagellates suggests marine or brackish conditions of deposition" (R. H. Tschudy, 1969, in Love, 1973, p. A33).

Subregional shoreline patterns suggest that if, indeed, the Harebell's "*Mytilus* zone" represents marine shoreline paleoenvironmental conditions, connection to the open sea probably would have been to the north into the Bearpaw sea. Such an interpretation would be reasonable in light of the probability of local subsidence associated with: (1) loads developed by the astonishingly thick and rapidly deposited coarse clastic sedimentary sheet of the Harebell Formation; and (2) tectonic loadings by the more westerly thrust sheets. On the other hand, Dr. William A. Cobban (personal communication, 1988) has suggested that his examination of the "*Mytilus*" fossils indicated a fresh-water occurrence of some unidentified form of pelecypod. If true, the hypothesized southerly lobe of the Bearpaw sea into northwestern Wyoming as shown in parts 24 to 25 of Figure 4 probably should be replaced by a shoreline pattern similar to that seen in parts 23 and 26 of Figure 4. In light of the important paleogeographic implications involved, perhaps this entire issue of the nature of the "*Mytilus* zones" in the Harebell Formation should be reopened.

Transgression of the Bearpaw sea was synchronous with the more general condition of gradual eastward shoreline progradation across most of Wyoming, Colorado, and New Mexico. It seems reasonable to suggest that the Bearpaw transgression of Montana and Alberta represents, in part, effects of subregional isostatic subsidence following deposition of the massive clastic loads that make up the Two Medicine and Judith River Formations and their marine equivalents.

Polygon 7 of Figure 6B encloses effects of shoreline regression of the Bearpaw sea (equivalent to regressive event "R9" of Kauffman, 1977). Present data suggest that shoreline progradation had a marked north-to-south component in addition to the dominant east-to-west component. As the local Bearpaw regression continued, the strand migrated to longitudes comparable to those in more southerly parts of the region (where the regression apparently was more gradual and regular).

Polygons 8 and 9 represent, respectively, transgressive and regressive shoreline migrations across southern Wyoming plus northwestern Colorado. The locally recognizable events resulted in deposition of the "Lewis" Shale, a name used quite improperly in this part of the Rocky Mountains. The transgressive/regressive interval represented by polygons 8 to 9 occurred during an interval equivalent to Kauffman's (1977) "R9" regression. As can be seen on maps 23 to 31 of Figure 4, the local "Lewis" sea was isolated from the shoreline bounding the Bearpaw sea. This marine inundation of southern Wyoming and northern Colorado was a highly localized, geologically late, and remarkably persistent event that occurred during a time regionally characterized by general regression of the Western Interior Seaway.

Geologic implications of shoreline migrational histories

Little reliable evidence exists for presence of extensive polar ice or continental glaciation during the Mesozoic. For that reason, variation in positions of marine shorelines at the temporal and geographic scales described in the present study should be evaluated in terms of rates of change in: (1) global sea levels due to evolution of the mid-oceanic ridges and oceanic basins; and (2) variously caused local, subregional, regional, or continental effects of tectonism, including differential subsidence and variations in sedimentary transport. The limited range of latitudes involved in the present study would seem to preclude recognition of important effects of eustacy caused by changing relative positions of continents and/or magmatic masses within the Earth's interior ("geoidal-eustacy" of Mörner, 1980).

Present data suggest a single, very slow, regional regression of the shoreline that was operative, as a minimum, throughout the entire interval encompassed by the study. Such a general regression is consistent with a previously calculated, slowly operating, global sea-level lowering during Late Cretaceous time due to the tectono-eustatic causes of changing volumes of mid-oceanic ridges (Pitman, 1978; Pitman and Golovchenko, 1983). Development of major, deep oceanic volcanic platforms, as suggested by Schlanger and others (1981), also could have affected global sea-level alterations at the time scale under present discussion. We hasten to point out, however, that Schlanger and others (1981) postulated global *trans*gression during the Campanian and early Maastrichtian.

A general regression such as suggested here could correspond either to parts of "first order" (durations of 200 to 300 m.y.) or "second order" (durations of 10 to 80 m.y.) eustatic cycles of Vail and others (1977). A gradual regression such as documented here also is consistent with theoretical predictions for Late Cretaceous lowering of global sea level due to opening of an Atlantic-type basin (Heller and Angevine, 1985). A gradual regional regression also could be consistent in interpretation with a general continental uplift. Finally, such a general regression could be interpreted using a multiplicity of combinations of rates of change in any of the above phenomena. Purely on the basis of intuition, we favor a working hypothesis of gradual global sea-level decrease due to topographic evolution of the world's Late Cretaceous system of mid-oceanic ridges and oceanic basins.

Present data also suggest the imposition of several local or subregional perturbations (expressed as geographically restricted shoreline transgressions, stillstands, and regressions) onto the general regressive trend. Such localized phenomena would be expected in light of subcontinental-scale thermal activation and migration known to have been underway during the Late Cretaceous (McGookey and others, 1972; Armstrong and Ward, 1989). Recognition of local differential subsidence and uplift also is consistent with effects of redistribution of depositional loads and isostatic rebounds adjacent to actively evolving foreland thrustbelts (see Jordan, 1981). Differential effects on subsidence and sediment accumulation more distant from edges of the thrust belt would be expected during Late Cretaceous time from thermal influences of a shallowly subducting oceanic lithospheric slab beneath the Western Interior (Cross and Pilger, 1978). Subregionally active, poorly understood stress fields (Cloetingh and others, 1985) associated with early phases of Laramide-style deformation (Berg, 1981) could result in local *apparent* sea-level fluctuations. Effects of any one, or any combination, of the above phenomena could account for asynchronous transgressions and regressions seen in the area of our study, in western and Arctic Canada (Jeletzky, 1978), and across the world (Matsumoto, 1980; Hubbard, 1988) during Late Cretaceous time.

Present data are inconsistent, however, with at least two other commonly held, geologically central interpretations of Late Cretaceous history of the Western Interior. For example, our study does not support the existence of geologically brief (1 to 10 m.y.) global sea-level fluctuations (e.g., the reputedly worldwide Claggett and Bearpaw "cyclothems" of Kauffman, 1977, 1979, 1984; Hancock and Kauffman, 1979) equivalent to "third order" global eustatic cycles of Vail and others (1977) and "short term" eustatic cycles of Haq and others (1987). Such rapid, worldwide sea-level fluctuations have been geologically unexplainable assuming absence of important Cretaceous glaciation (Pitman and Golovchenko, 1983; Miall, 1986). Secondly, our study does not support existence of a global sea-level highstand as suggested to have occurred during later phases of the Campanian by Vail and others (1977) and Haq and others (1987).

GENERAL CONCLUSIONS

1. First approximations of stratigraphic equivalents of North American Land-Mammal "Ages" (NALMAs; as defined by Lillegraven and McKenna, 1986) in the standard ammonite zonation of Upper Cretaceous marine strata of the North American West-

ern Interior (of Obradovich and Cobban, 1975) appear to be as follows (see Fig. 5):

Lancian—*Sphenodiscus (Coahuilites)* through *"Triceratops"* zones;
"Edmontonian" (not yet faunally defined)—*Didymoceras cheyennense* through *Baculites clinolobatus* zones;
Judithian—*Baculites* sp. (smooth; late form) through *Exiteloceras jenneyi* zones;
Aquilan—*Scaphites hippocrepis* through *Baculites asperiformis* zones.

2. Precision in geological communication about Upper Cretaceous strata in the North American Western Interior would be best served, at least until correlations become more securely refined, through use of existing provincial biostratigraphic terminologies, both for terrestrial and marine sections. Varying criteria have been applied toward placement of stage boundaries involving the European-based names of Santonian, Campanian, and Maastrichian to strata in the Western Interior. Unfortunate results have been unacceptably high levels of terminological confusion combined with inconsistency in geological correlation to the remainder of the world.

3. Mammalian faunas of the upper Fruitland and/or lower Kirtland Formations of the San Juan Basin of New Mexico are probably of "Edmontonian" age rather than of Judithian age, at least according to the biostratigraphic definition of the latter within the present chapter.

4. Considerable latitudinal taxonomic diversification occurred during the Judithian in the Western Interior. Comparative taxonomic information is inadequate to make comparable geographically based statements for other North American land-mammal "ages" during the Cretaceous.

5. Interpretation of changing patterns of ancient strandlines for the western shore of the Western Interior Seaway suggests a slow, general, regional regression that encompassed, at a minimum, the last 20 m.y. of the Cretaceous. We favor explanation by a slow decrease in global sea level related to topographic evolution of the world's Late Cretaceous system of mid-oceanic ridges and oceanic basins.

6. The overall pattern of slow, regional regression was superimposed by effects of local to subregional, asynchronous episodes of transgressions, stillstands, and regressions of the epeiric sea. The perturbations are best interpreted by principal influence of local and subregional tectonism, not global eustatic sea-level changes.

7. Interpreted patterns of shoreline evolution during the last 20 m.y. of the Cretaceous in the Western Interior are inconsistent with (A) the existence of geologically brief (1 to 10 m.y.) global fluctuations in sea level, and (B) the concept that the late Campanian was represented by an unusually high global sea level.

ACKNOWLEDGMENTS

The study was initiated through award to Lillegraven of a University of Wyoming Arts and Sciences Professorship, made possible by funding through the Chacey Kuehn Trust. The work also was supported in part by a National Science Foundation grant (EAR 82-05211) to Lillegraven. Mrs. Linda E. Lillegraven aided preparation of the manuscript in many ways. Dr. J. David Archibald helped with correlations for mammal-bearing localities in southeastern Montana and northwestern Colorado. Valuable discussions were held with Drs. Charles L. Angevine, Richard L. Armstrong, D. L. Blackstone, Jr., Donald W. Boyd, Brent H. Breithaupt, William A. Clemens, William R. Dickinson, William E. Frerichs, Paul H. Heller, J. David Love, Malcolm C. McKenna, and James R. Steidtmann. Special thanks are offered to Dr. William A. Cobban of the U.S. Geological Survey, Denver. Bill graciously reviewed all aspects of the work, provided unpublished data, and contributed generously from his wealth of experience on the Upper Cretaceous of the Western Interior. Finally, thanks are offered to Dr. Curt Teichert, who served as a second reviewer.

APPENDIX 1. SUMMARY INTERPRETIVE SITE-SPECIFIC DATA USED BY COMPUTER TO DEVELOP SHORELINE MAPS PRESENTED AS FIGURE 4. THE LAST 33 COLUMNS CODE ENVIRONMENT OF DEPOSITION AS: 3 = TERRESTRIAL; 2 = STRAND; 1 = MARINE; AND 0 = UNKNOWN (AND PROBABLY UNKNOWABLE). SOURCES OF DATA ARE FROM APPENDIX 2.

				Ammonite Zones						
Identifier	Loc. No.	N. Lat.	W. Long.	1-5	6-10	11-15	16-20	21-25	26-30	31-33
Canelo Hills, Arizona	1	31°45'	110°45'	33333	33333	33333	33333	33333	33333	333
SW San Juan Basin, New Mexico	2	35°45'	108°15'	33333	23333	33333	33333	33333	33333	333
Central San Juan Basin, New Mexico	3	36°15'	108°05'	11111	23211	11222	11111	22333	33333	333
North San Juan Basin, New Mexico	4	37°15'	107°30'	11111	23311	11111	11111	23333	33333	333
Black Mesa, Arizona	5	36°15'	110°15'	23333	33333	33333	33333	33333	33333	333
York Canyon, New Mexico	6	36°45'	105°00'	11111	11111	11111	11111	11111	23333	333
North Raton Basin, Colorado	7	37°47'	104°50'	11111	11111	11111	11111	11111	22333	333
Montrose Syncline, Colorado	8	38°21'	107°30'	11111	11111	11111	11111	12333	33333	333
East Book Cliffs, Colorado	9	39°13'	108°28'	11111	11111	11111	11111	33333	33333	333
Central Book Cliffs, Colorado	10	39°05'	109°35'	11111	11111	11111	12333	33333	33333	333
West Book Cliffs, Utah	11	39°32'	110°33'	11111	11111	22233	33333	33333	33333	333
Wasatch Plateau, Utah	12	39°31'	111°02'	12333	33333	33333	33333	33333	33333	333
Henry Mountains, Utah	13	38°00'	111°00'	11123	33333	33333	33333	33333	33333	333
Kaiparowits Plateau, Utah	14	37°14'	111°30'	33112	33333	33333	33333	33333	33333	333
Paunsaugunt Plateau, Utah	15	37°25'	112°20'	11123	33333	33333	33333	33333	33333	333
Kolob Plateau, Utah	16	37°30'	112°55'	33333	33333	33333	33333	33333	33333	333
Sharon Springs, Kansas	17	38°55'	101°30'	11111	11111	11111	11111	11111	11111	223
Pueblo, Colorado	18	38°20'	104°35'	11111	11111	11111	11111	11111	23333	333
Golden, Colorado	19	39°40'	105°15'	11111	11111	11111	11111	11111	11111	333
Middle Park, Colorado	20	40°07'	106°22'	11111	11111	11111	11111	11111	10000	003
Fish Creek, Colorado	21	40°25'	107°15'	11111	11111	11111	11331	23323	33111	333
Hamilton, Colorado	22	40°22'	107°35'	11111	11111	11111	33331	33333	33111	333
South Sand Wash Basin, Colorado	23	40°35'	107°50'	11111	11111	11111	33333	33333	33111	333
Piceance Creek, Colorado	24	39°48'	108°08'	11111	11111	11111	33331	33333	33333	333
Carbondale, Colorado	25	39°20'	107°15'	11111	11111	11111	11111	22333	33333	333
Redbird, Wyoming	26	43°17'	104°15'	11111	11111	11111	11111	11111	11111	333
Hiawatha Camp, Colorado	27	40°59'	108°33'	11111	11113	33333	33333	33333	33133	333
Vernal Coal Field, Utah	28	40°17'	109°18'	11111	11111	23333	33333	33333	33333	333
Coalville Coal Field, Utah	29	41°00'	111°29'	33333	33333	33333	33333	33333	33333	333
Dutch John, Utah	30	40°59'	109°15'	11111	11111	11233	33333	33333	33333	333
Carter, Wyoming	31	41°22'	110°27'	11123	33333	33333	33333	33333	33333	333
Burntfork, Wyoming	32	41°12'	109°53	11112	33333	33333	33333	33333	33333	333
Firehole, Wyoming	33	41°09'	109°26'	11111	11223	33333	33333	33333	33333	333
Vermilion Basin, Wyoming	34	41°02'	108°45'	11111	11113	33333	33333	33333	33123	333
Opal, Wyoming	35	41°39'	110°24'	11111	33333	33333	33333	33333	33333	333
Little America, Wyoming	36	41°30'	109°43'	11111	23333	33333	33333	33333	33333	333
Wilkins Peak, Wyoming	37	41°23'	109°30'	11111	12223	33333	33333	33333	33333	333
Baxter Basin, Wyoming	38	41°22'	109°05'	11111	11113	33333	33333	33333	33133	333
Bitter Creek, Wyoming	39	41°28'	108°38'	11111	11113	33333	33333	33333	32123	333
Red Desert, Wyoming	40	41°25'	108°28'	11111	11112	33333	33333	33333	21113	333
Muddy Creek, Wyoming	41	40°28'	107°40'	11111	11111	22233	33333	33333	11113	333
Seedskadee, Wyoming	42	41°44'	109°38'	00000	00000	00000	00000	00033	33333	333
Superior, Wyoming	43	41°47'	109°25'	00000	00000	00000	00000	00033	33133	333
Steamboat Mountain, Wyoming	44	41°50'	108°38'	00000	00000	00000	00000	00033	31123	333
SW Divide Basin, Wyoming	45	41°42'	108°20'	00000	00000	00000	00000	00033	21113	333
Central Divide Basin, Wyoming	46	42°05'	107°55'	11111	11113	33333	33333	33333	11233	333
Wamsutter, Wyoming	47	41°37'	107°58'	11111	11111	33333	33333	33333	11113	333
East Washakie Basin, Wyoming	48	41°18'	107°53'	11111	11111	33333	33333	33333	11000	000
Baggs, Wyoming	49	41°03'	107°44'	11111	11111	33333	33333	33333	11000	000
Slater, Colorado	50	40°52'	107°28'	11111	11111	33333	33333	33333	33110	000
Danforth Hills, Colorado	51	40°18'	108°08'	11111	11111	11111	33333	33333	33313	333

APPENDIX 1. SUMMARY INTERPRETIVE SITE-SPECIFIC DATA USED BY COMPUTER TO DEVELOP SHORELINE MAPS PRESENTED AS FIGURE 4. THE LAST 33 COLUMNS CODE ENVIRONMENT OF DEPOSITION AS: 3 = TERRESTRIAL; 2 = STRAND; 1 = MARINE; AND 0 = UNKNOWN (AND PROBABLY UNKNOWABLE). SOURCES OF DATA ARE FROM APPENDIX 2. (continued)

Identifier	Loc. No.	N. Lat.	W. Long.	Ammonite Zones 1-5	6-10	11-15	16-20	21-25	26-30	31-33
Meeker, Colorado	52	40°03'	107°57'	11111	11111	11111	33331	33333	33333	333
Sulphur Canyon, Utah	53	39°12'	109°22'	11111	11111	11111	11333	33333	33333	333
Prairie Canyon, Colorado	54	39°23'	108°59'	11111	11111	11111	11333	33333	33333	333
Big Salt Wash, Colorado	55	39°19'	108°40'	11111	11111	11111	11333	33333	33333	333
Corcoran Mine, Colorado	56	39°13'	108°34'	11111	11111	11111	11222	33333	33333	333
Rollins Mine, Colorado	57	38°52'	108°07'	11111	11111	11111	11112	33333	33333	333
Cedaredge, Colorado	58	38°51'	107°49'	11111	11111	11111	11111	33333	33333	333
Jensen, Utah	59	40°22'	109°31'	11111	11112	33333	33333	33333	33333	333
Dinosaur, Colorado	60	40°12'	108°55'	11111	11111	11112	33333	33333	33333	333
Blue Mountain, Colorado	61	40°15'	108°43'	11111	11111	11111	33333	33333	33333	333
Massadona, Colorado	62	40°14'	108°32'	11111	11111	11111	33333	33333	33333	333
Elk Springs, Colorado	63	40°22'	108°22'	11111	11111	11111	33333	33333	33333	333
Dickman Draw, Colorado	64	40°23'	108°04'	11111	11111	11111	33333	33333	30000	000
Rawlins, Wyoming	65	41°50'	107°26'	11111	11111	11122	33333	33333	11112	333
Lost Soldier, Wyoming	66	42°10'	107°35'	11111	11111	11111	33333	33333	11123	333
North Sinclair, Wyoming	67	41°58'	107°03'	11111	11111	11111	33233	33323	11111	333
NW Medicine Bow, Wyoming	68	41°55'	106°16'	11111	11111	11111	33133	22312	11111	333
SE Medicine Bow, Wyoming	69	41°48'	106°06'	11111	11111	11111	21111	11211	11111	333
Rock River, Wyoming	70	41°43'	105°58'	11111	11111	11111	11111	00011	11111	333
Hunter Canyon, Colorado	71	39°20'	108°30'	11111	11111	11111	11233	33333	33333	333
North Park, Colorado	72	40°31'	106°35'	11111	11111	11111	11111	11000	00000	000
South Sinclair, Wyoming	73	41°34'	107°06'	11111	11111	11111	33333	33333	00000	000
Riverton, Wyoming	74	42°54'	108°21'	11111	12333	33333	33333	33333	33333	333
Gas Hills, Wyoming	75	42°52'	108°11'	11111	11233	33333	33333	33333	33333	333
Fales Rocks, Wyoming	76	42°50'	107°14'	11111	11111	11112	33333	33332	33333	333
Alcova, Wyoming	77	42°37'	106°38'	11111	11111	11111	33111	11331	11111	333
Glenrock, Wyoming	78	42°52'	106°00'	11111	11111	11111	31111	11331	11113	333
Big Bend, Texas	79	29°06'	103°09'	11111	11111	11111	33333	33333	33333	333
Casper Canal, Wyoming	80	42°41'	106°26'	11111	11111	11111	33333	33331	11113	333
Salt Creek, Wyoming	81	43°10'	106°15'	11111	11111	11133	31333	33331	11113	333
North Fork, Wyoming	82	43°45'	106°42'	11111	11111	11133	31333	33331	12223	333
Nowater Creek, Wyoming	83	43°45'	107°30'	11111	11111	11133	33333	33331	13333	333
Zimmerman Butte, Wyoming	84	43°47'	107°49'	11111	11112	33333	33333	33333	33333	333
Roncco Mine, Wyoming	85	43°48'	108°22'	11111	33123	33333	33333	33333	33333	333
Cottonwood Creek, Wyoming	86	43°51'	108°28'	11111	33233	33333	33333	33333	33333	333
Bill, Wyoming	87	43°20'	105°26'	11111	11111	11111	11111	11121	11112	333
Elgin Creek, Wyoming	88	44°15'	106°51'	11111	11111	11133	33333	33331	13333	333
Parkman, Wyoming	89	44°57'	107°26'	11111	11111	11133	33333	33331	13333	333
Hardin, Montana	90	45°46'	107°36'	11111	11111	11111	11111	11111	11133	333
Mosby, Montana	91	46°53'	107°45'	11111	11111	11111	23333	11111	11133	333
Judith River, Montana	92	47°44'	109°33'	11122	23111	11123	33333	21111	11133	333
Forsyth, Montana	93	46°15'	106°43'	11111	11111	11111	11111	11111	11123	333
Lavina, Montana	94	46°12'	108°57'	11111	11111	11111	33333	21111	11133	333
Shawmut, Montana	95	46°15'	109°33'	11111	13111	11113	33333	31111	11133	333
Bruno Siding, Montana	96	46°17'	110°00'	11133	33111	32333	33333	31111	11133	333
Dearborn River, Montana	97	47°11'	112°12'	11333	33333	33333	33333	33333	33333	333
SE Glasgow, Montana	98	47°52'	106°22'	11111	11111	11111	11111	11111	11133	333
Sidney, Montana	99	47°38'	104°15'	11111	11111	11111	11111	11111	11333	333
McClusky, Montana	100	47°23'	100°11'	11111	11111	11111	11111	11111	11111	333
Glad Valley, South Dakota	101	45°23'	101°40'	11111	11111	11111	11111	11111	11111	333
Walhalla, North Dakota	102	48°53'	098°02'	11111	11111	11111	11111	11111	11133	333

APPENDIX 1. SUMMARY INTERPRETIVE SITE-SPECIFIC DATA USED BY COMPUTER TO DEVELOP SHORELINE MAPS PRESENTED AS FIGURE 4. THE LAST 33 COLUMNS CODE ENVIRONMENT OF DEPOSITION AS: 3 = TERRESTRIAL; 2 = STRAND; 1 = MARINE; AND 0 = UNKNOWN (AND PROBABLY UNKNOWABLE). SOURCES OF DATA ARE FROM APPENDIX 2. (continued)

Identifier	Loc. No.	N. Lat.	W. Long.	Ammonite Zones						
				1-5	6-10	11-15	16-20	21-25	26-30	31-33
Bonesteel, South Dakota	103	43°03'	098°43'	11111	11111	11111	11111	11111	11111	113
Landusky, Montana	104	47°43'	108°23'	11111	11111	11111	33111	11111	11133	333
South Glasgow, Montana	105	47°59'	106°41'	11111	11111	11111	11111	11111	11133	333
NW Glasgow, Montana	106	48°41'	106°48'	11111	11111	11111	11111	11111	11133	333
Wild Horse, Alberta	107	49°04'	110°46'	11111	23111	11133	33333	11111	11333	333
Lethbridge, Alberta	108	49°39'	112°43'	11113	33311	13333	33333	31111	33333	333
Legend, Alberta	109	49°34'	111°42'	11111	33111	11133	33333	31111	13333	333
Manyberries, Alberta	110	49°26'	110°30'	11111	13111	11113	33333	31111	11333	333
Cypress Hills, Saskatchewan	111	49°28'	109°58'	11111	11111	11111	33333	11111	11133	333
Shaunavon, Saskatchewan	112	49°24'	108°21'	11111	11111	11111	31111	11111	11133	333
Wood Mountain, Saskatchewan	113	49°15'	106°41'	11111	11111	11111	11111	11111	11133	333
Big Beaver, Saskatchewan	114	49°14'	105°23'	11111	11111	11111	11111	11111	11133	333
Radville, Saskatchewan	115	49°17'	104°16'	11111	11111	11111	11111	11111	11133	333
Champion, Alberta	116	50°11'	112°51'	11113	33111	11333	33333	31111	13333	333
Medicine Hat, Alberta	117	49°52'	110°39'	11111	11111	11111	33333	31111	11333	333
Hatton, Saskatchewan	118	50°08'	109°57'	11111	11111	11111	33333	11111	11133	333
Bateman, Saskatchewan	119	49°56'	106°59'	11111	11111	11111	11111	11111	11133	333
Tilley, Alberta	120	50°28'	111°39'	11111	13111	11133	33333	31111	11333	333
Hilda, Alberta	121	50°26'	110°21'	11111	11111	11111	33333	31111	11333	333
Success, Saskatchewan	122	50°21'	108°42'	11111	11111	11111	31111	11111	11133	333
Glen Kerr, Saskatchewan	123	50°33'	107°13'	11111	11111	11111	11111	11111	11133	333
Darmody, Saskatchewan	124	50°34'	106°21'	11111	11111	11111	11111	11111	11133	333
Regina, Saskatchewan	125	50°26'	104°52'	11111	11111	11111	11111	11111	11133	333
McLean, Saskatchewan	126	50°26'	104°02'	11111	11111	11111	11111	11111	11133	333
Gem, Alberta	127	50°57'	112°20'	11111	11111	11133	33333	31111	13333	333
Finnegan, Altberta	128	51°17'	112°01'	11111	11111	11133	33333	31111	11333	333
Blindloss, Alberta	129	51°07'	110°38'	11111	11111	11111	33333	31111	11333	333
Leader, Saskatchewan	130	50°50'	109°34'	11111	11111	11111	33333	11111	11133	333
Eatonia, Saskatchewan	131	51°15'	109°26'	11111	11111	11111	33333	11111	11133	333
Dunblane, Saskatchewan	132	51°09'	107°01'	11111	11111	11111	11111	11111	11133	333
Scapa, Alberta	133	51°49'	112°15'	11111	11111	11133	33333	31111	11333	333
Consort, Alberta	134	51°41'	110°35'	11111	11111	11111	33333	31111	11333	333
Coleville, Saskatchewan	135	51°39'	109°17'	11111	11111	11111	33333	11111	11133	333
Ruthilda, Saskatchewan	136	51°45'	108°11'	11111	11111	11111	31111	11111	11133	333
Hanley, Saskatchewan	137	51°28'	106°28'	11111	11111	11111	11111	11111	11133	333
Major, Saskatchewan	138	51°55'	109°39'	11111	11111	11111	33333	11111	11133	333
Donalda, Alberta	139	52°29'	112°23'	11111	11111	11133	33333	33111	33333	333
Provost, Alberta	140	52°17'	110°21'	11111	11111	11111	33333	11111	11333	333
Unita, Saskatchewan	141	52°21'	109°20'	11111	11111	11111	33333	11111	11133	333
Battleford, Saskatchewan	142	52°25'	107°53'	11111	11111	11111	31111	11111	11133	333
Saskatoon, Saskatchewan	143	52°06'	106°12'	11111	11111	11111	11111	11111	11133	333
Ryley, Alberta	144	53°15'	112°38'	11111	11111	11133	33333	33111	33333	333
McLaughlin, Alberta	145	52°52'	110°14'	11111	11111	11111	00000	11111	11333	333
Lloydminster, Saskatchewan	146	53°27'	109°25'	11111	11111	11111	30000	00000	00000	000
Cardston, Alberta	147	49°10'	113°07'	11333	33331	33333	33333	31113	33333	333
Pekisko, Alberta	148	50°22'	114°05'	11133	33111	33333	33333	31113	33333	333
Sunnyslope, Alberta	149	51°40'	113°18'	11111	11111	11333	33333	33113	33333	333
Lacombe, Alberta	150	52°36'	114°05'	11111	11111	11333	33333	33133	33333	333
Drayton Valley, Alberta	151	53°24'	115°23'	11111	11111	13333	33333	33333	33333	333
Whitecourt, Alberta	152	54°09'	116°39'	11111	11111	33333	33333	33333	33333	333
Valleyview, Alberta	153	54°45'	117°50'	11111	11111	33333	33333	33333	33333	333

APPENDIX 1. SUMMARY INTERPRETIVE SITE-SPECIFIC DATA USED BY COMPUTER TO DEVELOP SHORELINE MAPS PRESENTED AS FIGURE 4. THE LAST 33 COLUMNS CODE ENVIRONMENT OF DEPOSITION AS: 3 = TERRESTRIAL; 2 = STRAND; 1 = MARINE; AND 0 = UNKNOWN (AND PROBABLY UNKNOWABLE). SOURCES OF DATA ARE FROM APPENDIX 2. (continued)

				Ammonite Zones						
Identifier	Loc. No.	N. Lat.	W. Long.	1-5	6-10	11-15	16-20	21-25	26-30	31-33
Depuyer, Montana	154	48°08'	112°29'	11133	33333	33333	33333	31123	33333	333
Mount May, Alberta	155	54°07'	119°50'	11111	11113	33333	33333	33333	33333	333
Slave Lake, Alberta	156	55°26'	115°25'	11111	11111	13111	33333	33333	33333	333
Sundre, Alberta	157	51°52'	114°52'	11111	11111	33333	33333	33333	33333	333
Scobey, Montana	158	48°40'	105°41'	11111	11111	11111	11111	11111	11133	333
Havre, Montana	159	48°22'	109°42'	11112	23111	11113	33331	11111	11133	333
Trochu, Alberta	160	51°52'	112°55'	11111	11111	11133	33333	33113	33333	333
Roundup, Montana	161	46°41'	108°25'	11111	12111	11113	33333	11111	11133	333
Tabiona, Utah	162	40°22'	110°51'	00000	00000	00000	33333	33333	33333	333
Mobridge, South Dakota	163	45°39'	100°28'	11111	11111	11111	11111	11111	11111	113
Livingston, Montana	164	45°37'	110°37'	00000	13333	33333	33333	33333	33333	333
Columbus, Montana	165	45°28'	109°11'	11111	13111	13333	33333	33311	11333	333
Elk Basin, Wyoming	166	44°56'	108°49'	11111	13111	13333	33333	33311	33333	333
SE Yellowstone Park, Wyoming	167	44°00'	110°36'	00000	00000	00000	00000	00022	00000	000

APPENDIX 2. PRIMARY SOURCES OF INFORMATION USED IN CONSTRUCTION OF APPENDIX 1 AND PALEOSHORELINE MAPS (FIG. 4)

Alberta and British Columbia
Caldwell, 1968, 1982
Caldwell and North, 1984
Caldwell and others, 1978
Dodson, 1971
Gibson, 1977
Gill and Cobban, 1973
Given and Wall, 1971
Jeletzky, 1968, 1971
Jerzykiewicz and McLean, 1980
Lerbekmo, 1961
Lerbekmo and Coulter, 1985
Lillegraven, 1969
McCrory and Walker, 1986
McLean, 1971, 1977
Meijer Drees and Mhyr, 1981
Ower, 1960
Riccardi, 1983
Rice, 1981
Russell, 1940
Russell, 1987
Russell and Landes, 1940
Shaw and Harding, 1954
Stott, 1975
Williams and Burk, 1964

Saskatchewan and Manitoba
Caldwell, 1968, 1982
Caldwell and North, 1984
Forester and others, 1977
Furnival, 1946
Jeletzky, 1968
Lerbekmo and Coulter, 1985
McLean, 1971
McNeil and Caldwell, 1981
North and Caldwell, 1975
Reiskind, 1975
Riccardi, 1983
Williams and Baadsgaard, 1975

Montana
Cobban, 1951
Cobban and Reeside, 1952
Gautier, 1981
Gill and Cobban, 1966a, b, 1973
McLean, 1971
Nichols and others, 1982, 1985
Rice, 1981
Rice and Shurr, 1983
Rice and others, 1982
Roberts, 1972
Robinson and others, 1959, 1964
Ruppel and Lopez, 1984
Schultz and others, 1980
Tysdal and others, 1986
Williams and Burk, 1964

North Dakota
Gill and Cobban, 1961, 1965, 1966a, b, 1973
Witzke and others, 1983

Wyoming
Barwin, 1961
Bergstresser and Frerichs, 1982
Blackstone, 1975
Cobban, 1951
Dorr and others, 1977
Farabee and Canright, 1986
Frerichs, 1979, 1980
Frerichs and Adams, 1973
Gill and others, 1970
Gill and Cobban, 1966a, b, 1973
Kauffman, 1973
Keefer, 1965
Keefer and Rich, 1957
Lamerson, 1982
Lillegraven and McKenna, 1986
Lillegraven and Ostresh, 1988
Love, 1970, 1973
MacKenzie, 1975
Miller, 1977
Nichols and others, 1982
Oriel and Tracey, 1970
Reynolds, 1966, 1976
Rich, 1958
Robinson and others, 1959, 1964
Roehler, 1983
Schultz and others, 1980
Zapp and Cobban, 1962

South Dakota
Cobban, 1951
Gill and Cobban, 1973
Robinson and others, 1964
Rubey, 1930
Waage, 1968
Witzke and others, 1983

Utah
Bissell, 1952
Bowers, 1972
Cobban, 1951
Decourten, 1978
Doelling, 1972a, b, c, and d
Doelling and Graham, 1972 a, b
Eaton, 1987a
Fisher and others, 1960
Fouch and others, 1983
Gill and Hail, 1975
Gregory, 1951
Hansen, 1965
Isby and Picard, 1983
Keighin and Fouch, 1981
Lawrence, 1965
Lawton, 1985
Lohrengel, 1969
Matheny and Picard, 1985
Peterson, 1969a, b
Peterson and Kirk, 1977
Peterson and Ryder, 1975
Peterson and Waldrop, 1965
Peterson and others, 1980
Ryer, 1983
Spieker, 1946, 1949

Colorado
Armstrong, 1969
Baltz, 1965
Barnes and others, 1954
Bass and others, 1955
Billingsley, 1977
Brownfield and Johnson, 1986
Cobban, 1951, 1973
Dickinson, 1965
Dickinson and others, 1968
Dolly and Meissner, 1977
Fassett, 1976
Fisher and others, 1960
Frerichs, 1980
Frerichs and others, 1977
Gill and Cobban, 1966a
Gill and others, 1972
Gill and Hail, 1975
Goldstein, 1950
Hail, 1968
Hancock, 1925
Irwin, 1986
Izett and others, 1971
Johnson and May, 1980
Johnson and others, 1980
Kiteley, 1983
Kiteley and Field, 1984
Konishi, 1959
Lillegraven, 1987
Madden, 1985
Manfrino, 1984
Miller, 1977
Scott, 1964
Scott and Cobban, 1963, 1964, 1965
Siepman, 1986
Tweto, 1975
Weimer, 1976, 1983
Weimer and Tillman, 1980
Zapp and Cobban, 1960

Kansas
Cobban, 1951
Gill and others, 1972
Witzke and others, 1983

Arizona
Drewes, 1971
Hayes, 1970
Hayes and Drewes, 1978
Keith, 1978
Kluth, 1983
Molenaar, 1983a, b
O'Sullivan and others, 1972
Repenning and Page, 1956

New Mexico
Ash and Tidwell, 1976
Baltz, 1965
Brookins and Rigby, 1987
Brown, 1962
Cobban, 1951, 1973, 1976
Cobban and others, 1974
Cumella, 1983
Fassett, 1976, 1977, 1987
Fassett and Hinds, 1971
Flores and Tur, 1982
Klute, 1986
Lee, 1917
Lehman, 1985
Lucas and Schoch, 1982
Miller, 1983
Molenaar, 1977, 1983a, b
Orth and others, 1982
Peterson and Kirk, 1977
Pillmore, 1976
Pillmore and Maberry, 1976
Pillmore and others, 1984
Reeside, 1924
Siemers and King, 1974
Tschudy, 1973

Texas
Maxwell and others, 1967
Schiebout and others, 1987

REFERENCES CITED

Archibald, J. D., 1982, A study of Mammalia and geology across the Cretaceous-Tertiary boundary in Garfield County, Montana: University of California Publications in Geological Sciences, v. 122, 286 p.

——, 1987, Late Cretaceous (Judithian and Edmontonian) vertebrates and geology of the Williams Fork Formation, NW Colorado, *in* Currie, P. J., and Koster, E. H., eds., Fourth symposium on Mesozoic terrestrial ecosystems: Drumheller, Alberta, Tyrrell Museum of Palaeontology Occasional Paper 3, p. 7–11.

Archibald, J. D., and Clemens, W. A., 1984, Mammal evolution near the Cretaceous-Tertiary boundary, *in* Berggren, W. A., and Van Couvering, J. A., eds., Catastrophies and earth history; The new uniformitarianism: Princeton, New Jersey, Princeton University Press, p. 339–371.

Armstrong, R. L., 1969, K-Ar dating of laccolithic centers of the Colorado Plateau and vicinity: Geological Society of America Bulletin, v. 80, p. 2081–2086.

Armstrong, R. L., and Ward, P., 1989, Late Triassic to earliest Eocene magmatism in the North American Cordillera; Implications for the Western Interior Basin, *in* Caldwell, W.G.E., and Kauffman, E. G., eds., The Western Interior Basin: Geological Association of Canada Special Paper (in press).

Armstrong-Ziegler, J. G., 1978, A aniliid snake and associated vertebrates from the Campanian of New Mexico: Journal of Paleontology, v. 52, p. 480–483.

Ash, S. R., and Tidwell, W. D., 1976, Upper Cretaceous and Paleocene floras of the Raton Basin, Colorado and New Mexico: New Mexico Geological Society 27th Field Conference Guidebook, p. 197–203.

Baltz, E. H., 1965, Stratigraphy and history of Raton Basin and notes on San Luis Basin, Colorado–New Mexico: American Association of Petroleum Geologists Bulletin, v. 49, p. 2041–2075.

Barnes, H., Baltz, E. H., Jr., and Hayes, P. T., 1954, Geology and fuel resources of the Red Mesa area, La Plata and Montezuma counties, Colorado: U.S. Geological Survey Oil and Gas Investigations Map OM–149, scale 1:62,500.

Barwin, J. R., 1961, Stratigraphy of the Mesaverde Formation in the southern part of the Wind River Basin, Wyoming: Wyoming Geological Association 16th Annual Field Conference Guidebook, p. 171–179.

Bass, N. W., Eby, J. B., and Campbell, M. R., 1955, Geology and mineral fuels of parts of Routt and Moffat Counties, Colorado: U.S. Geological Survey Bulletin 1027–D, p. 143–250.

Berg, R. R., 1981, Review of thrusting in the Wyoming foreland: Contributions to Geology, University of Wyoming, v. 19, p. 93–104.

Bergstresser, T. J., and Frerichs, W. E., 1982, Planktonic foraminifera from the Upper Cretaceous Pierre Shale at Red Bird, Wyoming: Journal of Foraminiferal Research, v. 12, p. 353–361.

Billingsley, L. T., 1977, Stratigraphy of the Trinidad Sandstone and associated formations Walsenburg area, Colorado, *in* Veal, H. K., ed., Exploration frontiers of the Central and Southern Rockies: Denver, Colorado, Rocky Mountain Association of Geologists 1977 Symposium, p. 235–246.

Bissell, H. J., 1952, Geology of the Cretaceous and Tertiary sedimentary rocks of the Utah-Arizona-Nevada corner, *in* Guidebook to the Geology of Utah, Cedar City to Las Vegas, Nevada: Intermountain Association of Petroleum Geologists, no. 7, p. 69–78.

Blackstone, D. L., Jr., 1975, Late Cretaceous and Cenozoic history of the Laramie Basin region, southeast Wyoming, *in* Curtis, B. F., ed., Cenozoic history of the southern Rocky Mountains: Boulder, Colorado, Geological Society of America Memoir 144, p. 249–279.

Bowers, W. E., 1972, The Canaan Peak, Pine Hollow, and Wasatch formations in the Table Cliff region, Garfield County, Utah: U.S. Geological Survey Bulletin 1331–B, p. B1–B39.

Breithaupt, B. H., 1982, Paleontology and paleoecology of the Lance Formation (Maastrichtian), east flank of Rock Springs Uplift, Sweetwater County, Wyoming: Contributions to Geology, University of Wyoming, v. 21, p. 123–151.

Brookins, D. G., and Rigby, J. K., Jr., 1987, Geochronologic and geochemical study of volcanic ashes from the Kirtland Shale (Cretaceous), San Juan Basin, New Mexico, *in* Fassett, J. E., and Rigby, J. K., Jr., eds., The Cretaceous-Tertiary boundary in the San Juan and Raton basins, New Mexico and Colorado: Boulder, Colorado, Geological Society of America Special Paper 209, p. 105–110.

Brown, R. W., 1962, Paleocene flora of the Rocky Mountains and Great Plains: U.S. Geological Survey Professional Paper 375, 119 p.

Brownfield, M. E., and Johnson, E. A., 1986, A regionally extensive altered air-fall ash for use in correlation of lithofacies in the Upper Cretaceous Williams Fork Formation, northeastern Piceance Creek and southern Sand Wash basins, Colorado, *in* Stone, D. S., ed., New interpretations of northwest Colorado geology: Denver, Colorado, Rocky Mountain Association of Geologists, 1986 Symposium, p. 165–169.

Butler, R. F., and Lindsay, E. H., 1985, Mineralogy of magnetic minerals and revised magnetic polarity stratigraphy of continental sediments, San Juan Basin, New Mexico: Journal of Geology, v. 93, p. 535–554.

Caldwell, W.G.E., 1968, The Late Cretaceous Bearpaw Formation in the South Saskatchewan River Valley: Saskatchewan Research Council Geology Division Report 8, 86 p.

——, 1982, The Cretaceous system in the Williston Basin; A modern appraisal, *in* Christopher, J. E., and Kaldi, J., eds., Fourth International Williston Basin Symposium: Regina, Saskatchewan Geological Society Special Paper 6, p. 295–312.

Caldwell, W.G.E., and North, B. R., 1984, Cretaceous stage boundaries in the southern Interior Plains of Canada: Geological Society of Denmark Bulletin, v. 33, p. 57–69.

Caldwell, W.G.E., North, B. R., Stelck, C. R., and Wall, J. H., 1978, A foraminiferal zonal scheme for the Cretaceous system in the Interior Plains of Canada, *in* Stelck, C. R., and Chatterton, B.D.E., eds., Western and Arctic Canadian biostratigraphy: Geological Association of Canada Special Paper 18, p. 495–575.

Carpenter, K., 1979, Vertebrate fauna of the Laramie Formation (Maestrichtian), Weld County, Colorado: Contributions to Geology, University of Wyoming, v. 17, p. 37–49.

Clemens, W. A., 1961, A Late Cretaceous mammal from Dragon Canyon, Utah: Journal of Paleontology, v. 35, p. 578–579.

——, 1973a, Fossil mammals of the type Lance Formation, Wyoming; Part 3, Eutheria and summary: University of California Publications in Geological Sciences, v. 94, 102 p.

——, 1973b, The roles of fossil vertebrates in interpretation of Late Cretaceous stratigraphy of the San Juan Basin, New Mexico: Four Corners Geological Society Memoir, 1973, p. 154–167.

Clemens, W. A., and Lillegraven, J. A., 1986, New Late Cretaceous, North American advanced therian mammals that fit neither the marsupial nor eutherian molds, *in* Flanagan, K. M., and Lillegraven, J. A., eds., Vertebrates, phylogeny, and philosophy: Laramie, Contributions to Geology, University of Wyoming, Special Paper 3, p. 55–85.

Clemens, W. A., Lillegraven, J. A., Lindsay, E. H., and Simpson, G. G., 1979, Where, when, and what; A survey of known Mesozoic mammal distribution, *in* Lillegraven, J. A., Kielan-Jaworowska, Z., and Clemens, W. A., eds., Mesozoic mammals; The first two-thirds of mammalian history: Berkeley, University of California Press, p. 7–58.

Cloetingh, S., McQueen, H., and Lambeck, K., 1985, On a tectonic mechanism for regional sealevel variations: Earth and Planetary Science Letters, v. 75, p. 157–166.

Cobban, W. A., 1951, Scaphitoid cephalopods of the Colorado Group: U.S. Geological Survey Professional Paper 239, 42 p.

——, 1973, Significant ammonite finds in uppermost Mancos Shale and overlying formations between Barker Dome, New Mexico, and Grand Junction, Colorado, *in* Fassett, J. E., ed., Cretaceous and Tertiary rocks of the southern Colorado Plateau: Four Corners Geological Society Memoir, p. 148–153.

——, 1976, Ammonite record from the Pierre Shale of northeastern New Mexico: New Mexico Geological Society 27th Field Conference Guidebook, p. 165–169.

——, 1977, A new curved baculite from the Upper Cretaceous of Wyoming: U.S. Geological Survey Journal of Research, v. 4, p. 457–462.

Cobban, W. A., and Reeside, J. B., Jr., 1952, Correlation of the Cretaceous formations of the western interior of the United States: Geological Society of America Bulletin, v. 63, p. 1011–1044.

Cobban, W. A., Landis, E. R., and Dane, C. H., 1974, Age relations of upper part of Lewis Shale on east side of San Juan Basin, New Mexico: New Mexico Geological Society 25th Field Conference Guidebook, p. 279–282.

Cross, T. A., and Pilger, R. H., Jr., 1978, Tectonic controls of late Cretaceous sedimentation, western interior USA: Nature, v. 274, p. 653–657.

Cumella, S. P., 1983, Relation of Upper Cretaceous regressive sandstone units of the San Juan Basin to source area tectonics, *in* Reynolds, M. W., and Dolly, E. D., eds., Mesozoic paleogeography of the west-central United States: Rocky Mountain Paleogeography Symposium 2: Denver, Colorado, Rocky Mountain Section, Society of Economic Paleontologists and Mineralogists, p. 189–199.

Dalrymple, G. B., 1979, Critical tables for the conversion of K-Ar ages from old to new constants: Geology, v. 7, p. 558–560.

Decourten, F. L., 1978, Non-marine flora and fauna from the Kaiparowits Formation (Upper Cretaceous) of the Paria River Amphitheater, southwestern Utah: Geological Society of America Abstracts with Programs, v. 10, p. 102.

Dickinson, R. G., 1965, Geologic map of the Cerro Summit quadrangle, Montrose County, Colorado: U.S. Geological Survey Geological Quadrangle Map GQ–486, scale 1:24,000.

Dickinson, R. G., Leopold, E. B., and Marvin, R. F., 1968, Late Cretaceous uplift and volcanism on the north flank of the San Juan Mountains, Colorado, *in* Epis, R. C., ed., Cenozoic volcanism in the southern Rocky Mountains: Golden, Colorado School of Mines Quarterly, v. 63, p. 125–148.

Dodson, P., 1971, Sedimentology and taphonomy of the Oldman Formation (Campanian), Dinosaur Provincial Park, Alberta (Canada): Palaeogeography, Palaeoclimatology, Palaeoecology, v. 10, p. 21–74.

Doelling, H. H., 1972a, Alton coal field, *in* Doelling, H. H., and Graham, R. L., eds., Southwestern Utah coal fields; Alton, Kaiparowits Plateau and Kolob-Harmony: Salt Lake City, Utah Geological and Mineralogical Survey Monograph Series 1, p. 1–66.

——, 1972b, Kolob-Harmony coal fields, *in* Doelling, H. H., and Graham, R. L., eds., Southwestern Utah coal fields; Alton, Kaiparowits Plateau and Kolob-Harmony: Salt Lake City, Utah Geological and Mineralogical Survey Monograph Series 1, p. 251–333.

——, 1972c, Coalville coal field, *in* Doelling, H. H., and Graham, R. L., eds., Southwestern Utah coal fields; Alton, Kaiparowits Plateau and Kolob-Harmony: Salt Lake City, Utah Geological and Mineralogical Survey Monograph Series 1, p. 323–354.

——, 1972d, Henrys Fork coal field, *in* Doelling, H. H., and Graham, R. L., eds., Southwestern Utah coal fields; Alton, Kaiparowits Plateau and Kolob-Harmony: Salt Lake City, Utah Geological and Mineralogical Survey Monograph Series 1, p. 355–378.

Doelling, H. H., and Graham, R. L., 1972a, Kaiparowits Plateau coal field, *in* Doelling, H. H., and Graham, R. L., eds., Southwestern Utah coal fields; Alton, Kaiparowits Plateau and Kolob-Harmony: Salt Lake City, Utah Geological and Mineralogical Survey Monograph Series 1, p. 67–249.

——, 1972b, Vernal coal field, *in* Doelling, H. H., and Graham, R. L., eds., Eastern and northern Utah coal fields; Vernal, Henry Mountains, Sego, La Sal-San Juan, Tabby Mountain, Coalville, Henrys Fork, Goose Creek and Lost Creek: Salt Lake City, Utah Geological and Mineralogical Survey Monograph Series 2, p. 1–95.

Dolly, E. D., and Meissner, F. F., 1977, Geology and gas exploration potential, Upper Cretaceous and lower Tertiary strata, northern Raton Basin, Colorado, *in* Veal, H. K., ed., Exploration frontiers of the Central and Southern Rockies: Denver, Colorado, Rocky Mountain Association of Geologists 1977 Symposium, p. 247–270.

Dorr, J. A., Jr., Spearing, D. R., and Steidtmann, J. R., 1977, Deformation and deposition between a foreland uplift and an impinging thrust belt; Hoback Basin, Wyoming: Geological Society of America Special Paper 177, 82 p.

Drewes, H., 1971, Mesozoic stratigraphy of the Santa Rita Mountains, southeast of Tucson, Arizona: U.S. Geological Survey Professional Paper 658–C, 81 p.

Eaton, J. G., 1987a, Stratigraphy, depositional environments, and age of Cretaceous mammal-bearing rocks in Utah, and systematics of the Multituberculata (Mammalia) [Ph.D. thesis]: Boulder, University of Colorado, 308 p.

——, 1987b, The Campanian–Maastrichtian boundary in the Western Interior of North America: Newsletters on Stratigraphy, v. 18, p. 31–39.

Farabee, M. J., and Canright, J. E., 1986, Stratigraphic palynology of the lower part of the Lance Formation (Maestrichtian) of Wyoming: Sonder-Abdruck aus Palaeontographica Beitrage zur Naturgeschichte der Vorzeit, Abt. B, Bd. 199, p. 1–89.

Fassett, J. E., 1976, What happened during Late Cretaceous time in the Raton and San Juan basins—with some thoughts about the area in between: New Mexico Geological Society 27th Field Conference Guidebook, p. 185–190.

——, 1977, Geology of the Point Lookout, Cliff House, and Pictured Cliffs sandstones of the San Juan Basin, New Mexico and Colorado: New Mexico Geological Society 28th Field Conference Guidebook, p. 193–197.

——, 1987, The ages of the continental, Upper Cretaceous, Fruitland Formation and Kirtland Shale based on a projection of ammonite zones from the Lewis Shale, San Juan Basin, New Mexico and Colorado, *in* Fassett, J. E., and Rigby, J. K., Jr., eds., The Cretaceous–Tertiary boundary in the San Juan and Raton basins, New Mexico and Colorado: Boulder, Geological Society of America Special Paper 209, p. 5–16.

Fassett, J. E., and Hinds, J. S., 1971, Geology and fuel resources of the Fruitland Formation and the Kirtland Shale of the San Juan Basin, New Mexico and Colorado: U.S. Geological Survey Professional Paper 676, 76 p.

Fisher, D. J., Erdmann, C. E., and Reeside, J. B., Jr., 1960, Cretaceous and Tertiary formations of the Book Cliffs, Carbon, Emery, and Grand counties, Utah, and Garfield and Mesa counties, Colorado: U.S. Geological Survey Professional Paper 332, 80 p.

Flores, R. M., and Tur, S. M., 1982, Characteristics of deltaic deposits in the Cretaceous Pierre Shale, Trinidad Sandstone, and Vermejo Formation, Raton Basin, Colorado: The Mountain Geologist, v. 19, no. 2, p. 25–40.

Flynn, L. J., 1986, Late Cretaceous mammal horizons from the San Juan Basin, New Mexico: American Museum Novitates 2845, 30 p.

Forester, R. W., Caldwell, W.G.E., and Oro, F. H., 1977, Oxygen and carbon isotopic study of ammonites from the Late Cretaceous Bearpaw Formation in southwestern Saskatchewan: Canadian Journal of Earth Sciences, v. 14, p. 2086–2100.

Fouch, T. D., Lawton, T. F., Nichols, D. J., Cashion, W. B., and Cobban, W. A., 1983, Patterns and timing of synorogenic sedimentation in Upper Cretaceous rocks of central and northeast Utah, *in* Reynolds, M. W., and Dolly, E. D., eds., Mesozoic paleogeography of the west-central United States; Rocky Mountain paleogeography, Symposium 2: Denver, Colorado, Rocky Mountain Section, Society of Economic Paleontologists and Mineralogists, p. 305–336.

Fox, R. C., 1974, *Deltatheroides*-like mammals from the Upper Cretaceous of North America: Nature, v. 249, p. 392.

——, 1976, Cretaceous mammals (*Meniscoessus intermedius,* new species, and *Alphadon* sp.) from the lowermost Oldman Formation, Alberta: Canadian Journal of Earth Sciences, v. 13, p. 1216–1222.

——, 1981, Mammals from the Upper Cretaceous Oldman Formation, Alberta; V, *Eodelphis* Matthew, and the evolution of the Stagodontidae (Marsupialia): Canadian Journal of Earth Sciences, v. 18, p. 350–365.

——, 1984, A primitive, "obtuse-angled" symmetrodont (Mammalia) from the Upper Cretaceous of Alberta, Canada: Canadian Journal of Earth Sciences, v. 21, p. 1204–1207.

——, 1987, Patterns of mammalian evolution towards the end of the Cretaceous, Saskatchewan, Canada, *in* Currie, P. J., and Koster, E. H., eds., Fourth symposium on Mesozoic terrestrial ecosystems: Drumheller, Alberta, Tyrrell Museum of Palaeontology Occasional Paper 3, p. 96–100.

Fox, R. C., and Naylor, B. G., 1986, A new species of *Didelphodon* Marsh (Marsupialia) from the Upper Cretaceous of Alberta, Canada; Paleobiology and phylogeny: Neues Jahrbuch für Geologie und Paläontologie Abhandlungen, v. 172, p. 357–380.

Frerichs, W. E., 1979, Planktonic foraminifera from the Sage Breaks Shale, Centennial Valley, Wyoming: Journal of Foraminiferal Research, v. 9, p. 159–184.

——, 1980, Age of the western interior *Clioscaphites chouteauensis* zone: Journal of Paleontology, v. 54, p. 366–370.

Frerichs, W. E., and Adams, P., 1973, Correlation of the Hilliard Formation with the Niobrara Formation: Wyoming Geological Association 25th Annual Field Conference Guidebook, p. 187–192.

Frerichs, W. E., Pokras, E. M., and Evetts, M. J., 1977, The genus *Hastigerinoides* and its significance in the biostratigraphy of the Western Interior: Journal of Foraminiferal Research, v. 7, p. 149–156.

Furnival, G. M., 1946, Cypress Lake map-area, Saskatchewan: Geological Survey of Canada Memoir 242, 161 p.

Gautier, D. L., 1981, Petrology of the Eagle Sandstone, Bearpaw Mountains area, north-central Montana: U.S. Geological Survey Bulletin 1521, 54 p.

Gibson, D. W., 1977, Upper Cretaceous and Tertiary coal-bearing strata in the Drumheller–Ardley region, Red Deer River Valley, Alberta: Geological Survey of Canada Paper 76–35, 41 p.

Gill, J. R., and Cobban, W. A., 1961, Stratigraphy of lower and middle parts of the Pierre Shale, northern Great Plains: U.S. Geological Survey Professional Paper 424–D, p. D185–D191.

——, 1965, Stratigraphy of the Pierre Shale, Valley City and Pembina Mountain areas, North Dakota: U.S. Geological Survey Professional Paper 392–A, 20 p.

——, 1966a, The Red Bird section of the Upper Cretaceous Pierre Shale in Wyoming: U.S. Geological Survey Professional Paper 393–A, 73 p.

——, 1966b, Regional unconformity in Late Cretaceous Wyoming: U.S. Geological Survey Professional Paper 550–B, p. B20–B27.

——, 1973, Stratigraphy and geologic history of the Montana Group and equivalent rocks, Montana, Wyoming, and North and South Dakota: U.S. Geological Survey Professional Paper 776, 37 p.

Gill, J. R., and Hail, W. J., Jr., 1975, Stratigraphic sections across Upper Cretaceous Mancos Shale–Mesaverde Group boundary, eastern Utah and western Colorado: U.S. Geological Survey Oil and Gas Investigations Chart OC–68, 1 sheet.

Gill, J. R., Merewether, E. A., and Cobban, W. A., 1970, Stratigraphy and nomenclature of some Upper Cretaceous and lower Tertiary rocks in south-central Wyoming: U.S. Geological Survey Professional Paper 667, 53 p.

Gill, J. R., Cobban, W. A., and Schultz, L. G., 1972, Stratigraphy and composition of the Sharon Springs Member of the Pierre Shale in western Kansas: U.S. Geological Survey Professional Paper 728, 50 p.

Gingerich, P. D., Rose, K. D., and Krause, D. W., 1980, Early Cenozoic mammalian faunas of the Clark's Fork Basin–Polecat Bench area, northwestern Wyoming, *in* Gingerich, P. D., ed., Early Cenozoic paleontology and stratigraphy of the Bighorn Basin, Wyoming: Ann Arbor, University of Michigan Papers on Paleontology 24, p. 51–68.

Given, M. M., and Wall, J. H., 1971, Microfauna from the Upper Cretaceous Bearpaw Formation of south-central Alberta: Bulletin of Canadian Petroleum Geology, v. 19, p. 502–544.

Goldstein, A., 1950, Mineralogy of some Cretaceous sandstones from the Colorado Front Range: Journal of Sedimentary Petrology, v. 20, p. 85–97.

Gregory, H. E., 1951, The geology and geography of the Paunsaugunt region, Utah: U.S. Geological Survey Professional Paper 226, 116 p.

Hail, W. J., Jr., 1968, Geology of southwestern North Park and vicinity Colorado: U.S. Geological Survey Bulletin 1257, 119 p.

Hamilton, W., 1978, Mesozoic tectonics of the western United States, *in* Howell, D. G., and McDougall, K. A., eds., Mesozoic paleogeography of the western United States; Pacific Coast paleogeography Symposium 2: Los Angeles, California, Pacific Section, Society of Economic Paleontologists and Mineralogists, p. 33–70.

Hancock, E. T., 1925, Geology and coal resources of the Axial and Monument Butte quadrangles, Moffat County, Colorado: U.S. Geological Survey Bulletin 757, 134 p.

Hancock, J. M., and Kauffman, E. G., 1979, The great transgressions of the Late Cretaceous: Journal of the Geological Society of London, v. 136, p. 175–186.

Hansen, W. R., 1965, Geology of the Flaming Gorge area, Utah-Colorado-Wyoming: U.S. Geological Survey Professional Paper 490, 196 p.

Haq, B. U., Hardenbol, J., and Vail, P. R., 1987, Chronology of fluctuating sea levels since the Triassic: Science, v. 235, p. 1156–1167.

Hayes, P. T., 1970, Cretaceous paleogeography of southeastern Arizona and adjacent areas: U.S. Geological Survey Professional Paper 658–B, 42 p.

Hayes, P. T., and Drewes, H., 1978, Mesozoic depositional history of southeastern Arizona, *in* Callender, J. F., Wilt, J. C., Clemons, R. E., and James, H. L., eds., Land of Cochise; Southeastern Arizona: Albuquerque, New Mexico Geological Society, p. 201–207.

Heller, P. H., and Angevine, C. L., 1985, Sea-level cycles during the growth of Atlantic-type oceans: Earth and Planetary Science Letters, v. 75, p. 417–426.

Hubbard, R. J., 1988, Age and significance of sequence boundaries on Jurassic and Early Cretaceous rifted continental margins: American Association of Petroleum Geologists Bulletin, v. 72, p. 49–72.

Irwin, C. D., 1986, Upper Cretaceous and Tertiary cross sections, Moffat County, Colorado, *in* Stone, D. S., ed., New interpretations of northwest Colorado geology: Denver, Colorado, Rocky Mountain Association of Geologists 1986 Symposium, p. 151–156.

Isby, J. S., and Picard, M. D., 1983, Currant Creek Formation; Record of tectonism in Sevier–Laramide orogenic belt, north-central Utah: Contributions to Geology, University of Wyoming, v. 22, p. 91–108.

Izett, G. A., Cobban, W. A., and Gill, J. R., 1971, The Pierre Shale near Kremmling, Colorado, and its correlation to the east and the west: U.S. Geological Survey Professional Paper 684–A, p. A1–A19.

Jeletzky, J. A., 1968, Macrofossil zones of the marine Cretaceous of the western interior of Canada and their correlation with the zone and stages of Europe and the western interior of the United States: Geological Survey of Canada Paper 67–72, 66 p.

——, 1971, Marine Cretaceous biotic provinces and paleogeography of western and Arctic Canada; Illustrated by a detailed study of ammonites: Geological Survey of Canada Paper 70–22, 92 p.

——, 1978, Causes of Cretaceous oscillations of sea level in western and Arctic Canada and some general geotectonic implications: Geological Survey of Canada Paper 77–18, 44 p.

Jerzykiewicz, T., and McLean, J. R., 1980, Lithostratigraphical and sedimentological framework of coal-bearing Upper Cretaceous and lower Tertiary strata, Coal Valley area, central Alberta Foothills: Geological Survey of Canada Paper 79–12, 47 p.

Johnson, R. C., and May, F., 1980, A study of the Cretaceous–Tertiary unconformity in the Piceance Creek Basin, Colorado; The underlying Ohio Creek Formation (Upper Cretaceous) redefined as a member of the Hunter Canyon or Mesaverde Formation: U.S. Geological Survey Bulletin 1482–B, 27 p.

Johnson, R. C., May, F., Hansley, P. L., Pitman, J. K., and Fouch, T. D., 1980, Petrography, X-ray mineralogy, and palynology of a measured section of the Upper Cretaceous Mesaverde Group in Hunter Canyon, western Colorado: U.S. Geological Survey Oil and Gas Investigations Chart OC–91, 1 sheet.

Johnston, P. A., 1980, First record of Mesozoic mammals from Saskatchewan: Canadian Journal of Earth Sciences, v. 17, p. 512–519.

Johnston, P. A., and Fox, R. C., 1984, Paleocene and Late Cretaceous mammals from Saskatchewan, Canada: Palaeontographica, Abt. A, v. 186, p. 163–222.

Jordan, T. E., 1981, Thrust loads and foreland basin evolution, Cretaceous, western United States: American Association of Petroleum Geologists Bulletin, v. 65, p. 2506–2520.

Kauffman, E. G., 1973, A brackish water biota from the Upper Cretaceous

Harebell Formation of northwestern Wyoming: Journal of Paleontology, v. 47, p. 436–446.

——, 1977, Geological and biological overview: western interior Cretaceous basin: The Mountain Geologist, v. 14, p. 75–99.

——, 1979, Cretaceous, *in* Robison, R. A., and Teichert, C., eds., Treatise on invertebrate paleontology, Part A, Introduction; Fossilization (taphonomy), biogeography, and biostratigraphy: Boulder, Geological Society of America, p. 418–487.

——, 1984, Paleobiogeography and evolutionary response dynamic in the Cretaceous Western Interior Seaway of North America, *in* Westermann, G.E.G., ed., Jurassic–Cretaceous biochronology and paleogeography of North America: Geological Association of Canada Special Paper 27, p. 273–306.

Keefer, W. R., 1965, Stratigraphy and geologic history of the uppermost Cretaceous, Paleocene, and lower Eocene rocks in the Wind River Basin, Wyoming: U.S. Geological Survey Professional Paper 495–A, 77 p.

Keefer, W. R., and Rich, E. I., 1957, Stratigraphy of the Cody Shale and younger Cretaceous and Paleocene rocks in the western and southern parts of the Wind River Basin, Wyoming: Wyoming Geological Association 12th Annual Field Conference Guidebook, p. 71–78.

Keighin, C. W., and Fouch, T. D., 1981, Depositional environments and diagenesis of some nonmarine Upper Cretaceous reservoir rocks, Uinta Basin, Utah, *in* Ethridge, F. G., and Flores, R. M., eds., Recent and ancient nonmarine depositional environments; Models for exploration: Society of Economic Paleontologists and Mineralogists Special Publication 31, p. 109–125.

Keith, S. B., 1978, Paleosubduction geometries inferred from Cretaceous and Tertiary magmatic patterns in southwestern North America: Geology, v. 6, p. 516–521.

Kiteley, L. W., 1983, Paleogeography and eustatic-tectonic model of late Campanian Cretaceous sedimentation, southwestern Wyoming and northwestern Colorado, *in* Reynolds, M. W., and Dolly, E. D., eds., Mesozoic paleogeography of the west-central United States; Rocky Mountain paleogeography Symposium 2: Denver, Colorado, Rocky Mountain Section, Society of Economic Paleontologists and Mineralogists, p. 273–303.

Kiteley, L. W., and Field, M. E., 1984, Shallow marine depositional environments in the Upper Cretaceous of northern Colorado, *in* Tillman, R. W., and Siemers, C. T., eds., Siliciclastic shelf sediments: Society of Economic Paleontologists and Mineralogists Special Publication 34, p. 179–204.

Klute, M. A., 1986, Sedimentology and sandstone petrography of the upper Kirtland Shale and Ojo Alamo Sandstone, Cretaceous–Tertiary boundary, western and southern San Juan Basin, New Mexico: American Journal of Science, v. 286, p. 463–488.

Kluth, C. F., 1983, Geology of the northern Canelo Hills and implications for the Mesozoic tectonics of southeastern Arizona, *in* Reynolds, M. W., and Dolly, E. D., eds., Mesozoic paleogeography of the west-central United States; Rocky Mountain paleogeography Symposium 2: Denver, Colorado, Rocky Mountain Section, Society of Economic Paleontologists and Mineralogists, p. 159–171.

Konishi, K., 1959, Upper Cretaceous surface stratigraphy, Axial Basin and Williams Fork area, Moffat and Routt counties, Colorado, *in* Haun, J. D., and Weimer, R. J., eds., Symposium on Cretaceous rocks of Colorado and adjacent areas: Rocky Mountain Association of Geologists 11th Field Conference, Washakie, Sand Wash, and Piceance Basins, p. 67–73.

Lamerson, P. R., 1982, The Fossil Basin and its relationship to the Absaroka thrust system, Wyoming and Utah, *in* Powers, R. B., ed., Geologic studies of the Cordilleran thrust belt: Denver, Colorado, Rocky Mountain Association of Geologists, v. 1, p. 279–340.

Lawrence, J. C., 1965, Stratigraphy of the Dakota and Tropic Formations of Cretaceous age in southern Utah; Geology and resources of south central Utah: Utah Geological Society Guidebook 19, p. 71–91.

Lawton, T. F., 1985, Style and timing of frontal structures, thrust belt, central Utah: American Association of Petroleum Geologists Bulletin, v. 69, p. 1145–1159.

Lee, W. T., 1917, Geology of the Raton Mesa and other regions in Colorado and New Mexico: U.S. Geological Survey Professional Paper 101, p. 9–221.

Lehman, T. M., 1984, The multituberculate *Essonodon browni* from the Upper Cretaceous Naashoibito Member of the Kirtland Shale, San Juan Basin, New Mexico: Journal of Vertebrate Paleontology, v. 4, p. 602–603.

——, 1985, Depositional environments of the Naashoibito Member of the Kirtland Shale, Upper Cretaceous, San Juan Basin, New Mexico: New Mexico Bureau of Mines and Mineral Resources Circular 195, p. 55–79.

Lerbekmo, J. F., 1961, Stratigraphic relationship between the Milk River Formation of the southern Plains and the Belly River Formation of the southern Foothills of Alberta: Alberta Society of Petroleum Geologists Journal, v. 9, p. 273–276.

Lerbekmo, J. F., and Coulter, K. C., 1985, Late Cretaceous to early Tertiary magnetostratigraphy of a continental sequence; Red Deer Valley, Alberta, Canada: Canadian Journal of Earth Sciences, v. 22, p. 567–583.

Lillegraven, J. A., 1969, Latest Cretaceous mammals of upper part of Edmonton Formation of Alberta, Canada, and review of marsupial-placental dichotomy in mammalian evolution: Lawrence, University of Kansas Paleontological Contributions, Article 50 (Vertebrata 12), 122 p.

——, 1987, Stratigraphic and evolutionary implications of a new species of *Meniscoessus* (Multituberculata, Mammalia) from the Upper Cretaceous Williams Fork Formation, Moffat County, Colorado, *in* Martin, J. E., and Ostrander, G. E., eds., Papers in vertebrate paleontology in honor of Morton Green: Rapid City, Dakoterra, South Dakota School of Mines and Technology Special Paper 3, p. 46–56.

Lillegraven, J. A., and McKenna, M. C., 1986, Fossil mammals from the "Mesaverde" Formation (Late Cretaceous, Judithian) of the Bighorn and Wind River basins, Wyoming, with definitions of Late Cretaceous North American land-mammal "ages": American Museum Novitates 2840, 68 p.

Lillegraven, J. A., and Ostresh, L. M., Jr., 1988, Evolution of Wyoming's early Cenozoic topography and drainage patterns: National Geographic Research, v. 4, p. 303–327.

Lindsay, E. H., Butler, R. F., and Johnson, N. M., 1981, Magnetic polarity zonation and biostratigraphy of Late Cretaceous and Paleocene deposits, San Juan Basin, New Mexico: American Journal of Science, v. 281, p. 390–435.

Lohrengel, C. F., II, 1969, Palynology of the Kaiparowits Formation, Garfield County, Utah [abs.]: Geological Society of America Special Paper 121, p. 179.

Love, J. D., 1970, Cenozoic geology of the Granite Mountains area, central Wyoming: U.S. Geological Survey Professional Paper 495–C, 154 p.

——, 1973, Harebell Formation (Upper Cretaceous) and Pinyon Conglomerate (uppermost Cretaceous and Paleocene), northwestern Wyoming: U.S. Geological Survey Professional Paper 734–A, 54 p.

Lucas, S. G., and Schoch, R. M., 1982, Magnetic polarity zonation and biostratigraphy of Late Cretaceous and Paleocene continental deposits, San Juan Basin, New Mexico: American Journal of Science, v. 282, p. 920–927.

MacKenzie, M. G., 1975, Stratigraphy and petrology of the Mesaverde Group southern part of Big Horn Basin, Wyoming [Ph.D. thesis]: New Orleans, Louisiana, Tulane University, 156 p.

Madden, D. J., 1985, Description and origin of the lower part of the Mesaverde Group in Rifle Gap, Garfield County, Colorado: The Mountain Geologist, v. 22, p. 128–138.

Manfrino, C., 1984, Stratigraphy and palynology of the upper Lewis Shale, Pictured Cliffs Sandstone, and lower Fruitland Formation (Upper Cretaceous) near Durango, Colorado: The Mountain Geologist, v. 21, p. 115–132.

Matheny, J. P., and Picard, M. D., 1985, Sedimentology and depositional environments of the Emery Sandstone Member of the Mancos Shale, Emery and Sevier counties, Utah: The Mountain Geologist, v. 22, p. 94–109.

Matsumoto, T., 1980, Inter-regional correlation of transgressions and regressions in the Cretaceous period: Cretaceous Research, v. 1, p. 359–373.

Maxwell, R. A., Lonsdale, J. T., Hazzard, R. T., and Wilson, J. A., 1967, Geology of Big Bend National Park, Brewster County, Texas: Austin, University of Texas Bureau of Economic Geology Publication 6711, 320 p.

McCrory, V.L.C., and Walker, R. G., 1986, A storm and tidally-influenced prograding shoreline; Upper Cretaceous Milk River Formation of southern

Alberta, Canada: Sedimentology, v. 33, p. 47–60.

McGookey, D. P., Haun, J. D., Hale, L. A., Goodell, H. G., McCubbin, D. G., Weimer, R. J., and Wulf, G. R., 1972, Cretaceous system, *in* Mallory, W. W., and others, eds., Geologic atlas of the Rocky Mountain region: Denver, Colorado, Rocky Mountain Association of Geologists, p. 190–228.

McLean, J. R., 1971, Stratigraphy of the Upper Cretaceous Judith River Formation in the Canadian Great Plains: Saskatchewan Research Council Geology Division Report 11, 96 p.

——, 1977, Lithostratigraphic nomenclature of the Upper Cretaceous Judith River Formation in southern Alberta; Philosophy and practice: Bulletin of Canadian Petroleum Geology, v. 25, p. 1105–1114.

McNeil, D. H., and Caldwell, W.G.E., 1981, Cretaceous rocks and their foraminifera in the Manitoba escarpment: Geological Association of Canada Special Paper 21, 439 p.

Meijer Drees, N. C., and Mhyr, D. W., 1981, The Upper Cretaceous Milk River and Lea Park formations in southeastern Alberta: Bulletin of Canadian Petroleum Geology, v. 29, p. 42–74.

Miall, A. D., 1986, Eustatic sea level changes interpreted from seismic stratigraphy; A critique of the methodology with particular reference to the North Sea Jurassic record: American Association of Petroleum Geologists Bulletin, v. 70, p. 131–137.

Miller, F. X., 1977, Biostratigraphic correlation of the Mesaverde Group in southwestern Wyoming and northwestern Colorado, *in* Veal, H. K., ed., Exploration frontiers of the Central and Southern Rockies: Denver, Colorado, Rocky Mountain Association of Geologists 1977 Symposium, p. 117–137.

Miller, R. L., 1983, Subdivisions of the Menefee Formation and Cliff House Sandstone (Upper Cretaceous) in southwest San Juan Basin, New Mexico: U.S. Geological Survey Bulletin 1537–A, p. A29–A53.

Molenaar, C. M., 1977, Stratigraphy and depositional history of Upper Cretaceous rocks of the San Juan Basin area, New Mexico and Colorado, with a note on economic resources: New Mexico Geological Society 28th Field Conference Guidebook, p. 159–166.

——, 1983a, Major depositional cycles and regional correlations of Upper Cretaceous rocks, southern Colorado Plateau and adjacent areas, *in* Reynolds, M. W., and Dolly, E. D., eds., Mesozoic paleogeography of the west-central United States; Rocky Mountain paleogeography, Symposium 2: Denver, Colorado, Rocky Mountain Section, Society of Economic Paleontologists and Mineralogists, p. 201–224.

——, 1983b, Sedimentary facies and correlation of the Gallup Sandstone and associated formations, northwestern New Mexico, *in* Fassett, J. E., ed., Cretaceous and Tertiary rocks of the southern Colorado Plateau: Albuquerque, New Mexico, Four Corners Geological Society Memoir, p. 85–110.

Montellano, M., 1986, Mammalian fauna of the Judith River Formation (Late Cretaceous, Judithian) [Ph.D. thesis]: Berkeley, University of California, 184 p.

Mörner, N.-A., 1980, Relative sea-level, tectono-eustacy, geoidal-eustacy, and geodynamics during the Cretaceous: Cretaceous Research, v. 1, p. 329–340.

Nichols, D. J., Jacobson, S. R., and Tschudy, R. H., 1982, Cretaceous palynomorph biozones for the central and northern Rocky Mountain region of the United States, *in* Powers, R. B., ed., Geologic studies of the Cordilleran thrust belt: Denver, Colorado, Rocky Mountain Association of Geologists, v. 2, p. 721–733.

Nichols, D. J., Perry, W. J., Jr., and Haley, J. C., 1985, Reinterpretation of the palynology and age of Laramide syntectonic deposits, southwestern Montana, and revision of the Beaverhead Group: Geology, v. 13, p. 149–153.

North, B. R., and Caldwell, W.G.E., 1975, Foraminiferal faunas in the Cretaceous System of Saskatchewan, *in* Caldwell, W.G.E., ed., The Cretaceous System in the western interior of North America: Geological Association of Canada Special Paper 13, p. 303–331.

Obradovich, J. D., and Cobban, W. A., 1975, A time-scale for the Late Cretaceous of the western interior of North America, *in* Caldwell, W.G.E., ed., The Cretaceous System in the western interior of North America: Geological Association of Canada Special Paper 13, p. 31–54.

Oriel, S. S., and Tracey, J. I., Jr., 1970, Uppermost Cretaceous and Tertiary stratigraphy of Fossil Basin, southwestern Wyoming: U.S. Geological Survey Professional Paper 635, 53 p.

Orth, C. J., Gilmore, J. S., Knight, J. D., Pillmore, C. L., Tschudy, R. H., and Fassett, J. E., 1982, Iridium abundance measurements across the Cretaceous/Tertiary boundary in the San Juan and Raton basins of northern New Mexico, *in* Silver, L. T., and Schultz, P. H., eds., Geological implications of impacts of large asteroids and comets on the earth: Boulder, Colorado, Geological Society of America Special Paper 190, p. 423–433.

O'Sullivan, R. B., Repenning, C. A., Beaumont, E. C., and Page, H. G., 1972, Stratigraphy of the Cretaceous rocks and the Tertiary Ojo Alamo Sandstone, Navajo and Hopi Indian reservations, Arizona, New Mexico, and Utah: U.S. Geological Survey Professional Paper 521–E, 65 p.

Ower, J. R., 1960, The Edmonton Formation: Alberta Society of Petroleum Geologists Journal, v. 8, p. 309–323.

Peterson, F., 1969a, Four new members of the Upper Cretaceous Straight Cliffs Formation in the southeastern Kaiparowits region, Kane County, Utah: U.S. Geological Survey Bulletin 1274–J, 28 p.

——, 1969b, Cretaceous sedimentation and tectonism in the southeastern Kaiparowits region, Utah: U.S. Geological Survey Open-File Report OF69–202, 259 p.

Peterson, F., and Kirk, A. R., 1977, Correlation of the Cretaceous rocks in the San Juan, Black Mesa, Kaiparowits, and Henry basins, southern Colorado Plateau: New Mexico Geological Society 28th Field Conference Guidebook, p. 167–178.

Peterson, F., and Ryder, R. T., 1975, Cretaceous rocks in the Henry Mountains region, Utah and their relation to neighboring regions: Four Corners Geological Society 8th Field Conference Guidebook, p. 167–189.

Peterson, F., and Waldrop, H. A., 1965, Jurassic and Cretaceous stratigraphy of south-central Kaiparowits Plateau, Utah: Utah Geological Society and Intermountain Association of Petroleum Geologists Guidebook to the Geology of Utah 19, p. 47–69.

Peterson, F., Ryder, R. T., and Law, B. E., 1980, Stratigraphy, sedimentology, and regional relationships of the Cretaceous system in the Henry Mountains region, Utah, *in* Picard, M. D., ed., Henry Mountains symposium: Salt Lake City, Utah Geological Association Publication 8, p. 151–170.

Pillmore, C. L., 1976, Commercial coal beds of the Raton coal field, Colfax County, New Mexico: New Mexico Geological Society 27th Field Conference Guidebook, p. 227–247.

Pillmore, C. L., and Maberry, J. O., 1976, Depositional environments and trace fossils of the Trinidad Sandstone, southern Raton Basin, New Mexico: New Mexico Geological Society 27th Field Conference Guidebook, p. 191–195.

Pillmore, C. L., Tschudy, R. H., Orth, C. J., Gilmore, J. S., and Knight, J. D., 1984, Geologic framework of nonmarine Cretaceous–Tertiary boundary sites, Raton Basin, New Mexico and Colorado: Science, v. 223, p. 1180–1183.

Pitman, W. C., III, 1978, Relationship between eustacy and stratigraphic sequences of passive margins: Geological Society of America Bulletin, v. 89, p. 1389–1403.

Pitman, W. C., III, and Golovchenko, X., 1983, The effect of sealevel change on the shelfedge and slope of passive margins, *in* Stanley, D. J., and Moore, G. T., eds., The shelfbreak; Critical interface on continental margins: Society of Economic Paleontologists and Mineralogists Special Publication 33, p. 41–58.

Reeside, J. B., Jr., 1924, Upper Cretaceous and Tertiary formations of the western part of the San Juan Basin, Colorado and New Mexico: U.S. Geological Survey Professional Paper 134, p. 1–70.

Reiskind, J., 1975, Marine concretionary faunas of the uppermost Bearpaw Shale (Maestrichtian) in eastern Montana and southwestern Saskatchewan, *in* Caldwell, W.G.E., ed., The Cretaceous System in the western interior of North America: Geological Association of Canada Special Paper 13, p. 235–252.

Repenning, C. A., and Page, H. G., 1956, Late Cretaceous stratigraphy of Black

Mesa, Navajo and Hopi Indian reservations, Arizona: American Association of Petroleum Geologists Bulletin, v. 40, p. 255–294.

Reynolds, M. W., 1966, Stratigraphic relations of Upper Cretaceous rocks, Lamont–Bairoil area, south-central Wyoming: U.S. Geological Survey Professional Paper 550–B, p. B69–B76.

—— , 1976, Influence of recurrent Laramide structural growth on sedimentation and petroleum accumulation, Lost Soldier area, Wyoming: American Association of Petroleum Geologists Bulletin, v. 60, p. 12–33.

Riccardi, A. C., 1983, Scaphitids from the upper Campanian–lower Maastrichtian Bearpaw Formation of the western interior of Canada: Geological Survey of Canada Bulletin 354, p. 1–103.

Rice, D. D., 1981, Subsurface cross section from southeastern Alberta, Canada, to Bowdoin Dome area, north-central Montana, showing correlation of Cretaceous rocks and shallow, gas-productive zones in low-permeability reservoirs: U.S. Geological Survey Oil and Gas Investigations Chart OC–112, 1 sheet.

Rice, D. D., and Shurr, G. W., 1983, Patterns of sedimentation and paleogeography across the Western Interior Seaway during time of deposition of Upper Cretaceous Eagle Sandstone and equivalent rocks, northern Great Plains, *in* Reynolds, M. W., and Dolly, E. D., eds., Mesozoic paleogeography of the west-central United States; Rocky Mountain paleogeography Symposium 2: Denver, Colorado, Rocky Mountain Section, Society of Economic Paleontologists and Mineralogists, p. 337–358.

Rice, D. D., Shurr, G. W., and Gautier, D. L., 1982, Revision of Upper Cretaceous nomenclature in Montana and South Dakota: U.S. Geological Survey Bulletin 1529–H, p. H99–H104.

Rich, E. I., 1958, Stratigraphic relation of latest Cretaceous rocks in parts of Powder River, Wind River, and Big Horn basins, Wyoming: American Association of Petroleum Geologists Bulletin, v. 42, p. 2424–2443.

Rigby, J. K., Jr., and Wolberg, D. L., 1987, The therian mammalian fauna (Campanian) of Quarry 1, Fossil Forest study area, San Juan Basin, New Mexico, *in* Fassett, J. E., and Rigby, J. K., Jr., eds., The Cretaceous–Tertiary boundary in the San Juan and Raton basins, New Mexico and Colorado: Geological Society of America Special Paper 209, p. 51–79.

Roberts, A. E., 1972, Cretaceous and early Tertiary depositional and tectonic history of the Livingston area, southwestern Montana: U.S. Geological Survey Professional Paper 526–C, 120 p.

Robinson, C. S., Mapel, W. J., and Cobban, W. A., 1959, Pierre Shale along western and northern flanks of Black Hills, Wyoming and Montana: American Association of Petroleum Geologists Bulletin, v. 43, p. 101–123.

Robinson, C. S., Mapel, W. J., and Bergendahl, M. H., 1964, Stratigraphy and structure of the northern and western flanks of the Black Hills uplift, Wyoming, Montana, and South Dakota: U.S. Geological Survey Professional Paper 404, 134 p.

Roehler, H. W., 1983, Stratigraphy of Upper Cretaceous and lower Tertiary outcrops in the Rock Springs Uplift, Wyoming: U.S. Geological Survey Miscellaneous Investigations Series Map I–1500.

Rubey, W. W., 1930, Lithologic studies of fine-grained Upper Cretaceous sedimentary rocks of the Black Hills region: U.S. Geological Survey Professional Paper 165, p. 1–54.

Ruppel, E. T., and Lopez, D. A., 1984, The thrust belt in southwest Montana and east-central Idaho: U.S. Geological Survey Professional Paper 1278, 41 p.

Russell, L. S., 1940, Stratigraphy and structure, *in* Russell, L. S., and Landes, R. W., eds., Geology of the southern Alberta plains: Geological Survey of Canada Memoir 221, 128 p.

—— , 1952, Cretaceous mammals of Alberta: National Museum of Canada Bulletin, v. 126, p. 110–119.

—— , 1975, Mammalian faunal succession in the Cretaceous System of western North America, *in* Caldwell, W.G.E., ed., The Cretaceous System in the western interior of North America: Geological Association of Canada Special Paper 13, p. 137–161.

—— , 1987, Biostratigraphy and paleontology of the Scollard Formation, Late Cretaceous and Paleocene of Alberta: Royal Ontario Museum Life Sciences Contributions 147, 23 p.

Russell, L. S., and Landes, R. W., 1940, Geology of the southern Alberta plains: Canadian Geological Survey Memoir 221, p. 1–223.

Ryer, T. A., 1983, Transgressive-regressive cycles and the occurrence of coal in some Upper Cretaceous strata of Utah: Geology, v. 11, p. 207–210.

Sahni, A., 1972, The vertebrate fauna of the Judith River Formation, Montana: American Museum of Natural History Bulletin, v. 147, p. 321–412.

Schiebout, J. A., Rigsby, C. A., Rapp, S. D., Hartnell, J. A., and Standhardt, B. R., 1987, Stratigraphy of the Cretaceous–Tertiary and Paleocene–Eocene transition rocks of Big Bend National Park, Texas: Journal of Geology, v. 95, p. 359–375.

Schlanger, S. O., Jenkyns, H. C., and Premoli-Silva, I., 1981, Volcanism and vertical tectonics in the Pacific Basin related to global Cretaceous transgressions: Earth and Planetary Science Letters, v. 52, p. 435–449.

Schultz, L. G., Tourtelot, H. A., Gill, J. R., and Boerngen, J. G., 1980, Composition and properties of the Pierre Shale and equivalent rocks, northern Great Plains region: U.S. Geological Survey Professional Paper 1064–B, 114 p.

Scott, G. R., 1964, Geology of the Northwest and Northeast Pueblo Quadrangles, Colorado: U.S. Geological Survey Miscellaneous Geologic Investigations Map I–408, scale 1:24,000.

Scott, G. R., and Cobban, W. A., 1963, Apache Creek Sandstone Member of the Pierre Shale of southeastern Colorado: U.S. Geological Survey Professional Paper 475–B, p. B99–B101.

—— , 1964, Stratigraphy of the Niobrara Formation at Pueblo, Colorado: U.S. Geological Survey Professional Paper 454–L, 30 p.

—— , 1965, Geologic and biostratigraphic map of the Pierre Shale between Jarre Creek and Loveland, Colorado: U.S. Geological Survey Miscellaneous Geological Investigations Map I–439, 4 p. text, scale 1:48,000.

Shaw, E. W., and Harding, S.R.L., 1954, Lea Park and Belly River formations of east-central Alberta, *in* Clark, L. M., ed., Western Canada sedimentary basin; A symposium: American Association of Petroleum Geologists, p. 297–308.

Siemers, C. T., and King, N. R., 1974, Macroinvertebrate paleoecology of a transgressive marine sandstone, Cliff House Sandstone (Upper Cretaceous), Chaco Canyon, northwestern New Mexico: New Mexico Geological Society 25th Field Conference Guidebook, p. 267–277.

Siepman, B. R., 1986, Facies relationships in Campanian wave-dominated coastal deposits in Sand Wash Basin, *in* Stone, D. S., ed., New interpretations of northwest Colorado geology: Denver, Colorado, Rocky Mountain Association of Geologists 1986 Symposium, p. 157–164.

Sloan, R. E., and Russell, L. S., 1974, Mammals from the St. Mary River Formation (Cretaceous) of southwestern Alberta: Royal Ontario Museum Life Sciences Contributions 95, p. 1–21.

Sloan, R. E., Rigby, J. K., Jr., Van Valen, L. M., and Gabriel, D., 1986, Gradual dinosaur extinction and simultaneous ungulate radiation in the Hell Creek Formation: Science, v. 232, p. 629–633.

Smith, A. G., Hurley, A. M., and Briden, J. C., 1981, Phanerozoic paleocontinental world maps: Cambridge, Cambridge University Press, 102 p.

Spieker, E. M., 1946, Late Mesozoic and early Cenozoic history of central Utah: U.S. Geological Survey Professional Paper 205–D, p. D117–D161.

—— , 1949, Sedimentary facies and associated diastrophism in the Upper Cretaceous of central and eastern Utah: Geological Society of America Memoir 39, p. 55–81.

Stott, D. F., 1975, The Cretaceous System in northeastern British Columbia, *in* Caldwell, W.G.E., ed., The Cretaceous System in the western interior of North America: Geological Association of Canada Special Paper 13, p. 442–467.

Tschudy, R. H., 1973, The Gasbuggy core; A palynological appraisal, *in* Fassett, J. E., ed., Cretaceous and Tertiary rocks of the southern Colorado Plateau: Four Corners Geological Society Memoir, p. 131–143.

Tweto, O., 1975, Laramide (Late Cretaceous–early Tertiary) orogeny in the southern Rocky Mountains, *in* Curtis, B. F., ed., Cenozoic history of the southern Rocky Mountains: Geological Society of America Memoir 144, p. 1–44.

Tysdal, R. G., Marvin, R. F., and DeWitt, E., 1986, Late Cretaceous stratigraphy, deformation, and intrusion in the Madison Range of southwestern Montana: Geological Society of America Bulletin, v. 97, p. 859–868.

Vail, P. R., Mitchum, R. M., Jr., and Thompson, S., III, 1977, Seismic stratigraphy and global changes of sea level; Part 4, Global cycles of relative changes of sea level, *in* Payton, C. E., ed., Seismic stratigraphy; Applications to hydrocarbon exploration: American Association of Petroleum Geologists Memoir 26, p. 83–97.

Waage, K. M., 1968, The type Fox Hills Formation, Cretaceous (Maestrichtian), South Dakota; Part 1, Stratigraphy and paleoenvironments: New Haven, Connecticut, Yale University Peabody Museum of Natural History Bulletin 27, 175 p.

Weimer, R. J., 1976, Cretaceous stratigraphy, tectonics, and energy resources, western Denver Basin, *in* Epis, R. C., and Weimer, R. J., eds., Studies in Colorado field geology: Golden, Colorado School of Mines Professional Contributions 8, p. 180–227.

—— , 1983, Relation of unconformities, tectonics, and sea level changes, Cretaceous of the Denver Basin and adjoining areas, *in* Reynolds, M. W., and Dolly, E. D., eds., Mesozoic paleogeography of the west-central United States; Rocky Mountain paleogeography Symposium 2: Denver, Colorado, Rocky Mountain Section, Society of Economic Paleontologists and Mineralogists, p. 359–376.

Weimer, R. J., and Tillman, R. W., 1980, Tectonic influence on deltaic shoreline facies, Fox Hills Sandstone, west-central Denver Basin: Golden, Colorado School of Mines Professional Contributions 10, 131 p.

Whitmore, J. L., 1985, Fossil mammals from two sites in the Late Cretaceous Lance Formation in northern Niobrara County, Wyoming, *in* Martin, J. E., ed., Fossiliferous Cenozoic deposits of western South Dakota and northwestern Nebraska: Rapid City, Dakoterra, South Dakota School of Mines and Technology, v. 2, pt. 2, p. 353–367.

Williams, G. D., and Baadsgaard, H., 1975, Potassium-argon dates and Upper Cretaceous biostratigraphy in eastern Saskatchewan, *in* Caldwell, W.G.E., ed., The Cretaceous System in the western interior of North America: Geological Association of Canada Special Paper 13, p. 417–426.

Williams, G. D., and Burk, C. F., Jr., 1964, Upper Cretaceous, *in* McCrossan, R. G., and Glaister, R. P., eds., Geological history of western Canada: Calgary, Alberta Society of Petroleum Geologists, p. 169–189.

Wilson, R. W., 1983, Late Cretaceous mammals of western South Dakota: National Geographic Society Research Reports, v. 15, p. 749–752.

—— , 1987, Late Cretaceous (Fox Hills) multituberculates from the Red Owl local fauna of western South Dakota, *in* Martin, J. E., and Ostrander, G. E., eds., Papers in vertebrate paleontology in honor of Morton Green: Rapid City, Dakoterra, South Dakota School of Mines and Technology Special Paper 3, p. 118–122.

Wiltschko, D. V., and Dorr, J. A., Jr., 1983, Timing of deformation in Overthrust belt and foreland of Idaho, Wyoming, and Utah: American Association of Petroleum Geologists Bulletin, v. 67, p. 1304–1322.

Witzke, B. J., Ludvigson, G. A., Poppe, J. R., and Ravn, R. L., 1983, Cretaceous paleogeography along the eastern margin of the Western Interior Seaway, Iowa, southern Minnesota, and eastern Nebraska and South Dakota, *in* Reynolds, M. W., and Dolly, E. D., eds., Mesozoic paleogeography of the west-central United States; Rocky Mountain paleogeography Symposium 2: Denver, Colorado, Rocky Mountain Section, Society of Economic Paleontologists and Mineralogists, p. 225–252.

Wolfe, J. A., and Upchurch, G. R., Jr., 1987, North American nonmarine climates and vegetation during the Late Cretaceous: Palaeogeography, Palaeoclimatology, Palaeoecology, v. 61, p. 33–77.

Wood, H. E., 2nd, Chaney, R. W., Clark, J., Colbert, E. H., Jepsen, G. L., Reeside, J. B., Jr., and Stock, C., 1941, Nomenclature and correlation of the North American continental Tertiary: Geological Society of America Bulletin, v. 52, p. 1–48.

Zapp, A. D., and Cobban, W. A., 1960, Some Late Cretaceous strand lines in northwestern Colorado and northeastern Utah: U.S. Geological Survey Professional Paper 400–B, p. B246–B249.

—— , 1962, Some Late Cretaceous strand lines in southern Wyoming: U.S. Geological Survey Professional Paper 450–D, p. D52–D55.

Manuscript Submitted February 16, 1988
Manuscript Revised May 11, 1988
Manuscript Accepted by the Society June 12, 1989

Printed in U.S.A.

Geological Society of America
Special Paper 243
1990

Mammalian zonation near the Cretaceous-Tertiary boundary

J. David Archibald
Department of Biology, San Diego State University, San Diego, California 92182
Donald L. Lofgren
Department of Paleontology, University of California, Berkeley, California 94720

ABSTRACT

With the discovery and description of the Bug Creek faunas in 1965, it became necessary to reexamine the sequence of mammal ages (Lancian-Puercan) spanning the Cretaceous-Tertiary boundary. Bug Creek faunal assemblages have been viewed as being in part coeval with the Lancian assemblages or slightly younger. Because of the lack of Lancian sites above the Bug Creek–type sites and the discontinuous nature of the sediments preserving the latter type faunas, it appears that the Bug Creek faunas post-date Lancian faunas. Although the Bug Creek faunas cannot be well constrained biostratigraphically, the appearance of a number of new taxa and the continuation of these or closely related taxa into the Puercan is strong evidence for a biochronologic sequence of Lancian–Bug Creek–Puercan.

When the faunal contents of Lancian, Bug Creek, and earliest Puercan (Pu_1) sites are scrutinized, it is clear that important faunal introductions occur with the commencement of Bug Creek assemblages. In contrast, between Bug Creek and earliest Puercan (Pu_1) assemblages there are very few major faunal introductions and thus more faunal continuity. Therefore, the definition and concept of the Puercan Land Mammal Age is modified, and the Bug Creek faunas are formally defined and characterized as the *Protungulatum/Peradectes* Interval-Zone (Pu_0) of the Puercan Land Mammal Age. The *Protungulatum/Peradectes* Interval-Zone (Pu_0) postdates the (latest Cretaceous) Lancian Land Mammal Age and commences the (latest Cretaceous?–early Paleocene) Puercan Land Mammal Age. This interval-zone is defined as including faunas that occur during the time between the first appearance of the arctocyonid ungulate *Protungulatum* and the first appearance of the didelphid marsupial *Peradectes.*

Certain biochronological criteria (first appearances and "cladochronology") within this interval-zone strengthen the view that the original sequence (from oldest to youngest) of Bug Creek Anthills, Bug Creek West, and Harbicht Hill is correct. Therefore, three informal biochrons are recognized, the *Protungulatum/Mimatuta* (bk_1), *Mimatuta/Oxyprimus* (bk_2), and *Oxyprimus/Peradectes* (bk_3) biochrons.

Dinosaurs and Lancian mammals are found at all Pu_0 localities. Although the possibility of reworking cannot be completely dismissed, the abundance of Lancian mammals and the nature of dinosaur material at certain Pu_0 sites suggest some temporal overlap. Similarly, the stratigraphic placement of palynological change and an iridium anomaly relative to Pu_0 faunas remains equivocal.

Archibald, J. D., and Lofgren, D. L., 1990, Mammalian zonation near the Cretaceous-Tertiary boundary, *in* Bown, T. M., and Rose, K. D., eds., Dawn of the Age of Mammals in the northern part of the Rocky Mountain Interior, North America: Boulder, Colorado, Geological Society of America, Special Paper 243.

INTRODUCTION

The biotic turnover during the Cretaceous-Tertiary (K-T) transition remains one of the most widely debated episodes of extinction in earth history. These debates are all too often infused with much hyperbole and superficial data bases, while the obviously very basic field-oriented research is neglected. This is nowhere more true than in the case of the geochronologic framework that is used to argue pro or con for various extinction scenarios.

This chapter is not directly related to the debates surrounding the extinctions during the K-T transition (in fact, it will attempt to avoid the general issue), but it does impinge on the debate. Specifically, the chapter reexamines the biochronology/biostratigraphy of several critical terrestrial faunas, notably the Bug Creek Anthill faunal assemblage.

This review will be done explicitly from the perspective of the mammalian components of the faunas, including a clarification of the mammalian biochronology where possible. Following this section, the mammalian zonation is compared with various other faunal (dinosaurs), floral (pollen), and geologic (iridium) components that can be at least tentatively correlated to it.

HISTORY

The history of mammalian zonation in North America predates the present debate over K-T extinctions by many years. The first truly synthetic work involving zonation of the Cenozoic Era using mammalian faunas dates from the work of the Wood Committee (Wood and others) published in 1941. This has been greatly augmented and updated in the recent volume edited by Woodburne (1987). Both works were essentially restricted to the Cenozoic Era, although both made some mention of the terrestrial zonation of the latest Cretaceous in North America. The zonation of the latest Cretaceous is clearly identifiable as such from the time of the paleobotanical work of Dorf (1940) and later in various works of L. Russell (1964, 1975). This has been greatly expanded in a variety of works, most notably that of Lillegraven and others (1979) and Lillegraven and McKenna (1986).

The more traditional view as seen in Wood and others (1941) and Archibald and others (1987) has been that the boundary between the older Lancian and younger Puercan North American Land Mammal Ages and the K-T boundary is coincident. (For the sake of brevity, "North American Land Mammal Age" will be shortened to "Mammal Age.") In the case of Archibald and others (1987), this was viewed more as a result of historical accident rather than as a theoretical necessity.

With the discovery and description in 1965 of the Bug Creek Anthill faunal assemblage in eastern Montana (Bug Creek Anthills, Bug Creek West, and Harbicht Hill) by Sloan and Van Valen, it became necessary to reexamine the sequences of mammal ages encompassing the K-T boundary (Fig. 1). In this paper they briefly described the faunas from a series of localities that produced three different sorts of mammalian assemblages. The first sort yielded an assemblage of mammals that through informal usage has become known as "typically latest Cretaceous" in aspect. This refers to a normal Lancian fauna dominated by marsupials, ptilodontoid multituberculates, and proteutherians, as described more extensively by Clemens (1964, 1966, 1973) for the type Lance Formation in eastern Wyoming, by Lillegraven (1969) for the Scollard Formation in southern Alberta, and by Archibald (1982) for the type area of the Hell Creek Formation in northeastern Montana. The second assemblage has informally become known as "faunas of Paleocene aspect." This refers to the presence not only of Lancian mammals, but also mammals that appear to be most closely related to taxa first recorded in early Paleocene faunas. The third assemblage was represented by only one locality, Purgatory Hill, which yielded an early Paleocene (Puercan) mammalian fauna.

It is the second assemblage of Sloan and Van Valen (1965) that is of central concern to this paper. This assemblage consists of three localities, which according to the authors, could be placed in the ascending biostratigraphic order of Bug Creek Anthills, Bug Creek West, and Harbicht Hill. These authors recognized that the sites occurred in channel-fill deposits with the localities recorded as being 24, 18, and 12 m, respectively, below the base of the Hell Creek–Tullock formational boundary. This formational contact was also taken to be the Cretaceous-Tertiary boundary. Thus, the conclusion was that the Bug Creek faunal assemblage represented successive, fossil-bearing channels capped *within* the Cretaceous Hell Creek Formation.

Implicitly in their 1965 paper and explicitly in later papers (e.g., Van Valen, 1978; Sloan, 1987), one or both authors recognized a Bugcreekian Mammal Age separating the Lancian and Puercan (here including Mantuan) Mammal Ages.

Archibald (1981, 1982) described Lancian localities to the west in Garfield County that were positioned higher in the Hell Creek Formation than had been reported for the Bug Creek faunal assemblage by Sloan and Van Valen (1965). Especially important was the relatively faunally rich Flat Creek 5, occurring some 23 km to the southwest of Bug Creek (Fig. 1) and positioned only 5.2 m below the formational contact and presumed K-T boundary (as defined by the last appearance of dinosaurs). Based on this data, Archibald proposed that the Bug Creek Anthill (BCA) faunal assemblage, rather than postdating the Lancian Mammal Age, was probably correlative with the latter part of this age. It was argued that the BCA faunal assemblage preserving both "mammals of Paleocene aspect" and "typical latest Cretaceous mammals," was coeval with the youngest Lancian sites preserving only "typical latest Cretaceous mammals." The faunal differences were seen as being ecological rather than biostratigraphic, and hence these faunas were termed the Bug Creek and Hell Creek faunal-facies, respectively.

Subsequent work has suggested that the earlier interpretation of a faunal succession from the Lancian to the Bug Creek faunas is more likely correct. There are two lines of evidence that support this interpretation.

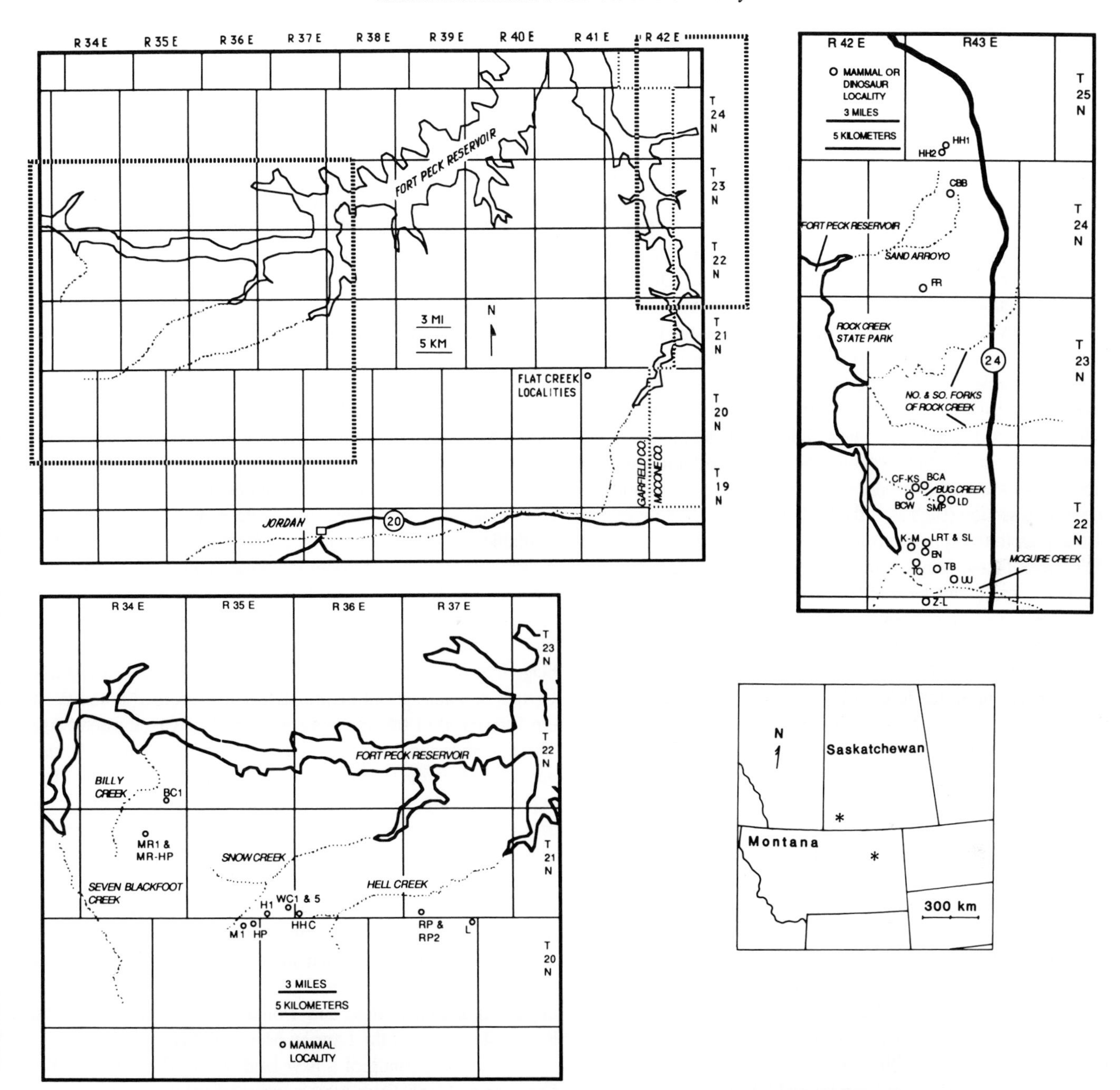

Figure 1. Maps showing the locations of sites of geological interest or of faunas referable to the Lancian Mammal Age or the *Protungulatum/Peradectes* (Pu_0) and the *Peradectes/Ectoconus* (Pu_1) Interval-Zones of the Puercan Mammal Age. Map in lower right shows general location of sites in McCone and Garfield Counties in Montana and in southwestern Saskatchewan. Map in upper left shows areas of interest in McCone County (right) and Garfield County (below), and the Flat Creek fossil sites and the Lerbekmo stratigraphic sampling site. The fossil localities and/or local faunas in Garfield County are (from west to east): McKeever Ranch 1 and McKeever Ranch–Harley's Point (MR1 & MR-HP): Billy Creek 1 (BC1), Morales 1 (M1); Herpijunk Promontory (HP); Hauso 1 (H1); Worm Coulee 1 and 5 (WC1 & 5); Hell's Hollow Channel (HHC); Rick's Place and Rick's Place 2 (RP & RP2). An additional location of geological interest in Garfield County is the Lerbekomo Site (L). The fossil localities and/or local faunas in McCone County are (from north to south): Harbicht Hill 1 and 2 (HH1 and HH2); Chris' Bone Bed (CBB); Ferguson Ranch (FR); Bug Creek Anthill (BCA); Carnosaur Flat and Ken's Saddle (CF-KS); Bug Creek West (BCW); Scmenge Point (SMP); Last Days B. A. (LD); Little Roundtop Channel (LRT); Second Level (SL); K-Mark 2 (K-M); Eagle Nest (EN); Shiprock, Tedrow Dinosaur Quarry, and Tedrow Quarry Local Fauna (TQ); Three Buttes (TB); Up Up the Creek (UU), and Z-Line (Z-L).

First, in an area ranging from 85 to 50 km to the west of the Bug Creek region, Dingus (1984, and personal communication) has shown that deposition in upper portions of the Hell Creek Formation and lower parts of the overlying Tullock Formation can be laterally discontinuous, as well one might expect in fluvial deposits. In some areas, such as at the Herpijunk Promontory microfossil site and the nearby Iridium Hill (Fig. 1), a series of lignites is present including the lowest iridium-bearing lignite (the so-called lower Z coal). In this area, in situ dinosaur remains occur to within 2 to 3 m of the iridium-bearing coal. Dingus was able to laterally trace some of the local coals of this Z-complex using various lithological criteria. In places such as the Lerbekmo site (Fig. 1) some 18 km to the east of Herpijunk Promontory, Dingus was able to show that the lowest iridium-bearing coal was absent and that the highest dinosaurs seemed to disappear just below the next highest coal. The simplest explanation is that the lowest coals in the first section (such as at Iridium Hill and Herpijunk) were either eroded away or never were deposited in the second section (such as at the Lerbekmo site). This reasoning also applies to the apparent contemporenity of the Bug Creek and Hell Creek faunal-facies. It now seems more likely that the two types of sites cannot be strictly correlated stratigraphically, but that some amount of rock is simply not represented through erosion or nondeposition above the stratigraphically highest Hell Creek faunal-facies sites.

Finally, for all the fieldwork that has been conducted in the Bug Creek area, no localities that produce exclusively "typical latest Cretaceous mammals" have been reported as occurring stratigraphically above localities producing "mammals of Paleocene aspect."

PURPOSE, RATIONALE, AND METHODOLOGY

Accepting the preceding arguments, we consider that the Bug Creek faunal assemblage in all likelihood postdates the Lancian Mammal Age, but predates the Puercan Mammal Age as both are currently defined and utilized. Whether, and if so how, this assemblage of faunas should be formally recognized raises several theoretical and practical issues.

One of the most problematic issues is that for now, the Bug Creek faunas cannot be well constrained biostratigraphically. As Figure 2 indicates, and as later discussed, the Bug Creek faunas (labeled as Pu_0 in Fig. 2) would appear to overlap biostratigraphically both the Lancian and Puercan mammal ages. There is, however, at present no evidence to support this view. Rather, the Bug Creek faunas as well as many of the overlying Puercan faunas occur in channel-fill deposits that cut older sediments. This has been discussed by a variety of recent authors (Fastovsky and Dott, 1986; Smit and others, 1987; Rigby and others, 1987). The seeming appearance then of potential overlap in Figure 2 is actually an artifact of our inability to determine where within the complex channeling sequence the Bug Creek faunas occur relative to Lancian sites in surrounding nonchannel and channel deposits. Also, as noted later, estimates of the biostratigraphic position for the Bug Creek faunal assemblage can be suggested using channel cappings, but such assessments may well underestimate the biostratigraphic extent of these faunas.

Our view then is that although the Bug Creek faunas cannot be well constrained biostratigraphically, there is no basis to suggest temporal overlap. In fact, to the contrary, the appearance of a number of new taxa in the Bug Creek faunas and the continuation of these or closely related taxa into the Puercan seems to us to be strong evidence for a biochronological sequence of Lancian–Bug Creek–Puercan. In fact, the superpositional relationship of Puercan over Bug Creek–type faunas is demonstrated by a sequence of sites in Saskatchewan (Fox, 1987). These sites, however, do not appear to represent as complete a faunal sequence as in the Bug Creek area.

Even accepting the Bug Creek faunal assemblage as being temporally intermediate between the Lancian and Puercan mammal ages, there remains the issue of if and how it should be formally recognized, and at what hierarchical level the faunas are best recognized. We believe that there is more than ample data available to formally recognize this faunal assemblage. The three possibilities include incorporating these faunas into the preexisting Lancian or Puercan mammal ages as an interval zone or recognizing a new Bugcreekian Mammal Age.

Of the three choices, incorporating the Bug Creek assemblage into the Lancian Mammal Age seems the least acceptable. Certain new lineages of mammals are first recorded in the Western Interior with the appearance of the Bug Creek assemblage. These include the first appearance of archaic ungulates (*Protungulatum, Mimatuta, Ragnarok*, and *Oxyprimus*) and the multituberculate family Eucosmodontidae (*Stygimys*). Other distinctive taxa also appearing at this time are a proteutherian, *Procerberus*, and another multituberculate, *Catopsalis* (see Table 1). Although Lancian and Bug Creek assemblages have many faunal elements in common, it seems that a clearly recognizable faunal change occurs between the two assemblages based on the first appearances. Therefore, we feel that incorporating the Bug Creek assemblage into the Lancian Age would needlessly obscure an important faunal change and would also excessively alter the historical concept of the Lancian Mammal Age.

As to the naming of a new land mammal age for the Bug Creek assemblage, there are a number of valid reasons why this might be appropriate. To this end, Archibald (1987a, 1987b) advocated the establishment of a new mammal age based on the Bug Creek assemblage. This was also the course suggested by Sloan (1987). Based on new information from the McGuire Creek area that is presented below, we feel that the naming of a new mammal age for the Bug Creek assemblage should be abandoned.

Based on theoretical and practical grounds, we feel that formal recognition of the Bug Creek assemblage can be best accommodated within the Puercan Mammal Age as an interval-zone. When the faunal contents of the Bug Creek and earliest Puercan (Pu_1) assemblages are closely scrutinized, it becomes clear that these assemblages have many faunal components in

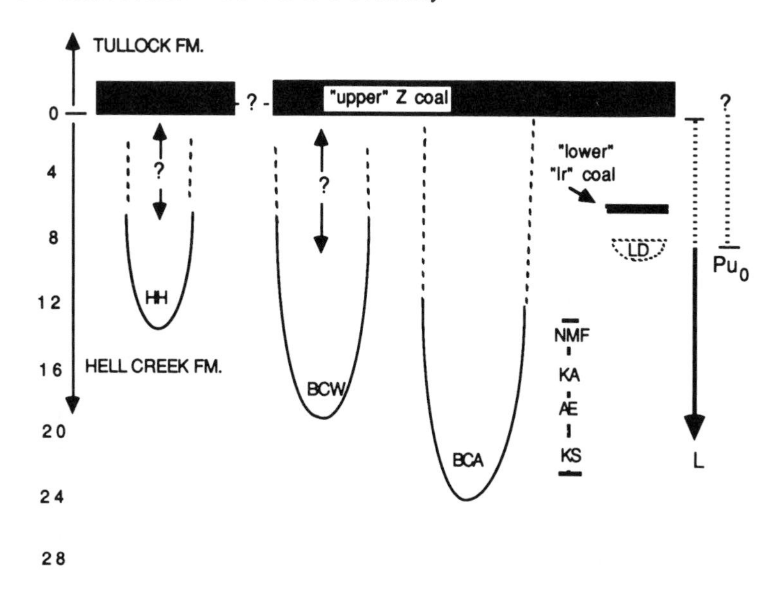

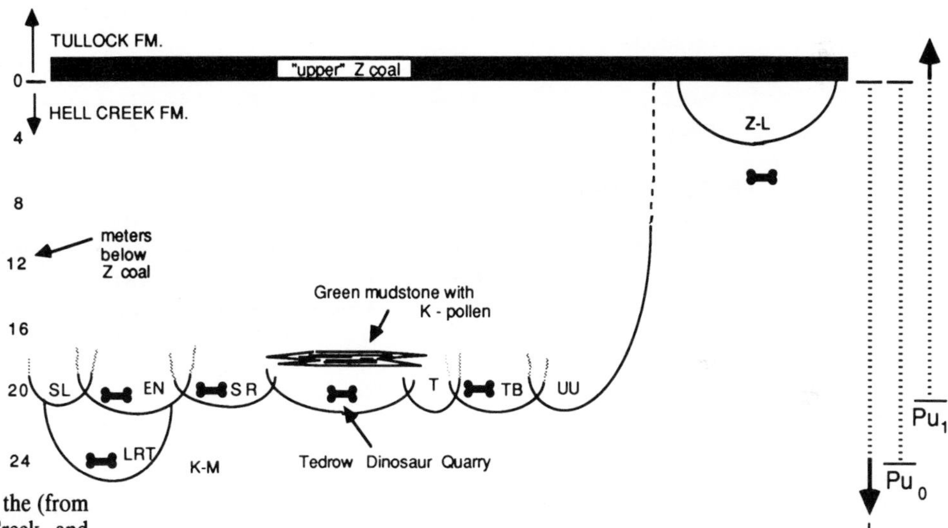

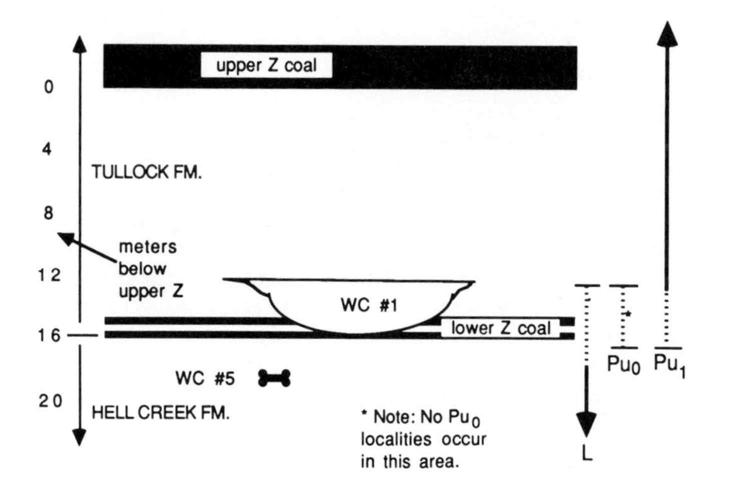

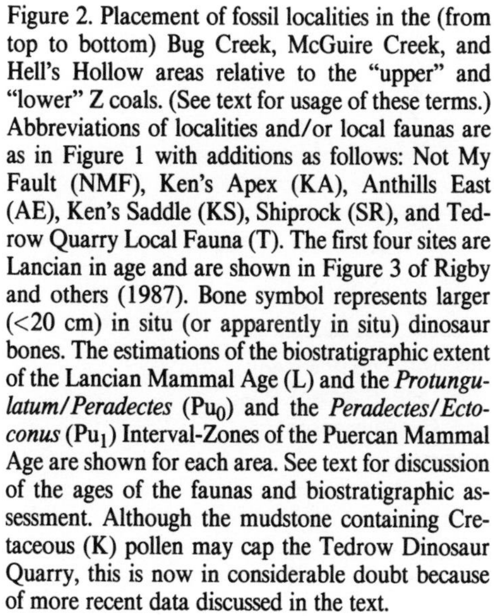

Figure 2. Placement of fossil localities in the (from top to bottom) Bug Creek, McGuire Creek, and Hell's Hollow areas relative to the "upper" and "lower" Z coals. (See text for usage of these terms.) Abbreviations of localities and/or local faunas are as in Figure 1 with additions as follows: Not My Fault (NMF), Ken's Apex (KA), Anthills East (AE), Ken's Saddle (KS), Shiprock (SR), and Tedrow Quarry Local Fauna (T). The first four sites are Lancian in age and are shown in Figure 3 of Rigby and others (1987). Bone symbol represents larger (<20 cm) in situ (or apparently in situ) dinosaur bones. The estimations of the biostratigraphic extent of the Lancian Mammal Age (L) and the *Protungulatum/Peradectes* (Pu_0) and the *Peradectes/Ectoconus* (Pu_1) Interval-Zones of the Puercan Mammal Age are shown for each area. See text for discussion of the ages of the faunas and biostratigraphic assessment. Although the mudstone containing Cretaceous (K) pollen may cap the Tedrow Dinosaur Quarry, this is now in considerable doubt because of more recent data discussed in the text.

TABLE 1.*

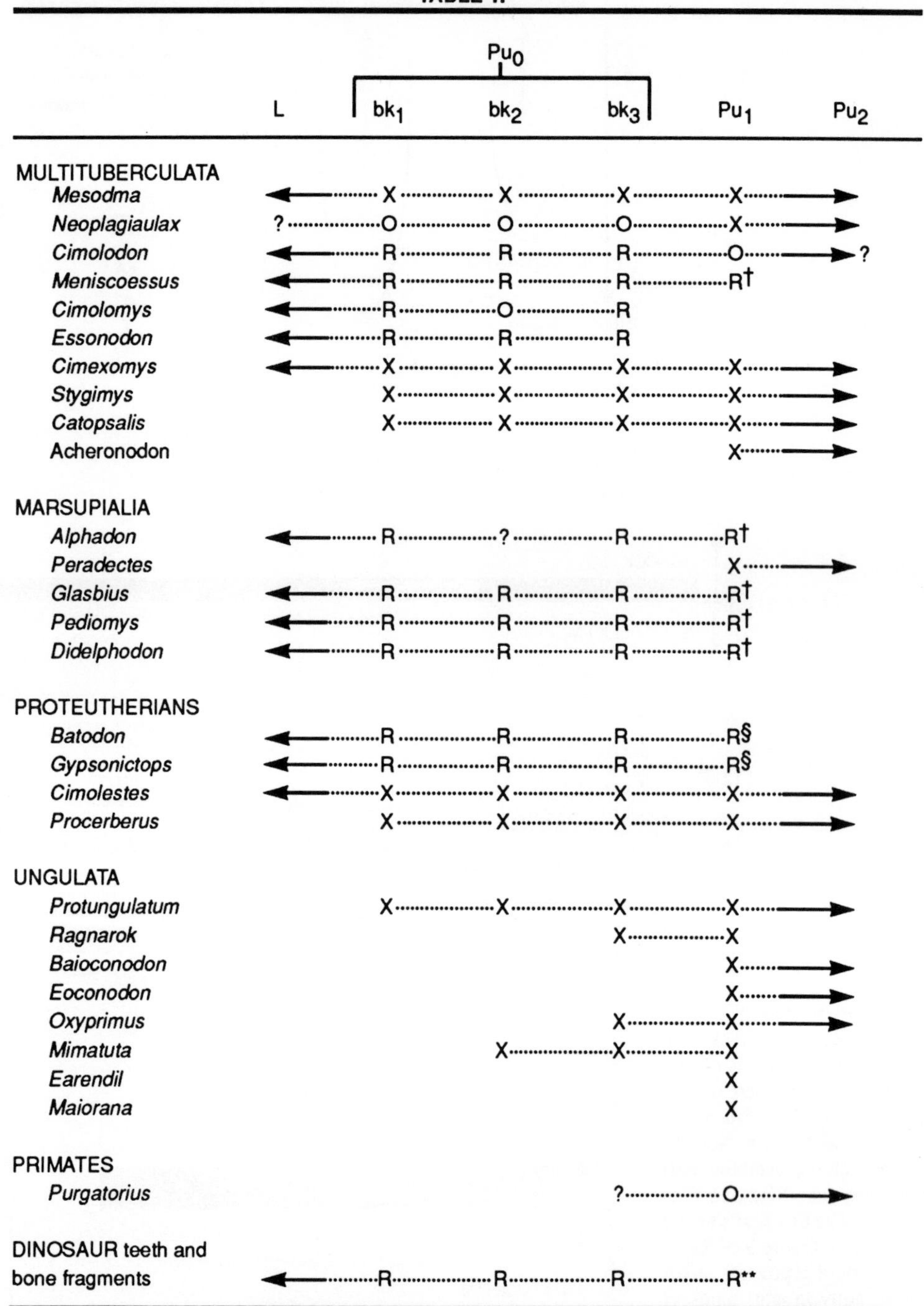

	L	Pu_0			Pu_1	Pu_2
		bk_1	bk_2	bk_3		
MULTITUBERCULATA						
Mesodma	←	X	X	X	X	→
Neoplagiaulax	?	O	O	O	X	→
Cimolodon	←	R	R	R	O	→ ?
Meniscoessus	←	R	R	R	R†	
Cimolomys	←	R	O	R		
Essonodon	←	R	R	R		
Cimexomys	←	X	X	X	X	→
Stygimys		X	X	X	X	→
Catopsalis		X	X	X	X	→
Acheronodon					X	→
MARSUPIALIA						
Alphadon	←	R	?	R	R†	
Peradectes					X	→
Glasbius	←	R	R	R	R†	
Pediomys	←	R	R	R	R†	
Didelphodon	←	R	R	R	R†	
PROTEUTHERIANS						
Batodon	←	R	R	R	R§	
Gypsonictops	←	R	R	R	R§	
Cimolestes	←	X	X	X	X	→
Procerberus		X	X	X	X	→
UNGULATA						
Protungulatum		X	X	X	X	→
Ragnarok				X	X	
Baioconodon					X	→
Eoconodon					X	→
Oxyprimus				X	X	→
Mimatuta			X	X	X	
Earendil					X	
Maiorana					X	
PRIMATES						
Purgatorius				?	O	→
DINOSAUR teeth and bone fragments	←	R	R	R	R**	

*Abbreviations: L, Lancian Mammal Age; Pu_0, *Protungulatum/Peradectes* Interval-Zone (Puercan); bk_1, bk_2, bk_3, informal biochrons in Pu_0; Pu_1, *Peradectes/Ectoconus* Internal-Zone (Puercan); Pu_2, *Ectoconus/Periptychus* Interval-Zone (Puercan); X, genus in the interval-zone; O, genus absent from interval zone but known before and after; R, genus argued by some to be reworked.

†Lancian genera common at some Pu_1 sites

§Lancian genera rare at some Pu_1 sites

**Dinosaur postcranial fragments/teeth present and common at some Pu_1 sites.

common (see Table 1). This is particularly true for a series of localities now under study by Lofgren in the McGuire Creek area immediately south of the Bug Creek area in McCone County, Montana (see Fig. 1). These sites, which will be outlined below, more than any described to date, indicate a gradation from the Bug Creek assemblage into earliest Puercan faunas (such as the Mantua Lentil Local Fauna). Hence, with this new information, if one were to name a new mammal age, it would be quite difficult to decide where to draw the faunal break. Thus, we have chosen to place the Bug Creek assemblage within the Puercan Mammal Age as a formal interval-zone and accordingly modify the definition and concept of the Puercan Mammal Age.

Although this changes the historical concept of the Puercan Land Mammal Age, we feel that the modifications are of minor extent and don't alter the original intent of the Wood Committee (Wood and others, 1941). In fact, the ages when proposed were designed to be open to some modification: "precise limits between successive ages are intended to be somewhat flexible and may presumably be modified in the light of later discoveries" (Wood and others, 1941, p. 6). It is reasonable to consider the Bug Creek discoveries from this perspective and therefore modify the Puercan to accommodate this new faunal interval.

A parallel example comes from the issue of whether or not to recognize "Mantuan" as a formal mammal age as advocated by Van Valen (1978). Most authors, including Archibald and others (1987), prefer to consider the "Mantuan" as a zone within the Puercan Mammal Age (Fig. 3). The basis for this decision is found in historical usage plus what we now know of the Mantua Lentil Local Fauna. The fauna from Mantua lentil was presented as a principal correlative of the Puercan Mammal Age by the Wood Committee (Wood and others, 1941). Thus, they regarded it as belonging to this age. Van Valen (1978) correctly emphasized that the Mantuan fauna is almost undoubtedly older than the type Puercan fauna of San Juan Basin, New Mexico. It does not follow though that one should elevate the Mantuan to the status of a mammal age. If this rationale were to be followed, one could similarly argue that all named faunal intervals within presently recognized mammal ages might be raised to the status of a mammal age. If a Mantuan Mammal Age is recognized, why not a Greybullian or a Lostcabinian Mammal Age? Formalized zonations, when warranted, serve the same purpose.

The methodology of definition and characterization followed here is that employed by Archibald and others (1987) and can be found in that paper. We have chosen to recognize the Bug Creek assemblage as an interval-zone within the Puercan Mammal Age. We follow the usage of the International Stratigraphic Guide (Hedberg, 1976) in naming and recognizing the beginning of an interval-zone based on the first appearance of a taxon and the first appearance of an unrelated taxon to simultaneously mark the end of the interval-zone in question and to begin the succeeding interval-zone. The only general theoretical construct to emphasize is that land-mammal ages and their various zones defined and characterized in Archibald and others (1987) are technically based on zonation generated from faunal aspects and hence are

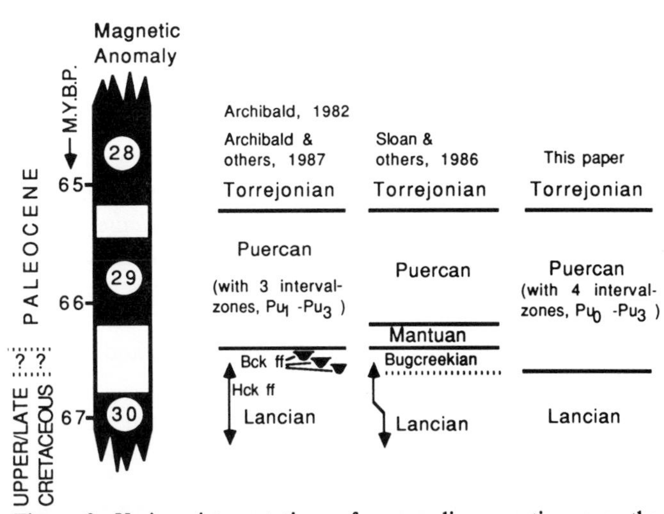

Figure 3. Various interpretations of mammalian zonation near the Cretaceous-Tertiary boundary in the Western Interior. The uncertainty regarding the K-T boundary refers only to the chronostratigraphic uncertainty surrounding the *Protungulatum/Peradectes* (Pu_0) Interval-Zone. As discussed in the text, some palynological correlations of Lofgren and Hotton (1988) suggested that most of this interval-zone could be Cretaceous in age, but more recent work by Hotton (personal communication) casts doubt on this interpretation.

biochronologically controlled. Without biostratigraphically defined zonation (e.g., a stage or sub-stage), the Bug Creek assemblage, as well as the mammal ages in Archibald and others (1987), are not ages in the sense of the North American Stratigraphic Code (1983). Archibald and others followed the tradition orginated by the Wood Committee (Wood and others, 1941), and followed by most subsequent authors on the subject, in utilizing faunal composition in dealing with mammal ages. The hope is that eventually a biostratigraphically bound zonation will emerge. This is beginning to occur for some mammal ages such as the Clarkforkian (Rose, 1981), which is more strictly a biostratigraphic unit. As will be discussed in this paper, the Bug Creek assemblages can begin to be placed in a biostratigraphic framework, although the exactness of this framework is far from what we would like it to be.

THE *PROTUNGULATUM/PERADECTES* INTERVAL-ZONE (PU_0)

Definition of the interval-zone and redefinition of the Puercan Mammal Age

The *Protungulatum/Peradectes* Interval-Zone is defined to include faunas that occur during the time between the first appearance of the arctocyonid ungulate, *Protungulatum*, and the first appearance of the didelphid marsupial, *Peradectes.* Although we prefer this preceding usage, reference to this faunal interval as the Bugcreekian Mammal Sub-Age is equivalent. In order to maintain stability of usage of informal abbreviations of the interval-zones, we refer to this interval-zone as Pu_0 and retain Pu_1 for the *Peradectes/Ectoconus* Interval-Zone as was done in Archibald and others (1987).

By incorporating this interval-zone within the Puercan Mammal Age, it becomes necessary to also redefine this mammal age. Hence, this age is now considered to commence with the appearance of the arctocyonid ungulate *Protungulatum* rather than with *Peradectes,* as was done in Archibald and others (1987). The Puercan Land-Mammal Age can now be redefined to include faunas that occur during the time between the first appearance of the archaic ungulate, *Protungulatum,* and the first appearance of the periptychid, *Periptychus.*

Biostratigraphic assessment

As mentioned in a previous section, the *Protungulatum/Peradectes* Interval-Zone cannot be as well constrained biostratigraphically as one would desire. This was, as noted, a result of the occurrence of these faunas within complex channel-fill deposits that cut into the underlying deposits of the Hell Creek Formation, which yield Lancian faunas. Maximum lowest biostratigraphic placement for the beginning and minimum highest biostratigraphic placement for the end of this interval-zone can be at least partially established and compared in three areas in northeastern Montana—the Bug Creek area, the McGuire Creek area, and the Hell's Hollow area (see Fig. 2).

For purposes of the following discussion it is necessary to clarify some stratigraphic terminology pertaining to the lignites in these areas. The term "Z coal" in McCone County dates from its usage in coal surveys by the U.S. Geological Survey (Collier and Knecthel, 1939) where it was employed to mark the formational contact between what are now called the Hell Creek and overlying Tullock Formations. The term "Z coal" was extended to the west into Garfield County in similar coal surveys (Rohrer, USGS, unpublished, in Archibald, 1982). This latter terminology was followed by Archibald (e.g., 1982), but it was realized that more than one lignite was present and thus the idea of a Z-complex was developed. The "lower Z" marks the formational boundary and approximates the K-T boundary (which was believed to lie about 2 to 3 m lower based on the highest remains of unreworked dinosaurs) and the "upper Z" marks the upper limit of this coal complex.

Although the terms Z coal, lower Z coal, and upper Z coal have been widely used throughout the region, a number of authors (e.g., Archibald, 1982; Fastovsky, 1987; Smit and others, 1987) have clearly indicated that direct correlations of these coals over this region are *not* possible. Thus, in the following discussions it must be kept in mind that lateral equivalency of the coals is not implied when the same sequential names are used. In order to emphasize this lack of lateral equivalency, the terms upper and lower will be placed in quotes when used outside of the type area of the Hell Creek Formation in Garfield County. Further, in McCone County it is the uppermost and thickest of the Z complex that has been identified as the formational boundary, while the K-T boundary is usually approximated at a lower, but not necessarily the lowest lignite in the Z complex (e.g., Smit and others, 1987), while in Garfield County both the formational and K-T boundaries are usually placed at the lowest of the Z complex (e.g., Archibald, 1982; see Fig. 2).

In the Bug Creek area in McCone County (Fig. 2), the formational contact has been generally agreed to occur at the base of the first laterally persistent lignite, which usually measures 1 to 2 m in thickness. According to Smit and others (1987), this formational coal varies from 3.5 to 11 m above a thin discontinuous lignite that approximates the K-T boundary based on a pollen change and the occurrence of an as-yet-unpublished iridium anomaly. As these authors note, and we concur, unreworked dinosaurs have not been found above this lower coal. These lignites are the "upper" and "lower" Z coal, respectively. The lowest stratigraphic placement of a Pu_0 (or Bugcreekian) site in the Bug Creek area remains the discovery site of Bug Creek Anthills positioned between 24 and 20 m below the formational contact (base of the "upper" Z). The channel capping is thought to be undetermined (Fastovsky and Dott, 1986) or to occur at or just above the "lower" Z coal (Smit and others, 1987). We favor the former interpretation; however, the biostratigraphically highest Lancian sites lateral to the channels nevertheless provide a maximum lowest stratigraphic placement for the Pu_0 sites. Rigby and others (1987) indicate a site (NMF in their Fig. 3) very close to Bug Creek Anthills that occurs slightly lower than 12 m below the "upper" Z. The highest definitively Lancian site in the general area with which we are familiar occurs in Russell Basin about 2 km east of Bug Creek Anthills. This site, Last Days B. A. locality, yielded a small sample of Lancian mammals and dinosaurs. It occurs about 2 m below the "lower" Z coal (K-T boundary) and about 6 m below the "upper" Z coal (formational contact). Thus, the maximum (lowest) stratigraphic occurrence (relative to sediments lateral to the channels) for the *Protungulatum/Peradectes* Interval-Zone in the Bug Creek area is just slightly less than 2 m below the "lower" Z coal, but of course could be higher (Fig. 2). No faunas of the next younger *Peradectes/Ectoconus* Interval-Zone (Pu_0) have been reported in the Bug Creek area, and thus no upper limit of the *Protungulatum/Peradectes* Interval-Zone (Pu_1) can be suggested in this area.

About 5 km south of the Bug Creek area, in the vicinity of McGuire Creek (Fig. 1), Lofgren has been studying faunas and their biostratigraphic setting for his Ph.D. research. The preliminary comments reported here and briefly noted in Lofgren and Hotton (1988), will be more fully discussed elsewhere by Lofgren. As in the Bug Creek area, the faunas usually come from channel deposits of considerable complexity. Only one unproductive locality in the upper part of the Hell Creek Formation in this area can with confidence be regarded as a Lancian site. This locality, K-Mark 2, occurs at about 24 m below the "upper" Z coal that, as in the Bug Creek area, marks the boundary between the Hell Creek and Tullock Formations (Fig. 2). This site is lateral to a channel deposit yielding the stratigraphically lowest Bug Creek–type fauna in the area. This site, Little Roundtop

Channel, is probably referable to Pu_0 (or Bugcreekian Sub-Age). As with the Lancian site, it occurs about 24 m below the "upper" Z formational contact, although the site is in a channel deposit cut by overlying channel deposits and thus is certainly capped at least somewhat biostratigraphically higher (Fig. 2). Lateral to the channeling, unreworked dinosaur remains in flood-plain deposits (e.g., below Z-line in Fig. 2) occur to within about 6 m of the "upper" Z and suggest that, if they could be found, the highest Lancian mammalian faunas (and highest Late Cretaceous pollen?) in this area would be at about this level. Although more equivocal than for the Last Days site in Russell Basin, these data suggest that the boundary between the Lancian and overlying Puercan (Pu_0 or Bugcreekian) occurs about 6 m below the Hell Creek–Tullock contact.

Unlike in the Bug Creek area, fortunately, it appears that the *Peradectes/Ectoconus* Interval-Zone (Pu_1) is well represented and can provide a highest possible placement for the underlying *Protungulatum/Peradectes* Interval-Zone (Pu_0). One small locality, Z-Line Local Fauna, occurring at the base of a channel-fill about 4 m below and capped by the upper Z coal is with little doubt assignable to Pu_1 based upon among other taxa, forms probably referable to *Peradectes* cf. *P. pusillus* and *Mimatuta minuial,* both of which appear in Pu_1 sites elsewhere (Archibald and others, 1987; and below).

Of even more potential interest are a series of localities that occur in a complex of channel deposits on the average of 20 m below the "upper" Z coal (Fig. 2). All of the localities—Second Level, Eagles Nest Area, Shiprock Local Fauna, Tedrow Quarry Local Fauna, Three Buttes Local Fauna, and Up Up the Creek—are faunally similar to Harbicht Hill (about 25 km north) and the slightly younger Hell's Hollow Local Fauna (about 85 km west; see faunal zonation below). Two of the sites, however, Shiprock Local Fauna and Up Up the Creek, have each also produced a few speciments referable to *Peradectes* cf. *P. pusillus,* the taxon that was used by Archibald and others (1987) to define the commencement of the Puercan. Also, the richest and best sampled of these two localities, Up Up the Creek, has also produced a few specimens of *Catopsalis alexanderi.* This taxon is known elsewhere (e.g., Hell's Hollow Local Fauna and its type locality near Denver) only from faunas referred to the *Peradectes/Ectoconus* Interval-Zone (Pu_1). It is a possibility that these two faunas are slightly younger than the other four in this channel complex. Given the lack of a clear channel capping (except for Three Buttes, Second Level, and Eagles Nest, which are capped by the "upper" Z) and relatively poor sampling at most of the sites, however, we favor the preliminary interpretation that all six sites are approximately the same age. We suspect that owing to their rarity, *Peradectes* and *Catopsalis alexanderi* have not yet been recovered from all the sites in the channel complex. Further collecting at these sites may provide an answer. In any case, it suggests that these sites are very close on either side of the Pu_0-Pu_1 boundary. These same six sites plus the probably biostratigraphically higher Z-line in addition to having Paleocene aspect mammals, also have yielded most of the typical latest Cretaceous or Lancian mammals known from Bug Creek Anthills proper. Especially common are the marsupials *Pediomys, Alphadon,* and *Glasbius,* which in combination represent about 50 percent of the abundance of the primitive ungulates at these sites. The only noticeable faunal distinction between the six lower sites and the higher Z-Line, is the absence of Lancian mammals and dinosaur bone fragments and teeth at the latter site.

The very gradational change from one interval-zone to another was, as noted earlier, a major reason that the first author (JDA) now agrees with the second author (DLL) that recognizing a separate mammal age for the Bug Creek faunal assemblage should at least for the foreseeable future be abandoned. Since Z-Line and the six lower sites are for the moment only known to be capped by the "upper" Z, this represents the lowest position ascertainable for the beginning of the *Peradectes/Ectoconus* Interval-Zone (Pu_1) and the highest position for the end of the *Protungulatum/Peradectes* Interval-Zone (Pu_0) in these sections, although a lower capping and hence earlier commencement for the Pu_1 interval-zone remains a very likely prospect.

Some 85 km to the west in Garfield County in Hell's Hollow (Archibald, 1982), localities can also be biostratigraphically arranged so as to estimate the limits of the *Protungulatum/Peradectes* Interval-Zone (Pu_0; Figs. 1 and 2). In this setting, however, no faunas have been identified as belonging to this interval-zone. Rather, Lancian faunas and overlying faunas referable to the *Peradectes/Ectoconus* Interval-Zone (Pu_1) permit estimates of the stratigraphic limitations of the *Protungulatum/Peradectes* Interval-Zone. It should be emphasized that there are no major unconformities in this section. Further, although this constitutes negative evidence, it does provide a crude estimate of the possible stratigraphic limitations of this interval-zone in the Hell's Hollow area.

In Hell's Hollow, a poorly sampled Lancian locality (Worm Coulee 5) and larger, scattered dinosaur postcranial and cranial fragments occur to within 1.8 to 3 m of the lower Z coal (Fig. 2). The stratigraphic uncertainty for Worm Coulee 5 arises because of reworking of this lignite both by the immediately superposed Hell's Hollow Channel and by post-Pleistocene erosion. For the sake of discussion this figure is rounded to 2.4 m. The immediately overlying Hell's Hollow Channel cuts (or maybe in its lower part is penecontemporaneous with) the lower Z coal. In this region, this lower coal not only approximates the K-T boundary as in McCone County, but also has traditionally been used to mark the formational contact (e.g., Archibald, 1982).

In this section, the lower Z has not been investigated for pollen or iridium, but it is laterally equivalent to a similar lignite only 4 km to the west that has yielded iridium (Alvarez, 1983) and that also approximates the K-T pollen boundary. Also two small vertebrate samples in the same area (Fig. 1), Morales 1 and Herpijunk Promontory, may be referable to the *Peradectes/Ectoconus* Interval-Zone ("mantuan" in Archibald, 1982). Both sites are in channel fills that cut the underlying Z coal from 0.8 to 1.4 m. Also in this area a rich, but mostly undescribed Lancian locality, Hauso 1, occurs about 12 to 15 m below the lower Z.

The capping of the Hell's Hollow Channel can be placed at about 12 m below the upper Z coal and at 3.7 m above the lower Z, while the actual fossil locality, Worm Coulee 1, occurs about 1.6 m below the channel capping (Fig. 2). Utilizing the channel capping, this places the lowest demonstrated position for the *Peradectes/Ectoconus* Interval-Zone at a similar 3.7 m above the lower Z coal and K-T boundary. Several sites averaging 15 km to the northwest (Billy Creek 1 and McKeever Ranch sites, see Fig. 1) either show stratigraphically (Billy Creek 1) or indicate faunally (the McKeever Ranch sites) that the *Peradectes/Ectoconus* Interval-Zone continues higher in the section for an as-yet-undetermined number of meters.

It is possible to utilize similar (but not the same) lithological criteria as presented in the foregoing discussion in order to provide minimum biostratigraphic constraints for the *Protungulatum/Peradectes* Interval-Zone (Pu_0) for the Bug Creek area, McGuire Creek area, and in Hell's Hollow. In the Bug Creek area (including Russell Basin), the lowest position for the inception of Pu_0 would be about 2 m below the "lower" Z, which marks the K-T boundary in the area, or 8 m below the "upper" Z, which marks the formational contact. In the McGuire Creek area, Pu_0 could commence as low as 24 m below the "upper" Z, but this seems unlikely as dinosaur remains in flood plains have been found to within about 6 m of the "upper" Z. In Hell's Hollow, Pu_0 could begin some 2 m below the lower Z, which marks the K-T and formational boundaries in this area. Thus, in both the Bug Creek area and Hell's Hollow the lowest position for the commencement for Pu_0 is about 2 m below the "lower" Z coal, while for the McGuire Creek area the data are too equivocal for an assessment. The upper limit for the Pu_0 is more poorly constrained. In the McGuire Creek area it appears that Pu_0 could persist to the base of the "upper" Z, while in Hell's Hollow it appears Pu_1 might supercede Pu_0 some 12 m below the upper Z (or about 4 m above the lower Z). No upper limit can be suggested for Pu_0 in the Bug Creek area.

Faunal composition and characterization

Various faunal lists for the *Protungulatum/Peradectes* Interval-Zone have been published by various authors, the most complete are found in Archibald (1982) and Archibald and Clemens (1984), and also can be extracted from Sloan's (1987) "phylogeny" of Paleocene mammals. The list of taxa at the end of this section provides an update of these lists of taxa to the generic level (Table 1). Taxa not identified at least to genus are excluded. The asterisk (*) indicates "typical latest Cretaceous" taxa, some of which may have been reworked into sediments bearing the BCA mammals. Although this is not the interpretation favored here, it is noted for completeness and discussed further under the section on Lancian mammals below.

As part of his recognition of a new "Bugcreekian age (stage)," Sloan (1987) stated that "[t]his newly defined age (stage) is typified by *Procerberus formicarum, Protungulatum donnae, P. gorgun, Mimatuta morgoth, Oxyprimus erikseni, Purgatorius ceratops, Stygimys kuszmauli,* and *Catopsalis joyneri,* all of which are restricted to this age (stage) so far as is presently known." All of these genera are known from younger sites (Archibald and others, 1987) in the Puercan Mammal Age. At the species-level, *Procerberus formicarium, Mimatuta morgoth,* and *Oxyprimus erikseni* have been reported; and *Stygimys kuszmauli* and *Protungulatum donnae* have been questionably reported from faunas of the *Peradectes/Ectoconus* Interval-Zone (Pu_1; Archibald, 1982) in the Hell's Hollow Local Fauna in Garfield County. Similarly, *Procerberus formicarum, Protungulatum donnae, P. gorgun, Mimatuta morgoth, Oxyprimus erikseni,* and *Stygimys kuszmauli* are all tentatively identified at one or more of the early Pu_1 sites (Shiprock, Up Up the Creek, and Z-Line) of Lofgren (unpublished manuscript) in the McGuire Creek area. This leaves only *Catopsalis joyneri* and *Purgatorius ceratops* as species that may be restricted to this faunal interval, but unfortunately the latter is very poorly known as it is currently represented by a single tooth purported to have come from Harbicht Hill.

The following are reported for the *Protungulatum/Peradectes* Interval-Zone (Pu_0):

First appearances: *Catopsalis, Mimatuta, Oxyprimus, Procerberus, Protungulatum, Purgatorius(?), Ragnarok, Stygimys.* (N.B.: Sloan and others, 1986; Smit and others, 1987, treated *Ragnarok* as a junior synonym of *Baioconodon.* Although probably correct, this has not been formally published by its author, M. Middleton.)

Last appearances: **Cimolomys, *Essonodon.*

Index fossils: none (at generic level).

Characteristic fossils: **Alphadon, *Batodon, Cimexomys, Cimolestes, *Cimolodon(?), *Didelphodon, *Glasbius, *Gypsonictops, *Meniscoessus, Mesodma, *Pediomys.*

Taxa absent but known before and after the *Protungulatum/Peradectes* Inteval-Zone: *Neoplagiaulax(?).*

Informal biochrons

In Archibald and others (1987), the mammal ages were subdivided into zones. The preferred method of zonation was the interval-zone concept as outlined in that paper. Our naming of the Bug Creek faunal assemblage as a new interval-zone—the *Protungulatum/Peradectes* Interval-Zone (Pu_0)—follows from this work. In addition, since its first description (Sloan and Van Valen, 1965) the Bug Creek faunal assemblage has been argued to represent at least three faunal levels.

A potentially serious problem, however, with attempting to subdivide this interval-zone lies in attempting to determine a biostratigraphic sequence *within* the interval-zone. The difficulty arises because the stratigraphic positions of all of the published Bug Creek–type sites have been questioned at one time or another. Some of these difficulties in stratigraphic placement and correlation, and possible faunal mixing, are discussed below. Fortunately, there are other ways to correlate faunas, namely by using the faunas themselves. This might at first seem to be circular, but in fact it is not. Biostratigraphic placement utilizing strati-

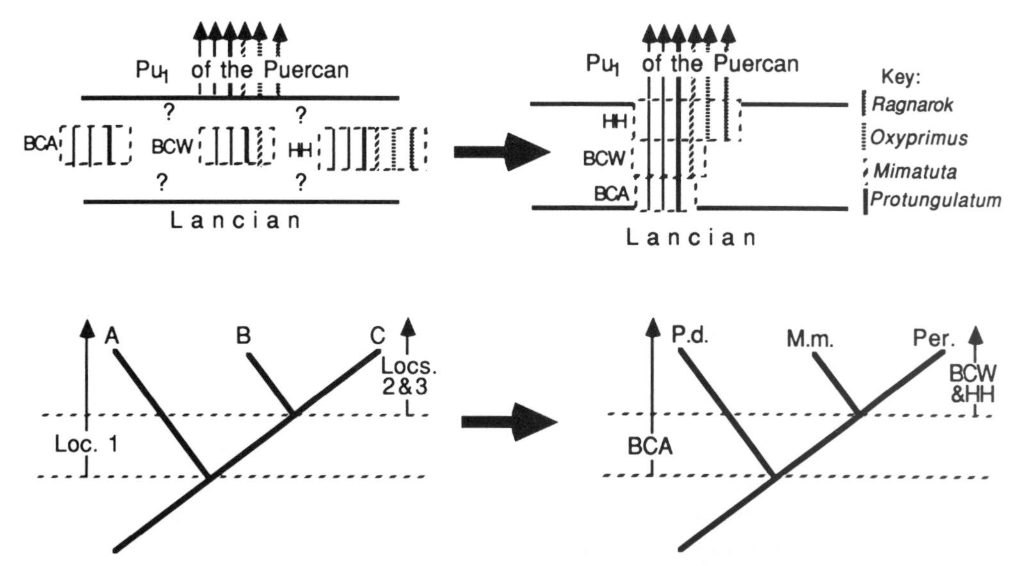

Figure 4. Lower half shows hypothetical (left) and actual (right) cases for the use of cladochronology in helping to determine the relative biochronologic ages of faunas/localities. Upper half shows how the relative ages of faunas of uncertain biostratigraphic position (left) can be resolved (right) using concurrent ranges of taxa that occur both in the unresolved and in the biostratigraphically resolved faunas. See text for explanation. Abbreviations are as follows: Bug Creek Anthills Local Fauna (BCA); Bug Creek West Local Fauna (BCW); Harbicht Hill Local Fauna (HH); *Protungulatum donnae* (P.d.); *Mimatuta morgoth* (M.m.); the remainder of the monophyletic clade, Periptychidae (Per.).

graphic sequencing of faunas is desirable, but in the absence of this or other independent means of correlation, the faunas can and often do provide the means of correlation.

In fact, biochronologic correlation (first appearances, last appearances, similar evolutionary grade, etc.) continues to be the single most important method of correlating faunas on a local, regional, and continental scale. Although this is a very basic principle it is frequently misunderstood. For example, Fastovsky and Dott (1986; Fastovsky, 1987) suggest that since the BCA locality is not chronostratigraphically and lithostratigraphically constrained, its age relative to other fossiliferous sites is not ascertainable. This is not correct, for if it were, one could not correlate any fossil-bearing beds on their faunal content alone (Archibald, 1986). Also, this is the single most important method of relative age correlation of sedimentary rocks. In the case of the Bug Creek–type sites, however, such biochronologic correlations must remain tentative. This is because such correlations are based almost entirely on the presence or absence of one or two taxa that could be affected significantly by sampling, taxonomical, and/or ecological biases. The effects these factors may have had on correlations (and biochronologic units) within the *Protungulatum/Peradectes* Interval Zone (Pu_0) are not known.

The consequences then for the *Protungulatum/Peradectes* Interval-Zone are that its subdivisions are based exclusively on faunal content without the benefit of accurate stratigraphic placement. Accordingly, we use the term "biochron" for the following sequencing of this interval-zone since we do not wish to imply stratigraphic control. Further, although we do feel that there are sequential faunal differences within this interval-zone, we do not feel that formal recognition of subdivisions is warranted based on our current knowledge.

One of the all-too-obvious results of not having good biostratigraphic control within the interval-zone is that, although internally consistent and coherent, the biochronological sequence cannot be very accurately correlated to other geologic scales. This is unfortunate, but such imprecise chronostratigraphic correlation is in fact quite common. Such is the case for correlation of the Paleocene/Eocene boundary with North American land mammal ages, where this boundary is correlated to various faunal or floral changes occurring at different times within the Clarkforkian or Wasatchian Mammal Age (see comments in Archibald and others, 1987 and references therein).

One way in which biochronology (and biostratigraphy) can become more precise and more explicitly testable is through the application of cladistic methodology. Hypotheses of relationship have already become commonplace in assessing biogeographic patterns and are now finding their way into studies of standing diversity, especially as it pertains to extinction patterns (Norell, 1987). Similarly, cladistic methodology can be useful in biochronologic correlation. The technique will be more fully explained elsewhere (Archibald, in preparation), but the general thesis is that cladograms of preferably more distantly related organisms can be compared to suggest the most parsimonious relative positioning of the localities at which the organisms occur.

For example (Fig. 4), three taxa, A, B, and C, have been analyzed cladistically, and B and C together form the sister taxon to A. If at localities 1, 2, and 3, taxon A is present only at locality 1 and taxa B and/or C are present at localities 2 and 3, one can argue that locality 1 is no older but could be younger than the most recent common ancestors of the sister taxa A and B plus C, while localities 2 and 3 are no older but could be younger than the most recent common ancestor of taxa B and C. In this simple case, then, the argument would be that locality 1 could be older relative to localities 2 or 3, and that the relative ages of localities 2 and 3 to each other cannot be resolved based upon these data.

The potential strength of this technique lies not in a single cladistic analysis such as this hypothetical example, but in comparing cladograms for as many different clades as possible. Through repeated comparisons of such cladograms one can establish whether there is congruency between cladograms regarding the relative positioning of the localities in question.

At present, there are only two published cladograms for species that make their appearance in the *Protungulatum/Peradectes* Interval-Zone. One includes the archaic ungulate family Periptychidae (Archibald and others, 1983) and the other examines the relationships within the multituberculate genus *Catopsalis* (Simmons and Maio, 1986). Unfortunately, in the latter case, only one species of *Catopsalis, C. joyneri,* occurs throughout the *Protungulatum/Peradectes* Interval-Zone and thus is not of particular use for biochronology. In the case for the periptychids (Fig. 4), *Mimatuta,* the sister taxon to remaining periptychids, occurs at Bug Creek West and Harbicht Hill, but not Bug Creek Anthills. The most likely sister taxon for all periptychids, *Protungulatum donnae,* occurs at all three localities. The argument then is that Bug Creek Anthills is no older but could be younger than the split between *Protungulatum donnae* and periptychids, while Bug Creek West and Harbicht Hill are no older but possibly younger than the split between *Mimatuta* and the remaining periptychids. This of course can be used only to support the view that Bug Creek Anthills is relatively older than the other two sites, but not how the other two sites relate to each other.

With further cladistic analysis of other taxa, a more precise and more explicitly testable zonation of the *Protungulatum/Peradectes* Interval-Zone can be established. This technique, which I term "cladochronology," has application beyond issues pertaining to the *Protungulatum/Peradectes* Interval-Zone.

Although cladochronological techniques cannot currently be applied to argue for the placement of Bug Creek West relative to Harbicht Hill, another method can provide such a relative biochronological placement. The method employs the concept of concurrent range zones, which have been used for many years. The major difference in using this concept biochronologically rather than biostratigraphically is that the overlapping occurrences of the taxa are used without direct reference to their biostratigraphic position. It is argued that if enough taxa demonstate overlap for a given suite of localities, the localities can be sequenced unambiguously. The statistical arguments for this are being considered elsewhere (Archibald and others, in progress).

In order to establish in which direction the biochronological sequence proceeds, biostratigraphically constrained faunas should bracket, overlie, or underlie the biostratigraphically ambiguous faunas (Fig. 4). Overlying faunas are arguably best, and they may provide the opportunity of using first appearances, which are theoretically better biochronological markers than last appearances (or extinctions).

In comparing Bug Creek West and Harbicht Hill, both faunas share the majority of their taxa including *Mimatuta.* This taxon continues into the overlying biostratigraphically constrained faunas of the *Peradectes/Ectoconus* Interval-Zone (Pu_1; Fig. 4). The taxa *Oxyprimus* and *Ragnarok,* however, are not present at Bug Creek West, but are present at Harbicht Hill and at the overlying biostratigraphically constrained faunas of the Pu_1 interval-zone. Thus, we argue that Harbicht Hill is relatively younger than Bug Creek West but is older than Pu_1 faunas because the latter faunas have produced specimens of *Peradectes,* which has not been reported from Harbicht Hill.

Protungulatum/Mimatuta *biochron (bk_1).* This informal biochron is recognized as including faunas that occur during the time between the appearance of *Protungulatum* and the appearance of *Mimatuta.*

The single most important fauna referable to this biochron is that recovered from the Bug Creek Anthills locality, including the discovery anthill and the nearby quarry sites (Fig. 1). Although first described over 20 years ago (Sloan and Van Valen, 1965), the fauna has never been fully described. Faunal lists, such as by Archibald (1982), Archibald and Clemens (1984), and Sloan and others (1986), that are drawn from various sources represent the most recent and thorough faunal lists for BCA. The faunal list in Table 1 is from these references. Faunal correlatives of BCA have been reported in the immediate vicinity (e.g., Ken's Apex in Sloan and others, 1986; Smit and others, 1987) but none have been described.

Of the sites referable to the *Protungulatum/Peradectes* Interval-Zone in the Bug Creek area, the BCA locality between about 24 and 20 m below the "upper" Z coal is topographically lowest relative to either the "lower" Z coal (the palynologically defined K-T boundary) or the "upper" Z coal (formational contact). This is one point on which all discussants of this locality seem to agree. The stratigraphic position of the locality is another matter. Some (Fastovsky and Dott, 1986; Rigby and others, 1987) feel that the capping of the channel bearing the site cannot be determined and hence the age of the locality is in doubt. Others (Smit and others, 1987) are only slightly more positive in suggesting that the channel cuts the "lower" Z coal and is Paleocene in age. They argue that the base of another channel (the Big Bugger) that has produced only Paleocene pollen may be equivalent to the top of the Bug Creek channel, and hence the Bug Creek channel could be Paleocene. Finally, there are those (Sloan and others, 1986; Sloan, 1987) who suggest the site lies within the Hell Creek Formation and is probably Cretaceous in age.

We have taken the more conservative stance and agree that the capping for the channel that bears the BCA site is equivocal (Fig. 2). Thus the channel could be capped as high as at the base of the "upper" Z coal. The biostratigraphic assessment suggests that the faunally older BCA site is capped closer to either side of the "lower" Z coal, than to the higher "upper" Z coal. This is indicated by the presence of sites referable to the next younger *Peradectes/Ectoconus* Interval-Zone at a comparable topographic position some 20 m below the "upper" Z coal in the McGuire Creek area and the presence of such sites at about 4 m above the lower coal in Hell's Hollow.

Even with this biostratigraphic uncertainty the cladochronological assessment given above suggests that the Bug Creek Anthills locality is (as was first assessed by Sloan and Van Valen, 1965) the relatively oldest fauna in the *Protungulatum/Peradectes* Interval-Zone. In addition to probably having only one archaic ungulate, *Protungulatum,* the site has yielded all genera (but possibly not species) of Lancian mammals except for Neoplagiaulax(?). Assuming that this relative age assessment is correct, *Catopsalis, Procerberus,* and *Stygimys* also first appear at BCA. Characterization of the *Protungulatum/Mimatuta* biochron (bk_1) is as follows:

First appearances: *Catopsalis, Procerberus, Protungulatum, Stygimys.*

Last appearances: none (at the generic level).

Index fossils: none (at generic level).

Characteristic fossils: **Alphadon, *Batodon, Cimexomys, Cimolestes, *Cimolodon, *Cimolomys, *Didelphodon, *Essonodon, *Glasbius, *Gypsonictops, *Meniscoessus, Mesodma, *Pediomys, Procerberus, Protungulatum, Stygimys.*

Taxa absent but known before and after the *Protungulatum/Mimatuta* biochron: Neoplagiaulax(?).

Mimatuta/Oxyprimus ***biochron (bk_2).*** This biochron is recognized as including faunas that occur during the time between the appearance of *Mimatuta* and the appearance of *Oxyprimus.*

The fauna that typifies this biochron is from the Bug Creek West site (BCW) of Sloan and Van Valen (1965), which lies about 1 km southwest of Bug Creek Anthills. One locality, Scmenge Point, has been reported as a faunal correlative of Bug Creek West (Sloan and others, 1986); however, the full mammalian faunal list has not yet been reported so comparisons cannot be made. Rigby and others (1987) place Scmenge Point about 2 km east of BCW. As with the *Protungulatum/Mimatuta* biochron, the fauna of BCW has never been fully described. Thus, the faunal list in Table 1 is from a composite of sources combined from Archibald (1982), Archibald and Clemens (1984), Sloan and others (1986), and references therein.

As with all other Bug Creek–type sites in the Bug Creek area, the capping for the channel that yields the Bug Creek West fauna has not been determined. In Sloan and Van Valen (1965) and most other later references, BCW is cited as occurring about 18 m below the "upper" Z coal. This appears to be close to the range of about 15 to 17 m shown in Figure 3 of Rigby and others (1987). In Figure 5 of the same paper, however, the position appears to be closer to 10 m below the "upper" coal. Lacking evidence to the contrary, we assume the latter figure is a *lapsus calami.*

Faunistically, the *Mimatuta/Oxyprimus* biochron is very similar to the *Protungulatum/Mimatuta* biochron. The former contains all the genera (and perhaps species) of the latter biochron with the exception of the appearance of *Mimatuta* in bk_2 and the absence of *Cimolomys,* which is probably sampling error as this taxon is reported from the next younger bk_3. As discussed above, bk_2 is argued to be relatively younger than bk_1 on cladochronological grounds. In the same section it was argued that bk_2 is older than bk_3 because bk_2 lacks *Oxyprimus,* which is found in bk_3 faunas and in faunas that are biostratigraphically assignable to the younger *Peradectes/Ectoconus* Interval-Zone (Pu_1).

First appearances: *Mimatuta.*

Last appearances: none (at the generic level).

Characteristic taxa: **Alphadon(?), *Batodon, Catopsalis, Cimexomys, Cimolestes, *Cimolodon, *Didelphodon, *Essonodon, *Glasbius, *Gypsonictops, *Mensicoessus, Mesodma, *Pediomys, Procerberus, Protungulatum, Stygimys.*

Taxa absent but known before and after the *Mimatuta/Oxyprimus* biochron: **Cimolomys, Neoplagiaulax(?).*

Oxyprimus/Peradectes ***biochron (bk_3).*** This biochron is recognized as including faunas that occur during the time between the appearance of *Oxyprimus* and the appearance of *Peradectes.*

The fauna that typifies this biochron is the Harbicht Hill locality, which lies aobut 22 km north of BCA (Fig. 1). Sloan and others (1986) report Harbicht Hill 1 and 2, which are assumed to be faunally equivalent although this is not specified.

One locality, Chris's Bone Bed, lying about 3 km south of Harbicht Hill, could be a faunal equivalent of Harbicht Hill or of Bug Creek West, while a second locality, Ferguson Ranch, could be a faunal equivalent of Harbicht Hill or the younger faunas of the *Peradectes/Ectoconus* Interval-Zone (Pu_1) such as the Hell's Hollow Local Fauna some 85 km to the west.

Chris's Bone Bed has yielded a few mammalian specimens (Lupton and others, 1980). One of these is *Protungulatum gorgun* that is first reported at Bug Creek West, but continues at Harbicht Hill (Van Valen, 1978) and at younger sites such as Ferguson Ranch (Sloan and others, 1986). This species also occurs at sites in the McGuire Creek area that may straddle both sides of the faunal boundary between the *Protungulatum/Peradectes* (Pu_0) and *Peradectes/Ectoconus* (Pu_1) interval-zones.

The Ferguson Ranch fauna has been mentioned without a full faunal listing in various papers (Sloan and others, 1986; Smit and others, 1987). Thus, it is somewhat difficult to assess its age relative to the criteria developed here. Sloan and others (1986) list the ungulates from this site and Smit and others (1987) note that the ungulates are at a similar evolutionary grade as those from the Hell's Hollow Local Fauna (Archibald, 1982). There is no mention, however, of the marsupial *Peradectes,* which is quite common at Hell's Hollow 85 km to the west of Ferguson Ranch and at the Z-Line locality in the McGuire Creek area 20 km south of Ferguson Ranch. We suspect that the taxon may be present but has yet to be recovered or recognized, and hence advisedly consider Ferguson Ranch to be referable to the

Peradectes/Ectoconus Interval-Zone (Pu_1) rather than to the *Oxyprimus/Peradectes* biochron (bk_3) of the *Protungulatum/ Peradectes* Interval-Zone (Pu_1).

In the McGuire Creek area, just to the south of Bug Creek (Fig. 1), several faunas belong either to the bk_3 biochron or to the next younger Pu_1 interval-zone. The most likely candidate for assignment to the bk_3 biochron is the Little Roundtop Channel locality (Fig. 2). This locality is in a channel deposit that is capped by a slightly higher channel complex containing faunas that may bracket the Pu_0 (bk_3)/Pu_1 faunal boundary. Based on a small sample, the fauna of Little Roundtop appears to match that of Harbicht Hill, except that *Pediomys* is reported from Little Roundtop but not Harbicht Hill.

The channel complex that overlies Little Roundtop has produced a series of localities and local faunas (Fig. 2) that are currently under study by Lofgren. These include: Second Level localities, Eagle Nest localities, Shiprock Local Fauna, Tedrow Quarry Local Fauna, Three Buttes Local Fauna, and Up Up the Creek Local Fauna (Fig. 1). All the sites are positioned about 20 m below the "upper" Z coal, but this lignite is known to form the channel capping for only the Second Level localities, Eagle Nest localities, and the Three Buttes Local Fauna.

As noted in the preceding two paragraphs, the faunas from these localities appear to be very close on either side of the Pu_0/Pu_1 boundary. Of these faunas, only Up Up the Creek has been quite extensively sampled using screen washing. The faunal list for this (and the other sites) is (are) very preliminary. It includes all of the genera and most of the species recognized at Harbicht Hill, but also includes a few specimens referable to *Peradectes* cf. *P. pusillus, Catopsalis alexanderi.* Both of these species have been reported only from the younger *Peradectes/Ectoconus* Interval-Zone (Pu_1), and thus suggest the referral of Up Up the Creek to this younger faunal interval.

Of the other less well-sampled faunas listed above, the Shiprock Local Fauna has also produced rare specimens referable to *Peradectes,* suggesting that this local fauna is comparable in age to Up Up the Creek. Further sampling of the other localities is required to determine whether reference to Pu_0 or Pu_1 is warranted.

In quite marked contrast to Harbicht Hill, Up Up the Creek and the other faunas listed above have produced relatively high percentages of Lancian mammals. Especially well represented are the multituberculate *Meniscoessus* and all the marsupial genera recognized from the Lancian Mammal Age. The marsupials are present in numbers nearly equaling that of the archaic ungulates. The eutherians, *Batodon* and *Gypsonictops,* are much rarer components of these faunas. The issue of whether the Lancian taxa are reworked is addressed later in the paper.

Another fauna possibly belonging to the *Oxyprimus/Peradectes* biochron (bk_3) is the "Medicine Hat Brick and Tile Quarry (Long Fall Horizon)" (Johnson and Fox, 1984; Fox, 1987). This fauna is from a locality in southern Saskatchewan (Fig. 1), and represents only one of two faunas of the *Protungulatum/Peradectes* Interval-Zone known from outside of eastern Montana. The presence of the same genera of archaic ungulates, especially *Oxyprimus,* at Harbicht Hill and Long Fall suggests that Long Fall is at least as young as the former site. The absence of *Peradectes* also suggests Long Fall is older than the *Peradectes/Ectoconus* Interval-Zone.

Before the discovery of possible Pu_1 faunas in the McGuire Creek area yielding *Alphadon* and *Pediomys,* the presence of these marsupials at Long Fall would also have suggested an older age than Pu_1. For now, the presence of such Lancian taxa at localities such as these does not appear to afford distinguishing between Pu_0 and Pu_1 localities.

Analyzing these and other components of this fauna, Johnson and Fox (1984, p. 215) conclude "that the Long Fall Horizon represents an interval in very latest Lancian time." Although we agree with the relative faunal placement of the fauna, we would now argue that Long Fall is best regarded as belonging to the Pu_0 interval-zone (bk_3 biochron). It should be noted that Sloan (1987) refers the Long Fall Horizon to his "Mantuan Age (Stage)." This usage is comparable to the *Peradectes/Ectoconus* Interval-Zone (Pu_1) of Archibald and others (1987) and is the usage followed by us. Lacking such taxa as *Peradectes,* we prefer, however, to refer Long Fall to the *Protungulatum/Peradectes* Interval-Zone (Pu_0).

As with the other faunas in the Pu_0 interval-zone, the bk_3 biochron lacks biostratigraphic constraint. The discovery site of Harbicht Hill is usually reported as being about 12 m below the "upper" Z coal (Fig. 2). Other referred faunas such as Little Roundtop Channel occur considerably lower at about 24 m below the "upper" Z coal. The best that can be noted for the present is that the faunas of the bk_3 biochron are poorly constrained biostratigraphically.

Faunally, the *Oxyprimus/Peradectes* biochron (bk_3) is quite similar to both the next older *Mimatuta/Oxyprimus* biochron (bk_2) and the next younger *Peradectes/Ectoconus* Interval-Zone (Pu_1). At the generic level, bk_3 differs from bk_2 in the first appearances of *Oxyprimus* and *Ragnarok,* and the questionable first appearance of the primate *Purgatorius.* Possible differences at the species level have yet to be fully explored. Whether the apparent last appearance in bk_3 of *Cimolomys* and *Essonodon* is real or a vagary of sampling remains to be seen following further sampling at younger Pu_1 sites in the McGuire Creek area. Within eastern Montana alone, the first appearances of *Peradectes* and the extremely rare multituberculate *Archeronodon,* and a variety of new species such as *Catopsalis alexanderi, Protungulatum mckeeveri, Ragnarok engdahli,* and *R. nordicum* all reinforce the faunal separation of bk_3 and Pu_1.

First Appearances: *Oxyprimus, Purgatorius(?), Ragnarok.*

Last Appearances: **Cimolomys,* *Essonodon.*

Index fossils: none (at the generic level).

Characteristic fossils. **Alphadon,* *Batodon, Catopsalis, Ci-*

*mexomys, Cimolestes, *Cimolodon, *Didelphodon, *Glasbius, *Gypsonictops, *Meniscoessus, Mesodma, Mimatuta, *Pediomys, Procerberus, Protungulatum, Stygimys.*

Taxa absent but known before and after the *Oxyprimus/Peradectes* biochron: *Neoplagiaulax(?).*

Problematical sites

There are three other published sites not in the vicinity of the Bug Creek area that can be referred to the *Protungulatum/Peradectes* Interval-Zone. Two of these sites appear not to be referable to this interval-zone, while the other is referable to the *Protungulatum/Peradectes* Interval-Zone, but not to a specific biochron.

The first of these sites is the poorly fossiliferous Rick's Place (and Rick's Place 2) located in Garfield County, Montana, about 65 km west of the Bug Creek area (Fig. 1). Archibald (1982) questionably assigned these sites to the Bug Creek faunal-facies based on the presence of fragmentary jaws of archaic ungulates and the apparent placement of the site within the Hell Creek Formation. Subsequent work by Clemens and students (personal communication) indicates that the sites are in channels downcutting from the Tullock Formation. Reference to the *Protungulatum/Peradectes* Interval-Zone is probably incorrect, but this cannot be unequivocally demonstrated. Sloan (1987) mentions this as a Bugcreekian fauna from Garfield County.

The second of these sites occurs in southern Wyoming and was reported by Breithaupt (1982). The site, UW locality V-79032, was referred to as Black Butte by Sloan (1987). The site has a normal Lancian fauna except for one dental remain that Breithaupt (1982) thought could belong to *Protungulatum* sp. Subsequently, Lillegraven (personal communication) indicated that reference to this taxon is questionable, and thus this site is in all probability not referable to the *Protungulatum/Peradectes* Interval-Zone.

Third, and finally, the Fr-1 site in southern Saskatchewan (Fig. 1) (Johnston, 1980; Johnston and Fox, 1984; Fox, 1987) can be confidently referred to the *Protungulatum/Peradectes* Interval-Zone, but reference to a specific biochron is equivocal. Fox (1987) notes that the fossil vertebrates indicate that Fr-1 is older than the Long Fall Horizon, also from southern Saskatchewan. As discussed above, the mammals from the Long Fall Horizon indicate it belongs to the *Oxyprimus/Peradectes* biochron and thus Fr-1 is best regarded as early to middle Pu_0 in age.

As with Pu_0 sites in eastern Montana, reference of the Fr-1 site to the Cretaceous or Paleocene is problematic. Fox (1987) argues it is Cretaceous based on the presence of dinosaur teeth and "typical latest Cretaceous mammals"; Smit and others (1987) argue for a Paleocene age because the locality occurs in sandstone of the Ravenscrag Formation that erodes an iridium-bearing coal seam (the Ferris-1). Aside from the issue of whether the site is Cretaceous or Paleocene based on sometimes conflicting definitions and issues of reworking, the site does *not* occur in a sandstone resembling the Paleocene Ravenscrag Formation, but rather comes from lithology indistinguishable from the Upper Cretaceous Frenchman Formation (Fox, personal communication).

Boundary between the **Protungulatum/Peradectes** *(Pu_0) and* **Peradectes/Ectoconus** *(Pu_1) Interval-Zones*

The faunas of the *Oxyprimus/Peradectes* biochron (bk_3) of the *Protungulatum/Peradectes* (Pu_0) Interval-Zone and the *Peradectes/Ectoconus* (Pu_1) Interval-Zone do not differ markedly from each other above the species level (Table 1). The only taxa that may last appear in the bk_3 biochron are *Cimolomys* and *Essonodon,* both of which are rare taxa in earlier faunas. There are nominally five genera that first appear in the Pu_1 interval-zone, although not at what are probably the earliest faunas referable to this interval-zone. These taxa are: *Peradectes, Baioconodon, Eoconodon, Earendil,* and *Maiorana.*

As argued by Middleton in his Ph.D. thesis (written communication to JDA) and referred to by Sloan and others (1986), *Ragnarok* is probably a junior synonym of *Baioconodon.* Hence, the latter taxon would replace the former as occurring at faunas of the bk_3 biochron and the Pu_1 interval-zone. *Earendil* and *Maiorana* are both known only from the Mantua Lentil Local Fauna (Van Valen, 1978) and it appears that the former taxon may be a synonymous with *Mimatuta,* a view shared by both Cifelli (1983) and Archibald (unpublished manuscript).

Of the two remaining taxa, only *Peradectes* is found at all or most of the faunas referable to the Pu_1 interval-zone and its first appearance defines the interval-zone. This marsupial is abundant in Hell's Hollow Local Fauna, Garfield County; common in the Z-Line Local Fauna, McGuire Creek area; uncommon in the Mantua Lentil Local Fauna, Bighorn Basin, northern Wyoming and the Alexander locality, Denver Basin, eastern Colorado; rare in the Up Up the Creek and Shiprock local faunas, McGuire Creek area; and unknown from Ferguson Ranch in the Bug Creek area, the Second Level, Eagles Nest, Tedrow Quarry, and Three Buttes local faunas in the McGuire Creek area, and the McKeever Ranch and Herpijunk localities in Garfield County.

Another taxon that appears to be useful in differentiating faunas assignable to either Pu_0 or Pu_1 is the multituberculate *Catopsalis.* In Pu_0 faunas, *C. joyneri* appears to be the only species of this genus that is present, while in Pu_1 faunas *C. alexanderi* is present, possibly in conjunction with *C. joyneri* at some sites. *C. alexanderi* is known from its type area in the Littleton Local Fauna of eastern Colorado (Middleton, 1982) as well as from the Mantua Lentil (Wyoming) and the Hell's Hollow (Montana) local faunas. Lofgren has also recovered a few specimens of this species from the Up Up the Creek Local Fauna in the McGuire Creek. With further study, additional taxa may provide better faunal separation between the Pu_0 (especially bk_3) and Pu_1 interval-zones.

A final comment on faunal composition pertains to the occurrence of dinosaur bones, bone fragments, and teeth relative to the Pu_0/Pu_1 faunal boundary. The issue of whether such material is reworked is discussed in a following section devoted to this topic; suffice it to note here how the pattern of occurrence of dinosaur material varies in the faunas. All faunas that we discussed for the bk_3 biochron have produced dinosaur teeth and bone fragments (Table 1). In contrast, some Pu_1 faunas produce dinosaur bone fragments and teeth and some do not. For example, the Mantua Lentil, Hell's Hollow, McKeever Ranch, Littleton, and Z-Line Local Faunas either very rarely produce a few worn dinosaur teeth or none at all. In contrast, at Ferguson Ranch and all the McGuire Creek area sites (except Z-Line), dinosaur teeth, bone fragments, and even bones can be quite common. Whether this represents a true extinction of dinosaurs within the *Peradectes/Ectoconus* Interval-Zone (Pu_1) or a reworking of dinosaur bones is at present a moot question.

The biostratigraphic aspects of the boundary between Pu_0/Pu_1 were considered earlier in the section dealing with biostratigraphic assessment of the Pu_0 interval-zone. The conclusion in this previous section was that although poorly constrained, the Pu_0 interval-zone ends and the Pu_1 interval-zone commences somewhere from about 4 m above the "lower" Z coal (or 12 m below the upper Z coal) to near the base of the "upper" Z coal.

CORRELATION OF GEOLOGIC/PALEONTOLOGIC PHENOMENA WITH THE *PROTUNGULATUM/PERADECTES* INTERVAL-ZONE

Beyond the interest of the biostratigrapher and paleomammalogist, the importance of understanding the *Protungulatum/-Peradectes* Interval-Zone and its zonation lies in how it relates to other paleontologic and geologic phenomena. It is of course in these other correlations where most of the controversies surrounding the *Protungulatum/Peradectes* Interval-Zone arise. The following is a brief review of some of these correlations.

Lancian mammals

One of the more intriguing aspects of the Bug Creek faunal assemblage when it was first described (Sloan and Van Valen, 1965) was that there was an apparent coincidence of what have become known as "typical latest Cretaceous mammals" and "Paleocene aspect mammals." As discussed previously in the section dealing with the history of these faunas, the former phrase refers to mammals that characterize the Lancian Mammal Age, such as marsupials, multituberculates, and proteutherian mammals, while the latter phrase refers to mammals that had heretofore only been described from unquestionable Paleocene faunas, specifically the genera *Protungulatum, Catopsalis, Stygimys,* and *Procerberus.* It was taken as given that this represented a true association of these mammals (Sloan and Van Valen, 1965).

Rumors, however, persisted that these were mixed faunas caused by reworking. This view was advocated formally by Smit and van der Kaars in 1984. They essentially argued that the Bug Creek channel deposits were of Paleocene age and had reworked sediments of Cretaceous age containing Lancian mammals. One of their main reasons for this argument was that the "Paleocene aspect mammals" were far more common in the faunas than were the Lancian mammals. This argument was countered by the fact that unevenness in species diversity is common in fossil as well as recent biotas (Archibald, 1987c). The example was given for the type Lance fauna, Wyoming, where the four most common taxa represent an impressive 58 percent of this fauna, in comparison to Bug Creek Anthills where the figure is a slightly smaller but similar 53 percent, indicating Bug Creek Anthills is not unusual in its faunal composition.

The discovery and description of two Pu_0 faunas in southern Saskatchewan, Fr-1 and the Long Fall Horizon (Johnston, 1980; Johnston and Fox, 1984; Fox, 1987), which include many "typical latest Cretaceous" mammals, reinforces the view that these mammals are not reworked. Especially compelling is the fact that at the Long Fall Horizon, the two marsupials, *Alphadon* sp. and *Pediomys elegans,* are individually more numerous than any of the species of archaic ungulate save for *Protungulatum.* Even more interesting is that in the best sampled of the slightly younger Pu_1 faunas in the McGuire Creek area, such as Up Up the Creek, these same marsupials plus *Glasbius* and *Didelphodon* are also quite common, nearly equaling the archaic ungulates in numbers of specimens. The only possible anomaly for this pattern is the Pu_1(?) Ferguson Ranch fauna. Strangely, this is the only fauna for which dinosaur teeth, but no Lancian mammals have been reported (e.g., Rigby and others, 1987). Whether this represents a true lack of Lancian mammals or whether the taxa simply have not been fully reported must await additional information regarding the fauna.

When it is realized that all Pu_0 and some Pu_1 faunas (excepting localities of one or a few isolated specimens) have yielded "typical latest Cretaceous" mammals the reworking issue begins to fade. At the present, this issue cannot unequivocally be resolved; the burden of proof, however, still rests with those who wish to demonstrate that some sort of mixing has occurred. In the preceding faunal listings and in Table 1, the suspect mammals have been so noted.

Dinosaurs

As with Lancian mammals, it was first thought that dinosaurs were unequivocal contemporaries of the Pu_0 faunas (Sloan and Van Valen, 1965). There was no reason to doubt this association as the Bug Creek area sites yielded dinosaur teeth and bone fragments along with the mammals. In light of some of the arguments concerning extinction, this association has been questioned (e.g., Smit and van der Kaars, 1984).

There would seem to be at least three important alternatives to consider regarding dinosaurs and sites referable to both the Pu_0 and Pu_1 interval-zones. Dinosaurs could have been extant (1) during the deposition of all Pu_0- and Pu_1-aged sediments in which they are found, (2) during the deposition of only some of the sediments in which they are found, or (3) during none of the deposition of sediment in which they are found.

The strongest sort of evidence arguing for association would be the co-occurrence of at least partially articulated or associated dinosaur remains and Pu_0-Pu_1 mammals. Almost equally strong would be the finding of an at least partially articulated dinosaur in sediments clearly overlying those containing a Pu_0 or Pu_1 mammalian fauna.

The first scenario (co-occurrence of Pu_0 or Pu_1 faunas and partially articulated or associated dinosaur remains) is now being studied by Lofgren in the McGuire Creek area. At all the Pu_1 (or Pu_0) localities occurring at about 20 m below the "upper" Z coal in this area (see Fig. 2), fragmentary dinosaur fragments and teeth are extremely common. This is especially the case for the area in which the Shiprock and Tedrow Quarry local faunas are recognized. Both faunas include larger (>20 cm) fragmentary dinosaur remains. Nearby, at the Tedrow Dinosaur Quarry, which is a small quarry site, associated remains (vertebrae, ribs, parts of the pelvis, etc.) of a ceratopsian were recovered along with Pu_0 mammals (*Stygimys* and *Ragnarok*).

The second scenario of dinosaur remains above Pu_0 (or Pu_1) faunas has been mentioned (Sloan and others, 1986) but not substantiated. When dinosaur specimens have been cited as occurring above Pu_0 localities it is always for localities whose stratigraphic position is suspect relative to the overlying Hell Creek–Tullock formational contact and the nearby K-T boundary (however it is defined). A typical example of this is the Bugcreekian-aged Chris's Bone Bed that is at the top of an isolated hill about 1 km from the nearest outcrop preserving the formational contact (Lupton and others, 1980). Sloan and others (1986) report that "many dinosaurs have been collected above this locality." Without some sort of sedimentary capping at the locality and with only topographic position as a guide, this argument, although possible, is weak.

Chris's Bone Bed does, however, provide data similar to that at McGuire Creek in suggesting that dinosaurs were extant at the time of deposition. This is in the form of larger, unarticulated parts of dinosaurs. Lupton and others (1980) provide a list and quarry map for such dinosaurian material. This evidence is not conclusive, but does suggest that, as at McGuire Creek, dinosaurs and Pu_0 (or Pu_1) mammals were at some point coeval.

The final type of evidence used to not only argue that Bug Creek–type mammals and dinosaurs were coeval, but that dinosaurs survived into the Paleocene, is the presence of shed dinosaur teeth at sites producing these types of mammals (Sloan and others, 1986; Rigby, 1985; Rigby and others, 1987). These authors reason that since the shed teeth are not very abraded, they are probably not reworked from older sediments. Argast and others (1987) subjected dinosaur and crocodilian teeth to mechanical abrasion and showed that there was little if any detectable abrasion, suggesting that arguments for lack of reworking based on lack of abrasion are equivocal at best.

Shed dinosaur teeth have been found rarely at other early Paleocene sites such as the Hell's Hollow local fauna (e.g., Archibald, 1982) and even at Eocene sites (J. I. Kirkland, written communication)! The point is, however, that the considerable abundance of shed dinosaur teeth at Pu_0 and some Pu_1 localities cannot be lightly dismissed without evidence to suggest that they are in fact reworked. Reworking becomes harder to accept in the face of the very probable association of larger dinosaur remains and Bug Creek–type mammals noted above. It remains probable that dinosaurs were contemporaries of Bug Creek–type mammals and possible that dinosaurs did survive into the Paleocene either in North America or elsewhere, but the evidence remains inconclusive.

An alternative view has been that the lack of dinosaurs (not including shed teeth) in approximately the upper 2 to 3 m of uppermost Cretaceous sediments in the Western Interior represents the extinction of dinosaurs before the deposition of iridium and the palynologically defined K-T boundary (Archibald and Clemens, 1982). The argument that this is a problem of statistical sampling (Alvarez, 1983) has been shown to be in error by McKenna (see Archibald, 1987c). There does remain the possibility that this "gap" is caused by some other factor. As Bryant has shown (see Archibald, 1987; Bryant and others, 1986), other vertebrates are almost totally lacking from this interval. One possible cause is the claim that carbonate is depleted in this interval (e.g., Retallack and Leahy, 1986).

For both "typical latest Cretaceous mammals" and dinosaurs at Bugcreekian sites there remains the possibility that they are reworked.

Finally, it should be emphasized that there is a tremendously rich fauna of non-dinosaurian lower vertebrates that has been recovered along with dinosaur and mammal remains at Pu_0 localities. Although they certainly represent part of the original biota as well as probably including some reworked material, the remains of these organisms, at least for the present, do not shed much light on issues surrounding the recognition of the *Protungulatum/Peradectes* Interval-Zone. For further information regarding these vertebrates, see Bryant (1988).

Iridium

An enrichment of iridium has been reported from a variety of marine and terrestrial sections, including sites in eastern Montana (e.g., Alvarez, 1983) and southern Saskatchewan (Nichols and others, 1986) near localities that have yielded faunas of the *Protungulatum/Peradectes* Interval-Zone. Sloan and others (1986) report the occurrence of iridium from the Bug Creek area. They report (as a personal communication from Orth, p. 632) " a slightly enriched zone of iridium just above a thin charcoal streak that varies from 0.6 m to a maximum of 1.8 m below the base of the lowest Z coal in the basin, 7 km east of SMP" (Scmenge Point). Without more information it is impossible to evaluate this reported enrichment. More importantly, as is the case with other stratigraphic correlations to the BCA faunal assemblage, the stratigraphic position of most if not all of the BCA localities is suspect as they cannot be accurately placed in the context of the formational boundary or the K-T boundary. Accordingly, unequivocal association of faunas of the *Protungulatum/Peradectes*

Interval-Zone and iridium has not been established in the Bug Creek area.

In Garfield County, immediately west of Hell's Hollow, the iridium-bearing lower Z coal appears to be cut by channeling that has yielded mammals at the Herpijunk locality (Fig. 1). This site, however, is almost certainly referable to the *Peradectes/Ectoconus* Interval-Zone, as is the case for the Hell's Hollow Local Fauna. Thus, here as in the Bug Creek area it cannot be determined whether the older *Protungulatum/Peradectes* Interval-Zone predates, straddles, or postdates the deposition of iridium.

The most that can be said with confidence is that iridium underlies mammalian faunas of the *Peradectes/Ectoconus* Interval-Zone and overlies dinosaur remains by several meters in Garfield County in the vicinity of Iridium Hill (e.g., Alvarez, 1983).

A recent report of an in situ dinosaur fragment about 1 m below an iridium-bearing clay in the type Lance Formation of Wyoming may be the closest association of iridium and apparently unreworked dinosaur (Bohor and others, 1987), but of course with no other means to bracket this interval it is difficult to speculate as to how much time this 1 m represents. Further, since the bone was reported as coming from a fluvial sandstone, it could well be reworked from somewhat higher or lower in the section.

Pollen

Palynology is probably one of the methods that could eventually unravel the stratigraphic confusion surrounding the Bug Creek faunal assemblage. For the present, however, there is much confusion surrounding palynological change in this assemblage. About 85 km west of the Bug Creek area, in the general vicinity of Hell Creek, it appears that one can find quite a dramatic palynological change (extinctions of some forms and a fern spore spike) quite closely associated (measured in centimeters) with an iridium anomaly (Hotton, personal communication; Smit and others, 1987).

Unfortunately, there are conflicting statements regarding palynological change in the Bug Creek area. Sloan and others (1986, p. 632) state, "Oltz recovered abundant well-preserved specimens of the diagnostic Cretaceous pollen grains *Aquillapollenites ampulus* and *A. delicatus* from the base of the combined Z coals *above* [our emphasis] both the BCA and BCW-SMP channels." Smit and others (1987, p. 68) do not agree, noting that they "did not find Cretaceous elements, like *Aquillapollenites* apart from some corroded grains. . . ." One paragraph following the first statement, Sloan and others (1986) appear to contradict themselves when they state that "[w]e interpret these data to mean BCA and BCW-SMP channels are of different ages, and that at least BCA is Cretaceous in age." If their first statement is accepted, all of the aforementioned sites would be Cretaceous in age if a palynofloral change is used to mark the Cretaceous-Tertiary boundary in these sediments.

Smit and others (1987) present a more detailed review of the palynology in eastern Montana. They note (p. 68), however, that "[t]he Cretaceous-Tertiary boundary interval was analyzed in detail at Herpijunk Promontory in Garfield County because the critical Cretaceous-Tertiary interval in the Bug Creek area is barren. Also a well-defined iridium anomaly which occurs within the lower Z-coal at a typical 'tonstein' layer in Garfield County is absent at Bug Creek."

This statement is rather troubling in view of any extinction scenario that purports to show a tight correlation in the Bug Creek area between a palynological change, an iridium anomaly, and various vertebrate extinctions, but more to the point, it suggests that these various "events" cannot as of yet be correlated to the Bug Creek–type faunas.

Even given this level of uncertainty, these authors suggest that all localities in the Bug Creek faunal assemblage are Paleocene in age (using the palynologically defined K-T boundary) except for Bug Creek Anthills proper for which the age assignment is uncertain. For the biochronologically next youngest site, Bug Creek West, these authors indicate that carbonaceous streaks in and above this locality contain Paleocene pollen while clay clasts in the locality contain Cretaceous pollen. The evidence suggests a Paleocene age for the upper part of this channel, but does not demonstrate a Paleocene age for the microvertebrates. These could have just as easily been reworked along with the hydrologically equivalent clay clasts of Cretaceous age. Hence, the BCW site, as well as other Bug Creek–sequence sites, could be Cretaceous in age.

This latter possibility would make sense if some collaborative work reported by Lofgren and Hotton (1988) proves to be correct. In the McGuire Creek area an entirely Cretaceous pollen assemblage was recovered from a mudstone lens that appeared to be 2 m above the base of the channel complex that bears the associated ceratopsian remains mentioned above under the discussion of dinosaurs (Fig. 2). This site, the Tedrow Dinosaur Quarry, appears to be lateral to and within the same channel complex that bears the Tedrow Quarry Local Fauna and less certainly the Shiprock Local Fauna. If a palynological change (and its close association to an iridium anomaly) is used to define the K-T boundary in the Western Interior, then these data would suggest a latest Cretaceous age for much of the *Protungulatum/Peradectes* Interval-Zone. This possibility has recently been complicated by the recovery of definitely Paleocene pollen from a unit that appears to be cut by the Tedrow Quarry channel deposits (Hotton, personal communication). If this is borne out by further work, it would appear that we still lack any faunas of the *Protungulatum/Peradectes* Interval-Zone that are definitively capped by latest Cretaceous pollen horizons. Thus, the evidence still favors a Paleocene assignment for this interval-zone.

For now the best that can be said of palynological data is that they may eventually help establish a Late Cretaceous or Paleocene age for some or all of the sites. Whether these data also can be used for more detailed biochronologic zonation remains to be demonstrated.

K-T boundary and the Hell Creek–Tullock contact

Determination of the positions of the Cretaceous/Tertiary boundary and the Hell Creek–Tullock formational contact relative to the localities that have produced the faunas of the *Protungulatum/Peradectes* Interval-Zone is of primary interest for those concerned with extinction scenarios. At least the K-T boundary determination, however, relies to a great extent upon the previously discussed faunal, floral, and geochemical data. Thus, correlating the K-T boundary with the *Protungulatum/Peradectes* Interval-Zone continues to be an elusive enterprise. Even if one is to make what still remains a conjectural correlation of an iridium anomaly and/or a palynological change in eastern Montana to the defining area for the K-T boundary area in northern Europe (Smit and others, 1987), one is still faced with the aforementioned problems of correlating these criteria to the Bug Creek–type faunas within eastern Montana. The presence of dinosaur teeth and/or bones at such sites in Montana and Saskatchewan are similarly problematic for use in correlation of the K-T boundary.

The assignment of some or all of the *Protungulatum/Peradectes* Interval-Zone to the Late Cretaceous or Paleocene is unresolved, but the evidence from both the McGuire Creek and Bug Creek areas is currently more suggestive of a Paleocene age. Similarly, the position of the *Protungulatum/Peradectes* Interval-Zone relative to the Hell Creek–Tullock formational contact remains equivocal. The evidence is mounting that at least some of the channel deposits containing these faunas are capped within the Tullock Formation in Garfield County if one utilizes the lower Z coal as the formational contact, but are wholly within the uppermost Hell Creek if one uses the "upper" Z coal in McCone County (Fig. 2).

ACKNOWLEDGMENTS

The authors thank Tom Bown, Richard Estes, Steve Walsh, and Mike Woodburne for reading and providing comments on the manuscript and Howard Hutchison for helpful discussions. Their assistance does not necessarily imply agreement on the contents of this paper. Financial support from NSF Grant EAR-84075507 and National Geographic Society Grant 252882 to J.D.A. and NSF Grant BSR-85-13253 to W. A. Clemens in support of D.L.L.'s portion of this research is gratefully acknowledged.

REFERENCES CITED

Alvarez, L. W., 1983, Experimental evidence that an asteroid impact led to the extinction of many species: Proceedings of the National Academy of Sciences, v. 80, p. 627–642.

Argast, S., Farlow, J. O., Gabet, R. M., and Brinkman, D. L., 1987, Transport-induced abrasion of fossil reptilian teeth; Implications for the existence of Tertiary dinosaurs in the Hell Creek Formation, Montana: Geology, v. 15, p. 927–930.

Archibald, J. D., 1981, The earliest known Paleocene mammal fauna and its implications for the Cretaceous–Tertiary extinction: Nature, v. 208, p. 650–652.

——, 1982, A study of Mammalia and geology across the Cretaceou–Tertiary boundary in Garfield County, Montana: University of California Publications in Geological Sciences, v. 122, p. 1–286.

——, 1986, Comment *on* 'Sedimentation, stratigraphy, and extinctions during the Cretaceous–Paleogene transition at Bug Creek, Montana': Geology, v. 14, p. 892–894.

——, 1987a, Latest Cretaceous and early Tertiary mammalian biochronology/biostratigraphy in the Western Interior: Geological Society of America Abstracts with Programs, v. 19, p. 258.

——, 1987b, The Bugcreekian Land Mammal Age; A reassessment: Journal of Paleontology Abstracts of Papers, v. 7, supp. to no. 3, p. 10A.

——, 1987c, Stepwise and non-catastrophic Late Cretaceous terrestrial extinctions in the Western Interior of North America; Testing observations in the context of an historical science in Les extinctions dans l'histoire des vertébrés (Extinctions in vertebrete history): Mémoires de la Société géologique de France, Nouvelle Série, no. 150, p. 45–52.

Archibald, J. D., and Clemens, W. A., 1982, Late Cretaceous extinctions: American Scientist, v. 70, p. 377–385.

——, 1984, Mammal evolution near the Cretaceous–Teriary boundary, *in* Berggren, W. A., and Van Couvering, J. A., eds., Catastrophes in earth history; The new uniformitarianism: Princeton, New Jersey, Princeton University Press, p. 339–371.

Archibald, J. D., Schoch, R. M., and Rigby, J. K., Jr., 1983, A new subfamily, Conacodontinae, and new species, *Conacodon kohlbergeri*, of the Periptychidae (Condylarthra, Mammalia): Postilla, no. 191, p. 1–24.

Archibald, J. D., Clemens, W. A., Gingerich, P. D., Krause, D. W., Lindsay, E. H., and Rose, K. D., 1987, First North American Land Mammal Ages of the Cenozoic Era, *in* Woodburne, M. O., ed., Cenozoic mammals of North America; Geochronology and biostratigraphy: Berkeley, University of California Press, p. 24–76.

Bohor, B. F., Triplehorn, D. M., Nichols, D. J., and Millard, H. T., Jr., 1987, Dinosaurs, spherules, and the "magic" layer; A new K-T boundary clay site in Wyoming: Geology, v. 15, p. 896–899.

Breithaupt, B. H., 1982, Paleontology and paleoecology of the Lance Formation (Maastrichtian) east flank of Rock Springs uplift, Sweetwater County, Wyoming: University of Wyoming Contributions to Geology, v. 21, p. 123–151.

Bryant, L. J., 1988, Non-dinosaurian vertebrates across the Cretaceous–Tertiary boundary in northeastern Montana: University of California Publications in Geological Sciences (in press).

Bryant, L. J., Hutchison, J. H., Clemens, W. A., and Archibald, J. D., 1986, Diversity changes in lower vertebrates (non-dinosaurian): Geological Society of America Abstracts with Programs, v. 18, p. 552.

Clemens, W. A., 1964, Fossil mammals at the type Lance Formation, Wyoming; Part I, Introduction and Multituberculata: University of California Publications in Geological Sciences, v. 48, p. 1–105.

——, 1966, Fossil mammals of the type Lance Formation, Wyoming; Part II, Marsupialia: University of California Publications in Geological Sciences, v. 62, p. 1–122.

——, 1973, Fossil mammals of the type Lance Formation, Wyoming; Part III, Eutheria and summary: University of California Publications in Geological Sciences, v. 94, p. 1–102.

Collier, A. J., and Knechtel, M., 1939, The coal resources of McCone County, Montana: U.S. Geological Survey Bulletin 905, 80 p.

Dingus, L., 1984, Effects of stratigraphic completeness on interpretations of extinction rates across the Cretaceous–Tertiary boundary: Paleobiology, v. 10, p. 420–438.

Dorf, E., 1940, Relationship between floras of the type Lance and Fort Union

Formations: Geological Society of America Bulletin, v. 51, p. 213–235.

Fastovsky, D. E., 1987, Paleoenvironments of vertebrate-bearing strata during the Cretaceous–Paleogene transition, eastern Montana and western North Dakota: Palaios, v. 2, p. 282–295.

Fastovsky, D. E., and Dott, R. H., Jr., 1986, Sedimentology, stratigraphy, and extinctions during the Cretaceous/Paleogene transition at Bug Creek, Montana, Geology, v. 14, p. 279–282.

Fox, R. C., 1987, Patterns of mammalian evolution towards the end of the Cretaceous, Saskatchewan, Canada, *in* Short Papers, Fourth Symposium on Mesozoic Terrestrial Ecosystems: Occasional Paper of the Tyrrell Museum of Palaeontology, no. 3, p. 7–11.

Hedberg, H. D., ed., 1976, International stratigraphic guide: New York, John Wiley and Sons, 200 p.

Johnston, P., 1980, First record of Mesozoic mammals from Saskatchewan: Canadian Journal of Earth Sciences, v. 17, p. 512–519.

Johnston, P., and Fox, R. C., 1984, Paleocene and Late Cretaceous mammals from Saskatchewan, Canada: Palaeontographica (A), v. 186, p. 163–222.

Lillegraven, J. A., 1969, Latest Cretaceous mammals of upper part of Edmonton Formation of Alberta, Canada, and review of marsupial-placental dichotomy in mammalian evolution: University of Kansas Paleontological Contribution, Art. 50 (Vert. 12), p. 1–122.

Lillegraven, J. A., Kielan-Jaworowska, K., and Clemens, W. A., 1979, Mesozoic mammals; The first two-thirds of mammalian history: Berkeley, University of California Press, 311 p.

Lillegraven, J. A., and McKenna, M. C., 1986, Fossil mammals from the "Mesaverde" Formation (Late Cretaceous, Judithian) of the Big Horn and Wind River Basins, with definitions of Late Cretaceous land mammal "ages": American Museum Novitates, no. 2840, p. 1–68.

Lofgren, D. L., and Hotton, C., 1988, Palynologically defined Cretaceous Bug Creek Facies channel, upper Hell Creek Formation, McCone County, northeast Montana: Geological Society of America Abstracts with Programs, v. 20, p. 428.

Lupton, C., Gabriel, D., and West, R. M., 1980, Paleobiology and depositional setting of a Late Cretaceous vertebrate locality, Hell Creek Formation, McCone County, Montana: Laramie, University of Wyoming Contributions to Geology, v. 18, p. 117–126.

Middleton, M. D., 1982, A new species and additional material of *Catopsalis* (Mammalia, Multituberculata) from the Western Interior of North America: Journal of Paleontology, v. 56, p. 1197–1206.

Nichols, D. J., Jarzen, D. M., Orth, C. J., and Oliver, P. Q., 1986, Palynological and iridium anomalies at Cretaceous–Tertiary boundary, south-central Saskatchewan: Science, v. 231, p. 714–717.

Norell, M. A., 1987, The phylogenetic determination of taxonomic diversity; Implications for terrestrial vertebrates at the K-T boundary: Journal of Vertebrate Paleontology Abstracts of Papers, v. 7, supplement to no. 3, p. 22A.

North American Stratigraphic Code, 1983: American Association of Petroleum Geologists Bulletin, v. 67, p. 841–875.

Retallack, G. J., and Leahy, G. D., 1986, Cretaceous–Tertiary dinosaur extinction: Science, v. 234, p. 1170–1171.

Rigby, J. K., Jr., 1985, Paleocene dinosaurs; The reworked sample question: Geological Society of America Abstracts with Programs, v. 17, p. 262.

Rigby, J. K., Jr., Newman, K. R., Smit, J., van der Kaars, S., Sloan, R. E., and Rigby, J. K., 1987, Dinosaurs from the Paleocene part of the Hell Creek Formation, McCone County, Montana: Palaios, v. 2, p. 296–302.

Rose, K. D., 1981, The Clarkforkian Land-Mammal Age and mammal faunal composition across the Paleocene–Eocene boundary: Ann Arbor, University of Michigan Papers in Paleontology, v. 26, p. 1–197.

Russell, L., 1964, Cretaceous non-marine faunas of northwestern North America: Royal Ontarion Museum Life Sciences, Contribution 61, p. 1–24.

—— , 1975, Mammalian faunal succession in the Cretaceous system of western North America, *in* Caldwell, W.G.E., ed., The Cretaceous System in the Western Interior of North America: Geological Association of Canada Special Paper 13, p. 137–161.

Simmons, N. B., and Maio, D., 1986, Paraphyly in *Catopsalis* (Mammalia: Multituberculata) and its biogeographic implications, *in* Flanagan, K. M., and Lillegraven, J. A., eds., Vertebrates, phylogeny, and philosophy: Laramie, Universit of Wyoming Contributions to Geology Special Paper 3, p. 87–94.

Sloan, R. E., 1987, Paleocene and latest Cretaceous mammal ages, biozones, magnetozones, rates of sedimentation, and evolution, *in* Fassett, J. E., and Rigby, J. K., Jr., eds., The Cretaceous–Tertiary boundary in the San Juan and Raton basins, New Mexico and Colorado: Geological Society of America Special Paper 209, p. 165–200.

Sloan, R. E., and Van Valen, L., 1965, Cretaceous mammals from Montana: Science, v. 148, p. 220–227.

Sloan, R. E., Rigby, J. K., Jr., Van Valen, L. M., and Gabriel, D., 1986, Gradual dinosaur extinction and simultaneous ungulate radiation in the Hell Creek Formation: Science, v. 232, p. 629–633.

Smit, J., and van der Kaars, S., 1984, Terminal Cretaceous extinctions in the Hell Creek area, Montana, compatible with catastrophic extinction: Science, v. 223, p. 1177–1179.

Smit, J., van der Kaars, S., and Rigby, J. K., Jr., 1987, Stratigraphic aspects of the Cretaceous–Tertiary boundary in the Bug Creek area of eastern Montana, U.S.A., *in* Les extinctions dans l'histoire des vertébrés (Extinctions in vertebrate history): Mémoires de la Société géologique de France, Nouvelle Série, no. 150, p. 53–73.

Van Valen, L., 1978, The beginning of the Age of Mammals: Evolutionary Theory, v. 4, p. 45–80.

Wood, H. E., II, Chaney, R. W., Clark, J., Colbert, E. H., Jepsen, G. L., Reeside, J. B., Jr., and Stock, C., 1941, Nomenclature and correlation of the North American continental Tertiary: Geological Society of America Bulletin, v. 52, p. 1–48.

Woodburne, M. O., 1987, Cenozoic mammals of North America; Geochronology and biostratigraphy: Berkeley, University of California Press, 576 p.

MANUSCRIPT ACCEPTED BY THE SOCIETY JUNE 12, 1989

Printed in U.S.A.

Geological Society of America
Special Paper 243
1990

The succession of Paleocene mammals in western Canada

Richard C. Fox
Laboratory for Vertebrate Paleontology, Departments of Geology and Zoology, University of Alberta, Edmonton, Alberta T6G 2E9, Canada

ABSTRACT

Although Paleocene mammals have been known from western Canada for nearly 70 years, only during the last 15 years have concerted efforts been made to discover, develop, and describe collections documenting their evolution. Whereas much remains to be done, a faunal succession has been reconstructed, based on collections from 41 mammalian local faunas in Alberta and Saskatchewan, ranging from early Puercan to late Tiffanian age (the first 8 to 10 m.y. of the Tertiary).

Latest Cretaceous mammalian local faunas in Saskatchewan show that evolution of progressive "Paleocene aspect" mammals, including condylarths, had begun during the North American Cretaceous as evidenced by the occurrence of fossils in stratigraphic settings free from the complexities that make uncertain the age of faunally similar assemblages in Montana.

The Ravenscrag Formation, southwestern Saskatchewan, yields the oldest (early Puercan) Tertiary mammals known from Canada (Rav W-1: MHBT Quarry, Pine Cree Park and Croc Pot localities), including the first discovery of the ancestral primate *Purgatorius* outside of eastern Montana, and the oldest unarguable carnivoran. Torrejonian mammals are known from the Coalspur Formation (Rocky Mountain Foothills; Diss locality) south of Edson, Alberta. Cochrane 2 (Porcupine Hills Formation, west of Calgary) has yielded unexpectedly diverse earliest Tiffanian mammals, and early Tiffanian mammals have recently been discovered in the Paskapoo Formation near Drumheller (Hand Hills West, lower level), and east of Innisfail, Alberta (Aaron's Locality). Middle Tiffanian localities occur in the Paskapoo Formation near Red Deer (DW 1 to 3, Mel's Place, Joffre Bridge Road Cut, Mammal Site No. 1, Erickson's Landing), and in the Hand Hills (Hand Hills West, upper level); the Police Point local fauna (Ravenscrag Formation), Alberta, appears to be late middle or early late Tiffanian in age. Late Tiffanian mammals at Roche Percée, Saskatchewan (Ravenscrag Formation), and Canyon Ski Lodge, Crestomere School, and Swan Hills, Alberta (Paskapoo Formation) conclude the Paleocene mammalian record known from Canada. Species lists for each locality are presented.

The early and middle Tiffanian mammalian record from Alberta fails to show a decline in species numbers seen at several American localities representing this interval: instead of the global cooling sometimes hypothesized to account for this decline, it now seems to be a result of biological or sedimentological events acting on a local scale, sampling error, or some combination of these factors.

Fox, R. C., 1990, The succession of Paleocene Mammals in western Canada, *in* Bown, T. M., and Rose, K. D., eds., Dawn of the Age of Mammals in the northern part of the Rocky Mountain Interior, North America: Boulder, Colorado, Geological Society of America, Special Paper 243.

INTRODUCTION

Mammals of Paleocene age were first discovered in western Canada in 1910, by Barnum Brown of the American Museum of Natural History. These specimens, which Simpson described in 1927, had been collected from a slump block from the Paskapoo Formation along the Red Deer River east of Red Deer, Alberta, at a locality known as Erickson's Landing. Over the next 50 years, a trickle of additional localities were discovered and a trickle of mammalian species was described from them—mostly by L. S. Russell, from fossils found in the vicinity of Calgary. D. E. Russell's (1967) summary of Paleocene mammalian faunas from North America included four major Paleocene mammal localities from Canada, all in Alberta: Calgary 2E, at a cutbank along the Elbow River and now within the city limits of Calgary; Cochrane 1 and 2, at outcrops along the Bow River near Cochrane, a village west of Calgary; and at Erickson's Landing. These sites were in rocks of the Paskapoo Formation (although in addition to these, two other finds—isolated teeth in the Rocky Mountain Foothills—were known from undivided strata of the Saunders Group); the oldest was thought to be Torrejonian, the youngest Tiffanian or perhaps Clarkforkian in age. The most diverse mammals had been found at Cochrane 2: L. S. Russell (1958) listed nine mammalian species, including three multituberculates, four "insectivores," and two condylarths from this locality. Up until 1967, all that had been published on Paleocene mammals from Canada was based on fewer than 50 specimens, most of which were isolated teeth or fragments of teeth.

The purpose of this paper is to list the results of work on Paleocene mammalian faunas from western Canada since D. E. Russell's (1967) summary—results obviously greatly increased in comparison to what had gone before—and to arrange all of the known occurrences of Paleocene mammals from Canada in biochronologic sequence. Knowledge of Canadian Paleocene mammals is now based on fossils from 41 localities/local faunas in Alberta and Saskatchewan, mostly as a consequence of collections made by field parties from the Laboratory of Vertebrate Paleontology, The University of Alberta (UALVP). These localities are in the Paskapoo, Porcupine Hills, Coalspur, and Ravenscrag Formations, continental deposits that accumulated during the Paleocene in the Alberta Syncline and, east of the Cypress Hills Arch, in the Williston Basin, where they occur in the Ravenscrag Formation. Figure 1 shows the overall distribution of localities yielding Paleocene mammals in Alberta and Saskatchewan, and Figure 2 shows the detailed distribution of localities clustering along the Blindman and Red Deer Rivers east of the city of Red Deer, Alberta; the reader should refer to these figures as required while progressing through the accounts of localities that follow. The richest of the localities is still Cochrane 2, having more than 80 species at present, based on approximately 1,700 identifiable specimens. The oldest occurrence is the Rav W-1 horizon at the Medicine Hat Brick and Tile Quarry, Saskatchewan; it is early Puercan in age. Several localities (Swan Hills Site 1, the Roche Percée localities, Canyon Ski Quarry, and Crestomere School) are late Tiffanian in age, probably falling within the *Plesiadapis churchilli* biochron as defined in the United States (although *P. churchilli* itself has not yet been found at Canyon Ski Quarry or Crestomere School); these are the youngest Paleocene mammal localities discovered in Canada to date. Figure 3 summarizes the biochronologic distribution of the major localities yielding Paleocene mammals included in this chapter.

Figure 1. Map of Alberta and Saskatchewan showing distribution of Paleocene mammal localities. 1, Medicine Hat Brick and Tile Quarry (Rav W-1 horizon); 2, Croc Pot; 3, Pine Cree Park; 4, R.C.A. Corehole 66-1 (Balzac); 5, Diss; 6, Calgary 2E and Dunbow Road; 7, Cochrane 1; 8, Nordegg; 9, Saunders Creek; 10, Cochrane 2; 11, Hand Hills East; 12, DW-1; 13, DW-2; 14, Mel's Place; 15, DW-3; 16, Burbank; 17, Joffre Bridge localities; 18, Erickson's Landing; 19, Hand Hills West; 20, Wintering Hills; 21, Police Point; 22, Swan Hills; 23, Crestomere School; 24, Canyon Ski Quarry and One-jaw Gap; 25, Roche Percée; 26, Tan-i-Bryn Rocks; 27, Forestry Trunk Road, south; 28, North Saskatchewan River. Aaron's Locality and Dickson Dam have been omitted from the map, having been discovered after its preparation.

Owing to its scope, much of the work reported here is still in progress and the taxonomy incomplete, as is reflected by the many tentative identifications of species in the lists that follow (nor has sampling at many sites been completed). However, the species that are important for estimating the age of these occurrences are for the most part fully identified; I have tried to draw attention to whatever uncertainties remain that might affect the resulting biochronology.

Most of the collections are of specimens found in situ and recovered either by underwater screening, or by on-site quarrying to better assure the salvage of little-damaged specimens. Surface collecting of loose specimens weathered free from their host rock has been negligible, since such specimens are encountered only

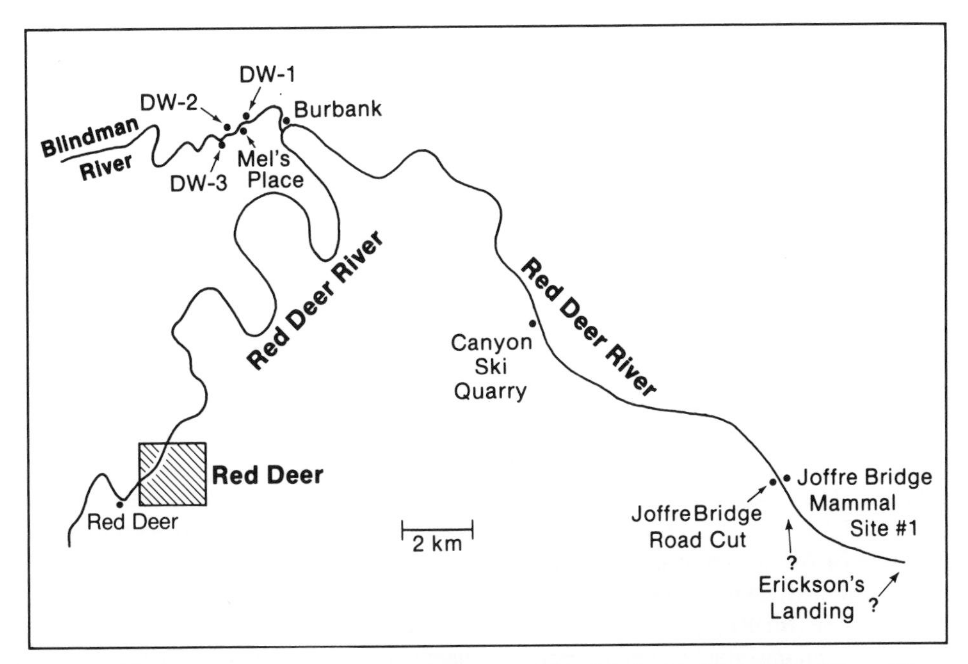

Figure 2. Map showing distribution of Paleocene mammal localities in the vicinity of Red Deer, Alberta.

rarely at best in the study area. The productive outcrops have been at roadcuts, railway cuts, mines, and eroded river banks. Most are laterally discontinuous with one another, sometimes over considerable distances, and none occur in badlands. Strata at the localities give no evidence of significant postdepositional disturbance, except at Diss, in the Foothills belt, where the beds are steeply inclined. Few localities have a vertical dimension sufficient to demonstrate a superpositional relation between strata containing mammals of different ages; the exceptions are the two horizons at the MHBT Quarry, the two horizons at Hand Hills West, and Canyon Ski Quarry relative to the Joffre Bridge localities, along the Red Deer River.

At present, correlation of the localities included here is based on their fossil content; neither magnetostratigraphic data nor radiometric dates are available as yet for calibration of these localities. However, palynofloral evidence, marker coals, an iridium anomaly, and radiometric dates have clearly defined the Cretaceous-Tertiary boundary in the study area (see, e.g., Jerzykiewicz and Sweet, 1986; Lerbekmo and Coulter, 1985; Lerbekmo and St. Louis, 1986; Lerbekmo and others, 1987). Additionally, Demchuck (1987) has determined a palynoflora stratigraphy for the Paleocene of central Alberta and correlatives elsewhere, which provides an independent biostratigraphic assessment for the ages of some of the localities included in this chapter.

The Paleocene mammals from Alberta and Saskatchewan are the most northerly mammals of this age discovered in the world, occurring between 49° and 55° North latitude and approximately 5° farther north in Paleocene latitude. As such, they might be expected to reflect the results of interchange with western Europe and East Asia via land bridges across the eastern and western Arctic, respectively. At present, the clearest evidence of a special faunal resemblance between Paleocene mammals from Canada and those from outside North America not previously known from local faunas in the United States is provided by the primate *Saxonella,* discovered only in the middle Tiffanian of Alberta and the Walbeck locality, German Democratic Republic (Fox, 1984a).

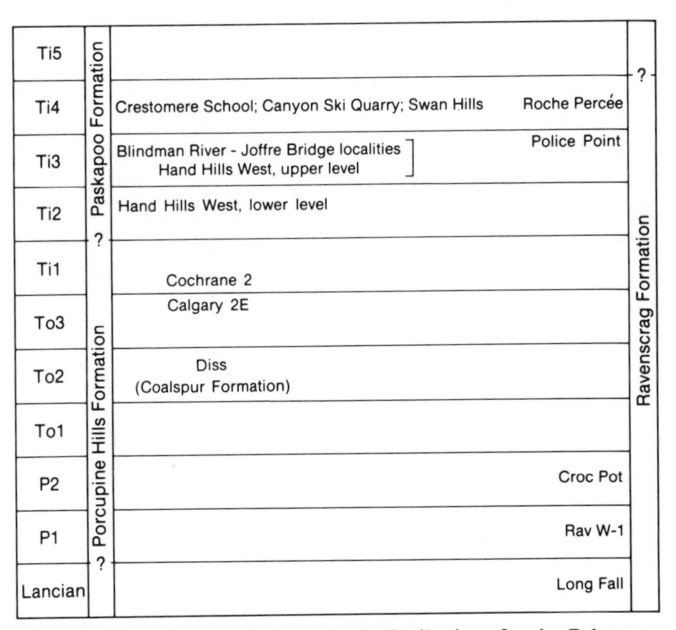

Figure 3. Chart showing biochronologic distribution of major Paleocene mammalian local faunas in Alberta and Saskatchewan. Diss may be To_1 and Hand Hills West, lower level, may be Ti_1 in age.

In the lists of species below, "cf." indicates tentative identification; a single or double asterisk signifies the type locality for species or genus, respectively. Legal descriptions and estimates of altitude are from topographic maps, produced by the Surveys and Mapping Branch, Department of Energy, Mines, and Resources, Ottawa.

PALEOCENE MAMMAL LOCALITIES IN WESTERN CANADA

FR-1: SASKATCHEWAN
Frenchman Formation, late Lancian (latest Cretaceous)
(Johnston, 1980; Fox, 1987, 1988)

The Fr-1 locality is latest Cretaceous in age, yielding the first Mesozoic mammals known from Saskatchewan (Johnston, 1980). It is located about 3.5 km northwest of the hamlet of Ravenscrag, in NE¼S24,T6,R24,W3, at approximately 1,025 m above sea level. Along with mammals from the Long Fall horizon at the Medicine Hat Brick and Tile (MHBT) Quarry listed below, the mammals from Fr-1 provide an important datum for characterizing mammalian evolution during the latest Cretaceous and, hence, for recognition of the beginning of mammalian evolution in the Paleocene; they are included here on that account. The Fr-1 mammals occur in greenish gray deposits typical of the Frenchman Formation in southwestern Saskatchewan, and in association with the unreworked remains of dinosaurs; the locality is below the level of the Ferris (= No. 1) Coal Zone, which marks both the contact between the Frenchman and Ravenscrag Formations in the region, and the Cretaceous-Tertiary boundary (Lerbekmo, 1985). The local fauna from Fr-1 includes evidence that at least two lineages of condylarths had evolved prior to the end of the Cretaceous, thereby decoupling the evolution of progressive mammals of "Paleocene aspect" from the terminal extinction event at the end of the Mesozoic (see Smit and van der Kaars, 1984). In the list below, the superscript "1" indicates species added since Johnston's (1980) study was completed.

Multituberculata: *Mesodma* sp.; *Cimolodon,* cf. *C. nitidus*[1]; *Catopsalis,* cf. *C. joyneri*[1]; *Meniscoessus,* cf. *M. robustus*[1];
Marsupicarnivora: unidentified genus and species.
Condylarthra: *Protungulatum,* cf. *P. donnae;* ?Periptychidae, unidentified genus and species[1].
Eutheria, *incertae sedis,* Palaeoryctidae: *Cimolestes,* cf. *C. cerberoides.*

MEDICINE HAT BRICK AND TILE QUARRY: SASKATCHEWAN
Ravenscrag Formation
(Johnston and Fox, 1984; Fox, 1987, 1988

The Medicine Hat Brick and Tile (MHBT) Quarry is located on the southern edge of the Cypress Hills Plateau, 3.5 km west northwest of Ravenscrag, Saskatchewan, in NW¼S23,T6, R24,W3 (not "W6," as in Johnston and Fox, 1984), at about 1,044 m above sea level. In the quarry, strata of the Ravenscrag Formation have been exposed as overburden during commercial excavation for ceramic clays of the Upper Cretaceous Battle and Whitemud Formations. Mammals have been found at two horizons, Long Fall (latest Cretaceous) and Rav W-1 (Puercan).

I discovered the first mammals at the MHBT Quarry in September 1975, leading to the collection of large samples of fossiliferous rock in 1976 to 1978 for underwater screening. These mammals—from the Rav W-1 horizon (Johnston and Fox, 1984)—were found to be concentrated in clay lenses beneath indurated sandstone ledges in loosely consolidated, pale brown to buff, low-angled, fluviatile, silty sandstones. The sandstone sequence probably represents channel and overbank deposits of small meandering or braided streams. The mammalian remains are isolated teeth, jaw fragments with teeth, and disarticulated postcranial bones, in association with fossils of *Amia, Lepisosteus,* amphibians, small reptiles (turtles, lizards, champsosaurs, crocodiles), and birds; *Lepisosteus* fossils are particularly abundant. The mammals at Rav W-1 indicate an early Puercan age (here I include the "Mantuan," recognized by some as a land-mammal age earlier than the Puercan, as but part of the Puercan), and at present are the oldest Cenozoic mammals known from Canada. Especially noteworthy among the Rav W-1 mammals are *Purgatorius,* the single record of this earliest primate outside of eastern Montana, and a carnivoran, only the second occurrence of this important order from rocks as old as Puercan.

Mammals at a second horizon—"Long Fall"—were found by P. A. Johnston in July 1979. The mammals at Long Fall are accompanied by fossils of *Myledaphus, Amia, Lepisosteus,* salamanders, turtles, champsosaurs, teiid lizards, crocodiles, and dinosaurs. The Long Fall horizon is several meters below Rav W-1, at the base of brown, steeply cross-bedded sandstones, which apparently represent laterally accreted point-bar deposits (Lerbekmo, 1985); Long Fall is separated from Rav W-1 by a disconformity and is late Lancian (latest Cretaceous) in age (Johnston and Fox, 1984; Fox, 1987, 1988).

In the lists for Long Fall and Rav W-1, species discovered since the completion of Johnston and Fox (1984) are indicated by superscript "1".

Long Fall horizon: late Lancian (latest Cretaceous)

Multituberculata: *Mesodma thompsoni; Cimolodon nitidus*[1]; Microcosmodontinae, unidentified genus and species[1]; *Stygimys,* new species[1]; *Catopsalis,* cf. *C. joyneri*[1]; *Catopsalis,* new species; *Meniscoessus,* cf. *M. robustus*[1]; *Cimolomys,* cf. *C. gracilis*[1]; *Cimexomys minor*[1]; *Cimexomys,* cf. *C. hausoi.*
Marsupicarnivora: *Alphadon* sp.; *Pediomys elegans.*
Condylarthra: *Protungulatum,* cf. *P. donnae; Oxyprimus,* cf. *O. erickseni; Baioconodon* (=*Ragnarok*) sp.; *Mimatuta* sp.; ?Hyopsodontidae, unidentified genus and species.
Eutheria, order *incertae sedis,* Palaeoryctidae: *Cimolestes,* cf. *C. cerberoides*[1]; *Cimolestes,* cf. *C. stirtoni*[1]; *Procerberus,* cf. *P. formicarum.* **Leptictidae:** *Gypsonictops illuminatus*[1].

Rav W-1 horizon: early Puercan (P_1)

Multituberculata: *Ptilodus,* cf. *P. tsosiensis; Mesodma thompsoni; Mesodma,* new species[1]; *Ectypodus* sp.; *Parectypodus armstrongi**; *Parectypodus* sp.; *Neoplagiaulax kremnus**; *Xyronomys* sp.; Neoplagiaulacidae, new genus and species[1]; *Microcosmodon arcuatus**; *Stygimys camptorhiza**; *Taeniolabis* sp.; *Cimeximys,* cf. *C. minor*[1]; *Cimexomys,* cf. *C. hausoi.*
Primates: *Purgatorius* sp.
Carnivora, family *incertae sedis:* new genus and species[1].
Condylarthra: *Carcinodon aquilonius; Oxyclaenus corax**; *Loxolophus schizophrenus**; *Baioconodon,* cf. *B. denverensis; ?Eoconodon* sp.; *Anisonchus,* cf. *A. oligistus; Oxyacodon,* cf. *O. agapetillus*[1]; *Bubogonia saskia***; *?Litaletes* sp.; *Litomylus orthronepius**.
Eutheria, order *incertae sedis,* Palaeoryctidae: *Cimolestes,* cf. *C. simpsoni*[1]; *Procerberus,* new species[1]; Palaeoryctidae, unidentified genus and species[1]. **Leptictidae:** *Prodiacodon,* new species[1].

Comments

My interpretations of the ages of the Long Fall and Rav W-1 mammals differs from that of two recent commentators, J. F. Lerbekmo (1985) and R. E. Sloan (1987). Lerbekmo thought that Rav W-1 and Long Fall are included in the same point-bar deposits, are the same age, and that both are Paleocene. According to his view, the Cretaceous aspect of the Long Fall assemblage is owing to Paleocene erosion and redeposition of Cretaceous vertebrates from the beds beneath. However, Lerbekmo's interpretation is mistaken: Long Fall and Rav W-1 show no mammalian species in common other than those already known to range from Late Cretaceous to Paleocene. More particularly, the "Paleocene" species from Long Fall differ from, and are more primitive than, the "Paleocene" species from Rav W-1 (nor do "Cretaceous" species occur at Rav W-1). Further, the dark shales of the Battle Formation that underlie the Ravenscrag strata at the Quarry have yielded no vertebrate fossils there, nor would they be expected to do so; the Battle has been found to be barren of vertebrate fossils throughout its considerable areal extent elsewhere, in Alberta, as well as in Saskatchewan. Hence, there is neither the opportunity for, nor evidence of, temporally significant reworking.

Sloan (1987) recognized that the taxonomic differences between the mammals at Long Fall and Rav W-1 were valid indicators of a significant difference in age between them; he believed, however, that both are Paleocene, with Long Fall being early Mantuan (an interval he accepts as a land-mammal age) and Rav W-1, later Puercan. However, Sloan's interpretation is no more consistent with the evidence than is Lerbekmo's: the ray *Myledaphus bipartitus,* the teiid lizard *Chamops segnis,* the multituberculates *Cimolodon nitidus* and *Cimolomys,* cf. *C. gracilis,* the marsupials *Alphadon* sp. and *Pediomys elegans,* and the placental *Cimolestes,* cf. *C. cerberoides,* all indicate a Cretaceous age, independently of the occurrence of dinosaurs at Long Fall. In fact, Sloan (1987, p. 174) believed *Cimolestes cerberoides* to be an index fossil of the Cretaceous, so by that criterion alone his assignment of a Paleocene age to these beds is mistaken. Accordingly, it remains true that "the MHBT Quarry is the only locality now known at which latest Cretaceous and early Paleocene mammals are found in superposition, bracketing the K-T boundary, as Johnson and Fox (1984) reported originally" (Fox, 1987, p. 99).

PINE CREE PARK: SASKATCHEWAN
Ravenscrag Formation, early Puercan (P_1)
(Russell, 1974; Van Valen, 1978; Johnston and Fox, 1984)

This locality, which L. S. Russell and party discovered in 1971, is in Pine Cree Park, Saskatchewan, a small community park at the southeast end of the Cypress Hills, approximately 16 km northeast of the MHBT Quarry ("near the SW corner of sect. 2, twp. 8, rge. 21, W. 3rd mer., about two miles northwest of Southfork village" [Russell, 1974, p. 8]). The fossils come from an outcrop of the Ravenscrag Formation, at a cutbank along a tributary of Swift Current Creek. Originally, one mammalian jaw with teeth was found at the site: the holotype of the arctocyonid *Carcinodon aquilonius* Russell 1974 (see Van Valen, 1978; and Johnston and Fox, 1984, for additional discussion of relationships of this species). In 1977, J. E. Storer, Saskatchewan Museum of Natural History, Regina, collected two neoplagiaulacid multituberculate teeth from the site (personal communication, July 1987).

The beds at the cutbank are brownish to gray sandstones and shales containing gastropod and bivalve shells, gar scales, and the bones of small fish, reptiles, and occasionally, mammals (Russell, 1974; personal observation by author); tetrapod fossils are not abundant at the site. L. S. Russell (1974) placed the Ravenscrag beds at Pine Cree Park approximately 14 m above the No. 3 coal seam, a level that would be stratigraphically above Rav W-1 and Croc Pot (q.v.), both of which are inferred to be below the level of the No. 2 seam, while still Paleocene (see Lerbekmo, 1985). In June 1987, Gao Keqin and I identified the only extensive coal locally near to Pine Cree Park as the No. 1 (Ferris) coal seam, seen at the contact between the Frenchman and Ravenscrag Formations along the section road leading from Southfork village to the entrance of the park; the No. 3 seam does not crop out here. If that is the case, *C. aquilonius* from Pine Cree occurs low in the Ravenscrag, and is likely early Puercan in age, in agreement with the occurrence of *C. aquilonius* at the MHBT Quarry, Rav W-1 horizon (Johnston and Fox, 1984).

Multituberculata: Neoplagiaulacidae, unidentified genus and species.
Condylarthra: *Carcinodon aquilonius**.

CROC POT: SASKATCHEWAN

Ravenscrag Formation, late Puercan (P_2)

(Fox, in preparation)

"Croc Pot" is the name given to a third locality in the Ravenscrag Formation near the Cypress Hills, approximately 2 km west of the MHBT Quarry, in SE¼S22,T6,R24,W3, at about 1,052 m above sea level. The fossils occur in fluviatile, loosely consolidated, variegated gray and buff to pale and dark brown, silty sandstones and shales, poorly exposed on a steep, grassy slope on the southern edge of the Cypress Hills Plateau. The first mammals at this locality were discovered by P. A. Johnston in August 1979. Small samples from the fossiliferous beds were taken in 1980 for underwater screening in the laboratory; these samples have proven to be richly fossiliferous. The mammals occur as isolated teeth, jaw fragments with teeth, and skeletal parts, in association with disarticulated fossils of *Amia, Lepisosteus,* and small amphibians and reptiles; crocodilians appear to be the most common vertebrates in the local fauna. The occurrence of *Taeniolabis taoensis* at this locality indicates a late Puercan age, within the *T. taoensis* subzone of Sloan (1987), and hence, younger than Rav W-1.

Multituberculata: cf. *Prochetodon* sp; *Mesodma formosa; Mesodma thompsoni; Mesodma* sp.; *Parectypodus,* new species; *Xyronomys* sp.; Neoplagiaulacidae, unidentified genus and species 1; Neoplagiaulacidae, unidentified genus and species 2; Microcosmodontinae, unidentified genus and species; *Stygimys* sp.; *Catopsalis* sp.; *Taeniolabis taoensis; Cimexomys,* cf. *C. hausoi.*
Condylarthra: *Loxolophus* sp., near *L. schizophrenus; Baioconodon,*cf. *B. denverensis; Desmatoclaenus* sp.; Arctocyonidae, unidentified genus and species.
Eutheria, *incertae sedis,* Palaeoryctidae: *Cimolestes* sp. **Leptictidae:** *Prodiacodon* sp.

R.C.A. COREHOLE 66-1 (BALZAC): ALBERTA

Porcupine Hills Formation (Carrigy, 1971), ?Puercan (P_2?)

(Fox, 1968; Van Valen, 1978; Johnston and Fox, 1984)

Core from this well, drilled by the Research Council of Alberta in 1966, at a site in Lsd 2, S25,T26,R2,W5, 9.7 km west of Balzac, Alberta (Carrigy, 1971, p. 101), yielded a well-preserved mammalian dentary with teeth, described as the holotype of the arctocyonid *Prothryptacodon albertensis* Fox 1968. Van Valen (1978) and Johnston and Fox (1984) further discussed the relationships of this species. The rock containing the fossil is a hard greenish siltstone, recovered from a depth of 220 m, from what is probably the Porcupine Hills Formation. Johnston and Fox (1984) concluded that the likely age of this fossil is Puercan, instead of Torrejonian, as Fox (1968) had originally supposed. This locality is the "Balzac West" locality of Savage and D. E. Russell (1983) and "Alberta Well 66-1" locality of Sloan (1987).

Condylarthra: *Prothryptacodon albertensis**.

DUNBOW ROAD: ALBERTA

Porcupine Hills Formation (Carrigy, 1971), ?Puercan

(Fox, in preparation)

The Dunbow Road locality is at a small commercial quarry (the DeWinton Quarry of Demchuck, 1987) exposing brown sandstones, siltstones, and shales, excavated for aggregate used in the manufacture of bricks (J. F. Lerbekmo, personal communication, 1987). It lies immediately west of Provincial Highway 2, 10 km south of Calgary, in E½S36,T21,R1,W5, at approximately 1,098 m above sea level. The quarry was brought to my attention as a possible site for earliest Paleocene mammals by A. R. Sweet (see McIntyre and others, 1984). In 1984, I located a seam containing broken molluscan shell at the quarry, which was sampled for underwater screening in the laboratory. The only mammalian fossil recovered was a single therian premolar; it possibly is from a condylarth, but no more precise identification is possible at present. Two more samples were collected from the Dunbow quarry in July 1987, but these have not been processed at this writing.

Mammalia: unidentified genus and species.

DISS: ALBERTA

Coalspur Formation (Jerzykiewicz, 1985), Torrejonian

(Fox, 1983a, in preparation)

The Diss locality is at a small railroad cut exposing the Coalspur Formation (Jerzykiewicz, 1985; Jerzykiewicz and McLean, 1980), 2 km southeast of Diss, along the Canadian National tracks leading to the Sterco-Luscar coal mine at Coal Valley; the locality is approximately 1,372 m above sea level, at SW¼S8,T48,R20,W5, in the Foothills region southeast of Robb. The first mammal recognized from this site is the impression of a multituberculate p4, which R.R.G. Williams found in July 1978, while quarrying for fossil fish with M.V.H. Wilson from the UALVP. Subsequent collections were made in 1981, 1982, 1984, 1985, and 1986. The mammals occur in a grayish green to brownish siltstone and are well preserved; specimens include an associated skull, lower jaw, and incomplete postcranium of the primitive eutherian *Pararyctes* (Fox, 1983a; in that paper, the Diss locality is erroneously cited as being 10 km south of Edson, Alberta). The mammal bed has also yielded an articulated skull of the batrachosauroidid salamander *Opisthotriton.* Although clearly diagnostic mammalian species have yet to be found at the site, the Diss local fauna is probably Torrejonian in age, an estimate that is a compromise between the range of *Palaechthon* (Torrejonian and Tiffanian) and the occurrence of the new species of *Parectypodus,* otherwise known only at Rav W-1 (early Puercan). From palynological evidence, Jerzykiewicz and Sweet (1986) thought the Coalspur Formation was no younger than early Paleocene; Demchuck (1987), however, has shown the upper parts of the formation to be correlative with the early Torrejonian.

Multituberculata: *Ptilodus,* new species; *Baiotomeus,* new species (see Krause, 1987a); *Mimetodon,* new species; *Parectypodus,* new species; *Neoplagiaulax,* cf. *N. nanophus; Neoplagiaulax,* cf. *N. nelsoni;* Neoplagiaulacidae, unidentified genus and species 1; Neoplagiaulacidae, unidentified genus and species 2.
Lipotyphla: new genus and species.
Primates: *Palaechthon* sp.
?Creodonta: unidentified genus and species.
Carnivora: *Simpsonictis,* cf. *S. jayanneae.*
Condylarthra: *Colpoclaenus* sp.; *Promioclaenus* sp. Condylarthra, unidentified genus and species 1; Condylarthra, unidentified genus and species 2.
Eutheria, order *incertae sedis,* Palaeoryctidae: *Pararyctes,* new species (Fox, 1983a). **Pantolestidae:** *Propalaeosinopa,* new species. **Pentacodontidae:** *Aphronorus,* new species.

CALGARY 2E (INCLUDING CALGARY 7E OF RUSSELL, 1929): ALBERTA
Porcupine Hills Formation (Carrigy, 1971), late Torrejonian (To_3)
(Russell, 1926, 1929, 1932, 1958; D. E. Russell, 1967; Krause, 1978; Youzwyshyn, in preparation)

This locality, which L. S. Russell discovered in 1919, is recorded as being on the east bank of the Elbow River at Calgary, Lsd 3,S4,T24,R1,W5, and in the Paskapoo (now the Porcupine Hills) Formation. It is the source of the first Paleocene mammal from Canada (*Catopsalis calgariensis*) to be recognized as such (Russell, 1926). By 1958, the locality had evidently been destroyed by encroachment of commercial development accompanying the growth of Calgary, and/or by weathering, although Russell (1958) reported being still able to locate fossils in loose blocks of sandstone in the river bed (and see Russell and Churcher, 1972). No more recently collected fossils from this locality have been recorded (Krause, 1978); accordingly, the list below is from Russell (1958), with appropriate taxonomic changes that have accrued since (G. P. Youzwyshyn, personal communication, February, 1987).

Russell (1958, p. 98), in reviewing earlier work (see Russell, 1932), considered the faunule "to be slightly older than the Cochrane faunules [q.v.] and more or less intermediate between the typical Tiffanian and typical Torrejonian." Krause (1978) thought a "very tentative" Torrejonian age for Calgary 2E was most appropriate (which is accepted here), while Sloan preferred a late Torrejonian (Sloan, 1970) or earliest Tiffanian (Sloan, 1987) age, the latter correlative with the Douglass Quarry local fauna, Montana (Krause and Gingerich, 1983). In his 1987 paper, however, Sloan (Fig. 1) unaccountably shows "Calgary" (= Calgary 2E) as younger than "Cochrane" (= Cochrane 1 plus 2?), which is incorrect. The occurrence together of *Baiotomeus,* cf. *B. douglassi, Eucosmodon molestus, Catopsalis calgariensis,* and *Tetraclaenodon,* cf. *T. puercensis* indicates a late Torrejonian age.
Multituberculata: *Baiotomeus,* cf. *B. douglassi; Neoplagiaulax,* new species, cf. *N. macrotomeus;* Neoplagiaulacidae, unidentified genus and species; *Anconodon cochranensis; Anconodon gidleyi; Eucosmodon molestus; Catopsalis calgariensis*.*
Primates: *Pronothodectes* sp. (= "Microsyopinae, *incertae sedis*": Krause, 1978)
Carnovora: *Protictis* sp.
Condylarthra: *Claenodon,* cf. *C. montanensis; Tetraclaenodon,* cf. *T. puercensis;* cf. *Tetraclaenodon* sp.

NORDEGG: ALBERTA
Paskapoo Formation (Tozer, 1956), Tiffanian
(Russell, 1932)

Russell (1932) described and illustrated an upper molar identified as ?*Chriacus* sp., collected from "the north Bank of North Saskatchewan river, near mile 113, Canadian National Railways, Nordegg Branch," apparently in Lsd 1,S30,T39, R10,W5. This tooth demonstrated the Paleocene age of the beds in which it occurred, which at the time were mapped as Cretaceous.
Condylarthra: ?*Chriacus* sp.

SAUNDERS CREEK: ALBERTA
Paskapoo Formation (Tozer, 1956), early Tiffanian (Ti_1)
(Russell, 1948; Simons, 1960)

Russell's (1948) description of a pantodont upper molar from a site 4.8 km east of Saunders Creek, Alberta, at an outcrop along the Nordegg Branch of the Canadian National Railway, provided further confirmation of the Paleogene age of the "Upper Saunders Beds" in the Rocky Mountain Foothills. Russell (1948) identified the species represented by this tooth as *Pantolambda* sp., and believed it to be indicative of a middle Paleocene (Torrejonian) age. Simons (1960) referred this specimen to *Caenolambda,* from the late Torrejonian or early Tiffanian, while Sloan (1970, 1987) gave an early Tiffanian age for this locality ("Saunder's Creek" in Sloan, 1987, Fig. 1).
Pantodonta: *Caenolambda* sp.

COCHRANE 1: ALBERTA
Porcupine Hills Formation (Carrigy, 1971), early Tiffanian (Ti_1)
(Russell, in Rutherford, 1927; Russell, 1958; D. E. Russell, 1967; Gingerich, 1982; Youzwyshyn, in preparation)

R. L. Rutherford and L. S. Russell discovered the Cochrane 1 locality in 1926. It is cited (e.g., in Russell, 1958) as being at Cochrane, Alberta, on the north bank of the Bow River, in Lsd 7,S4,T26,R4,W5 (see Rutherford, 1927); the fossils came from rocks now recognized as part of the Porcupine Hills Formation, but earlier included within the Paskapoo Formation. Russell (1958) stated that the only identifiable mammalian remains known from this locality were found in 1926; I know of no more

recent collections made after Russell's (1958) summary, and in fact, field parties from the UALVP have been uniformly unsuccessful in relocating this site. Hence, my list of mammals from Cochrane 1 follows Russell (1958), with updated taxonomic changes (from G. P. Youzwyshyn, personal communication, February, 1987); the occurrence of *Ectocion collinus* at Cochrane 1 is indicative of an early Tiffanian age (see Gingerich, 1982).
Multituberculata: *Anconodon cochranensis**.
Condylarthra: *Ectocion collinus**.
Eutheria, order *incertae sedis*, Pantolestidae: *Propalaeosinopa septentrionalis**.

COCHRANE 2: ALBERTA
Porcupine Hills Formation (Carrigy, 1971), early Tiffanian (Ti_1)

(Russell, in Rutherford, 1927; Russell, 1929, 1932 1958; D. E. Russell, 1967; Krause, 1978; Gingerich, 1982; Youzwyshyn, 1988; Fox and Youzwyshyn, in preparation)

This locality, in the Porcupine Hills Formation 0.5 km east of Cochrane, is one of the more important localities yielding Paleocene mammals in western Canada. The fossils come from beds at a southward-facing outcrop in SE¼S1,T26,R4,W5, at about 1,059 m above sea level; the outcrop overlooks the Bow River and the main transcontinental line of the Canadian Pacific Railroad. R. L. Rutherford and L. S. Russell discovered the locality in 1926; it is here called "Cochrane 2," a name intended to replace the names "Cochrane 11" (as in, e.g., Rutherford, 1927; Russell, 1928; D. E. Russell, 1967; and Krause, 1978), and "Cochrane II" (e.g., as in Russell, 1929; and Gingerich, 1982). Present ambiguities about the name of this locality are such as to confuse even knowledgeable workers, such as Savage and Russell (1983, p. 53), who have listed "Cochrane 2" and "Cochrane 11" as separate localities having different ages, which is incorrect.

Early papers concerning the Cochrane 2 mammals include Rutherford (1927) and Russell (1929, 1932, 1958). From this early work, Russell (e.g., 1965) believed the Cochrane mammals to be Clarkforkian (latest Paleocene) in age, while Sloan (1970, Fig. 5) thought them to be middle Tiffanian.

Field parties from UALVP have made a number of collections at Cochrane 2, beginning in 1969. Mammals have been found at only one horizon, about 17 m above the CPR tracks, in calcareous gray-greenish siltstones and shales containing gastropod and bivalve shells, coalified fragments of wood, and occasional well-rounded pebbles. The mammalian fossils occur within an interval that varies between a few centimeters and 1 m, along about 30 m of strike, beginning at the west end of the outcrop. L. S. Russell (1929, p. 167) stated that the mammal horizon at Cochrane 2 is "about 300 feet, stratigraphically, above that of locality 1"; this cannot be checked independently until Cochrane 1 is relocated. The UALVP collecting at Cochrane 2 has been by quarrying and splitting the fossiliferous rock on site in the attempt to recover jaws, followed by underwater screening of the residual quarried rock in the laboratory, for isolated teeth. D. W. Krause (1978), who made significant collections at Cochrane 2 for UALVP during 1972 to 1977, described the primates from the site and concluded that they indicated a "latest Torrejonian or latest Tiffanian . . . age but, provisionally . . . the former" (p. 1269). Krause's (1978) estimate that the mammals from the Porcupine Hills Formation at Cochrane were older than those from the Paskapoo Formation in the Red Deer River valley contradicted Carrigy's (1971) contention from lithostratigraphy that the Porcupine Hills Formation was in fact younger than the Paskapoo. Demchuck's (1987) review of the palynostratigraphy of these same beds shows that Krause (1978) was correct, as does the work on mammals from the two formations subsequent to Krause's.

G. P. Youzwyshyn is currently completing a review of the mammals from Cochrane 2 (based on approximately 1,700 identifiable specimens) for his M. Sc. research at The University of Alberta (UA): he has determined that the Cochrane 2 local fauna is earliest Tiffanian in age (from the occurrence of *Nannodectes intermedius, Plesiadapis praecursor, Elphidotarsius russelli,* and *Ectocion collinus;* and see Krause, 1978, Gingerich, 1982), approximately correlative with the Douglass Quarry local fauna, Montana (see Krause and Gingerich, 1983). (For completeness, it should be noted here that Sloan's (1987) assessment of the "Cochrane" locality [locality number unspecified] as older than Calgary 2E is erroneous.)

The Cochrane 2 local fauna includes over 80 mammalian species (Fox and Youzwyshyn, in progress; as compared to the nine species cited in the most recent prior summaries), an unexpectedly large number for the early parts of the Tiffanian: the Tiffanian is an interval thought to show a dramatic decline in mammalian diversity from a Torrejonian high as a consequence of global cooling from a subtropical to warm-temperature regime (see, e.g., Sloan, 1970; Gingerich, 1976, 1980; Gingerich and others, 1980; Rose, 1981a, b; Krause, 1984). If such an event took place, it occurred only after the time that the Cochrane 2 local fauna represents (assuming that a change to a more temperate climate would depress mammalian species diversity); as shown below, however, the high number of mammalian species from middle Tiffanian localities along the Blindman River, near Red Deer, Alberta, makes that alternative unlikely as well. Continuing research by D. W. Krause (Krause and Maas, 1987; Krause, personal communication, May, 1988) at Douglass Quarry, Montana, also has failed to reveal a depression in mammalian species numbers in the early Tiffanian.

The list below is the consequence of my own review of the mammals from Cochrane 2; for Cochrane 2, "cf." means compare with."
Multituberculata: *Ptilodus,* new species 1 (see Krause, D. W., unpublished Ph.D. thesis); *Ptilodus,* new species 2 (see Krause, 1982 thesis); *Ptilodus,* new species; *Baiotomeus,* new species; Ptilodontidae, unidentified genus and species; *Mesodma pygmaea* (see Sloan, 1987); *Mimetodon silberlingi; Ectypodus,* cf. *E. szalayi; Ectypodus* sp.; *Parectypodus sinclairi; Parectypodus,* cf. *P. sylviae; Parectypodus,* new species; *Neoplagiaulax,* cf. *N. hunteri;*

Neoplagiaulax nelsoni; Neoplagiaulacidae, unidentified genus and species; *Anconodon cochranensis; Anconodon gidleyi; Anconodon* sp.; Eucosmodontinae, unidentified genus and species; *Acheronodon,* new species; *Pentacosmodon* sp.
Marsupicarnivora: *Peradectes* sp.
Lipotyphla: *Litocherus,* cf. *L. lacunatus; Litocherus* sp.; Erinaceidae, unidentified genus and species; *"Leptacodon" mumusculum; Leptacodon,* cf. *L. tener; Leptacodon,* cf. *L. packi;* cf. *Nyctitherium,* new genus and species; *Limaconyssus* sp.; Nyctitheriidae, unidentified genus and species.
Dermoptera: *Elpidophorus,* cf. *E. elegans.*
Primates: ?*Palaechthon* sp.; *Plesiolestes,* cf. *P. sirokyi; Ignacius fremontensis; Ignacius,* cf. *I. frugivorus;* ?Pronothodectes sp.; *Nannodectes intermedius; Plesiadapis praecursor; Plesiadapis,* cf. *P. anceps; Elphidotarsius russelli*; Carpodaptes,* cf. *C. hazelae; Picrodus silberlingi.*
Carnivora: *Simpsonictis pegus; Simpsonictis,* cf. *S. tenuis;* Viverravidae, new genus and species.
Creodonta: unidentified genus and species.
Condylartha: ?*Oxyprimus* sp.; *Chriacus pelvidens; Chriacus,* cf. *C. baldwini; Thryptacodon orthogonius;* ?Oxyclaeninae, new genus and species; *Colpoclaenus,* cf. *C. keeferi; Claenodon,* cf. *C. montanensis;* ?*Claenodon* sp.; ?*Mimotricentes* sp.; *Desmatoclaenus* sp.; *Ectocion collinus* (Gingerich, 1982); *Litomylus dissentaneus; Litomylus,* new species; Apheliscinae, new genus and species; Mesonychidae, unidentified genus and species; ?Periptychidae, unidentified genus and species.
Pantodonta: unidentified genus and species.
?Rodentia, cf. Mimotonidae (Li and Ting, 1985): unidentified genus and species.
Eutheria, order *incertae sedis,* Palaeoryctidae: *Palaeoryctes,* cf. *P. punctatus; Pararyctes pattersoni; Pararyctes,* new species; Didelphodontinae, unidentified genus and species; Palaeoryctidae, unidentified genus and species 1; Palaeoryctidae, unidenticied genus and species 2; Palaeoryctidae, unidentified genus and species 3; "palaeoryctoid", new genus and species. **Leptictidae:** *Prodiacodon, cf. P. puercensis; Prodiacodon concordarcensis; Prodiacodon furor; Myrmecoboides montanensis.* **Pantolestidae:** *Propalaeosinopa septentrionalis; Paleotomus senior; Paleotomus,* new species. **Pentacodontidae:** *Bisonalveus browni;* Pentacodontidae, new genus and species. **Mixodectidae:** *Eudaemonema,* cf. *E. cuspidata.* **Apatemyidae:** *Jepsenella,* cf. *J. praepropera.* Eutheria, unidentified genus and species (cf. *Thylacaelurus*).

AARON'S LOCALITY: ALBERTA
Paskapoo Formation, early Tiffanian (Ti_1 or Ti_2)
(Fox, in preparation)

This locality was discovered by W. Roberts, from the Museum of Zoology, UA, and his son Aaron in June 1987; it is at a small roadcut, 42 km east of Innisfail, Alberta, in NW¼S15,T35, R24,W4, at about 876 m above sea level. The fossils occur with clay galls in a hard, brownish channel sandstone, and show evidence of postmortem abrasion. Samples taken from the site for underwater screening in the laboratory have been partly processed at this writing, with six mammalian taxa having been recognized; two of these, *Ignacius fremontensis* and cf. *Bisonalveus browni* suggest an early Tiffanian age.
Multituberculata: *Ptilodus,* cf. *Ptilodus,* new species 1 of Krause (1982);
Primates: *Ignacius fremontensis; Elphidotarsius,* cf. *E. wightoni.*
?Creodonta: unidentified genus and species.
Eutheria, order *incertae sedis,* Pentacodontidae: *Bisonalveus,* cf. *B. browni.*

HAND HILLS LOCALITIES: ALBERTA
Paskapoo Formation (Carrigy, 1971), Tiffanian
(Fox, in preparation)

The Hand Hills are an erosional remnant plateau located approximately 25 km northeast of Drumheller, Alberta. They are capped by gravels that range from Miocene to Pleistocene. In the summer of 1981, D. B. Schowalter collected several isolated mammalian teeth from bentonitic shales poorly exposed in a small roadcut on the north slope of the Hand Hills below the level of the gravels; in 1983, several more teeth and tooth fragments from this site were collected by Schowalter and Dr. B. G. Naylor, Tyrrell Museum of Palaeontology. These collections are housed in UALVP. In 1984, I visited the site but found no additional fossils.

This locality is now known as Hand Hills East; it is in SW¼S14,T30,R17,W4, at about 1,044 m above sea level. The mammals from Hand Hills East are not diagnostic as to age, but owing to the nearness of the locality to localities immediately to the west that are rich in mammals and are at a similar altitude (see below), Hand Hills East is undoubtedly Tiffanian.

In 1984, I discovered a rich bed of vertebrate fossils, including mammals, in a large roadcut approximately 1.7 km west of Hand Hills East. At this site, in SE¼S21,T30,R17,W4, at approximately 1,052 m above sea level, the road cuts through pale gray to buff shales and siltstones beneath the capping gravels. The mammals occur in a thin brown to dark gray shale layer in association with broken molluscan shells, and consist of isolated teeth and tooth fragments, and jaw fragments with teeth; in this layer, fossils of *Amia* and *Champsossaurus* are especially abundant. The productive layer is near the top of the roadcut, and is scoured down into the paler shales beneath, producing a shallow, but temporally important, disconformity. Large samples were taken from this layer for underwater screening in 1984 to 1986, and it has been named "Hand Hills West, upper level"; from the occurrence of *Plesiadapis rex,* the mammals from this layer appear to be middle Tiffanian in age.

In August 1986, I discovered a second layer at Hand Hills West containing molluscan shells and vertebrate bone, in a gray shale approximately 22 m below the first, along the same roadcut. A small sample taken for underwater screening showed the presence of mammals in this second layer, now called "Hand Hills West, lower level"; the primates indicate an early Tiffanian age.

A large sample was collected from this layer in 1987 for underwater screening in the laboratory, but has not been processed at this writing.

The two layers at Hand Hills West are especially important in being in superposition, an uncommon relationship among the mammalian occurrences discovered so far in the Paleocene of western Canada. Demchuck's (1987) palynostratigraphic estimate of the age of the shales below the disconformity is consistent with the early Tiffanian age indicated by the mammals from the lower level.

Hand Hills East: Tiffanian

Multituberculata: unidentified genus and species.

Hand Hills West, lower level: early Tiffanian (Ti_1 or Ti_2)

Multituberculata: *Neoplagiaulax,* new species.
Primates: *Elphidotarsius russelli; Picrodus silberlingi.*
Eutheria, order incertae sedis, Palaeoryctidae: *Pararyctes* sp.

Hand Hills West, upper level: middle Tiffanian (Ti_3)

Multituberculata: *Ptilodus,* new species 2 of Krause (1982); *Baiotomeus* sp.; *Neoplagiualax,* cf. *N. hazeni; Neoplagiaulax,* new species; *Ectypodus,* cf. *E. powelli; Microcosmodon woodi.*
Marsupicarnivora: *Peradectes elegans.*
Lipotyphla: *Litocherus,* cf. *L. notissimus; Litocherus,* cf. *L. zygeus; Leptacodon,* cf. *L.* packi.
Dermoptera: *Elpidophorus elegans.*
Primates: Microsyopidae, unidentified genus and species; *Ignacius frugivorus; Plesiadapis rex; Nannodectes simpsoni; Carpodaptes hazelae; Elphidotarsius wightoni; Zanycteris palaeocena; Saxonella,* new species.
Carnivora: *Raphictis,* cf. *R. gausion; Raphictis* sp.
Condylarthra: *Ectocion wyomingensis.*
Eutheria, order *incertae sedis,* Palaeoryctidae: *Palaeoryctes,* cf. *P. punctatus; Pararyctes pattersoni; Pararyctes,* new species. Pantolestidae: *Propalaeosinopa septentrionalis.* **Pentacodontidae:** *Bisonalveus* sp. **Apatemyidae:** unidentified genus and species.

BLINDMAN RIVER, LOCALITIES DW-1, MEL'S PLACE, DW-2, DW-3: ALBERTA
Paskapoo Formation, middle Tiffanian (Ti_3)
(Fox, 1984a to d; in preparation)

These localities are outcrops of the Paskapoo Formation along successive cutbanks of the Blindman River ("Paskapoo" is the Cree name given the Blindman River [Russell, 1965]), southeast of the village of Blackfalds, Alberta. Figures 1 and 2 show their distributions on regional and local scales. D. Wighton discovered DW-1, 2, and 3 in 1977; M. Fox discovered Mel's Place in 1981. They are located in S13–14, T39,R27,W4, at an altitude of about 838 m above sea level.

At the Blindman River localities, the Paskapoo Formation consists mostly of sombre gray-greenish, brown, and black shales, mudstones, siltstones, sandstones, and limestones, together representing deposition associated with slowly flowing streams, shallow lakes, and swamps. The mammalian bones are normally disarticulated, but they show no evidence of depositional abrasion, so were probably not transported far from place of death to final burial. Some show tooth marks, evidence of (prolonged?) subaerial exposure before burial. All of the mammalian fossils collected from these localities have been from species of small body size, with the largest represented by rare fossils of *Cyriacotherium* (Pantodonta) and of condylarths such as *Colpoclaenus* and *Ectocion.* Two extensively preserved multituberculate skeletons have been found at DW-2, in addition to several incomplete mammalian skulls. Many of the remaining identifiable mammals are represented by incomplete jaws with teeth, but preservation of the anterior parts of the dentition is not uncommon, a rarity for small mammals of this age. Prominent among the nonmammalian vertebrates at these localities are fossils belonging to a necrosaurid lizard.

DW-1, the first of the Blindman sites that Wighton found, appears now to be exhausted: it evidently consisted of a small lens of concentrated, well-preserved remains. It was worked extensively in 1978 and 1979, mostly by the author. DW-2 has yielded a large and important collection of exceptionally preserved mammalian fossils; it was quarried extensively in 1978 to 1984, and 1986 to 1987. Mel's Place has been little sampled: access to it is difficult at all but the lowest river levels. Mammals collected from DW-3 include only the few teeth and tooth fragments that Wighton found on discovery of the site: the fossils occur in a loose sandstone containing clay galls and molluscan shells, and are probably best sampled by underwater screening (yet to be undertaken). At DW-1 and 2, and Mel's Place, the mammals occur in shales that are hand quarried to best assure recovery without breakage. The Blindman River localities are within sight of one another, and can be correlated directly, by reference to the same lignites and mollusc-rich beds as markers. The occurrence of what appears to be a primitive stage in the upper molar evolution of *Plesiadapis rex* at DW-2 indicates that the Blindman localities are probably early middle Tiffanian in age. A preliminary list of mammalian species at DW-2 was published in Fox (1984c), which the present list is intended to replace. Figures 4 to 6 show representative specimens from DW-2, illustrating the exceptional preservation of many specimens of small mammals from this site.

DW-1

Multituberculata: *Ptilodus,* new species 2 of Krause (1982); *Mimetodon silberlingi.*
Marsupicarnivora: ?*Peradectes* sp.
Lipotyphla: *Litocherus,* cf. *L. zygeus;* "litolestine," new genus and species; Erinaceidae, new genus and species; *Leptacodon packi.*

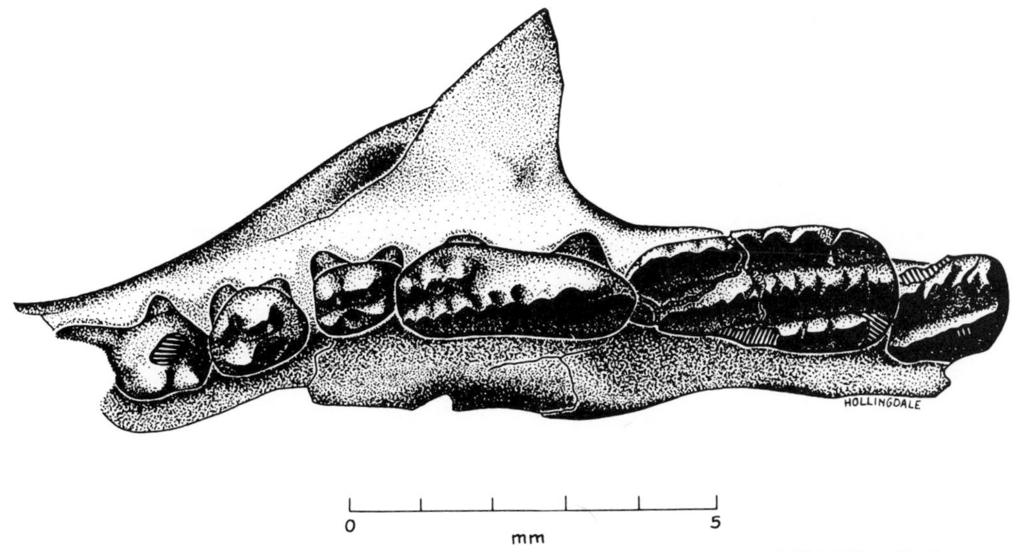

Figure 4. *Neoplagiaulax*, cf. *N. hunteri*: undescribed and uncatalogued specimen in UALVP collections, showing incomplete left maxillary with P1 to 4, M1 to 2; from DW-2, Paskapoo Formation, Alberta.

Dermoptera: *Elpidophorus elegans.*
Primates: *Elphidotarsius wightoni; Carpodaptes hazelae; Carpodaptes cygneus* (cf. Swan Hills species).
Condylarthra: *Colpoclaenus keeferi.*
Eutheria, order *incertae sedis*, Pentacodontidae: *Bisonalveus,* new species.

Mel's Place

Primates: *Carpodaptes hazelae.*

DW-2

Multituberculata: *Ptilodus,* new species 2 of Krause (1982); *Prochetodon,* new species; *Mesodma pygmaea; Ectypodus,* species 1; *Ectypodus,* species 2; *Neoplagiaulax,* cf. *N. hunteri; Neoplagiaulax,* cf. *N. hazeni; Neoplagiaulax,* new species 1-3; *Microcosmodon woodi.*
Marsupicarnivora: *Peradectes elegans.*
Lipotyphla: *Litocherus,* cf. *L. notissimus; Litolestes ignotus; Diacocherus minutus;* Erinaceidae, new genus and species; ?Erinaceidae, unidentified genus and species 1; ?Erinaceidae, unidentified genus and species 2; ?Erinaceidae, unidentified genus and species 3; *Leptacodon packi.*
Dermoptera: *Elpidophorus elegans.*
Primates: *Micromomys fremdi*;* Microsyopidae, unidentified genus and species 1; Microsyopidae, unidentified genus and species 2; *Ignacius frugivorius; Plesiadapis rex?; Plesiadapis* sp.; *Elphidotarsius wightoni; Carpodaptes hazelae; Saxonella, new species; Zanycteris palaeocena.*
Carnivora: *Protictis paralus; Raphictis,* cf. *R. gausion.*
Condylarthra: *Colpoclaenus keeferi; Metachriacus,* new species; *Claenodon,* cf. *C. ferox;* Arctocyonidae, new genus and species 1; Arctocyonidae, new genus and species 2. ?Hyopsodontidae, new genus and species.
Pantodonta: *Cyriacotherium,* cf. *C. argyreum.*
Eutheria, order *incertae sedis*, Palaeoryctidae: *Pararyctes,* new species 1; *Pararyctes,* new species 2. **Leptictidae:** ?*Prodiacodon,* new species. **Pantolestidae:** *Propalaeosinopa albertensis; Propalaeosinopa septentrionalis.* **Pentacodontidae:** *Bisonalveus,* new species; *Aphronorus* sp. **Mixodectidae:** *Eudaemonema,* new species. **?Apatemyidae:** unidentified genus and species.

DW-3

Multituberculata: *Neoplagiaulax,* cf. *N. hazeni.*
Primates: *Carpodaptes hazelae.*
Eutheria, order *incertae sedis*, Pentacodontidae: *Bisonalveus,* new species.

RED DEER: ALBERTA
Paskapoo Formation, middle Tiffanian (Ti_3)
(Russell, 1929)

This locality, in Paskapoo beds at Lsd 1 and 2, S18,T38, R27,W4, is listed by L. S. Russell (1929, p. 170) as being "about 1 mile west of the town of Red Deer, on the right bank of the Red Deer River"; a mammalian trackway was collected from the locality. Included here by default is an undescribed plesiadapid P3 that L. S. Russell collected in 1929, from "1 mile above Red Deer"; the specimen is in the UALVP collections. In 1987, an attempt was made to relocate the source(s) of these specimens, but it was unsuccessful. The beds in the vicinity are at about 869 m above sea level, higher in altitude than middle Tiffanian beds to the east of Red Deer, but lower than the late Tiffanian horizon

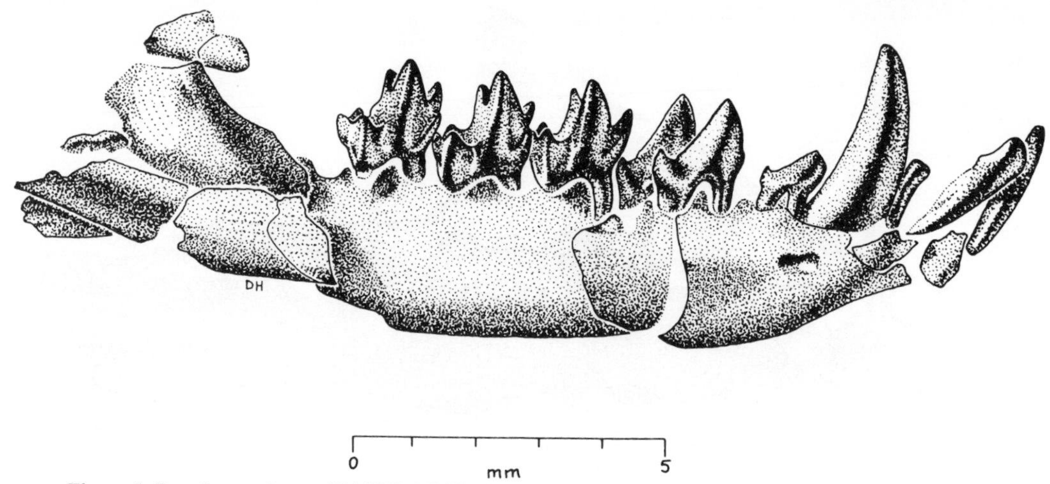

Figure 5. *Peradectes elegans,* UALVP 16270, right dentary containing il to 2, 4, c, pl to 3, ml to 3; from DW-2, Paskapoo Formation, Alberta (see Fox, 1983b).

at Canyon Ski Quarry (q.v.); at present, then, a middle Tiffanian age seems preferrable, but without fossil evidence in its support.
Primates: Plesiadapidae, unidentified genus and species.
Mammalia: unidentified genus and species (trackway).

BURBANK: ALBERTA
Paskapoo Formation, middle Tiffanian (Ti₃)
(Fox, in preparation)

"Burbank" is the name given to an extensive series of outcrops at the confluence of the Blindman and Red Deer Rivers, 1.2 km from the Blindman River localities, and in S13,T39,R27,W4, approximately 838 m above sea level. Most of the outcrop at Burbank consists of poorly bedded gray siltstones and shales that contain well-preserved leaf impressions and insect remains (Taylor and Stockey, 1984); the gray beds are capped by a more resistant, cliff-forming brown sandstone. E. Speirs collected the first mammalian fossil from Burbank in the summer of 1977. The specimen is a well-preserved dentary with teeth of *Propalaeosinopa albertensis,* and was the first Paskapoo mammal to be discovered in the area. Mammals are not common at this locality, in contrast to its abundant plant and insect remains.
Condylarthra: *Colpoclaenus keeferi.*
?Pantodonta: unidentified genus and species.
Eutheria, order *incertae sedis,* Pantolestidae: *Propalaeosinopa albertensis; Propalaeosinopa septentrionalis.*

JOFFRE BRIDGE ROADCUT: ALBERTA
Paskapoo Formation, middle Tiffanian (Ti_3)
(Fox, in preparation)

Approximately 12 km east of Red Deer and south of the Joffre Bridge across the Red Deer River, relocation of Provincial Highway 11 in 1978 produced a roadcut exposing highly fossiliferous beds in the Paskapoo Formation. This roadcut is located in NW¼S13,T38,R26,W4. In 1978, D. Wighton first found mammals at this locality, at a horizon now named "Joffre Bridge Roadcut, lower level," at about 818 m above sea level. From 1978 to the present, E. Speirs has quarried for fossils at the Joffre Bridge Roadcut, but mostly at a horizon 20 m above the lower level, collecting many valuable plant and insect specimens (Taylor and Stockey, 1984) and discovering rich deposits of teleost fishes (under study by M.V.H. Wilson, UALVP). This upper level has yielded several isolated mammalian jaws and, in 1986, the skull and postcranium of a large pantodont, probably *Titanoides,* also found by Speirs: the isolated jaws occur in an indurated black shale containing abundant molluscan shells, whereas the pantodont was excavated from poorly lithified brown mudstones and siltstones directly above the mollusc layer.

Speirs and field parties from the UALVP have worked the lower level intermittently, from 1979 to 1982, and in 1986. Here isolated teeth, jaw fragments with teeth, and the isolated bones of small mammals are found in a black shale only a few centimeters thick; the shale is directly beneath a massive sandstone that appears devoid of vertebrate fossils.

From the primates occurring at the lower level, its age is probably middle Tiffanian. Mammals from the upper level are consistent with this age, but are represented so far only by long-ranging species that themselves are not diagnostic. A peculiarity of the beds at the Joffre Bridge Roadcut is the apparent absence of *Bisonalveus,* one of the most abundant therians at the Blindman River localities and Hand Hills West, upper level, which are approximately correlative with the Roadcut beds.

Joffre Bridge Roadcut, lower level

Multituberculata: *Ptilodus,* new species 1 of Krause (1982); *Prochetodon,* new species; *Mesodma pygmaea; Mimetodon silberlingi; Neoplagiaulax,* cf. *N. hunteri; Neoplagiaulax,* cf. *N. hazeni; Neoplagiaulax,* new species.

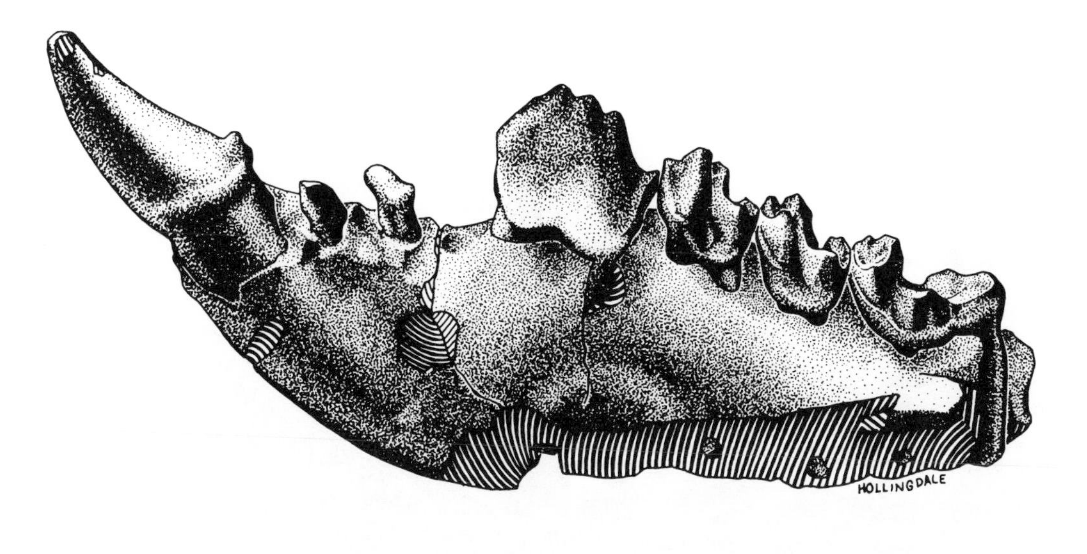

Figure 6. *Elphidotarsius wightoni*, UALVP 21002 (holotype), incomplete left dentary having il to 2, c, p4, ml to 3; from DW-2, Paskapoo Formation, Alberta (see Fox, 1984c). Bar equals 5 mm.

Marsupicarnivora: *Peradectes elegans.*
Lipotyphla: *Litocherus,* cf. *L. notissimus; Litocherus,* cf. *L. zygeus; Litolestes ignotus;* Erinaceidae, new genus and species. Lipotyphla, unidentified genus and species.
Dermoptera: *Elpidophorus elegans.*
Primates: Microsyopidae, unidentified genus and species; *Ignacius frugivorus; Carpodaptes hazelae.*
Condylarthra: Arctocyonidae, unidentified genus and species; *Ectocion wyomingensis.*
Pantodonta: unidentified genus and species.
Eutheria, order *incertae sedis,* Palaeoryctidae: *Pararyctes pattersoni;* cf. *Stilpnodon* sp.; **Palaeoryctidae, unidentified genus and** species. Pantolestidae: *Propalaeosinopa albertensis; Propalaeosinopa septentrionalis.*

JOFFRE BRIDGE ROADCUT, UPPER LEVEL

Lipotyphla: *Litocherus,* cf. *L. zygeus.*
Pantodonta: cf. *Titanoides* sp.

JOFFRE BRIDGE, MAMMAL SITE NO. 1: ALBERTA
Paskapoo Formation, middle Tiffanian (Ti3)
(Fox, in preparation)

Directly across the Red Deer River from the Joffre Bridge Roadcut is another locality, "Joffre Bridge, Mammal Site No. 1," in a high river bank exposing gray shales and siltstones of the Paskapoo Formation; this bank is located in NW¼S13 and SW¼S24,T38,R26,W4, at about 830 m above sea level. The fossils occur at approximately 22 m above river level, in a dark gray to brown or black shale rich in broken molluscan shells. The shell bed lies immediately beneath a thin, fissile black shale, which in turn is beneath a lignite. *Amia* bones and scales and *Champsosaurus* bones are common in the shell bed; mammalian fossils (jaws with teeth, skull parts) are usually well preserved (see Fig. 7), but are rare. D. W. Krause found the first mammals at this site in July 1974, and he collected additional specimens in 1977. Beginning in 1978, E. Speirs quarried here with good success, as have field parties from UALVP since 1979.

Multituberculata: *Ptilodus,* new species 1 of Krause (1982); *Prochetodon,* new species;
Marsupicarnivora: unidentified genus and species.
Lipotyphla: *Litocherus,* cf. *L. notissimus; Litocherus,* cf. *L. zygeus; Diacocherus minutus.*
Dermoptera: *Elpidophorus elegans.*
Pantodonta: *Titanoides* sp.
Eutheria, order *incertae sedis,* Pantolestidae: *Propalaeosinopa albertensis.*

ERICKSON'S LANDING: ALBERTA
Paskapoo Formation, middle Tiffanian (Ti3)
(Simpson, 1927; Krause, 1978)

In 1910, a field party from the American Museum of Natural History collected a small sample of mammalian remains from a slump block on the Red Deer River at Erickson's Landing ("about 20 miles below the town of Red Deer"; Simpson, 1927, p. 1). These fossils provided the first record of Paleocene mammals in Canada (although this was not recognized until later; Simpson, 1927), but the exact place at which they were found has been in doubt (see Krause, 1978, p. 1266).

L. S. Russell (personal communication, May, 1988) has kindly informed me that he visited the American Museum locality in 1924 and photographed it in 1928, when remains of the

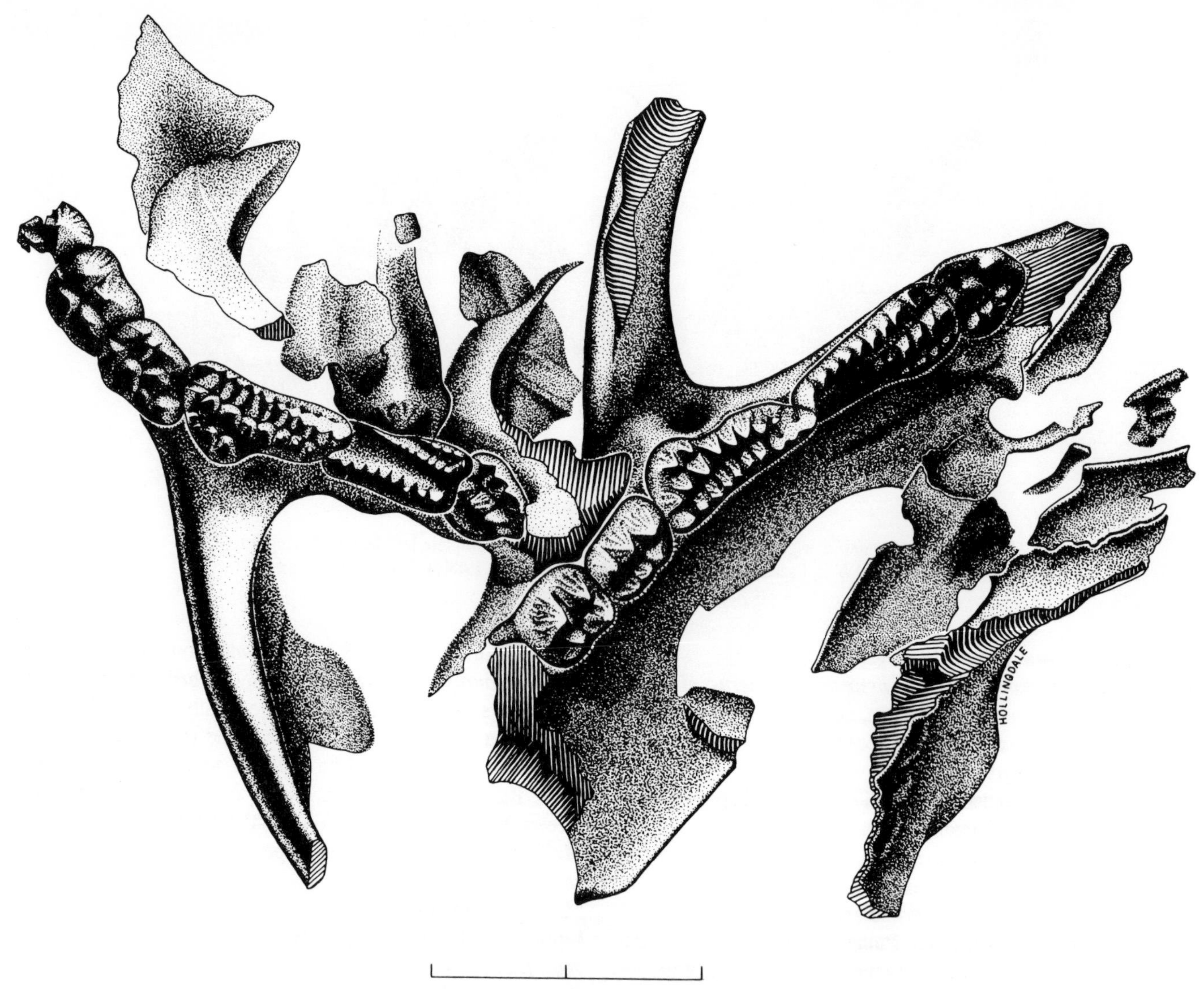

Figure 7. *Ptilodus,* new species 1 of Krause (1982), undescribed incomplete skull showing left and right Pl to 4, M1 to 2 (left P1 partly hidden, lingual to right M1); from Joffre Bridge, Mammal Site No. 1, Paskapoo Formation, Alberta. Bar equals 2 mm.

slumped shell layer were still in place. His photographs show the slump to have been at the river bank containing Joffre Bridge Mammal Site No. 1, but he points out that there is no way to determine from which of two shell beds cropping out there the slump block was derived; consequently, I have listed the Erickson's Landing mammals separately from those at Mammal Site No. 1. Simpson (1927) described the original faunule, which included the holotypes of two new genera and species. The occurrence of *Plesiadapix rex* in this faunule indicates a middle Tiffanian age (Gingerich, 1976; Krause, 1978).
Primates: *Plesiadapis rex.*
Dermoptera: *Elpidophorus elegans***.*
Condylarthra: *Ectocion* sp.
Eutheria, order *incertae sedis,* Pantolestidae: *Propalaeosinopa albertensis***.*

JOFFRE BRIDGE EAST
Paskapoo Formation, middle Tiffanian (Ti_3)
(Fox, in preparation)

In the summer of 1987, D. B. Schowalter collected an incomplete shearing blade of an unidentified multituberculate from a mollusc layer in the Paskapoo Formation east of Joffre Bridge, in NE¼S13,T38,R25,W4; the site is approximately 861 m above sea level, exposed on the bank to the north of Highway 11 as it descends to the Joffre Bridge. In altitude, the layer at Joffre Bridge East approaches that at Canyon Ski Quarry (q.v.) about 8 km to the northwest, and on that basis might be equivalent in age, but I have listed it with the other Joffre Bridge localities until a difference in age can be shown.
Multituberculata: unidentified genus and species.

JOFFRE PIPELINE: ALBERTA

Paskapoo Formation, middle Tiffanian (Ti_3)

(Fox, in preparation)

This locality was discovered during excavation in the Paskapoo Formation for a natural gas pipeline crossing the Red Deer River 4.5 km downstream from the Joffre Bridge; it is located southeast of the village of Joffre, in SE¼S9,T38,R25,W4, at about 823 m above sea level. E. Speirs collected the mammals—three teeth/tooth fragments—in August 1986, and they are catalogued in the collections of the Tyrrell Museum of Palaeontology, Drumheller, Alberta. Owing to the nearness of the Joffre Bridge localities, which are at a similar altitude, the fossils at Joffre Pipeline are probably middle Tiffanian in age.
Condylarthra: *Colpoclaenus* sp.

POLICE POINT: ALBERTA

Ravenscrag Formation, middle or late Tiffanian (Ti_3 or Ti_4)

(Krishtalka, 1973; Krause, 1978)

In 1973, L. Krishtalka described a new local fauna of Paleocene mammals from the Cypress Hills, Alberta, based on collections made by R. C. Fox and assistants in 1966, 1967, and 1969 (a very large collection was made for the UALVP in 1970 by D. E. O'Brien, but remains undescribed). The Police Point locality (locality UAR-1 of Krishtalka) is in Lsd 16,S15,T8,R1,W4, at 1,333 m above sea level, and approximately 12 km east of the village of Elkwater, Alberta (Krishtalka [1973] named the local fauna from this locality the "Cypress Hills local fauna," a name best suppressed owing to the fact that there are several localities in the Cypress Hills that have yielded fossil mammals, from Late Cretaceous to Miocene in age; the name "Police Point local fauna" is favored here as an unambiguous substitute, and is already in use in the literature in reference to this locality). The fossils at Police Point occur directly beneath the Oligocene(?) Cypress Hills Conglomerate, in poorly lithified, highly bentonitic shales, and in association with broken molluscan shells. Krishtalka (1973) assigned a Tiffanian age to the Police Point mammals, comparable to the age of the Scarritt Quarry local fauna, Montana. Krause (1978) reviewed the primates from Police Point and concluded that they indicated a middle Tiffanian age (p. 1269), one younger than either Erickson's Landing or Scarritt (Krause, 1978, Fig. 11), while Sloan (1987, Fig. 1) indicated that Police Point is older than Erickson's Landing in the middle Tiffanian, which is incorrect. In my opinion, the plesiadapid at Police Point is best identified as *Plesiadapis,* cf. *P. churchilli,* instead of *P. fodinatus* (Krishtalka, 1973) or *Plesiadapis* sp. (Krause, 1978). If so, and if the succession of *Plesiadapis* species is reliable for zonation of this part of the Paleocene, then Police Point is younger than previously believed, either late in the middle Tiffanian or early in the late Tiffanian.
Multituberculata: *Neoplagiaulax hunteri; Mesodma pygmaea; Parectypodus sinclairi; Neoplagiaulax hunteri.*
Marsupicarnivora: *Peradectes protinnominatus* (Krishtalka and Stucky, 1983).
Lipotyphla: *Litocherus notissimus; Litolestes* sp.; *Leptacodon tener; Leptacodon packi;* cf. *Nyctitherium* sp.
Dermoptera: *Elpidophorus elegans.*
Primates: *Ignacius* sp.; *Plesiadapis,* cf. *P. churchilli; Carpodaptes,* cf. *C. cygneus.*
Condylarthra: *Ectocion* sp.
Eutheria, order *incertae sedis,* Palaeoryctidae: *Pararyctes pattersoni;* Palaeoryctidae, unidentified genus and species. **Pantolestidae:** *Propalaeosinopa albertensis.*

CANYON SKI QUARRY: ALBERTA

Paskapoo Formation, late Tiffanian (Ti_4)

(Krause, 1978; Fox, in preparation)

The Canyon Ski Quarry locality is located immediately east of Red Deer, in the Paskapoo Formation exposed at the top of the Red Deer River valley, in N½S29,T38,R26,W4, at about 921 m above sea level; the fossiliferous layer crops out at a cut for a road descending to the Canyon Ski Lodge, a resort on the floor of the valley. D. W. Krause discovered this locality in August, 1976; Krause (1978) recorded the occurrence there of the primate *Carpodaptes cygneus,* as yet the only mammal described from the site.

At Canyon Ski Quarry, the mammalian fossils occur in a grayish-brown shale in association with broken mulluscan shells; the productive layer is about 0.5 m thick. The concentration of mammalian fossils in the layer appears to be low, and underwater screening of the shale is the only collecting technique that has been employed and that seems practical for the future. This locality is important because it is one of the few from Canada that can be shown to be in superposition to other occurrences of Paleocene mammals—in this instance, to those at Joffre Bridge (q.v.) approximately 7 km downstream. From the occurrence of *Carpodaptes cygneus,* Krause (1978, p. 1269) believed that the Canyon Ski Quarry local fauna is probably late Tiffanian in age.
Multituberculata: *Neoplagiaulax* sp.
Lipotyphla: *Litolestes ignotus; Diacocherus meizon; Leptacodon packi.*
Primates: *Ignacius frugivorus; ?Nannodectes* sp.; *Carpodaptes cygneus.*

ONE-JAW GAP: ALBERTA

Paskapoo Formation, late Tiffanian (Ti_4?)

(Fox, in preparation)

This locality, discovered by K. Soehn and the author in July 1987, is at a large roadcut approximately 3 km northeast of Canyon Ski Quarry, in NW¼S34,T38,R26,W4, at about 975 m above sea level. An incomplete mammalian dentary with at least one molar and an isolated upper molar, both unidentified at this writing, were collected from this cut, which from its altitude, is

probably at least as high stratigraphically as is the Canyon Ski Quarry directly across the Red Deer River.
Mammalia: unidentified genus and species.

CRESTOMERE SCHOOL: ALBERTA
Paskapoo Formation, late Tiffanian (Ti_4)
(Fox, in preparation)

In 1975, A. R. Sweet sent to D. W. Krause, then a graduate student at UA, several mammalian tooth fragments during preparation of rock samples for pollen analysis; the fossils were from a roadcut exposing pale gray to brownish shales of the Paskapoo Formation, approximately 22 km west of Ponoka, Alberta. The roadcut is in NE¼S26,T42,R28,W4, on Secondary Highway 792, about 1 km south of Provincial Highway 53 and Crestomere School, at about 899 m above sea level. In 1984, as a consequence of the discovery of jaws by G. P. Youzwyshyn, a field party from UALVP opened quarries at this site, which were worked still more intensively in 1986, and further in 1987. The fossils occur in several layers over an interval of about 0.5 m, in bentonitic to silty greenish-gray, brown, and dark gray shales; molluscan shells occur in some of the layers that contain mammals. Isolated teeth, well-preserved jaws with teeth, and occasional incomplete skulls of small mammals have been found at this important site. The occurrence of *Carpodaptes,* cf. *C. hobakensis,* suggests that the Crestomere School local fauna is late Tiffanian in age, which is consistent with its altitude. The brief list below is highly misleading in that most specimens that have been collected from Crestomere School are unprepared at this writing and, hence, unidentified.
Lipotyphla: *Litolestes ignotus.*
Primates: *Carpodaptes,* cf. *C. hobackensis.*
Condylarthra: *Ectocion wyomingensis.*

SWAN HILLS SITE 1: ALBERTA
Paskapoo Formation, late Tiffanian (Ti_4)
(Russell, 1967; Krishtalka, 1973; Krause, 1978; Gingerich, 1986; Stonley, in preparation)

L. S. Russell (1967) described the first mammals from the Paskapoo Formation of the Swan Hills; the specimens had been collected from an artificial exposure at a well site in Lsd4,S16, T67,R10,W5, approximately 1,067 m above sea level, and are catalogued (a total of 31 specimens in 15 species) in the collections of the Royal Ontario Museum, Toronto. The Swan Hills mammals, at nearly 55° North, are the most northerly Paleocene mammals currently known; L.S. Russell (1967) believed them to be late Paleocene (Tiffanian) in age.

In 1970, D. E. O'Brien made a collection for UALVP from site 1; G. J. Stonley, as part of his M.Sc. research at UA on the Swan Hills mammals, made additional collections from site 1 in 1986 and 1987. In 1986, Stonley discovered a second locality in the Paskapoo (called here "Swan Hills site 2") that yields mammals; it is located in Lsd12,S17,T67,R10,W5, at approximately 1,067 m above sea level. (Site 2 of Stonley should not be confused with locality 2 of L. S. Russell (1967), which is located elsewhere and has not yielded mammals.) The list below is of species that Stonley has identified at site 1; the mammals from site 2 are unidentified at this writing, but it is not anticipated that they will differ in age from those at site 1. The occurrence of *Plesiadapis churchilli* at site 1 indicates a late Tiffanian age (Krause, 1978).
Multituberculata: *Ptilodus kummae; Prochetodon foxi (Krause, 1987b); Mesodma pygmaea; Neoplagiaulax hunteri; Neoplagiaulax,* cf. *N. hazeni; Microcosmodon* sp.
Lipotyphla: *Litocherus lacunatus; Litolestes ignotus; Diacocherus meizon; Leptacodon tener;* Nyctitheriidae, *incertae sedis.*
Dermoptera: *Elpidophorus,* new species.
Primates: *Ignacius frugivorus; Plesiadapis churchilli; Carpodaptes cygneus**; Plesiadapidae, *incertae sedis.*
Carnivora: *Protictis,* cf. *P. laytoni.*
Condylarthra: *Claenodon* sp.; *Ectocion osbornianus; Phenacodus,* cf. *P. primaevus.*
Pantodonta: *Titanoides,* cf. *T. primaevus.*
Eutheria, *incertae sedis,* Palaeoryctidae: *Pararyctes* sp. **Pantolestidae:** *Propalaeosinopa* sp.

ROCHE PERCÉE (SEVERAL LOCALITIES): SASKATCHEWAN
Ravenscrag Formation, late Tiffanian (Ti_4)
(Krause, 1977, 1978; Krause and Fox, in preparation)

D. W. Krause is responsible for a large and important collection of fossil mammals from Roche Percée, made mostly during the summers 1972 to 1974 (Krause, 1977). The fossils come from several localities in the Ravenscrag Formation, extensively exposed in southeastern Saskatchewan as a consequence of strip mining for thermal coal. Systematic descriptions of the multituberculates have been published (Krause, 1977), but none of the remainder of the local fauna, which is under study by Krause and the author. The occurrence of *Plesiadapis churchilli* indicates a late Tiffanian age for the Roche Percée local fauna (Krause, 1978).
Multituberculata: *Ptilodus kummae**; *Prochetodon foxi* (Krause, 1987b); *Mesodma pygmaea; Mimetodon silberlingi; Ectypodus,* cf. *E. powelli; Neoplagiaulax hunteri; Neoplagiaulax,* cf. *N. hazeni; Microcosmodon conus.*
Lipotyphla: *Litocherus lacunatus; Litolestes,* cf. *L. ignotus; Diacocherus,* cf. *D. meizon; Leptacodon packi.*
Primates: *Micromomys vossae**; *Ignacius frugivorus; Plesiadapis churchilli; Carpodaptes cygneus.*
Carnivora: *Protictis,* cf. *P. dellensis; Raphictis,* cf. *R. gausion; Raphictis,* new species 1; *Raphictis,* new species 2;
Condylarthra: *Chriacus* sp.; *Thryptacodon australis; Ectocion osbornianus; Phenacodus primaevus.*
Pantodonta: *Titanoides* sp.; *Cyriacotherium,* cf. *C. argyreum* (Rose and Krause, 1982).
Eutheria, order *incertae sedis,* Palaeoryctidae: *Pararyctes* sp. **Pantolestidae:** *Propalaeosinopa* sp.

POORLY KNOWN LOCALITIES

The localities below are Paleocne in age from their regional geologic setting, but biostratigraphically useful fossils have yet to be obtained from them. Some of these localities—Tan-i-Bryn Rocks, for example—hold little prospect of yielding any further mammalian material; others, such as Wintering Hills and Dickson Dam, are listed here because they have been discovered only very recently, with no opportunity to make extensive collections. It is likely that diagnostic fossils will be found with the advent of larger samples from these sites.

TAN-I-BRYN ROCKS: ALBERTA
Porcupine Hills Formation (Carrigy, 1971), ?Tiffanian

This occurrence, on the property of Dr. H. Sutmoller east of Carstairs, Alberta, in NW¼S23,T29,R29,W4, is an articulated but incomplete skeleton, lacking the skull, of a dog-sized mammal; it was collected by Gao Keqin, S. Sutmoller, and the author in 1986. Unfortunately, the skeleton was extensively damaged in preparation at the UALVP.
Mammalia: unidentified genus and species.

FORESTRY TRUNK ROAD, SOUTH: ALBERTA
?Porcupine Hills Formation, Paleocene
(Fox and Naylor, in preparation)

This specimen is an incomplete maxillary having five teeth, catalogued in the collections of the Tyrrell Museum of Palaeontology, Drumheller, Alberta, and as yet unidentified. It comes from a cutbank on the Little Red Deer River, west of Water Valley, Alberta, a locality otherwise unidentified (B. G. Naylor, personal communication, September, 1987).
Mammalia: unidentified genus and species

NORTH SASKATCHEWAN RIVER: ALBERTA
Paskapoo Formation, ?Tiffanian

The piece of shale containing this specimen was found in 1973 as float in glacial gravels along the North Saskatchewan River west of Genesee, and is catalogued in the collections of the Tyrrell Museum of Palaeontology. It was found in S15,T5, R5,W5, but a search by J. E. Storer, then of the Provincial Museum and Archives of Alberta, failed to reveal a source outcrop in the vicinity.
Mammalia: *Bisonalveus* sp.

POPLAR RIVER MINE: SASKATCHEWAN
Ravenscrag Formation, ?early Tiffanian

This occurrence, from spoil piles at the Poplar River Mine, near Coronach, south-central Saskatchewan (N½S5,T2,R27,W2), is a vertebra from an unidentified mammal collected by T. Tokaryk, Saskatchewan Museum of Natural History, in 1984. J. Hartman's analysis of fresh-water gastropods at the site suggests an early Tiffanian age (J. E. Storer, personal communication, July, 1987).

WINTERING HILLS: ALBERTA
Paskapoo Formation (Carrigy, 1971)
(Fox, in preparation)

The Wintering Hills are an erosional remnant located about 20 km south and slightly east of Drumheller. They are capped by gravels of Pleistocene age (J. Burns, Provincial Museum and Archives of Alberta, personal communication, September, 1987). In August 1987, D. B. Schowalter discovered a fossiliferous deposit containing mammalian fossils at a roadcut on the north slope of the Wintering Hills; the mammalian fossils occur in a brown channel sandstone along with bones of non-mammalian vertebrates and broken molluscan shells. This locality is approximately 9 km south and east of East Coulee, in S½S33,T26, R18,W4, and is approximately 998 m above sea level.
Multituberculata: unidentified genus and species.
Carnivora: *Protictis* sp.

DICKSON DAM: ALBERTA
Paskapoo Formation
(Fox, in preparation)

The Dickson Dam impounds the Red Deer River approximately 20 km west of Innisfail; construction of the dam was completed in 1984. In September 1987, L. A. Lindoe, UALVP, found mammals at the dam site, at an outcrop adjacent to the spillway north of the dam; this locality is in NE¼S33,T35, R2,W5, and is at approximately 930 m above sea level.
Multituberculata: *Ptilodus,* new species 2 of Krause (1982).
Condylarthra: Arctocyonidae, unidentified genus and species.
Pantodonta: *Cyriacotherium* sp. Pantodonta, unidentified genus and species, but substantially larger than *Cyriacotherium.*

CONCLUSIONS

This chapter is the first attempt in 20 years (since D. E. Russell, 1967) to compile and arrange in chronologic sequence all of the Paleocene mammalian local faunas known from Canada (summarized in Fig. 3), and to list the species that they contain. The oldest of these local faunas is early Puercan (P_1), and the youngest late Tiffanian (Ti_4), a biochronologic range that encompasses an interval of from eight to ten million years (Savage and Russell, 1983, Fig. 3-22; Sloan, 1987, Fig. 7). The fossils of interest come from the Paskapoo, Porcupine Hills, Coalspur, and Ravenscrag Formations, in the Alberta Syncline and Williston Basin, Alberta and Saskatchewan.

The primary basis for age determination and correlation of these local faunas has rested on their containing taxa that define Paleocene Land-Mammal Ages (LMAs) and their subdivisions in the Western Interior of the United States: the American succession is much the richer and more intensively studied, providing thereby an appropriate standard for comparison. However, opportunities to check the succession established in this paper by independent means within the Alberta and Williston basins have been few, and a chronostratigraphy for the beds containing the succession has yet to be defined: for example, in only three instances, the MHBT Quarry, Hand Hills West, and Canyon Ski Quarry/Joffre Bridge localities, are strata of different ages—as determined by their mammalian content—in superpositional relationship with each other. For none of these local faunas are radiometric dates available (datable bentonites have yet to be found in proximity to the mammalian occurrences), nor has a magnetostratigraphy been determined for any but the earliest horizons under consideration here (see Lerbekmo, 1985, for the magnetostratigraphy of the K-T boundary in Alberta and Saskatchewan). For these reasons, Demchuck's (1987) pioneering attempt to develop a palynostratigraphic zonation of the continental Paleocene of the central Alberta Basin has provided welcome corroboration for the correlations of a number of the mammalian local faunas, and his recognition of pollens equivalent to those defining the P6 zone in the USA (Nicholls and Ott, 1978) suggests that latest Tiffanian or Clarkfordian beds are present in the Alberta Syncline, holding the possibility of discovery of mammals of that age there.

That the Paleocene mammalian local faunas from western Canada should show close taxonomic resemblance with correlatives in the United States, especially those from the northern tier of eastern Rocky Mountain and western plains states, is scarcely surprising in view of their close geographic proximity. Nonetheless, the Canadian local faunas have unique components that add significant new information about North American mammalian evolution during the Paleocene, as enumerated below.

1. Locality Fr-1 and the Long Fall mammals at the MHBT Quarry, although not Paleocene in age, are important to understanding the sources of the North American Paleocene mammalian fauna. The work of Sloan and Van Valen (1965) and of Archibald (1982) in eastern Montana appeared to indicate that progressive, "Paleocene-aspect" mammals occurred in the Late Cretaceous, before the onset of the great wave of extinction at the end of the period. A careful reexamination of the strata in question, however, has suggested that the age of these assemblages—latest Cretaceous or early Paleocene—cannot be determined (Fastovsky and Dott, 1986). Nonetheless, Fr-1 and Long Fall, although clearly Late Cretaceous in age, contain many of the same progressive groups, especially condylarths, that occur in the problematic Montana sites (Bug Creek Anthills, Bug Creek West, and Harbicht Hill), in corroboration of the claim that the latter are from the Late Cretaceous after all. More importantly, the Saskatchewan sites by themselves show that the evolution of "Paleocene aspect" mammals preceded the terminal Mesozoic extinctions.

2. The discovery of an undoubted carnivoran at the Rav W-1 horizon, MHBT Quarry, is the firmest evidence available of the existence of this major mammalian order in rocks as old as early Paleocene. (The only record of comparable age for the Carnivora is a p4 from the Puercan of New Mexico [McIntyre, 1966], but agreement has not been universal that the specimen pertains to a carnivoran: Flynn and Galiano [1982] cite only a questionable early Paleocene occurrence for the order, for example, and do not discuss this specimen.)

3. The discovery of *Purgatorius,* the earliest primate, at the Rav W-1 horizon, MHBT Quarry, is the first extension of the distribution of this mammal beyond the vicinity of the Fort Peck Reservoir, in eastern Montana. The Saskatchewan *Purgatorius* is at least as old as *Purgatorius* from the Paleocene of Montana (Johnston and Fox, 1984). (Many specialists appear to believe that the latest Cretaceous record of *Purgatorius*—based on a single tooth—is spurious, owing to contamination from Paleocene deposits during the washing and screening process.)

4. The discovery of species-rich local faunas from the early and middle Tiffanian of Alberta is inconsistent with the long-held thesis that the beginning of the North American Tiffanian saw a decline in mammalian species diversity in response to global cooling; the Alberta sites reflect no such decline, which is particularly surprising given the higher latitudes of the Alberta localities and the expectation that the effects of depressed temperatures might be felt more strongly there. The record of early Tiffanian mammals from Alberta is consistent with results of parallel research in correlative beds in Montana, which also fail to reveal a decline in mammalian species diversity in the early Tiffanian (Krause and Maas, 1987, and in preparation). However, the species richness of early middle Tiffanian sites in Alberta make unlikely a later and more rapid change in Tiffanian climates, as hypothesized by Krause and Maas (1987).

5. The discovery of a new species of the little known primate *Saxonella* in the middle Tiffanian of Alberta is the clearest indication to date of a European and North American Paleocene mammal that has not been found in the Paleocene of the United States. *Saxonella* previously had been known only from the Walbeck locality of the German Democratic Republic, which is perhaps a Ti_3 correlative (Sloan, 1987). It may be that *Saxonella* was restricted to a northerly distribution, thereby accounting for its absence in the extensively sampled Tiffanian of the United States.

ACKNOWLEDGMENTS

This summary of the succession of Paleocene mammals from western Canada stems from a suggestion made to me about the utility of such an undertaking by Dr. A. R. Sweet (Institute of Sedimentology and Petroleum Geology, Geological Survey of Canada), while we both were visiting the newly opened Tyrll Museum of Palaeontology in October, 1985. I also thank Drs. K. D. Rose (Johns Hopkins University) and T. M. Bown (U.S. Geological Survey, Denver) for their invitation to present this paper at the Boulder symposium. For important locality information, I thank Dr. A. R. Sweet; Dr. J. E. Storer (Saskatchewan

Museum of Natural History, Regina); Dr. B. G. Naylor and A. G. Newman (Tyrrell Museum of Palaeontology, Drumheller); and D. B. Schowalter (Delia, Alberta). G. P. Youzwyshyn and G. J. Stonley (UA) kindly allowed me to cite the results of their work before its completion, and have brought information of relevance to my attention. Drs. D. W. Krause (SUNY, Stony Brook) and L. S. Russell (Royal Ontario Museum, Toronto) provided helpful comments in their review of this paper for publication. D. Hollingdale prepared the drawings of specimens.

The advances that have been made in knowledge of Paleocene mammals from western Canada over the past 20 years are mostly owing to work in the field; in this regard, the following deserve special recognition for field discoveries that have greatly enriched the research collections of Paleocene mammals at UALVP, and hence the basis for present understanding of mammalian evolution during the Paleocene of northern North America: G. Denomme (Morinville, Alberta); Dr. P. A. Johnston (Tyrrell Museum of Paleontology); Gao Keqin (Institute of Vertebrate Paleontology and Paleoanthropology, Beijing, and UA); Dr. D. W. Krause; E. Speirs (Red Deer, Alberta); D. Wighton (UA). Scholars using the UALVP collection of Paleocene mammals are in their special debt.

REFERENCES CITED

For this chapter, I have attempted to cite original publications instead of citing the most recent summary of the subject that includes those publications. For example, rather than cite Savage and D. E. Russell (1983) as my source for distribution of Paleocene mammals from Canada, I have cited the original papers themselves. I have done this partly in the belief that original authors have the clearer right to citation and any credit that it may bring (although citation of textbooks and comparable summaries is unarguably easier). In this I depart from what seems to be increasingly common practice in North American vertebrate paleontology.

Archibald, J. D., 1982, A study of Mammalia and geology across the Cretaceous-Tertiary boundary in Garfield County, Montana: University of California Publications in Geological Sciences, v. 122, p. 1–286.

Carrigy, M. A., 1971, Lithostratigraphy of the uppermost Cretaceous (Lance) and Paleocene strata of the Alberta plains: Research Council of Alberta Bulletin 27, p. 1–161.

Demchuck, T. D., 1987, Palynostratigraphy of Paleocene strata of the central Alberta Plains [M.Sc. thesis]: Edmonton, University of Alberta, 151 p.

Fastovsky, D. E., and Dott, R. H., Jr., 1986, Sedimentology, stratigraphy, and extinctions during the Cretaceous-Paleocene transition at Bug Creek, Montana: Geology, v. 14, p. 279–282.

Flynn, J. J., and Galiano, H., 1982, Phylogeny of early Tertiary Carnivora, with a description of a new species of *Protictis* from the middle Eocene of northwestern Wyoming: American Museum of Natural History Novitates 2725, p. 1–64.

Fox, R. C., 1968, A new Paleocene mammal (Condylarthra: Arctocyonidae) from a well in Alberta, Canada: Journal of Mammalogy, v. 49, p. 661–664.

—— , 1983a, Evolutionary implications of tooth replacement in the Paleocene mammal *Pararyctes:* Canadian Journal of Earth Sciences, v. 20, p. 12–22.

—— , 1983b, Notes on the North American Tertiary marsupials *Herpetotherium* and *Peradectes:* Canadian Journal of Earth Sciences, v. 20, p. 1565–1578.

—— , 1984a, First North American record of the Paleocene primate *Saxonella:* Journal of Paleontology, v. 58, p. 892–894.

—— , 1984b, The dentition and relationships of the Paleocene primate *Micromomys* Szalay, with description of a new species: Canadian Journal of Earth Sciences, v. 21, p. 1262–1267.

—— , 1984c, A new species of the Paleocene primate *Elphidotarsius* Gidley; Its stratigraphic position and evolutionary relationships: Canadian Journal of Earth Sciences, v. 21, p. 1268–1277.

—— , 1984d, *Melaniella timosa,* n. gen. and sp., an unusual mammal from the Paleocene of Alberta, Canada: Canadian Journal of Earth Sciences, v. 21, p. 1335–1338.

—— , 1987, Patterns of mammalian evolution towards the end of the Cretaceous, Saskatchewan, Canada, *in* Currie, P. J., and Koster, E. H., eds., Fourth Symposium on Mesozoic Terrestrial Ecosystems: Tyrrell Museum of Palaeontology Occasional Paper 3, p. 96–100.

—— , 1988, The Wounded Knee local fauna and mammalian evolution near the Cretaceous-Tertiary boundary, Saskatchewan, Canada: Palaeontographica (A) (in press).

Gingerich, P. D., 1976, Cranial anatomy and evolution of early Tertiary Plesiadapidae (Mammalia, Primates): Ann Arbor, University of Michigan Papers on Paleontology 15, p. 140.

—— , 1980, *Tytthaena parrisi,* oldest known oxyaenid (Mammalia, Creodonta) from the late Paleocene of western North America: Journal of Paleontology, v. 54, p. 570–576.

—— , 1982, Paleocene "*Meniscotherium semicingulatum*" and the first appearance of Meniscotheriidae (Condylartha) in North America: Journal of Mammalogy, v. 58, p. 488–491.

—— , 1986, Systematic position of *Litomylus? alphamon* Van Valen (Mammalia: Insectivora); Further evidence for the late Paleocene age of Swan Hills Site-1 in the Paskapoo Formation of Alberta: Journal of Mammalogy, v. 60, p. 1135–1137.

Gingerich, P. D., Rose, K. D., and Krause, D. W., 1980, Early Cenozoic mammalian faunas of the Clark's Fork Basin–Polecat Bench area, northwestern Wyoming, *in* Gingerich, P. D., ed., Early Cenozoic paleontology and stratigraphy of the Bighorn Basin, Wyoming: Ann Arbor, University of Michigan Papers on Paleontology 24, p. 51–68.

Jerzykiewicz, T., 1985, Stratigraphy of the Saunders Group in the central Alberta Foothills; A progress report: Geological Survey of Canada Paper 85–1B, Current Research, part B, p. 247–258.

Jerzykiewicz, T., and McLean, J. R., 1980, Lithostratigraphic and sedimentological framework of coal-bearing Upper Cretaceous and lower Tertiary strata, Coal Valley area, central Alberta foothills: Geological Survey of Canada Paper 79–12, 47 p.

Jerzykiewicz, T., and Sweet, A. R., 1986, The Cretaceous-Tertiary boundary in the central Alberta Foothills; 1, Stratigraphy: Canadian Journal of Earth Sciences, v. 23, p. 1356–1374.

Johnston, P. A., 1980, First record of Mesozoic mammals from Saskatchewan: Canadian Journal of Earth Sciences, v. 17, p. 512–519.

Johnston, P. A., and Fox, R. C., 1984, Paleocene and Late Cretaceous mammals from Saskatchewan, Canada: Palaeontographica (A), Bd. 186, p. 163–222.

Krause, D. W., 1977, Paleocene multituberculates (Mammalia) of the Roche Percée local fauna, Ravenscrag Formation, Saskatchewan, Canada: Palaeontographica (A), Bd. 159, p. 1–36.

—— , 1978, Paleocene primates from western Canada: Canadian Journal of Earth Sciences, v. 15, p. 1250–1271.

—— , 1982, Evolutionary history and paleobiology of early Cenozoic Multituberculata (Mammalia), with emphasis on the Family Ptilodontidae [Ph.D. thesis]: Ann Arbor, University of Michigan, 555 p.

—— , 1984, Mammalian evolution in the Paleocene; Beginning of an era, *in* Broadhead, T. W., ed., Mammals; Notes for a short course: Knoxville, University of Tennessee Department of Geological Sciences Studies in Geology 8, p. 87–109.

——, 1987a, *Baiotomeus,* a new ptilodontid multituberculate (Mammalia) from the middle Paleocene of western North America: Journal of Paleontology, v. 61, p. 595–603.

——, 1987b, Systematic revision of the genus *Prochetodon* (Multituberculata, Mammalia) from the late Paleocene and early Eocene of western North America: Ann Arbor, University of Michigan Museum of Paleontology Contributions, v. 27, p. 221–236.

Krause, D. W., and Gingerich, P. D., 1983, Mammalian fauna from Douglass Quarry, earliest Tiffanian (late Paleocene) of the eastern Crazy Mountain Basin, Montana: Ann Arbor, University of Michigan Museum of Paleontology Contributions, v. 26, p. 157–196.

Krause, D. W., and Maas, M. C., 1987, Composition and diversity of mammalian faunas across the Torrejonian-Tiffanian boundary in western North America: Geological Society of America Abstracts with Programs, v. 19, p. 287.

Krishtalka, L., 1973, Late Paleocene mammals from the Cypress Hills, Alberta: Lubbock, Texas Tech University Special Publications of the Museum 2, p. 1–77.

Krishtalka, L., and Stucky, R. K., 1983, Paleocene and Eocene marsupials of North America: Carnegie Museum Annals, v. 52, p. 205–227.

Lerbekmo, J. F., 1985, Magnetostratigraphic and biostratigraphic correlations of Maestrichtian to early Paleocene strata between south-central Alberta and southwestern Saskatchewan: Canadian Petroleum Geology Bulletin, v. 33, p. 213–226.

Lerbekmo, J. F., and Coulter, K. C., 1985, Late Cretaceous and early Tertiary magnetostratigraphy of a continental sequence; Red Deer River valley, Alberta, Canada: Canadian Journal of Earth Sciences, v. 22, p. 567–583.

Lerbekmo, J. F., and St. Louis, R. M., 1986, The terminal Cretaceous iridium anomaly in the Red Deer River valley, Alberta, Canada: Canadian Journal of Earth Sciences, v. 23, p. 120–124.

Lerbekmo, J. F., Sweet, A. R., and St. Louis, R. M., 1987, The relationship between the iridium anomaly and palynofloral events at three Cretaceous-Tertiary boundary localities in western Canada: Geological Society of America Bulletin, v. 99, p. 325–330.

Li Chuan-kei, and Ting Su-yin, 1985, Possible phylogenetic relationship of Asiatic eurymylids and rodents, with comments on mimotonids, *in* Luckett, W. P., and Hartenberger, J.-L., eds., Evolutionary relationships among rodents; A multidisciplinary analysis: New York, Plenum Press, p. 35–58.

MacIntyre, G. T., 1966, The Miacidae (Mammalia, Carnivora); Part 1, The systematics of *Ictidoppapus* and *Protictis:* American Museum of Natural History Bullletin, v. 131, p. 115–210.

McIntyre, D. J., Sweet, A. R., and Wall, J. H., 1984, Palynology and micropaleontology of the Foothills belt near Calgary: 6th International Palynological Conference Guidebook, Field Excursion no. 3, 73 p.

Nicholls, D. J., and Ott, H. L., 1978, Biostratigraphy and evolution of the *Momipites-Caryapollenites* lineage in the early Tertiary in the Wind River Basin, Wyoming: Palynology, v. 2, p. 93–112.

Rose, K. D., 1981a, The Clarkforkian Land-Mammal Age and mammalian faunal composition across the Paleocene-Eocene boundary: Ann Arbor, University of Michigan Papers on Paleontology 26, 189 p.

——, 1981b, Composition and species diversity in Paleocene and Eocene mammal assemblages; An empirical study: Journal of Vertebrate Paleontology, v. 1, p. 367–388.

Rose, K. D., and Krause, D. W., 1982, Cyriacotheriidae, a new family of early Tertiary pantodonts (Mammalia) from western North America: Proceedings of the American Philosophical Society, v. 126, p. 26–50.

Russell, D. E., 1967, Le Paléocène continental d'Amérique du Nord: Muséum National d'Histoire Naturelle, série C, t. 16, 99 p.

Russell, L.S., 1926, A new species of the genus *Catopsalis* Cope from the Paskapoo Formation of Alberta: American Journal of Science, v. 12, p. 230–234.

——, 1928, A new fossil fish from the Paskapoo beds of Alberta: American Journal of Science, v. 15, p. 103–107.

——, 1929, Paleocene vertebrates from Alberta: American Journal of Science, v. 17, p. 162–178.

——, 1932, New data on the Paleocene mammals of Alberta, Canada: Journal of Mammalogy, v. 13, p. 38–54.

——, 1948, A Middle Paleocene mammal tooth from the foothills of Alberta: American Journal of Science, v. 246, p. 152–156.

——, 1958, Paleocene mammal teeth from Alberta: National Museum of Canada Bulletin 147, p. 96–103.

——, 1965, The continental Tertiary of western Canada, *in* Folinsbee, R. E., and Ross, D. M., eds., Vertebrate paleontology in Alberta: University of Alberta Department of Geology Bulletin 2, p. 41–52.

——, 1967, Palaeontology of the Swan Hills area, north-central Alberta: Royal Ontario Museum Life Sciences Contributions 71, p. 1–31.

——, 1974, Fauna and correlation of the Ravenscrag Foramtion (Paleocene) of southwestern Saskatchewan: Royal Ontario Museum Life Sciences Contributions 102, p. 1–52.

Russell, L. S., and Churcher, C. S., 1972, Vertebrate paleontology, Cretaceous to Recent, interior plains, Canada: 24th International Geological Congress, Montreal, Quebec, Excursion A21, p. 1–46.

Rutherford, R. L., 1927, Geology along the Bow River between Cochrane and Kananaskis, Alberta: Scientific and Industrial Research Council of Alberta Report 17, p. 1–29.

Savage, D. E., and Russell, D. E., 1983, Mammalian paleofaunas of the world: Reading, Addison-Wesley, 432 p.

Simons, E. L., 1960, The Paleocene Pantodonta: American Philosophical Society Transactions, pt. 6, v. 50, p. 3–98.

Simpson, G. G., 1927, Mammalian fauna and correlation of the Paskapoo Formation of Alberta: American Museum of Natural History Novitates 268, p. 1–10.

Sloan, R. E., 1970, Cretaceous and Paleocene terrestrial communities of western North America, *in* North American Paleontological Convention Proceedings: Lawrence, Allen Press, Inc., Part E, p. 427–453.

—— 1987, Paleocene and latest Cretaceous mammal ages, biozones, magnetozones, rates of sedimentation, and evolution, *in* Fassett, J. E., and Rigby, J. K., Jr., eds., The Cretaceous-Tertiary boundary in the San Juan and Raton basins, New Mexico and Colorado: Geological Society of America Special Paper 209, p. 165–200.

Sloan, R. E., and Van Valen, L., 1965, Cretaceous mammals from Montana: Science, v. 223, p. 220–227.

Smit, J., and Van der Kaars, S., 1984, Terminal Cretaceous extinctions in the Hell Creek area, Montana; Compatible with catastrophic extinction: Science, v. 223, p. 1177–1179.

Taylor, T. N., and Stockey, R. A., 1984, Field trip guide, 2nd International Organization of Paleobotany Conference, Edmonton, Alberta, 19–22 August 1984: International Organization of Paleobotany, 54 p.

Tozer, E. T., 1956, Uppermost Cretaceous and Paleocene non-marine molluscan faunas of western Alberta: Geological Survey of Canada Memoir 280, p. 1–125.

Van Valen, L., 1978, The beginning of the Age of Mammals: Evolutionary Theory, v. 4, p. 45–80.

Youzwyshyn, G. P., 1988, Late Paleocene mammals from near Cochrane, southwestern Alberta [M.Sc. thesis]: Edmonton, University of Alberta, 345 p.

Manuscript Accepted by the Society June 12, 1989

Printed in U.S.A.

Geological Society of America
Special Paper 243
1990

The biogeographic origins of late Paleocene–early Eocene mammalian immigrants to the Western Interior of North America

David W. Krause
Department of Anatomical Sciences, State University of New York, Stony Brook, New York 11794-8081
Mary C. Maas*
Department of Anthropology, State University of New York, Stony Brook, New York 11794-4364

ABSTRACT

South America, Central America, the southeastern United States, Arctic Canada, Europe, Asia, and Africa all have been suggested as possible or probable biogeographic sources for taxa that appeared in the Western Interior of North America during the late Paleocene and early Eocene. Recent compilations of the geographic and temporal distributions of Paleocene and Eocene mammals and new data, derived primarily from recent collections from early Tiffanian (late Paleocene) quarries in the Crazy Mountains Basin of south-central Montana, permit tests of these hypotheses, particularly those involving a southern New World origin.

Significant first appearances of mammalian higher taxa in the Western Interior occur in the earliest Tiffanian, late Tiffanian, earliest Clarkforkian, and earliest Wasatchian. Those that appear in the earliest Tiffanian probably were derived from late Torrejonian forms in the same region. It appears, therefore, that there was not a pronounced geographic shift in North American mammalian faunas across the Torrejonian-Tiffanian boundary as suggested in some southern New World origin hypotheses. It has been suggested that Palaeanodonta, Dinocerata, and Notoungulata (represented by Arctostylopidae), which appear in the late Tiffanian in the Western Interior, originated in South America, but the evidence is inconclusive and highly controversial. New higher taxa that appear in the Western Interior at the beginning of the Clarkforkian, particularly Rodentia and Tillodontia, probably originated in Asia and dispersed across Beringia. Most of the suprageneric taxa that first appear at the beginning of the Wasatchian in the Western Interior (Perissodactyla, Artiodactyla, Adapidae, Omomyidae, and Hyaenodontidae) also probably appeared in Asia and Europe at essentially the same time; there is no evidence for heterochrony. Recent paleontological discoveries and paleogeographic evidence suggest that the ultimate origins of some or all of these taxa lay in either Africa or the Indian subcontinent. The latter biogeographic source has not been seriously considered previously.

*Present address: Department of Biological Anthropology and Anatomy, Box 3170, Duke University Medical Center, Durham, North Carolina 27710.

Krause, D. W., and Maas, M. C., 1990, The biogeographic origins of late Paleocene–early Eocene mammalian immigrants to the Western Interior of North America, *in* Bown, T. M., and Rose, K. D., eds., Dawn of the Age of Mammals in the northern part of the Rocky Mountain Interior, North America: Boulder, Colorado, Geological Society of America, Special Paper 243.

INTRODUCTION

The late Paleocene and early Eocene of the Western Interior of North America have been characterized by a series of major changes in mammalian faunal composition. The first representatives of several modern orders of mammals appeared in North America during this interval of time, presumably as immigrants, and several important archaic taxa (e.g., Multituberculata, Plesiadapiformes) were severely diminished or became extinct. Shortly after their appearance in North America, many of the immigrant taxa became very abundant and taxonomically rich; they dominated terrestrial ecosystems and were responsible for a profound restructuring of early Tertiary mammalian communities by displacing or replacing resident taxa. Not surprisingly, considerable attention has been paid to the possible centers of origin of mammalian higher taxa first appearing in the late Paleocene and early Eocene of the Western Interior—South America, Central America, the southeastern United States, Arctic Canada, Europe, Asia, and Africa all have been suggested. Thus, almost every possible alternative has been put forward.

Here we review the evidence for and test previous hypotheses concerning biogeographic origins of supraspecific taxa of mammals first appearing in the Western Interior (1) at the beginning of the Tiffanian Land-Mammal Age (late Paleocene)—several genera; (2) in the late Tiffanian—Palaeanodonta, Dinocerata, and Arctostylopidae; (3) at the beginning of the Clarkforkian Land-Mammal Age (latest Paleocene–earliest Eocene)—Rodentia, Tillodontia, and Coryphodontidae (Pantodonta); and (4) at the beginning of the Wasatchian Land-Mammal Age (early Eocene)—Perissodactyla, Artiodactyla, Adapidae (Primates), Omomyidae (Primates), Hyaenodontidae (Creodonta), ?Didymoconidae ("Condylarthra"), and Didelphini (Marsupialia). Recent compilations of the geographic and temporal distributions of Paleocene and Eocene mammals facilitate such tests (e.g., Li and Ting, 1983; Savage and Russell, 1983; Russell and Zhai, 1987; Archibald and others, 1987; Stucky, unpublished data). In addition, data derived from new collections of early Tiffanian (late Paleocene) localities in the Crazy Mountains Basin of Montana are used to test some of these hypotheses, particularly those involving a southern New World origin.

BACKGROUND

The following is a brief, historical overview of hypotheses concerning centers of origin of Tiffanian, Clarkforkian, and Wasatchian mammals that have been put forward during the last 20 or so years. The purported sites of origin and dispersal routes are graphically depicted in Figure 1, a paleogeographic map of approximately late Paleocene–early Eocene time.

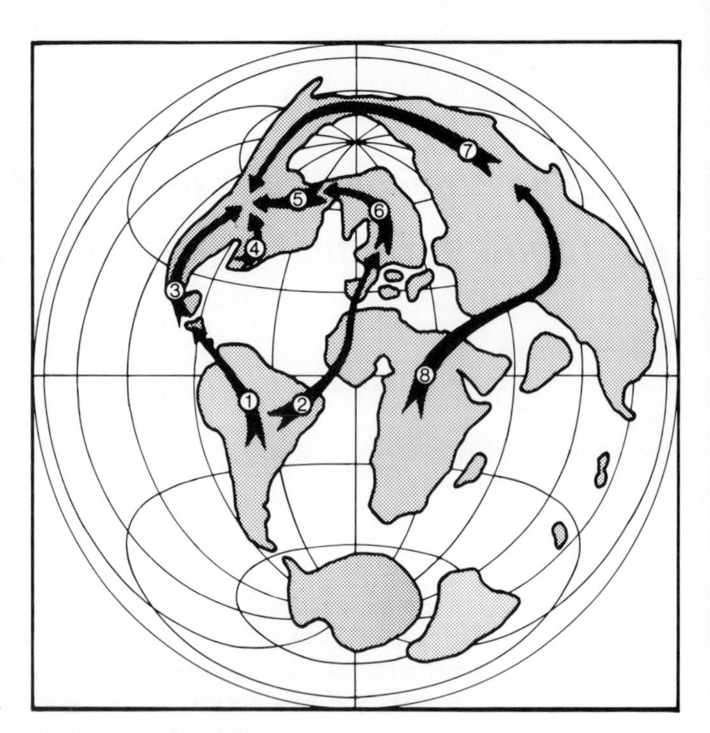

Figure 1. Paleogeographic map of approximately late Paleocene–early Eocene time with previously hypothesized migration routes of mammals into the Western Interior. Numbers on arrows indicating migratory routes correspond to numbers in the text.

Route 1. South America to the Western Interior via Central America

Marshall and de Muizon (1988) have argued recently that a series of four mammalian dispersal events from South to North America via Central America (Fig. 1, Route 1) took place in the Late Cretaceous and Paleocene. The first two purportedly occurred during the Late Cretaceous and early Paleocene and therefore are largely irrelevant to this analysis. The third, during the late Tiffanian, was examined in greater detail by Gingerich (1985). According to Gingerich, Dinocerata, Notoungulata (represented by Arctostylopidae), and Edentata dispersed from South to North America in the late Tiffanian as climates began to ameliorate (see also Webb, 1985). The fourth dispersal event of Marshall and de Muizon (1988) occurred in the latest Paleocene and involved only the marsupial family Didelphidae (= Didelphini of most authors; e.g., Clemens, 1979; Crochet, 1979, 1980; Crochet and Sigé, 1983; Krishtalka and Stucky, 1983a, 1983b), which is represented in North America at the beginning of the Wasatchian by the genus *Peratherium* (North American forms of which have been placed in *Herpetotherium* by Crochet, 1977, 1980; but which are retained in *Peratherium* by Krishtalka and Stucky, 1983a, 1983b). Neither of these dispersal events appears to have had a significant impact on the established mammalian faunas of the Western Interior; none of the immigrant taxa became very abundant or speciose.

Route 2. South America to the Western Interior via Africa, Europe, Greenland, and Arctic Canada

Contrary to Marshall and de Muizon (1988), Crochet and Sigé (1983) tentatively suggested that didelphid marsupials of the tribe Didelphini may have reached the Western Interior from South America by way of Africa, Europe, Greenland, and Arctic Canada (Fig. 1, Route 2).

Route 3. Central America to the Western Interior

Sloan, in 1969, and later Gingerich (1976, 1977a, 1977b) suggested an intriguing biogeographic hypothesis to account for mammalian faunal dynamics from the middle Paleocene through early Eocene in the Western Interior. Employing evidence from both mammalian systematics and paleoclimatic reconstructions, Sloan and Gingerich hypothesized that subtropically and tropically adapted Torrejonian (middle Paleocene) species of the Western Interior were displaced southward into Central America (or southern North America; Fig. 1, Route 3) at the beginning of the cooler Tiffanian; more temperate-adapted species therefore dominated the northern faunas during the Tiffanian. Sloan and Gingerich further suggested that, coincident with the climatic amelioration that began toward the end of the Tiffanian and continued into the Clarkforkian and Wasatchian, southern species dispersed northward again and displaced or replaced the more temperate-adapted Tiffanian species. Both Sloan and Gingerich noted that forms ancestral to taxa that appear for the first time in Clarkforkian and Wasatchian faunas have, for the most part, not been identified in Tiffanian faunas from the Western Interior; they suggested, however, that suitable ancestors for some of these Clarkforkian and Wasatchnian taxa are known from the Torrejonian. One of the logical extensions and predictions of the southern New World origin hypothesis (and, in this case, the Central America hypothesis) is that early Tiffanian mammals appearing for the first time in the Western Interior should not exhibit strong affinities with known late Torrejonian mammals from the same region but should instead represent descendants of unknown more northerly faunas.

Gingerich progressively modified his views concerning the Central American origin of Western Interior Paleogene faunas as new collections, particularly from both North America and Asia, were accumulated during the late 1970s and 1980s. Even in 1976 he acknowledged that some taxa (e.g., adapids and hyaenodontids) may have had their origins elsewhere (e.g., Africa) and that they dispersed to North America across high-latitude land bridges from Asia or Europe. Gingerich and Rose (1977) suggested that only the first wave of immigrants, those that arrived in the Clarkforkian, came from Central America while Wasatchian immigrants came from northern continents as the high-latitude land bridges opened up with climatic warming (see below). Later, Gingerich (1980a, 1980b, 1985) restricted the number of taxa that arrived in the Western Interior from Central (or South) America to those that appear in the late Tiffanian; he suggested that the Clarkforkian immigrants had arrived from Asia and the Wasatchian newcomers from Europe (see below).

Morris (1966) had earlier speculated that elements of the Wasatchian fauna of the Western Interior had a southern New World origin based primarily on the presence of a morphologically primitive species of *Hyracotherium*, a Wasatchian indicator genus, in the Punta Prieta fauna of Baja California, Mexico (see also Schiebout, 1981). Morris (1966, 1968) favored either a Tiffanian or Clarkforkian age for this fauna, as did Sloan (1969, 1987), Ferrusquia-Villafranca (1978), and Schiebout (1979). Flynn and Novacek (1984) and Novacek and others (1987), however, with the benefit of additional collections, have convincingly demonstrated that the Punta Prieta fauna is of Wasatchian age.

Route 4. Southeastern United States to the Western Interior

Schiebout (1979, 1981) suggested that the United States Southeast was a distinct biogeographic province during the Paleocene (Fig. 1, Route 4), separated from the Western Interior by the Cannonball Sea, a remnant of the Late Cretaceous Mid-Continental Sea that is depicted as extending northward from the Gulf of Mexico to at least as far as southern Saskatchewan and Manitoba. Sloan (1987) estimated that the Cannonball Formation is of Puercan through early or middle Tiffanian age. Schiebout predicted that, when found, Paleocene and early Eocene faunas of the U.S. Southeast will indicate stronger affinities with those of Europe than do those of the Western Interior. Following Wood (1977), who suggested that the Southeast was the site of origin for Rodentia, Schiebout (1981, p. 386) opined that "the southeast, as well as Mexico or even Central America, must be considered as a source region for immigrants" to the Western Interior in the early Eocene.

Route 5. Arctic Canada to the Western Interior

Hickey and others (1983, 1984) proposed that Wasatchian mammals of the Western Interior and Europe were derived from Arctic Canada (Fig. 1, Route 5). This hypothesis was based on the assignment of a late Paleocene age to a fauna from Ellesmere Island that contains mammalian taxa with Eocene affinities.

Route 6. Europe to the Western Interior via Greenland and Arctic Canada

Strong similarities between North American Wasatchian and European Sparnacian (= early part of Ypresian Standard Stage of Cavelier and Pomerol, 1986) faunas have long been noted (e.g., Simpson, 1947; Kurten, 1966; Sloan, 1969; Savage, 1971; Szalay and McKenna, 1971; McKenna, 1975a; Savage and Russell, 1983; Flynn, 1986) but there appears to be considerable disagreement concerning the direction(s) of dispersal between the two continents. Godinot (1978, 1981, 1982), Hooker (1980), and Gingerich (1980a) speculated that the direction of dispersal was from east to west (Fig. 1, Route 6; although Godinot explicitly recognized that disperal was possible in both directions). This was based on the discovery that several genera common to both North America and Europe were represented by more primitive species in Europe. Savage and Russell (1983, Fig. 4-8, see also

Fig. 4-37) illustrated what, in their view, were all possible dispersal routes of early Eocene mammals between holarctic continents. They indicated direct dispersal from North America to Europe, from North America to Asia, from Europe to Asia, and from Asia to Europe; none of the routes went from Europe to North America or from Asia to North America. More recently, Gingerich (1986), upon discovering a still more primitive Wasatchian fauna in North America than had been recognized previously, suggested that the direction of dispersal between North America and Europe was from west to east and that the new members of the North American fauna had arrived from Asia. Sloan (1987) agreed, stating, for instance, that *Hyracotherium* dispersed from Asia to North America to Europe. Gingerich (1986) further proposed that at least some of the Asian emigrants had their origins in Africa.

Route 7. Asia to the Western Interior via Beringia

Gingerich (1980a) suggested that the allochthonous members of the Clarkforkian fauna of the Western Interior had Asian affinities, a view also espoused by Webb (1985; Fig. 1, Route 7). Later, Gingerich (1986) added that elements of the Wasatchian fauna probably also were derived from Asia (see also Webb, 1985). Rose (1981a), on the other hand, although recognizing that intermingling of mammalian faunas between the Western Interior and Asia had occurred at the beginning of the Clarkforkian, believed that the evidence was too inconclusive to specify the direction of dispersal between North America and Asia. Other authors (e.g., Russell and Dashzeveg, 1986) have noted that early Eocene mammals of Asia do not show marked endemism but, rather, that they show close affinities with North American and European forms. According to Sloan (1987), times of Paleocene and early Eocene dispersal between the Western Interior and Asia were throughout the Torrejonian, late (but not latest) Tiffanian (Ti4), early (but not earliest) Clarkforkian (Cf2), and earliest Wasatchian (Wa1). There are a number of internal inconsistencies in Sloan's presentation concerning absolute age dates for the boundaries of his faunal zones and the taxa he lists as immigrants at those zonal boundaries. Nonetheless, he suggests that among the Torrejonian through earliest Wasatchian immigrants to North America are *Deltatherium, Pantolambda, Cyriacotherium, Esthonyx,* and *Hyracotherium.* Immigrants to Asia from (or through) North America include mesonychids, mioclaenids, notoungulates (i.e., arctostylopids), and *Prodinoceras.* Sloan (1987, p. 187) claims: "In each of these cases the complete prior evolution of the ancestors of the immigrating species is well documented in the continent from which it emigrated."

Route 8. Africa to the Western Interior via Asia and Beringia

Gingerich (1976, 1977a, 1980a, 1980b, 1981c, 1982b, 1986) speculated that several, perhaps all, of the mammalian faunal elements that arrived in the Western Interior at the beginning of the Wasatchian had their ultimate origins in Africa (Fig. 1, Route 8). These mammals include artiodactyls, perissodactyls, hyaenodontid creodonts, and primates of modern aspect. With regard to the latter, he (1986) suggested that the dispersal route was by way of Arabia, South Asia, and Central Asia to North America and then, from North America, finally, to Europe. He explicitly discounted dispersal directly from Africa to Europe. In any case, no one has suggested a direct dispersal route from Africa to North America across the opening Atlantic (as has been suggested between Africa and South America to account for the appearance of platyrrhine primates and caviomorph rodents in the South American Oligocene; e.g., Hoffstetter, 1972, 1974; Lavocat, 1974, 1980; Tarling, 1980).

Finally, we must note that hypotheses of the biogeographic origins of late Paleocene–early Eocene mammalian faunas of the Western Interior have been, and continue to be, profoundly influenced by biochronologic correlations. The status of the Clarkforkian Land-Mammal Age is an important case in point. Although defined as a land-mammal age by H. E. Wood and others (1941), the validity of the Clarkforkian was questioned by R. C. Wood (1967). It was not until 1980, after extensive collecting in the northern Bighorn Basin by Gingerich and Rose, that it was formally redefined (Rose, 1980, 1981a). During the intervening years (1967 to 1981), many of the hypotheses concerning the origins of North American late Paleocene–early Eocene mammals were put forward. During these years also, many collections of Paleocene and Eocene mammals were made in the People's Republic of China (see Li and Ting, 1983; Russell and Zhai, 1987). These collections, both in North America and in Asia, had, and continue to have, a fundamental influence on changing views concerning the centers of origin of North American late Paleocene–early Eocene mammalian faunas and the timing of biogeographic events.

METHODOLOGY

In assessing the relative merits of previous hypotheses for the biogeographic origins of mammalian immigrants to the Western Interior during the late Paleocene and early Eocene, many lines of evidence can be evaluated. These include (1) aspects of faunal turnover, (2) indices of faunal resemblance, (3) paleogeographic reconstructions, (4) paleoclimatic patterns, and (5) phylogenetic relations. The first of these is relevant to the identification of times at which there were high origination rates of mammalian genera in the Western Interior; it does not specify whether the originations were owing to immigrations from other biogeographic provinces or to in situ evolution. The second line of evidence provides an indication of the relative likelihood of interchange between the Western Interior and particular biogeographic provinces, while the third and fourth provide physical evidence for the potential for interchange. The fifth line of evidence can provide an indication of the direction(s) of dispersal (from or to the Western Interior from other biogeographic provinces). One additional source of information that is of potentially great importance is the paleobiogeographic patterns of the nonmammalian continental biota (see Briggs, 1987). This evidence is not considered in detail here, in part because it is beyond the scope of this study but also because the biostratigraphic and biochronologic data for most continental plants and nonmammalian animals are not as detailed as they are for mammals and, as a consequence, synchroneity cannot be established.

Faunal turnover

This factor is evaluated, in part, to determine if indeed there were significant turnovers in the Western Interior mammalian fauna during the late Paleocene and early Eocene, specifically at the Torrejonian-Tiffanian, Tiffanian-Clarkforkian, and Clarkforkian-Wasatchian boundaries. Such is, of course, to be expected since the boundaries were established on that basis. Nonetheless, such a test is appropriate since the four North American land-mammal ages delimiting these boundaries were named and characterized more than 45 years ago (Wood and others, 1941). In light of the considerable amount of new information regarding each of these ages, which has resulted in their subdivision into a series of faunal zones (e.g., Gingerich, 1975, 1976, 1983a; Sloan, 1987; Archibald and others, 1987; Krishtalka and others, 1987), it is important to determine if indeed the magnitudes of the turnovers between land-mammal ages are greater than those between the faunal zones that lie entirely within a land-mammal age. If the turnovers between North American land-mammal ages are greater than within land-mammal ages, then one can proceed to determine the cause(s) of the turnovers. If the turnovers appear to be owing to a high number of originations, the possibility of simultaneous immigration of taxa must be considered. However, whether the high originations are owing to in situ evolution or to immigration from other biogeographic provinces can only be determined by examining additional lines of evidence (see below).

Indices of faunal resemblance

Faunal resemblance indices (FRIs) can serve as a retrospective indicator of the *relative* likelihood of biotic interchange between different biogeographic provinces. Where possible and relevant, we have calculated FRIs to determine overall similarity between approximately coeval Paleocene and early Eocene mammalian faunas in the Western Interior and those in contiguous biogeographic provinces. It must be stressed that this does not indicate *direction* of dispersal, but is only a quantitative indication of the relative degree of interchange that might have taken place.

Paleogeography

Paleogeographic data also can be used to provide an indication of the relative likelihood of biotic interchange between biogeographic provinces. Recent geophysical studies have resulted in a plethora of paleogeographic reconstructions of continental positions. These serve to provide a data base for the assessment of interchange across land bridges (whether physically continuous/ discontinuous or temporally intermittent/persistent), between North America and neighboring landmasses such as South America, Europe, and Asia at particular times during the late Paleocene and early Eocene. Also, paleogeographic reconstructions that include potential barriers to dispersal, such as epicontinental seas or mountain ranges, need to be considered.

Paleoclimate

Paleoclimatic data also can be utilized to evaluate the likelihood of interchange between biogeographic provinces. For instance, according to southern New World origin hypotheses, there should be evidence for a climatic cooling event beginning approximately at the Torrejonian-Tiffanian boundary that might have resulted in a southward shift of faunas. There also should be evidence for a climatic warming event beginning in the Clarkforkian (or perhaps the late Tiffanian) and continuing into the Wasatchian that allowed southern New World elements to disperse northward again into the Western Interior. In the case of interchange between either North America and Europe or between North America and Asia, there should be evidence for equable climates to allow passage across high-latitude land bridges.

Phylogenetic relations

An assessment of phylogenetic relations, if the fossil record is adequate to the task, is necessary to determine whether the new taxa that appear in the Western Interior were derived from taxa that lived earlier in the same region or from taxa that lived earlier in another biogeographic province. If the latter is the case, then an indication of the direction of dispersal is provided. One of the most serious limitations of this methodology lies in the correlation of strata in different biogeographic provinces. Intercontinental (and inter-regional) age correlations during the Paleocene and early Eocene are poorly known; they often are based on the dispersing mammals themselves; thus, the probability of circular reasoning is high. As recently stated by Wing (1984, p. 441): "Too often paleobiologists are forced into the logically tenuous position of studying purportedly diachronous events in the same group of organisms that they use for correlation." For instance, if the first appearances of the same taxon (or taxa) in two different areas is used to provide an age estimate for one of the areas, then that taxon cannot be used as evidence for the direction of dispersal between the two areas. In this study we have attempted to review, where possible, the most recent literature concerning the ancestry (or at least more primitive sister group) of each of the supraspecific mammalian taxa that appear in the Western Interior at the beginnings of the Tiffanian, Clarkforkian, and Wasatchian Land-Mammal Ages.

RESULTS

Faunal turnover

We have compiled appearance and disappearance rates of mammalian genera for each of 15 faunal zones of the Puercan, Torrejonian, Tiffanian, and Clarkforkian Land-Mammal Ages, as well as for the first two zones of the Wasatchian Land-Mammal Age (Fig. 2). These zones generally conform to those recognized recently by Archibald and others (1987; see also Sloan, 1987) except that four zones, rather than only three, are recognized here for the Torrejonian. The last Torrejonian zone (To3) of Archibald and others (1987) is here divided into two (To3 and To4), as was done previously by Gingerich (1975, 1976, 1983a). Utilization of the latter scheme allows for more consistency in *types* of zones for the time interval considered primarily here (late Torrejonian through earliest Wasatchian). Almost all of the zones are lineage zones (To3 of Archibald and others, 1987, is an interval zone), where the successive first appearances of taxa defining each of the zones are thought to form parts of phylogenetic lineages. In contrast, the three zones of the Puercan Land-Mammal Age and the first two of the Torrejonian conform to the definition of interval zones in that they are defined and limited by the successive first appearances of unrelated taxa. The last two Torrejonian zones, all of the Tiffanian zones, the zone straddling the Tiffanian-Clarkforkian boundary (Ti6-Cf1), the second Clarkforkian zone, and the first two Wasatchian zones (Wa0 and Wa1 of Gingerich, 1983a, 1986, 1987) are lineage zones. The lineage zone straddling the Tiffanian-Clarkforkian boundary has been divided into two interval subzones, Ti6 and Cf1, by Archibald and others (1987). The third and youngest Clarkforkian zone is an abundance, or acme, zone.

We have calculated appearance and disappearance rates (number of mammalian genera per million years) for each of the 16 zones (zones Wa0 and Wa1 are combined since only a preliminary list for the Wa0 fauna is available; Gingerich, 1982a, 1986, 1987), despite the fact that the average longevity of each genus is considerably longer than one faunal zone. Harper (1975) has pointed out correctly that rates based on such a relationship may not be reliable. However, this criticism does not affect our analysis appreciably because the zones do not differ greatly in length. Using the time scale of Berggren and others (1985) and assuming equal durations for zones within each land-mammal age, those in the Puercan are estimated to be 0.6 Ma, in the Torrejonian 0.7 Ma, in the Tiffanian and Clarkforkian 0.55 Ma, and in the Wasatchian 0.8 Ma. Although Sloan (1987) has provided a detailed report on absolute ages of Paleocene land-mammal ages, his scale (based on absolute ages reported by Lowrie and Alvarez, 1981) is not adopted here. In part this is because his report is restricted to the Paleocene (i.e., no age estimates are provided for zones Cf2, Cf3, Wa0, and Wa1), but also because his age estimates for Paleocene faunal zones are based on relative rates of sedimentation, which involve many assumptions, and because of inconsistencies between the geomagnetic time scale he used and what we believe to be the more widely adopted Decade of North American Geology (DNAG) scale utilized by Berggren and others (1985).

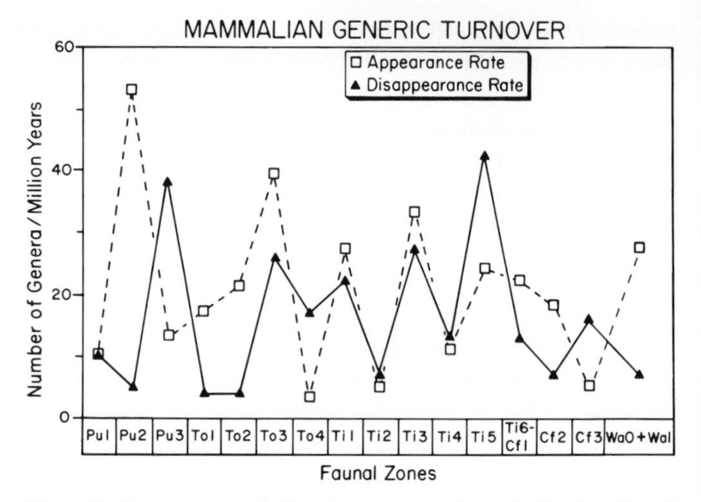

Figure 2. Appearance and disappearance rates for early Tertiary faunal zones. Rates calculated as number of genera per million years. Puercan through Clarkforkian faunal zones from Archibald and others (1987); Wasatchian faunal zones from Gingerich (1983a, 1986). Time scale from Berggren and others (1985).

The heights of the peaks and valleys of the rate curves must be evaluated independently for each zone in light of completeness estimates for individual faunal zones (Table 1). For instance, the high disappearance rate in zone Pu3 may be an artifact since the completeness estimate for zone To1 is only 67 percent compared to an average for the first 14 zones (Pu1 through Cf3) of 83 percent. By comparison, the completeness estimates in zones Pu2 and Pu3 are 93 and 84 percent respectively, indicating that some confidence can be placed in rate estimates for disappearances in Pu2 and rate estimates for appearances in Pu3.

The compilations of generic first and last appearances include range extensions provided by our discoveries in the Crazy Mountains Basin, south-central Montana, and therefore differ slightly from those of Archibald and others (1987). The data reveal several interesting patterns, some of which do not pertain directly to this analysis and which we will present elsewhere; the pattern most relevant to this study is that of appearance and disappearance rates. A theoretically expected pattern obtains: appearance rates are roughly equivalent to or greater than disappearance rates for all of the 16 faunal zones except four. These four zones, in which disappearance rates are considerably greater than appearance rates, are Pu3, To4, Ti5, and Cf3, that is, the last faunal zone of each of the four land-mammal ages. At the beginning of each of the land-mammal ages, appearance rates are invariably greater than disappearance rates. In other words, the land-mammal age boundaries are the only places where there is a crossing over from relatively high disappearance rates to relatively high appearance rates, precisely the times at which one would expect significant faunal turnover. That being the case, one

TABLE 1. GENERIC COMPLETENESS ESTIMATES FOR FAUNAL ZONES OF THE PUERCAN, TORREJONIAN, TIFFANIAN, AND CLARKFORKIAN LAND-MAMMAL AGES

	Total genera reported for zone*	Genera known before and after but not during zone	Estimated total genera†	Completeness estimates
Pu1	15	4	19	79
Pu2	42	3	45	93
Pu3	41	8	49	84
To1	26	13	39	67
To2	36	15	51	71
To3	71	3	74	96
To4	43	15	58	74
Ti1	53	7	60	88
Ti2	37	14	51	71
Ti3	63	2	65	97
Ti4	44	12	56	79
Ti5	59	3	62	95
Ti6-Cf1	41	11	52	79
Cf2	51	4	55	93
Cf3§	43	11	54	80

Note: Completeness estimates for each faunal zone were calculated according to the following formula: Total number of genera/estimated number of genera X 100.

*Includes genera known before, during, and after the faunal zone, first appearances, and last appearances.

†Includes genera known before, during and after the faunal zone, first appearances and last appearances, and genera known before and after but not during the faunal zone.

§Cf3 completeness estimates calculated using Wasatchian generic occurrences from Archibald and others (1987).

can proceed to test whether those taxa that appear at the beginnings of the North American land-mammal ages are owing to in situ evolution or to immigration from other biogeographic provinces.

Indices of faunal resemblance

Age correlations of Paleocene and early Eocene strata among holarctic continents are not firmly established; they are particularly bedeviled by the lack of independently corroborated data (Berggren and others, 1985). Correlations between North America and Europe are based on different types of data (e.g., marine plankton, terrestrial mammals, pollen), which, in most cases, do not agree (Savage and Russell, 1983; Wing, 1984; Berggren and others, 1985). Correlations between Paleocene faunas of North America and Asia (and between Europe and Asia) are based almost entirely on fossil mammals; there are as yet no adequate marine data to test mammal-based correlations independently (Russell and Zhai, 1987). These are serious and limiting problems that will be considered in the discussion section. For the present purpose of broad-level comparisons between Paleocene and early Eocene faunas, the following represents our understanding of tentative and approximate correlations of land-mammal ages or "ages" and, in some cases, particular faunas, between the three northern continents. Wasatchian (North America), Ypresian (Europe), and Bumbanian (Asia) ages appear to be roughly equivalent (Savage and Russell, 1983; Russell and Zhai, 1987); at the very least, they overlap significantly. The Dormaal Reference Level faunas (early Sparnacian) of Russell and others (1982b), however, are considered by Gingerich (1976), Gingerich and Rose (1977), Rose (1980), and Godinot (1981, 1982) to be equivalents of the late Clarkforkian of North America (but see Conclusions section where we argue that the Dormaal Reference Level faunas are probably early Wasatchian equivalents). The Clarkforkian, at least the early Clarkforkian, is considered to be equivalent to the late Cernaysian (Europe) and to the late Nongshanian (Asia) (Rose, 1980; Savage and Russell, 1983; Archibald and others, 1987; Russell and Zhai, 1987). The late Tiffanian (North America) appears to be equivalent to the early Cernaysian (Rose, 1980; Savage and Russell, 1983; Archibald and others, 1987). An early or middle Tiffanian equivalent appears to be represented by the Walbeck fauna of Germany (Gingerich, 1976; Savage and Russell, 1983; Archibald and others, 1987; Sloan, 1987). The early Nongshanian appears to be equivalent to the late Tiffanian (Russell and Zhai, 1987; Sloan, 1987). The only Torrejonian (North America) equivalent in Europe appears to be represented by the Hainin fauna of Belgium (Savage and Russell, 1983; Archibald and others, 1987). Finally, the Shanghuanian (Asia) appears to be equivalent to the Torrejonian and early Tiffanian (Russell and Zhai, 1987; Sloan, 1987).

Table 2 provides a tabulation of Faunal Resemblance Indices (FRIs) for genera and families contained in composite and, in some cases, single faunas of Paleocene through early Eocene age from North America, Europe, and Asia. FRIs are calculated as a percentage of taxa (usually genera or families) shared between two faunas divided by the total number of taxa in the smaller fauna. FRIs between any of the northern continents and Central America cannot be calculated simply because no mammals of Paleocene or early Eocene age are known from Central America. Similarly, FRIs between any of the northern continents and South America were not calculated since it appears that the Didelphidae was the only family shared between South and North America during this interval (if the Peradectini are elevated to familial status, as suggested by Marshall [1987], then they too are represented on both continents). No known mammalian genera were shared between South and North America during the Paleocene and early Eocene (Savage and Russell, 1983).

Composite faunas for North America were compiled from Archibald and others (1987) and Stucky (unpublished data), for Asia from Russell and Zhai (1987), and for Europe from Savage and Russell (1983), the latter being the same source employed by Flynn (1986) in his calculations of FRIs for early Eocene mammalian faunas. The Hainin fauna was compiled from Vianey-

TABLE 2. FAUNAL RESEMBLANCE INDICES (FRI) FOR PALEOCENE–EARLY EOCENE COMPOSITE FAUNAS BETWEEN HOLARCTIC CONTINENTS

North America/Europe	Hainin	Walbeck	Cernaysian	Dormaal*	Sparnacian
Wasatchian	....	....	....	54/88	52/90
Clarkforkian	....	....	16/65	29/62	31/60
Tiffanian	....	23/88	24/76	....	....
Torrejonian	28/72	....	....	....	....

North America/Asia	Shanghuanian	Nongshanian	Bumbanian
Wasatchian	....	....	30/71
Clarkforkian	....	9/22	....
Tiffanian	....	7/26	....
Torrejonian	0/18	....	....

Europe/Asia	Shanghuanian	Nongshanian	Bumbanian
Sparnacian	....	....	22/47
Dormaal*	....	....	18/47
Cernaysian	....	4/6	....
Walbeck	....	0/0	....
Hainin	0/0	....	....

Note: Numerator represents index based on genera; denominator represents index based on families.
*Dormaal includes the reference level faunas of Dormaal, Erquelinnes, Meudon, Pourcy, Palette, Rians, Kyson, and Silveirinha.

Liaud (1979), Sudre and Russell (1982), Crochet and Sigé (1983), and Savage and Russell (1983); the Walbeck fauna from Savage and Russell (1983); and the Dormaal Reference Level fauna from Russell and others (1982b), with additions from Antunes and Russell (1981), Godinot and others (1978, 1987), Hooker and Insole (1980), and Godinot (1981, 1984). Only comparisons indicated by the tentative age correlations (e.g., Sparnacian with Bumbanian and not Sparnacian with either Shanghuanian or Nongshanian) are shown in Table 2. In instances where overlap between ages or faunas is a possibility or probability (e.g., Sparnacian with both late Clarkforkian and Wasatchian or Nongshanian with both late Tiffanian and early Clarkforkian), FRIs were calculated to account for each comparison and are shown in the table.

Despite imprecision in correlation, a more or less consistent pattern is revealed by comparison of the North American, European, and Asian mammalian faunas. Whether at the familial or generic level, throughout the Paleocene and early Eocene, North American and European faunas were more similar to each other than were North American and Asian faunas, which in turn, were more similar to each other than were European and Asian faunas. This can be taken as an indication of the relative likelihood of dispersal among the three holarctic continents at roughly equivalent times during the Paleocene and early Eocene. The mammalian faunas of Europe and Asia appear to have been almost completely isolated from each other during the Paleocene; the only taxon that appears to have been indisputably shared between the two continents is the mesonychid *Dissacus.* In the early Eocene, a sizeable proportion of the mammalian fauna was apparently shared between Europe and Asia, but whether this was owing to direct dispersal between the two continents across or around the Obik Sea, as suggested by Savage and Russell (1983), or whether North America served as an intermediate corridor cannot be determined from these data.

Interestingly, these data have some bearing on a suggestion by Gingerich (1985, p. 124), who viewed the early and middle Paleocene as an interval of worldwide cosmopolitanism in mammalian faunas, the late Paleocene as an interval of endemism, and the early Eocene as an interval of reestablished cosmopolitanism. There are no definitely identified early Paleocene mammalian faunas from continents other than North America, but the FRIs calculated here (Table 2) suggest that, at least among northern continents, the middle Paleocene faunas generally were not as similar to one another as were late Paleocene faunas. Compared to the late Paleocene, the early Eocene faunas were indeed relatively cosmopolitan in distribution. These data therefore tentatively suggest a trend for increasing cosmopolitanism among northern continents from the middle Paleocene to the early Eocene.

Finally, although FRIs are not calculated here because of limited data, an assessment can be made of the similarities of the few known Paleocene Mammalia of the southeastern United States with those from elsewhere. This can be used as a test of Schiebout's hypothesis that the United States Southeast was isolated from the rest of the continent during most of the Paleocene by the Cannonball Sea, was faunally more intimately connected with Europe, and potentially served as a center of origin for later mammals of the Western Interior. Only two localities in the southeastern United States have yielded Paleocene mammals. Simpson (1932) described a remarkably complete specimen of a new species of the periptychid "condylarth" genus *Anisonchus,* recovered from an oil well drill core in Louisiana; he tentatively ascribed a Torrejonian age to the locality. *Anisonchus* is elsewhere known only from the Western Interior. Periptychids also have been reported from the Paleocene of Asia (Russell and Zhai, 1987) but their reference to that family is questionable (Russell, personal communication, 1988). No periptychids of any kind are known from the Paleocene of Europe (Savage and Russell, 1983), although an early Eocene periptychid from England has been described (Hooker, 1979). Schoch (1985) recently described a small collection of late Paleocene mammals from South Carolina. The only identifiable specimens are of the taeniodont *Ectoganus gliriformis lobdelli* and *Mingotherium,* a new genus placed in a new family whose relationships to other eutherians are obscure. *Ectoganus gliriformis lobdelli* elsewhere is known only from the Tiffanian and Clarkforkian of the Western Interior and taeniodonts as a whole are restricted to North America (Schoch, 1986). Schoch (1985, p. 8) speculated that *Mingotherium* "may have been part of a late Paleocene–early Eocene dinoceratan/xenungulate/pseudictopid/*Mingotherium* (Uintatheriamorpha) radiation in Asia and the Americas." Whatever the affinities of *Mingotherium,* the shared presence of *Anisonchus* and *Ectoganus gliriformis lobdelli* suggests that the strongest affinities of the Paleocene mammalian faunas of the U.S. Southeast were with the Western Interior, and not with Europe as Schiebout predicted. More Paleocene mammalian faunas from the southeastern United States are clearly necessary to confirm or refute this tentative conclusion.

Paleogeography

McKenna (1975a, 1983a, 1983b), Adams (1981), Barron and others (1981), Savage and Russell (1983), Ziegler and others (1983), Parrish (1987), and Briggs (1987), among others, have recently provided summaries of Late Cretaceous and early Tertiary paleogeography, and the following is distilled largely from those works, except as noted. For reference, a simplified Lambert equal-area paleogeographic map of late Paleocene–early Eocene continental positions and distributions of epicontinental seas is presented in Figure 1.

North America was connected to Europe during the late Paleocene and earliest Eocene via two land bridges, one attached to Europe north of the Baltic Sea (the DeGeer Bridge) and the other attached south of it (the Thulean Bridge). The DeGeer Bridge may have provided a potential dispersal route during the Danian (early/middle Paleocene) as well. However, the European end of the route, Fennoscandia, may have been isolated from the more southerly portions of Europe by marine barriers (McKenna, 1983b). The Thulean Bridge would presumably have provided a more direct connection between western Europe and North America, albeit for a more limited duration (late Paleocene to early Eocene). The land connection between Europe and North America began to separate during the early Eocene.

At least one land bridge, across the Bering Straits, also connected North America with northeastern Asia and probably was emergent throughout the Paleocene, in fact, throughout most of the Cenozoic. McKenna (1983a) argued that the paleolatitude of this region (Beringia) was higher than the land bridges between North America and Europe. He (1983a, p. 464) therefore concluded "that in general the North Atlantic connections would have been more hospitable than was the case at the north end of the Pacific" (but see Parrish, 1987). An estimated late Paleocene paleolatitude of 83° ± 9°N has been reported from the central Alaska Cantwell Formation (Hillhouse and Gromme, 1982), which must have been part of or adjacent to the Beringia corridor. It is estimated that the Eocene paleolatitude of Ellesmere Island, at the North American end of the DeGeer Route, was 75.9° ± 3.5°N (McKenna, 1983a). The more southerly North Atlantic connection, the Thulean Bridge, was at approximately 65°N (McKenna, 1983b).

Connections between North and South America are not well documented and are the subject of much current controversy (e.g., papers in Bonini and others, 1984). Most authors have suggested that the proto–Greater Antilles were situated between the two continents in the Late Cretaceous but that they then moved northeastward, creating a large gap. Anderson and Schmidt (1983) and Donnelly (1985), however, suggested that, rather than forming a proto–Greater Antilles, Central America remained more or less in situ as North and South America moved away from one another. In any case, any dispersal of mammals between the two continents during the early Tertiary seems to have taken place by sweepstakes; no major corridor was utilized until the Great American Biotic Interchange in the late Pliocene.

The hypothesis that the southeastern United States may have been separated biogeographically from the Western Interior during the Paleocene by a northward extension of the Gulf of Mexico, the Cannonball Sea (Schiebout, 1979, 1981), is not supported by additional data from marine fossil invertebrates (Lemke, 1960; Stanley, 1965; Cvancara, 1966; Feldman, 1972; Marincovich and others, 1985). The strongest similarities of Cannonball Formation invertebrates appear to lie with European and particularly Arctic Ocean Paleocene forms rather than with species from the nearer Midway Group of the Gulf of Mexico. The Cannonball Sea is therefore considered by Marincovich and others (1985) to have been a southward extension of the early Tertiary circumboreal sea rather than a northward extension from the Gulf of Mexico. If true, there is little or no current paleogeographic support for the concept of a southeastern United States refugium in which Paleocene mammals evolved in isolation east of a marine barrier, separating it from the Western Interior.

Europe and western Asia were separated from eastern Asia during the late Paleocene and throughout the Eocene by the Obik Sea (Turgai Straits), which transgressed from the north and joined the Arctic and Tethys Seas. The Obik Sea regressed in the early Oligocene. The Tethys Sea lay north of Africa, Arabia, and the Indian subcontinent and south of Europe and Asia. Recent paleomagnetic evidence suggests that the Indian subcontinent collided with Asia in the early Eocene (e.g., Curray and others, 1982; Allegre and others, 1984; Besse and others, 1984; Patriat and Achache, 1984), considerably earlier than dated by most workers previously.

A direct connection between Europe and Africa was present in the Late Cretaceous (Burchfiel and Royden, 1982) but possibly not again until the Oligocene or Miocene (Berggren and Hollister, 1977). Nonetheless, several workers (Cappetta and others, 1978, 1987; Hartenberger and others, 1985; Gheerbrant, 1987) have argued that late Paleocene and early Eocene mammals with Euramerican affinities from north Africa indicate that dispersal between the two continents was possible during this interval. Southern Europe and northern Africa were closely approximated to one another during the Paleocene and Eocene (e.g., Azzaroli, 1981; Dercourt and others, 1985), and intervening islands may have served as stepping stones (for interpretations of possible routes, see Gheerbrant, 1987). Europe itself was little more than a series of islands at this time, highly subject to eustatic sea-level changes (Aubry, 1985). Current geophysical evidence alone, however, cannot yet serve as a decisive factor in evaluating the potential for dispersal between Africa and Europe, between Africa plus Arabia and Asia, or between Africa plus Arabia and the Indian subcontinent.

Paleoclimate

Paleoclimatic data for the late Paleocene and early Eocene pertain directly to several of the biogeographic hypotheses for the origin of Clarkforkian and Wasatchian mammalian faunas. First, climatic effect has a major role in explanations for the hypothesized southward movement of mammals at the Torrejonian-Tiffanian transition, and their subsequent northward dispersal in the late Tiffanian through early Wasatchian (e.g., Gingerich, 1976, 1985). Second, climatic evidence is important to assess the plausibility of northern dispersal corridors, between North America and Europe as well as between North America and Asia, at the Tiffanian-Clarkforkian and Clarkforkian-Wasatchian transitions (e.g., Godinot, 1982).

Data from paleobotany (e.g., Dorf, 1964; Wolfe and Hopkins, 1967; Wolfe, 1978, 1980, 1985, 1987a, 1987b; Hickey, 1977, 1980), marine stable oxygen isotope ratios (e.g., Savin, 1977; Buchardt, 1978; Wolfe and Poore, 1982; Shackleton, 1984; Boersma, 1984), distributions of calcareous plankton (Haq and others, 1977), and palynology (e.g., Krutzsch, 1967; Leffingwell, 1971; Boulter, 1984) have been used to infer Paleocene and early Eocene climates. The various data sets are in general agreement that there was a trend from warm temperatures and tropical or subtropical climates in the middle Paleocene, to relatively cool temperatures and temperate climates at the beginning of the late Paleocene, and then a return to warm, subtropical climates commencing in the latter part of the late Paleocene and continuing into the early Eocene (for a recent brief summary see Gingerich, 1985). From leaf floras, it can be inferred that the cooler period was also associated with a greater mean annual temperature range, and thus less equable, more seasonal climates (Wolfe, 1978). In addition, Crowley and others (1986) have suggested that climatic conditions characterized by pronounced seasonal fluctuations in temperature prevailed during much of the early Cenozoic, with the exception of the early Eocene. The issue of interpretation of broad climatic trends from paleobotanical data is complicated, however, by the suggestion that the predominantly deciduous character of some, particularly northern, Paleocene megafloras may in fact reflect a brief low-temperature interval near the Cretaceous-Tertiary boundary (Wolfe and Upchurch, 1986). This low-temperature excursion may have decimated the mesothermal, broad-leaved evergreen flora of the Northern Hemisphere; thus, the deciduous vegetation throughout the Paleocene could simply reflect in part the previous history of the flora rather than an overall decline in temperatures that continued throughout most of the Paleocene (Wolfe and Upchurch, 1986; Wolfe, 1987a, 1987b). Reports of regional variation in late

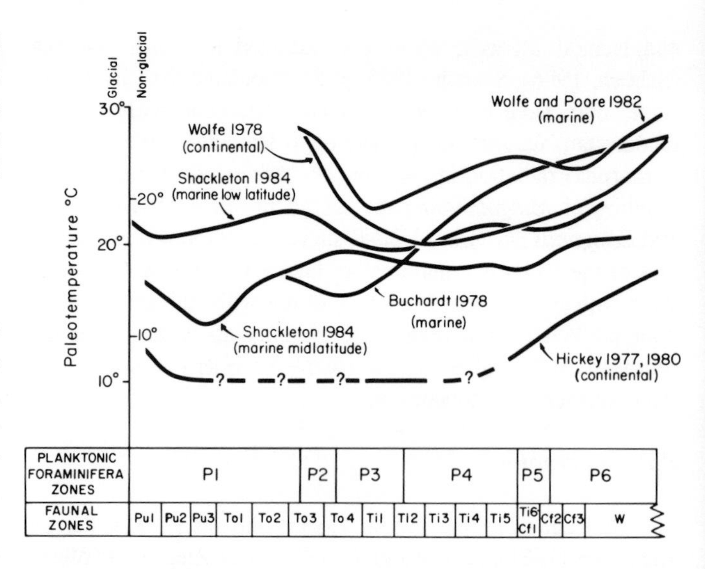

Figure 3. Early Tertiary paleotemperature curves, based on stable oxygen isotope ratios and paleobotanical data. Time scale from Berggren and others (1985).

Paleocene vegetation and climate further complicate paleoclimatic interpretations (e.g., Wolfe, 1985; Wing, 1987).

There is, in fact, disagreement among the various data sets about the precise timing of the coolest interval of the middle and late Paleocene, as measured by mean annual temperature (Fig. 3; correlations between continental and marine zones are provisionally taken from Berggren and others, 1985). Wolfe and Poore (1982) presented oxygen isotope data from Atlantic deep sea cores that show the cooling trend beginning early in plankton zone P2 and the warming trend beginning in the middle of plankton zone P3. According to Berggren and others (1985), these time periods coincide with the middle Torrejonian and the earliest Tiffanian, respectively. This is in general agreement with the temperature trend shown by North Sea continental shelf data (Buchardt, 1978), which indicate that the cool interval began in the Danian, between plankton zones NP4 and NP6 and that warming commenced by the early Thanetian. These intervals correlate approximately with the middle Torrejonian and the middle Tiffanian, respectively (Berggren and others, 1985). Similarly, Pacific Ocean low-latitude surface temperatures, compiled by Shackleton (1984) from a variety of sources and plotted on the time scale of Berggren and others (1985), show a cooling trend beginning approximately at 64 Ma and a warming trend beginning approximately at 61 Ma. These ages correlate with the early Torrejonian and early-to-middle Tiffanian (Berggren and others, 1985). However, midlatitude Atlantic Ocean surface temperatures, also compiled by Shackleton (1984) from a variety of sources, show the cool interval beginning later and extending later. Wolfe's (1978) frequently cited paleobotanical data place the coolest interval in the mid-Tiffanian, with the cooling trend beginning in the late Torrejonian. In Hickey's (1980) leaf flora–

based curve, there are gaps in the Torrejonian and in the late Tiffanian. Haq and others (1977) identified a cooler episode approximately at the middle-late Paleocene boundary, based on distributions of calcareous plankton.

The inconsistencies among the different data sets are undoubtedly a reflection of different sources of information. For instance, stable isotopes based on ocean temperatures give an overall view of global climate, whereas leaf floras more directly reflect continental paleotemperatures (Parrish, 1987). The inconsistencies also may reflect the different geographic positions from which the samples were taken. Buchardt's (1978) stable isotope data come from moderate-depth benthonic organisms collected from continental-shelf deposits in the North Sea, Wolfe and Poore's (1982) isotope data are from North Atlantic deep sea cores, Shackleton (1984) presents both South Atlantic and Pacific surface temperature data, Wolfe's (1978) Paleocene leaf flora localities are from the southeastern United States, and Hickey's (1977, 1980) from the Bighorn and Williston basins of the northwestern United States. Only Hickey's data, therefore, are derived from the same geographic area as the mammalian faunas that have been studied from the Western Interior, but Hickey had no data for most of the Torrejonian and for the late Tiffanian, two of the more crucial intervals relevant to this analysis. Furthermore, the floral assemblages analyzed by Hickey apparently are biased in that they represent only streamside vegetation (Wolfe, 1985). Most of the recent leaf flora studies for the Paleocene have focused on biotic changes across the Cretaceous-Tertiary boundary, and provide little detailed information regarding later Paleocene climatic fluctuations (e.g., Wolfe and Upchurch, 1986; Upchurch and Wolfe, 1987). Wolfe and Upchurch (1987) and Wolfe (1985; personal communication, 1987) are developing some leaf flora data from the Powder River Basin and from Alberta, but as yet, this information is preliminary.

There is, fortunately, another potential source of paleoclimatic inference that can be applied to the late Paleocene and early Eocene transitions in mammalian faunas of the Western Interior. Studies of broad-scale species-distribution patterns suggest that differences in mammalian species diversity, both the number of species and their relative abundance, may reflect ecological differences. For instance, in modern mammalian communities, differences in numbers of species have been associated with topographic and latitudinal gradients: there are fewer species in temperate mammalian faunas than in tropical mammalian faunas (e.g., Simpson, 1964; Fleming, 1973; Wilson, 1974; Andrews and others, 1979; McCoy and Connor, 1980). Bats contribute in no small measure to higher tropical—relative to temperate—species richness (Fleming, 1973; Wilson, 1974) but, even when bats are subtracted, a direct relation between latitude and species richness is maintained (e.g., McCoy and Connor, 1980; Eisenberg, 1981). Similar associations have been noted in other vertebrate groups (e.g., MacArthur, 1965; Karr, 1971; Rogers, 1975; Schall and Pianka, 1978). There are some important exceptions, most notably between tropical and temperate Australian faunas; nevertheless, the pattern of increased species richness from temperate to tropical environments has been clearly demonstrated for a variety of extant vertebrates, particularly in the Western Hemisphere (Schall and Pianka, 1978; Eisenberg, 1981). As with species richness, the evenness with which species are distributed in a fauna (i.e., their relative abundances) may also reflect ecological factors. On a global scale, species are more evenly distributed in tropical than in temperature climates (e.g., Pianka, 1983).

The explanation for the general pattern of greater species richness and equability in tropical than temperate environments probably involves interaction of many factors, including area effect, spatial heterogeneity, climatic stability and predictability, diversity and stability of food sources, and historical factors (e.g., Flessa, 1975; Osman and Whitlatch, 1978; McCoy and Connor, 1980; Glanz, 1982; Pianka, 1983; Schum, 1984; Van Devender, 1986). In particular, instability and/or unpredictability of climate or of primary production is associated with lower species diversity, including both richness and relative abundance (Pianka, 1983). Thus, when a fauna is dominated by a single species and other species are extremely rare, this may reflect instability in the community (Hutchinson, 1961). Rose (1981a) also suggested that low equability in the fauna may be associated with environmental unpredictability, although he noted that, in fossil communities, uneven patterns of species abundance may also reflect taphonomic factors (Rose, 1981a, 1981b; and see below). However, the association between environmental factors and animal diversity in extant communities provides the rationale for extrapolation of ecological parameters, such as climatic trends, from diversity patterns of past biological communities.

There has been considerable discussion as to how the concept of diversity is best applied to species assemblages. The use of diversity measures in ecology has been reviewed by Hurlbert (1971), Peet (1974), and Whittaker (1972, 1977), among others. Rose (1981a, 1981b) discussed the use of diversity measures particularly with reference to their application to fossil communities and recommended the use of heterogeneity indices, designed to measure both richness and evenness. Of diversity indices, the heterogeneity index of Simpson (1949), the Shannon-Weaver heterogeneity formulation (e.g., Peet, 1974), and the equitability index of Whittaker (1972, 1977) are most commonly applied (Fig. 4).

For finite samples, the Simpson index is:

$$L = \text{sum of } [n_i (n_i - 1)]/[N(N - 1)]$$

where n_i is the number of individuals in species *i*, and N the total sample size (Peet, 1974). Peet (1974) describes the Simpson index, which measures the probability that two individuals selected at random from a sample will be of the same species, as the weighted mean of proportional abundances.

The Shannon index is based on information theory and reflects the assumption that heterogeneity is a measure of uncer-

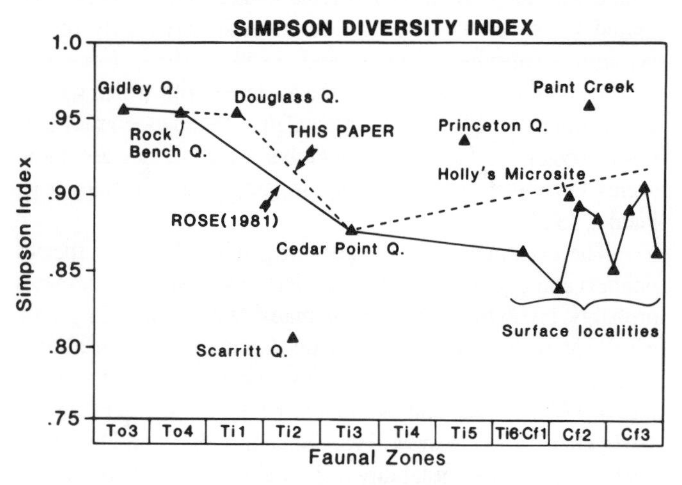

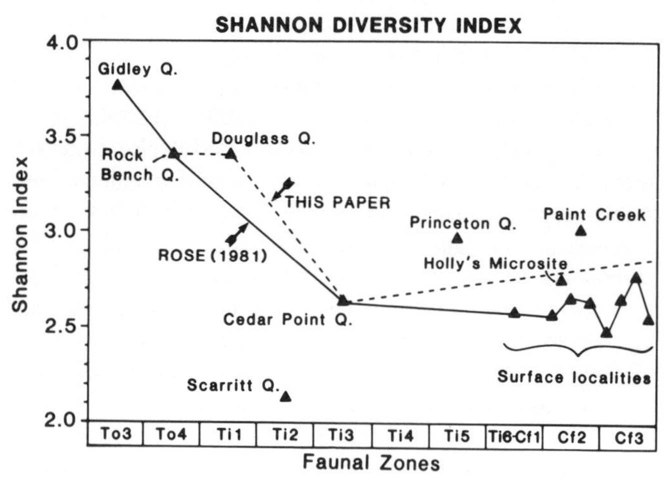

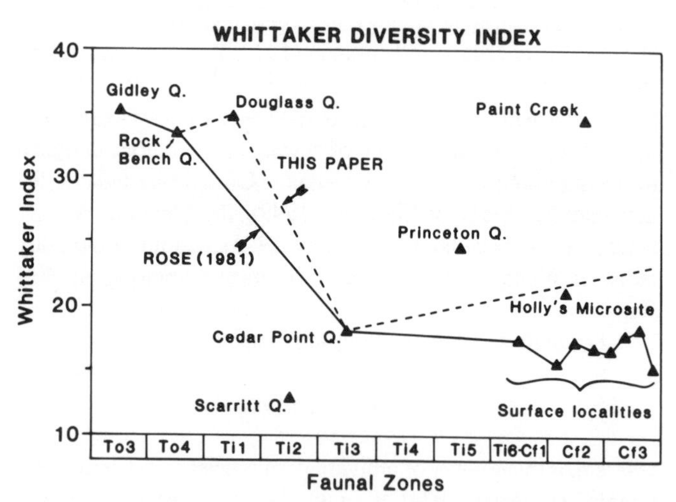

Figure 4. Diversity indices for late Torrejonian (To3) through late Clarkforkian (Cf3) mammalian faunas. Top, Simpson Index; middle, Shannon Index; bottom, Whittaker Index.

tainty regarding the identity of a species selected at random from a population (Peet, 1974). The Shannon index,

$$H' = -\text{ sum of } p_i \text{ Log } p_i$$

where p_i is the frequency of the ith species, is frequently estimated using $h' = -$ sum of (n_i/N) Log (n_i/N), a practice that introduces some additional error (Peet, 1974).

Whittaker (1972, 1977) suggested an alternative measure of diversity that he described as being based more directly on the concept of equitability:

$$E = S/\text{Log } p_1 - \text{Log } p_s$$

where S is the number of species, p_1 the frequency of the most abundant species, and p_s the frequency of the least abundant species.

As noted above, the fossil record presents some unique difficulties in terms of diversity analysis (e.g., Grayson, 1973, 1978; Graham, 1976, 1986; Lasker, 1976; Rose, 1981a, 1981b). Many of these reflect input from taphonomic factors that may irrevocably alter such information as relative abundance of species. In part because of this, many studies of diversity in the fossil record restrict themselves to examination of species-richness patterns. Rose (1981a, 1981b) discussed potential biases in relative abundance of species, and thus measures of species evenness, associated with different taphonomic processes, collection procedures, and sample sizes. He concluded that analysis of species evenness is valid for fossil assemblages if samples could be shown to be comparable in size, collection methods, and depositional histories. Consequently, in his analysis of faunal change between the middle Paleocene and early Eocene, Rose (1981a, 1981b) concentrated on large, quarried assemblages that he felt were roughly equivalent in taphonomic history. Of necessity, however, he was forced to include what he believed to be less representative faunal assemblages, gleaned from surface localities, to represent certain time periods, and had little data for other periods, particularly the earliest Tiffanian.

Rose's (1981a, 1981b) analysis of changes in species-diversity patterns showed relatively high diversity in the Torrejonian, followed by a decline in diversity in the Tiffanian, and then a return to relatively high diversity in the Wasatchian (Simpson, Whittaker, and Shannon indices). He noted a similar pattern in species richness. He interpreted these patterns as consistent with change from a stable, predictable environment to an unpredictable environment, and back to a stable environment in the early Eocene, based on analogy with modern abundance distributions (MacArthur, 1960; Hutchinson, 1961). Rose argued that mammalian species diversity patterns, when considered in conjunction with paleobotanical evidence (e.g., Wolfe, 1978), could be interpreted as indicative of a change from a subtropical-adapted to warm-temperate–adapted fauna from the Torrejonian to Tiffanian, and back to a subtropical-adapted fauna in the Wasatchian.

Although Rose (1981a, 1981b) had two large samples (Gidley and Rock Bench Quarries) for the late Torrejonian, he had only one (Scarritt Quarry) for the early Tiffanian. He, in fact, cautioned that the sample from Scarritt Quarry was small and probably biased taphonomically and therefore that it did not provide a reliable estimate for species diversity. He suspected that the same situation obtained at the late Tiffanian Princeton Quarry and did not include those two quarries in his inferred climatic curve, which is indicated in Figure 4 by solid lines. Rose also suspected the Clarkforkian surface-collected samples were biased in favor of large mammals. However, lacking any screen-washed or quarried samples, he used the surface locality diversity indices as approximations of paleoclimate. He could not find evidence in mammalian diversity indices to confirm the pronounced climatic warming in the Clarkforkian that Hickey (1977, 1980) and Wolfe (1978) had found and he appropriately noted this discrepancy.

Recent collecting efforts in the Crazy Mountains Basin of south-central Montana have significantly augmented previous collections from Douglass Quarry (Ti1) and Scarritt Quarry (Ti2) (see preliminary faunal lists in Tables 3 and 4). Current collections from these two quarries are now about as large as that previously known from Gidley Quarry (To3) and can be used to fill in some of the gaps in our knowledge of species diversity changes during the Paleocene. In addition, two Clarkforkian localities collected more intensively since Rose's work—Holly's Microsite (a quarry) and Paint Creek (a surface-collected and screen-washed locality)—can be used to assess more accurately Clarkforkian species diversity (see faunal lists in Krause, 1986).

An independent assessment of taphonomic bias can be obtained by a comparison of weighted molar size averages and size ranges (Fig. 5) and weighted distributions of molar size (Fig. 6) for mammalian species from the various localities. Weighted averages for a fauna are computed from the average size for each species, weighted by the number of individuals (MNI) represented for that species. Weighted means, ranges, and distributions for assemblages where it is suspected that taphonomic factors have biased the faunal composition can be compared with assemblages where taphonomic factors are thought not to have appreciably altered the faunal composition. Among Paleocene quarry assemblages, Gidley Quarry (To3), Rock Bench Quarry (To4), and Cedar Point Quarry (Ti3) are from similar depositional regimes, were collected in a similar manner, and were considered by Rose (1981a) to be relatively unbiased representations of faunal composition.

The much larger collection from Scarritt Quarry confirms Rose's suspicion that large species are not adequately represented. The same is true for Princeton Quarry. Douglass Quarry, on the other hand, does not appear to be biased in weighted molar size estimates; its distribution is similar to those from Gidley, Rock Bench, and Cedar Point Quarries. At least one of the two new Clarkforkian sites, Holly's Microsite, appears to be biased for small size. The other, Paint Creek, is a relatively small sample and therefore also may be skewed. On the other hand, the Clarkforkian surface localities are biased, relative to the Paleocene quarry assemblages, in favor of large-sized individuals.

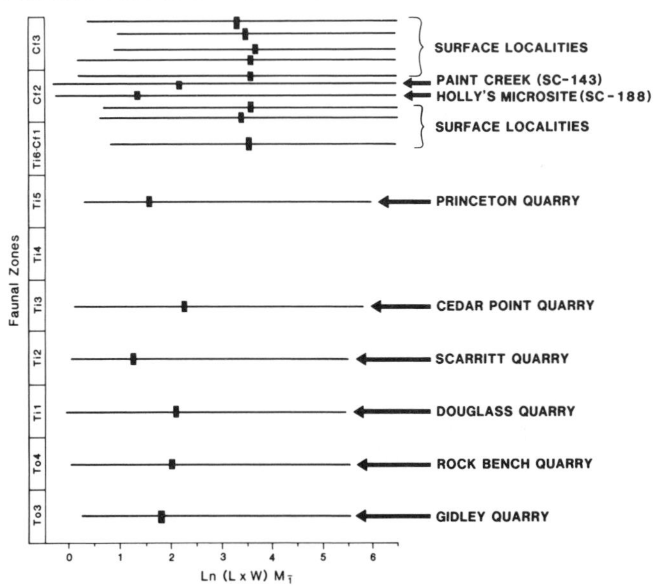

Figure 5. Means (vertical bars) and ranges (horizontal lines) of weighted measures of tooth size in major collections of middle Paleocene through early Eocene mammals in the Western Interior of North America. Gidley, Rock Bench, Cedar Point, and Gidley Quarries consist of coarse-grained, clay-clast conglomerates; Princeton Quarry, Scarritt Quarry, and Holly's Microsite consist of fine-grained mudstones; and Paint Creek is a screen-washed locality that also was intensively surface collected. Note similarities among collections based on similarity in sedimentology and collection history.

Following Rose (1981a, 1981b), we computed three diversity indices for each quarry sample. The sample from Douglass Quarry is very diverse, according to any of the three diversity measures used (dashed lines in Fig. 4). It is as diverse or more so than Rock Bench Quarry in zone To4. Forty-eight species have been identified and these are distributed quite evenly. Another recently well-sampled Ti1 zone locality, Cochrane 2 in Alberta, also contains a very species-rich mammalian fauna (Fox, 1987, and this volume). The species richness of both these localities is consistent with that of modern tropical mammalian faunas, and the species evenness at Douglass Quarry is consistent with general ecological stability. These two localities seem to demonstrate therefore that the pronounced late Paleocene cooling event did not occur at the beginning of the Tiffanian as previously suspected. Furthermore, given that the Cedar Point Quarry sample (Ti3) does not appear to be strongly biased taphonomically (i.e., skewed) and that it shows very low diversity (Rose, 1981a), the cooling event that paleobotanical evidence indicates must have been present somewhere between the early and middle Tiffanian may have been more abrupt than previously thought. It must be cautioned, however, that another recently sampled middle Tiffan-

TABLE 3. PRELIMINARY MAMMALIAN FAUNAL LIST FOR DOUGLAS QUARRY (Ti1), CRAZY MOUNTAINS BASIN, MONTANA

Taxon	Total/MNI	Frequency/MNI
Order MULTITUBERCULATA		
Family PTILODONTIDAE		
Ptilodus new species 1	28/192	0.135
Ptilodus new species 2*	3/19	0.014
Family NEOPLAGIAULACIDAE		
*Mesodma pygmaca**	1/3	0.005
Mimetodon? sp.*	2/2	0.010
Neoplagiaulax hunteri	8/24	0.038
*Neoplagiaulax nelsoni**	1/1	0.005
Neoplagiaulax new species*	7/15	0.034
Family CIMOLODONTIDAE		
Anconodon cochranensis	6/15	0.029
*Anconodon lewisi**	12/29	0.058
Family EUCOSMODONTIDAE*		
*Microcosmodon woodi**	1/3	0.005
Genus and species indet.	2/4	0.010
Order Marsupialia*		
Family Didelphidae*		
Paradactes cf. P. pauli	1/1	0.005
Order PROTEUTHERIA		
Family PALAEORYCTIDAE		
Acmeodon sp.*	4/10	0.019
Palaeoryctidae indet., sp. 1	1/2	0.005
Palaeoryctidae indet., sp. 2*	1/2	0.005
Family PENTACODONTIDAE		
Bisonalyeus browni	16/73	0.077
New genus and species	4/18	0.019
Family PANTOLESTIDAE		
Paleotomus senior	2/7	0.010
Propalaeosinopa diluculi	9./32	0.043
Family LEPTICTIDAE*		
*Myrmecoboides montanensis**	1/2	0.005
*Prodiacodon concordiarcensis**	1/10	0.005
*Prodiacodon furor**	1/4	0.005
Genus and species indet.*	1/1	0.005
Family indeterminate*		
Genus and species indet.*	2/2	0.010
Order LIPOTYPHLA		
Family NYCTITHERIIDAE		
Leptacodon munusculum	3/5	0.014
Order PRIMATES?		
Family PLESIADAPIDAE		
Nannodectes intermedius	10/114	0.048
Plesiadapis praecursor	7/86	0.034
Family PAROMOMYIDAE*		
Ignacius sp*	1/2	0.005
Family CARPOLESTIDAE*		
*Elphidotarsius russelli**	2/8	0.010
Family PICRODONTIDAE*		
*Picrodus silberlingi**	1/1	0.005
Family indeterminate*		
Genus and species indet.*	1/3	0.005
Order DERMOPTERA*		
Family PLAGIOMENIDAE*		
*Elpidophorus elegans**	1/5	0.005
Order CARNIVORA		
Family VIVERRAVIDAE		
Protictis new species	1/1	0.005
Simpsonictis sp*	2/5	0.010
Order MESONYCHIA*		
Family MESONYCHIDAE*		
Dissacus sp.*	1/5	0.005
Order "CONDYLARTHRA"		
Family HYOPSODONTIDAE		
Litomylus dissentaneus	4/16	0.019
New genus and species*	6/13	0.029
Family PHENACODONTIDAE		
Ectocion collinus	20/238	0.096
Phenacodus bisonensis	5/59	0.024
Family ARCTOCYONIDAE		
Claenodon montanensis	3/12	0.014
Claenodon? sp.	2/7	0.010
Colpoclaenus keeferi?	1/5	0.005
Chriacus orthogonius?	2/7	0.010
Chriacus pelvidens	4/21	0.019
Chriacus sp.	5/23	0.024
Mimotricentes fremontensis	4/47	0.019
Thryptacodon cf. *T. demari*	5/39	0.024
Order PANTODONTA		
Family PANTOLAMBDIDAE		
Titanoides sp.	2/12	0.010
Totals	208/1205	1.005

*Taxa added since Krause and Gingerich (1983).

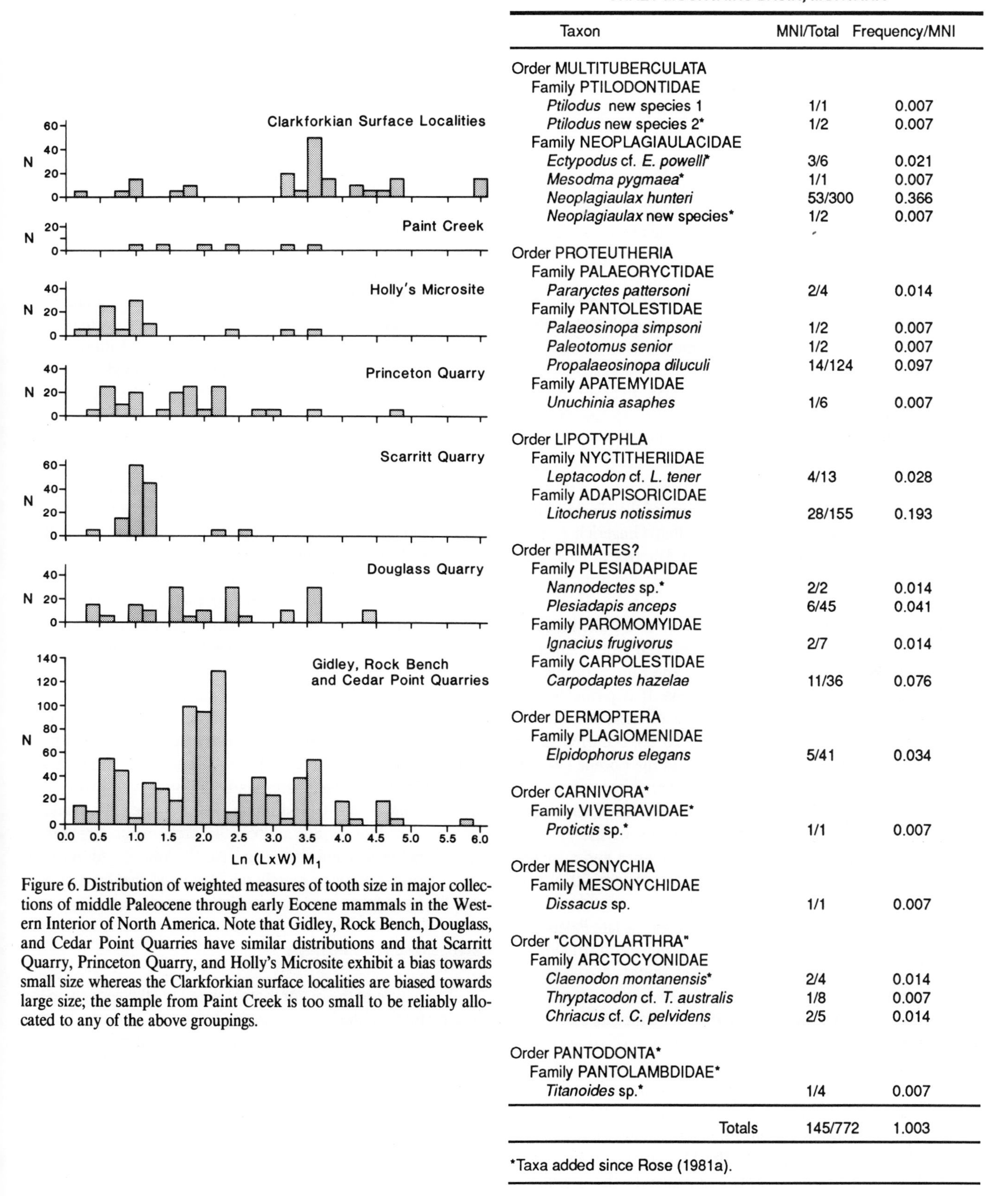

Figure 6. Distribution of weighted measures of tooth size in major collections of middle Paleocene through early Eocene mammals in the Western Interior of North America. Note that Gidley, Rock Bench, Douglass, and Cedar Point Quarries have similar distributions and that Scarritt Quarry, Princeton Quarry, and Holly's Microsite exhibit a bias towards small size whereas the Clarkforkian surface localities are biased towards large size; the sample from Paint Creek is too small to be reliably allocated to any of the above groupings.

TABLE 4. PRELIMINARY MAMMALIAN FAUNAL LIST FOR SCARRITT QUARRY (Ti2), CRAZY MOUNTAINS BASIN, MONTANA

Taxon	MNI/Total	Frequency/MNI
Order MULTITUBERCULATA		
Family PTILODONTIDAE		
Ptilodus new species 1	1/1	0.007
Ptilodus new species 2*	1/2	0.007
Family NEOPLAGIAULACIDAE		
Ectypodus cf. *E. powelli**	3/6	0.021
*Mesodma pygmaea**	1/1	0.007
Neoplagiaulax hunteri	53/300	0.366
Neoplagiaulax new species*	1/2	0.007
Order PROTEUTHERIA		
Family PALAEORYCTIDAE		
Pararyctes pattersoni	2/4	0.014
Family PANTOLESTIDAE		
Palaeosinopa simpsoni	1/2	0.007
Paleotomus senior	1/2	0.007
Propalaeosinopa diluculi	14/124	0.097
Family APATEMYIDAE		
Unuchinia asaphes	1/6	0.007
Order LIPOTYPHLA		
Family NYCTITHERIIDAE		
Leptacodon cf. *L. tener*	4/13	0.028
Family ADAPISORICIDAE		
Litocherus notissimus	28/155	0.193
Order PRIMATES?		
Family PLESIADAPIDAE		
Nannodectes sp.*	2/2	0.014
Plesiadapis anceps	6/45	0.041
Family PAROMOMYIDAE		
Ignacius frugivorus	2/7	0.014
Family CARPOLESTIDAE		
Carpodaptes hazelae	11/36	0.076
Order DERMOPTERA		
Family PLAGIOMENIDAE		
Elpidophorus elegans	5/41	0.034
Order CARNIVORA*		
Family VIVERRAVIDAE*		
Protictis sp.*	1/1	0.007
Order MESONYCHIA		
Family MESONYCHIDAE		
Dissacus sp.	1/1	0.007
Order "CONDYLARTHRA"		
Family ARCTOCYONIDAE		
*Claenodon montanensis**	2/4	0.014
Thryptacodon cf. *T. australis*	1/8	0.007
Chriacus cf. *C. pelvidens*	2/5	0.014
Order PANTODONTA*		
Family PANTOLAMBDIDAE*		
Titanoides sp.*	1/4	0.007
Totals	145/772	1.003

*Taxa added since Rose (1981a).

ian local fauna, from the DW2 Locality in Alberta, is much more species-rich than is Cedar Point Quarry (Fox, this volume). Finally, when species lists and MNI for the new collections from Holly's Microsite and Paint Creek are each combined with those for surface localities of roughly the same stratigraphic level (see Rose, 1981a), the resultant diversity indices are higher than those for the surface localities and Holly's Microsite, but lower than those computed for Paint Creek (although it must be cautioned that the anomalous estimate for Paint Creek may be an artifact of small sample size). Nonetheless, our data suggest that Clarkforkian mammalian diversity probably was greater than previously calculated by Rose (1981a, 1981b). This in turn suggests warmer and/or more predictable climates than revealed by Rose's data, a conclusion that is in more general agreement with Hickey's paleobotanical evidence.

Overall, the paleoclimatic trends do not support the hypothesis that climatic cooling at the beginning of the late Paleocene (Ti1) led to a southward displacement of the mammalian fauna and its replacement by a less diverse temperate-adapted fauna. Furthermore, Clarkforkian mammalian diversity patterns and at least some of the paleobotanical data indicate that the subsequent climatic trend toward warmer and more equable climates began earlier than previously thought. This provides little support for the hypothesis that more diverse, tropical-adapted faunas dispersed northward at the beginning of the Clarkforkian. In fact, it was partly on this evidence that Gingerich (1985) postulated dispersal from South to North America of palaeanodonts, notoungulates, and dinoceratans at an earlier time: the late Tiffanian.

The paleoclimatic evidence also pertains to the likelihood of mammalian dispersal via land bridges between North America and Asia and/or between North America and Europe during the middle-to-late Paleocene and early Eocene. It is probable that broad-leaved evergreen forests, indicative of mesothermal climates (mean annual temperature 13° to 20°C), occurred as far north as 60°N in the Paleocene and 61°N in the Eocene (Wolfe, 1980), and up to 65°N to 70°N during warm intervals of the Eocene (Wolfe, 1985; Spicer and others, 1987). However, except for the late Paleocene–early Eocene connection via the Thulean Bridge, land connections between North America and adjacent continents would have been farther north, between 70°N and 80°N (Crowley and others, 1986, Fig. 3).

High-latitude paleobotanical evidence for the late Paleocene and early Eocene offers few details for habitat reconstruction of the potential Palearctic-Nearctic dispersal routes. A vegetation with no modern analog, termed Polar Broad-Leaved Deciduous Forest, was present in areas north of 70° during the late Paleocene and early Eocene (Wolfe, 1985, 1987a). This vegetation type is typical of floral assemblages from the Paleocene and early Eocene of Ellesmere Island and the early Eocene of Spitsbergen, as well as the Cantwell Formation in central Alaska, and the southern Alaskan Chickaloon Formation. Ellesmere Island and Spitsbergen were part of or adjacent to the DeGeer Route, and the Cantwell and Chickaloon Formations occur in the presumed vicinity of the Beringia route. Wolfe (1985) has suggested that the large leaf size characteristic of the Polar Broad-Leaved Deciduous Forests was in part due to long growing-season days combined with high temperatures.

In addition, it is argued on the basis of high diversity of some late Paleocene Alaskan floras (Chickaloon Formation and lower Tolstoi Formation assemblages) that there was a poleward migration of broad-leaved deciduous taxa from the middle latitudes, particularly along streams (Spicer and others, 1987). If this was the case, these corridors could have provided congenial habitats for dispersal of midlatitude mammals. It should be noted that the paleolatitude of several parts of present-day Alaska is controversial. In particular, it has been argued that terranes containing southern and peninsular Alaskan floras actually were at middle latitudes during the early Tertiary (Stone and Packer, 1979; Stone and others, 1982; Bruns, 1983). However, these arguments have been challenged on the basis of paleomagnetic evidence for Wrangellia and the peninsular area of Alaska (Hillhouse and Gromme, 1983), and geologic and paleontologic evidence for the more eastern Yakutat block (Plafker, 1984; Wolfe and McCoy, 1984).

In any event, it is generally agreed that climate did not constitute an impassable barrier to Paleocene and early Eocene Palearctic-Nearctic mammalian dispersal. Nevertheless, the northerly position of all corridors connecting these regions most likely would have exercised a temperature-induced filtering effect on dispersals of midlatitude mammals (e.g., Godinot, 1982; McKenna, 1983a, 1983b). Because of its more southerly position, the Thulean Bridge, which was in position during the late Paleocene and early Eocene, may have been somewhat more hospitable to certain taxa. This would have enhanced the probability of dispersal between North America and Europe, in comparison to the probability of dispersal between Asia and North America. If, as indicated by paleotemperature curves, climates were cooler during the late Paleocene than the early Eocene, then the more southerly North Atlantic corridor between North America and Europe would have been particularly important for dispersals at the beginning of the Clarkforkian. In any case, the available paleoclimatic data do not address the problem of *direction* of dispersal between continents (e.g., North America to Asia, or Asia to North America), merely the likelihood that land bridges habitable by at least some mammalian taxa did exist.

Phylogenetic relationships

The nature of the major mammalian faunal turnovers in the Western Interior during the middle Paleocene to early Eocene can be discerned, in part, by examining the composition of the mammalian faunas on either side of the turnover boundaries and, particularly, by examining the phylogenetic relationships of the taxa that appear in the first faunal zone after each boundary. A determination of the most likely ancestor (or more primitive sister taxon) of the taxon first appearing after a major faunal turnover boundary potentially can lead to inference of the geographic source of that taxon, whether in the same biogeographic province or a different one. If several unrelated taxa appear to have the same allochthonous center of origin, one can infer that a significant dispersal event occurred at that boundary. The following is an accounting of the published views of various workers on the relationships of each of the supraspecific taxa that appear at the beginnings of each of the late Paleocene through early Eocene land-mammal ages (i.e., Torrejonian-Tiffanian, Tiffanian-Clarkforkian, and Clarkforkian-Wasatchian).

Torrejonian-Tiffanian turnover. Based on data from Archibald and others (1987) and from recent collections from Douglass Quarry in the Crazy Mountains Basin, the supraspecific taxa listed in Table 5 are currently recognized as first appearing in the Western Interior at the beginning of the Tiffanian Land-Mammal Age. All of the higher taxa that first appear in zone Ti1 are genera, whereas both of the later land-mammal age boundaries (i.e., Tiffanian-Clarkforkian and Clarkforkian-Wasatchian) are marked by the first appearances of families and even orders, in addition to genera. The Ti1 first-appearance records for at least two genera, *Pararyctes* and *Diacodon,* are questionable (Archibald and others, 1987). Furthermore, several genera are poorly known or not well-studied (such as *Bisonalveus, Titanoides,* and at least one new genus from Douglass Quarry in the Crazy Mountains Basin). However, those that are represented by good material and have been studied recently appear to be descended from Torrejonian (mostly late Torrejonian) genera known from the Western Interior. Thus, *Plesiadapis* and *Nannodectes* are thought to be derived from *Pronothodectes* (Gingerich, 1976; Sloan, 1987), *Carpodaptes* from *Elphidotarsius* (Simpson, 1928, 1937; Sloan, 1969; Rose, 1975a), *Stelocyon* and *Colpoclaenus* from *Oxyclaenus* (Van Valen, 1978; Sloan, 1987), *Thryptacodon* from *Prothryptacodon* (Sloan, 1969, 1987; Van Valen, 1978), and *Ectocion* and *Phenacodus* from *Tetraclaenodon* or something near it (Radinsky, 1966; Sloan, 1969, 1987; West, 1976; Van Valen, 1978). One does not have to appeal to a different geographic source for these taxa, either from farther north in North America than has previously been sampled, as implied in southern New World origin hypotheses, or from another biogeographic province.

TABLE 5. HIGHER MAMMALIAN TAXA RECORDED AS FIRST APPEARANCES AT THE BEGINNING OF THE TIFFANIAN LAND-MAMMAL AGE

1. *Pararyctes** (Proteutheria, Palaeoryctidae)
2. *Bisonalveus* (Proteutheria, Pentacodontidae)
3. *Diacodon** (Lipotyphla, Erinaceomorpha incertae sedis)
4. *Plesiadapis* (Primates?, Plesiadapidae)
5. *Nannodectes* (Primates?, Plesiadapidae)
6. *Carpodaptes* (Primates?, Carpolestidae)
7. *Stelocyon*† ("Condylarthra," Arctocyonidae)
8. *Colpoclaenus* ("Condylarthra," Arctocyonidae)
9. *Thryptacodon* ("Condylarthra," Arctocyonidae)
10. New genus ("Condylarthra," Hyopsodontidae)
11. *Ectocion* ("Condylarthra," Phenacodontidae)
12. *Phenacodus* ("Condylarthra," Phenacodontidae)
13. *Titanoides* (Pantodonta, Titanoideidae)

*Record questionable–see Archibald and others (1987).
†Although previously reported as being derived from a latest Torrejonian (Gingerich, 1978) horizon, the type and only locality of *Stelocyon* is more likely of earliest Tiffanian age (Butler and others, 1987).

TABLE 6. HIGHER MAMMALIAN TAXA RECORDED AS FIRST APPEARANCES AT THE BEGINNING OF THE CLARKFORKIAN LAND-MAMMAL AGE

1. Order Tillodontia - *Esthonyx*
2. Order Rodentia - *Acritoparamys, Franimys*
3. Family Coryphodontidae (Pantodonta) - *Coryphodon*
4. *Protentomodon* (Pentacodontidae, Proteutheria)
5. *Leipsanolestes* (Insectivora, Erinaceidae)
6. *Planetetherium* (Dermoptera?, Plagiomenidae)
7. *Haplomylus* ("Condylarthra," Hyopsodontidae)
8. *Apheliscus* ("Condylarthra," Hyopsodontidae)
9. *Prosthecion* ("Condylarthra," Phenacodontidae)

Tiffanian-Clarkforkian turnover. Based on information compiled primarily by Gingerich and Rose (e.g., Gingerich and Rose, 1977; Rose, 1980, 1981a) in the Clark's Fork Basin for over a decade, as well as additions from Archibald and others (1987), the higher taxa listed in Table 6 are known to appear at the beginning of the Clarkforkian Land-Mammal Age. The list includes, in addition to several genera, the appearance of two orders (Rodentia, Tillodontia) and one family (Coryphodontidae). Each of these higher taxa were once thought to be derived from North American Torrejonian forms: Rodentia from Plesiadapiformes (Wood, 1962, 1977; McKenna, 1969; Van Valen and Sloan, 1966; Van Valen, 1966, 1971; Szalay, 1968), Tillodontia from the arctocyonid "condylarth" *"Claenodon" procyonoides* or a species very close to it (Van Valen, 1963; see also Sloan, 1969), and Coryphodontidae from ?*Pantolambda* (Cope, 1883; Patterson, 1934; although Gingerich [1976] also cites Simons [1960, p. 6] for this attribution of ancestry, Simons states that "the assumption that they [Coryphodontidae] are descended from any of the known earlier groups of pantodonts is not indicated by present evidence").

Recent discoveries in Paleocene deposits of the People's Republic of China document the presence of a group of mammals, the Eurymyloidea, that most probably includes the ancestors, or near ancestors, of rodents (Li, 1977; Gingerich and Gunnell, 1979; Hartenberger, 1977, 1980; Dawson and others, 1984; Korth, 1984; Li and Ting, 1985; Luckett and Hartenberger, 1985; Novacek, 1985; Dashzeveg and Russell, 1988). It is therefore no longer necessary to suggest, by elimination, that rodents originated in Central America (Sloan, 1969; Gingerich, 1976) or in the southeastern United States (Wood, 1977; Schiebout, 1981). Tillodonts also are now regarded to be Asian in origin (Chow and Wang, 1979; Gingerich and Gunnell, 1979; Sloan, 1987). Although the specific group from which tillodonts were derived has not been determined with any degree of certainty, affinities with, or at least strong similarities to, pantodonts have been suggested (e.g., Chow and Wang, 1979; Cifelli, 1983; Huang and Zheng, 1987; Marshall and de Muizon, 1988). In any case, a diversity of tillodonts has been discovered recently in the Shanghuanian of Asia (Li and Ting, 1983; Russell and Zhai, 1987), and their appearance in North America at a later time suggests a dispersal event from the former to the latter.

The biogeographic origin of North American coryphodontid pantodonts is obscure. Coryphodontids are represented in North America (and Europe) only by *Coryphodon* but were much more diverse in the early Eocene of Asia, where they are represented by *Eudinoceras, Metacoryphodon, Hypercoryphodon, Asiocoryphodon,* and *Heterocoryphodon* (Lucas and Tong, 1988). Pantodonts as a whole are absent in the Cernaysian of Europe (Savage and Russell, 1983) but are very diverse in the Shanghuanian of China (Russell and Zhai, 1987); Asia, therefore, appears to be a potential source for the order (Webb, 1985). So too, however, is South America, as evidenced by the recent discovery of a possible pantodont in purported Late Cretaceous (probably Paleocene) deposits of Bolivia (de Muizon and Marshall, 1987; Marshall and de Muizon, 1988). However, although South America is therefore a potential site of origin for the order as a whole, there is no evidence to suggest that Coryphodontidae per se originated on that continent. According to Lucas (1984; Lucas and Tong, 1988), *Pantolambda* (and *Caenolambda*) is the primitive sister taxon of *Coryphodon,* the most primitive and possibly ancestral coryphodontid. As noted by Patterson (1939), however, *Pantolambda* is so primitive relative to *Coryphodon* that the two forms can be allied only indirectly. Nonetheless, Lucas (1984) supported a North American origin for *Coryphodon* because (1) all known Asian pantodonts discovered since Patterson's (1939) work are even further removed phylogenetically from *Coryphodon,* (2) the oldest record of *Coryphodon* is from the Western Interior, and (3) the most primitive species of *Coryphodon, C. proterus,* is from the Western Interior. Even so, a strictly ancestor-descendant relationship between *Pantolambda* (or *Caenolambda*) and *Coryphodon* is not defensible on present evidence, and no structural intermediates from the fairly well-sampled late Tiffanian of the Western Interior have turned up in the 50 years since Patterson's statement. As such, currently available information does not strongly indicate a clear choice for determining the biogeographic origins of *Coryphodon* and coryphodontids, a view now also espoused by Lucas (personal communication, 1987).

Although Miacidae (sensu Flynn and Galiano, 1982; Gingerich, 1983b) first appear in North America in the middle Clarkforkian (Cf2; Rose, 1980, 1981a) and therefore not at the Tiffanian-Clarkforkian boundary, this may be a sampling artifact since zone Cf1 is relatively poorly known (Table 1). Gingerich (1976) also listed this family as being among the potential wave of immigrants into North America from Central America but more recently he has suggested a European or Asian origin (Gingerich, 1983b). In any case, there is no known potential ancestor for *Uintacyon,* the first known North American miacid, in the Paleocene of the Western Interior.

The other well-established first appearances in the Clarkforkian of the Western Interior, all of them genera, are either too poorly known to permit even speculations concerning ancestry (*Protentomodon, Leipsanolestes, Prosthecion* [this form may not even be a valid genus; see Rose, 1981a; Kihm, 1984]), or are thought to have ancestors in the late Tiffanian of the same region (*Apheliscus, Haplomylus, Planetetherium*). Of the latter category, *Apheliscus* is thought to be derived from *Phenacodaptes* (e.g., Gazin, 1959; Sloan, 1969, 1987; Van Valen, 1978; Rose, 1981a), *Haplomylus* from *Haplaletes* (Van Valen, 1978; Sloan, 1987), and *Planetetherium* from *Elpidophorus* (Sloan, 1969) or something near *Elpidophorus* (Rose, 1975b). The previously suspected links of these Clarkforkian first appearances with either Tiffanian Central American taxa or with Torrejonian forms of the Western Interior therefore appear to be unsupported in light of more recent evidence. Instead, all the Clarkforkian first appearances appear to have been derived from Tiffanian Western Interior and Asian forms, or possibly even European forms.

Clarkforkian-Wasatchian turnover. Based on data compiled by Rose (1981a—50 m level of northern Bighorn Basin and 46 m [150 ft] level of southeastern Bighorn Basin), with additions and modifications from Gingerich (1981b), Bown and Schankler (1982), Gingerich and Rose (1982), Krause (1982), Gingerich (1982a, 1983b, 1986), Gunnell (1985), and Bown and Rose (1987), the supraspecific taxa listed in Table 7 are known to appear in the Western Interior at (or near) the beginning of the Wasatchian Land-Mammal Age (Wa0 and Wa1 faunal zones of Gingerich, 1983a, 1986, 1987). The taxa from both Wa0 and Wa1 faunal zones are included here simply because a complete faunal list for the former was not available for this analysis. However, a preliminary list of the mammalian fauna from one of the two localities included in this zone, UM locality SC-67, was published by Gingerich (1982a): *Meniscotherium priscum, Arfia* sp., *Diacodexis metsiacus, Didymictis* sp., *Ectocion osbornianus, Ectocion parvus, Hyopsodus* sp., *Hyracotherium grangeri, Oxyaena platypus, Paramys* sp., and *Viverravus* sp. To this can be added *Cantius torresi* (Gingerich, 1986) and *Phenacodus* sp. (Gingerich, 1987).

Two orders (Perissodactyla, Artiodactyla), four families (Adapidae, Omomyidae, Hyaenodontidae, ?Didymoconidae), one tribe (Didelphini), and several genera of mammals are present in the early Wasatchian (Wa0 + Wa1) that were not present at Clarkforkian or earlier horizons. (Meniscotheriidae previously also was considered an early Wasatchian indicator taxon [Gingerich, 1982a], but a specimen of *Meniscotherium priscum* was recently discovered at a late Clarkforkian horizon in the Bighorn Basin [Gingerich, personal communication, 1988]). As with the Clarkforkian first appearances, most of the early Wasatchian first appearances were thought to have ancestral links with Torrejonian (or Puercan) forms at the time that Sloan (1969) and Gingerich (1976) proposed a Central American origin for Wasat-

TABLE 7. HIGHER MAMMALIAN TAXA RECORDED AS FIRST APPEARANCES AT THE BEGINNING OF THE WASATCHIAN LAND-MAMMAL AGE (Wa0 and Wa1)

1. Order Perissodactyla - *Homogalax, Hyracotherium*
2. Order Artiodactyla - *Diacodexis*
3. Family Adapidae (Primates) - *Cantius*
4. Family Omomyidae (Primates) - *Teilhardina, Tetonius*
5. Family Hyaenodontidae (Creodonta) - *Arfia, Prolimnocyon, Prototomus, Tritemnodon*
6. Family ?Didymoconidae ("Condylarthra") - *Wyolestes*
7. Tribe Didelphini (Marsupialia, Didelphidae, Didelphinae) - *Peratherium*
8. *Didelphodus* (Proteutheria, Palaeoryctidae)
9. *Apatemys* (Proteutheria, Apatemyidae)
10. *Macrocranion* (Insectivora, Dormaaliidae)
11. *Scenopagus* (Insectivora, Dormaaliidae)
12. *Talpavoides* (Insectivora, Dormaaliidae)
13. *Eolestes* (Insectivora, Erinaceidae)
14. *Parapternodus* (Insectivora, Apternodontidae)
15. *Miacis* (Carnivora, Miacidae)
16. *Microparamys* (Rodentia, Paramyidae)

chian mammals. Thus, perissodactyls were thought to be derivable from the phenacodontid "condylarth" *Tetraclaenodon* or something near *Tetraclaenodon* (Radinsky, 1966; Sloan, 1969; Van Valen, 1978). (Morris [1968] and, more recently, Van Valen [1978] and Sloan [1987] have, in addition, entertained the notion that perissodactyls may have been derived from the arctocyonid "condylarth" *Desmatoclaenus.*) Similarly, artiodactyls were thought to be derivable from arctocyonids like *Tricentes* (or *Chriacus*) (Van Valen, 1971, 1978) or *Thryptacodon* (Sloan, 1969), adapids and omomyids from primitive plesiadapiforms (e.g., Sloan, 1969; Gingerich, 1976), hyaenodontids from palaeoryctids (Van Valen, 1966), *Hyopsodus* from *Litaletes* (Gingerich, 1976), and so on.

The ancestries of many of the above taxa are still far from well established, but a number of the previously suggested relationships indicating a Central American origin are questionable. For instance, Rose's (1982, 1985a, 1985b, 1987a) recent work on the postcrania of arctocyonids such as *Chriacus* and the earliest artiodactyls suggests that they were specializing in different directions. Thewissen and others (1983) recently described a species of *Diacodexis,* the earliest known artiodactyl genus, from late early Eocene deposits in Pakistan that is apparently more primitive than other species of the genus from Europe and North America and therefore tentatively suggested an Asian origin for the Artiodactyla. The same species was discovered in the early middle Eocene of India by Kumar and Jolly (1986), who also favored an Asian origin for artiodactyls. Fischer (1985, 1986) recently resurrected the notion that perissodactyls have close affinities with hyracoids, thus suggesting an Old World origin (see also McKenna, 1975b; Prothero and Manning, 1987; Wible, 1986; MacFadden, 1988). Gingerich (1976, 1980a, 1981a, 1982b, 1986) and Franzen (1987) argued, as had McKenna (1967) earlier, that adapids and/or omomyids originated in Africa, although Szalay and Li (1986) suggested a possible Asian origin for euprimates. Interestingly, possible omomyids have recently been found in the late Paleocene (Cappetta and others, 1987; Gheerbrant, 1987) and early Eocene (Hartenberger and others, 1985) of northern Africa but these materials have yet to be described. Gingerich (1976, 1980a, 1980b) argued for an African origin of Hyaenodontidae because of their Oligocene diversity on that continent. Since Gingerich's work, creodonts have been identified from the late Paleocene Adrar Mgorn locality (Cappetta and others, 1987; Gheerbrant, 1987) and from the early Eocene El Kohol locality (Mahboubi and others, 1983, 1984a, 1984b, 1986) of northern Africa. The Adrar Mgorn creodont has yet to be described but is listed as a ?proviverrine hyaenodontid by Cappetta and others (1987) and Gheerbrant (1987). The single, fragmentary specimen of a creodont from El Kohol has also not been described and has not been assigned to any family. Direct ancestors of the above taxa (Artiodactyla, Perissodactyla, Adapidae, Omomyidae, and Hyaenodontidae) still remain elusive. In all cases, however, the most recent evidence seems to support an Old World origin for these taxa even though there are no strong biochronological data to suggest that any of them appear anywhere else before they do in the Western Interior (see discussion section).

Didymoconids are listed as first appearing in the early Wasatchian of the Western Interior because of the recent discovery of a new genus, *Wyolestes,* which Gingerich (1981b) placed in a new subfamily, Wyolestinae. He also assigned the poorly known late Eocene Asian genus *Mongoloryctes* to the subfamily and placed the family in the suborder Mesonychia. Additional specimens of *Wyolestes* have since been discovered in western North America (Gingerich, 1982d; Novacek and others, 1987). Didymoconids are elsewhere known only from Asia (Shanghuanian through Tabenbulukian [late Oligocene]), and in some considerable diversity (Russell and Zhai, 1987). Gingerich (1981b, p. 532) stated that the "systematic diversity and temporal range of Didymoconidae in Asia strongly suggests that Asia was their geographical center of origin and radiation." *Wyolestes* is morphologically much more primitive than known Asian didymoconids, even those from the Shanghuanian and Nongshanian, and Gingerich (1981b) therefore suggested an early derivation from a *Yantanglestes*-like mesonychid ancestor. However, convergent similarities in molar morphology between *Wyolestes* and Asian didymoconids cannot yet be ruled out. This, plus a poor understanding of relationships among Asian didymoconids, precludes definitive statements concerning the biogeographic origins of *Wyolestes.* A comprehensive review of the family is clearly in order and is being undertaken by Jin Meng.

Marsupials of the tribe Didelphini appear for the first time in the Western Interior in the early Wasatchian. Didelphini are represented in some diversity from the Late Cretaceous and Paleocene of South America (e.g., Marshall, 1987; Marshall and de

Muizon, 1988), thus suggesting that South America was the center of origin for the tribe. North American didelphine Didelphini are placed in two genera, *Peratherium* and *Herpetotherium*, by Krishtalka and Stucky (1983b), although Crochet (1977, 1980) had earlier placed all known North American species of the former into the latter genus. All other North American early Tertiary marsupials are placed in the didelphine tribe Peradectini. Marshall (1987) has elevated the Peradectini to familial status and, by so doing, has argued that *Peratherium* is the first representative of the family Didelphidae in the Western Interior. Despite these problems of systematics (see also Fox, 1983), the ancestry of *Peratherium* is not known. Crochet (1979, 1980) clearly favored a South American origin for North American Didelphini but more recently, Crochet and Sigé (1983) have argued that entry into the Western Interior may not have been directly from South America but rather via a South America–Africa–Greenland–Arctic Canada route. Russell and others (1982a) favor dispersal from North America to Europe.

Most of the genera that represent first appearances in the early Wasatchian belong to the Proteutheria and Insectivora. The relationships of most of these are poorly known; their familial assignments are contentious and their ancestor-descendant relationships are nearly impossible to document on currently available evidence. One possible exception is *Apatemys*, which, although thought by Sloan (1969) to be derived from *Labidolemur*, has been more recently studied by Gingerich (1982c) and is considered by him to have emigrated from Europe. This assessment is based primarily on a greater diversity of apatemyids in the Sparnacian of Europe than in the Wasatchian of North America rather than on the identification of a specific ancestral form. In any case, no particular affinities with southern New World forms, or with North American Torrejonian forms, are implicated. Although specific ancestor-descendant sequences for the various rodent genera that appear at the beginning of the Wasatchian also have not been determined, they apparently can be derived from known stocks in the Clarkforkian of the Western Interior (Korth, 1984).

It appears therefore, as with Clarkforkian first appearances, that one does not have to appeal to Torrejonian forms of the Western Interior to explain the first appearances of higher taxa in the early Wasatchian. Many of the early Wasatchian first appearances appear to be immigrants from other continents and most likely entered North America by way of northern rather than southern routes. *Peratherium*, representing the tribe Didelphini, appears to be the only likely exception.

DISCUSSION

The results cited above can be used to evaluate each of the previously proposed biogeographic hypotheses to account for new taxa that appear in the Western Interior of North America in the late Paleocene and early Eocene (routes are shown on Fig. 1).

Route 1. South America to the Western Interior via Central America

No mammalian genera or families are unequivocally known to have been shared between South America and the Western Interior during the middle Paleocene through early Eocene. Palaeanodonts, dinoceratans, and notoungulates (represented by arctostylopids), all of which appear in the Western Interior during the late Tiffanian, are thought by Gingerich (1985) to have their origins in South America (see also Webb, 1985). None of the North American representatives of these groups, however, have undoubted affinities with South American forms.

Palaeanodonts are particularly problematic since their fossil record is restricted to the Paleocene, Eocene, and Oligocene of North America and the Oligocene of Europe and since the fossil records of their purported relatives are spotty at best. Palaeanodonts may be related to either Xenarthra (New World armadillos, anteaters, and sloths) or Pholidota (Old World pangolins), both, or neither (e.g., Emry, 1970; Rose, 1978, 1979, 1987b, and personal communication, 1988). As such, the origin of Xenarthra and the origin of Palaeanodonta may be unrelated questions. Furthermore, Rose (1978) postulated, and McKenna (1980) agreed, that palaeanodonts were derived from pantolestoids. Matthew (1915) proposed an origin of Xenarthra from paleanodonts or the ancestor of palaeanodonts, a possibility addressed by others and still maintained recently by Simpson (1978), who briefly reviewed the history of work on the phylogeny and paleobiogeography of Xenarthra. Simpson (1978), however, concluded that "in the absence of clear contrary evidence" the Xenarthra probably originated in South America. The more recent discoveries of a pholidotan and a supposed xenarthran in the middle Eocene of Europe (Storch, 1978, 1981) and a possible edentate (Ernanodontidae) in the late Paleocene of China (Ting, 1979; Radinsky and Ting, 1984) and perhaps Mongolia (Badamgarav and Reshetov, 1985) further confound the issue since none of these higher taxa have been found in the better sampled Paleocene and Eocene deposits of the Western Interior. On the other hand, the recent discovery of purported xenarthrans (Sudamericidae; Scillato-Yané and Pascual, 1984, 1985; Scillato-Yané, 1986) and xenarthran-like mammals (Gondwanatheriidae; Bonaparte, 1986a, 1986b, 1987) in the middle Paleocene and Late Cretaceous, respectively, of Patagonia may be taken as evidence in support of a South American center of origin for the group (but see Van Valen, 1988). Again, however, this is not necessarily an argument in support of a South American origin for palaeanodonts; their place of origin remains elusive but seems most likely to have been North America on present evidence (Rose, personal communication, 1988).

No concensus has been reached concerning the biogeographic origin of dinoceratans either. McKenna (1980), Gingerich (1985), and others have suggested a possible southern origin because of similarities in cheektooth (particularly M^3) morphology between *Carodnia* (a Paleocene xenungulate) and dinoceratans. Schoch and Lucas (1985), however, suggested that dinoceratans, as well as the South American Xenungulata, arose from an "anagalid' (*Pseudictops*-like) common ancestor in Asia. Sloan (1987, p. 179) stated that *Probathyopsis* (considered a junior synonym of *Prodinoceras* by Schoch and Lucas [1985], as originally suggested by Rose [1981a]), the earliest known North American dinoceratan, may have been derived from either South America (by way of *Carodnia*) or Asia.

The case for notoungulates is even more controversial. Until recently (Cifelli and others, 1987, 1989), the family Arctostylopidae, a strictly Asian and North American group, generally was considered a constituent of the Notoungulata. Non-arctostylopid notoungulates are restricted to South America. When Arctostylopidae was still considered to be a family of Notoungulata, a diversity of opinion existed regarding the paleobiogeography of notoungulates, with workers favoring different origins: South America (Simpson, 1951, 1965, 1978, 1980; Szalay and McKenna, 1971; Marshall and others, 1983; Gingerich, 1985; Webb, 1985; Marshall and de Muizon, 1988), Central America (Gingerich and Rose, 1977), or Asia (Matthew and Granger, 1925; Matthew, 1928; Patterson, 1958; Patterson and Pascual, 1972; Zheng, 1979) (reviewed more fully by Cifelli and others, 1989). Cifelli and others (1989), however, in a detailed systematic treatment of the Arctostylopidae, recognized divergent specializations in arctostylopids and notoungulates, elevated the former to ordinal rank (Arctostylopida), considered Arctostylopida and Notoungulata to be sister taxa, and suggested (p. 39) that "a common ancestor of the two groups would have been exceedingly primitive and, probably, not exclusive." They also suggested (p. 39) that a "common notoungulate/arctostylopid ancestor . . . might have been sufficiently primitive to have given rise to many other orders of mammals." As such, the Arctostylopida are considered by Cifelli and others (1989) to constitute an Asian radiation that dispersed to North America, possibly in the late Paleocene.

Didelphini, which appear in the Western Interior at the beginning of the Wasatchian, probably also had a South American origin and may have made their way to North America via Central America. However, Crochet and Sigé (1983) have recently suggested that dispersal was by way of Africa, Europe, Greenland, and Arctic Canada (see below).

Whatever the case, none of the taxa that may have arrived in the Western Interior from South America in the late Paleocene and early Eocene were part of a major wave of immigrants. Also, none of these taxa became common or speciose elements in the Western Interior after their arrival.

Route 2. South America to the Western Interior via Africa, Europe, Greenland, and Arctic Canada

As part of Crochet and Sigé's (1983) argument for a circuitous route for the entry of Didelphini from South America to the Western Interior via Africa, Europe, Greenland, and Arctic Canada (Fig. 1, Route 2), they cited a later appearance of the tribe in North America than in Europe. They recognized *Herpetotherium edwardi* (placed in *Peratherium* by most other workers; see Krishtalka and Stucky, 1983b) from the late Wasatchian as the earliest occurrence of Didelphini in North America and *Peratherium constans* from the Silveirinha (Portugal) and Dormaal (Belgium) localities, both of which are generally considered to be at a Clarkforkian-equivalent level, as the earliest occurrences of the tribe in Europe (the latter species also has been recently identified at Palette, France, another Dormaal Reference Level fauna; Godinot and others, 1987). *Peratherium,* however, is known from the early Wasatchian (Wa1) in the Western Interior (*P. mcgrewi* from the Bighorn Basin—see Bown, 1979; placed in *P. innominatum* by Krishtalka and Stucky, 1983b). Moreover, as argued below, the Silveirinha, Dormaal, and Palette localities are probably earliest Wasatchian, and not Clarkforkian, age equivalents. As such, the European occurrences of *Peratherium* are not necessarily earlier than the North American occurrences of the genus. It therefore seems unparsiminious, on the basis of this evidence, to suggest a route other than directly from South to North America via Central America.

Route 3. Central America to the Western Interior

In light of recent collections of Paleocene mammals from holarctic continents, particularly North America and Asia, we can find little or no support for a Central American origin (Fig. 1, Route 3) for new taxa that appear in the Western Interior during the late Paleocene and early Eocene. Earliest Tiffanian first occurrences of supraspecific taxa in the Western Interior can be derived from late Torrejonian forms from the same area. The ancestry of Clarkforkian first occurrences appears to lie either in the late Tiffanian of the Western Interior or with Asian forms (although a European origin for some cannot be ruled out). Some Wasatchian first occurrences (mainly genera) were apparently derived from Clarkforkian forms of the Western Interior, whereas Didelphini probably originated in South America. However, most suprageneric taxa (Perissodactyla, Artiodactyla, Omomyidae, Adapidae, Hyaenodontidae) were shared between North America, Europe, and Asia at the beginnings of the Wasatchian, Ypresian (sensu Cavelier and Pomerol, 1986), and Bumbanian, although their ultimate origins remain obscure (see section on Europe below). In any case, there is no particular indication that Central America served as a refugium in which any of these taxa evolved prior to their dispersal into the Western Interior. Any appeals to Central America as a center of origin for Clarkforkian and Wasatchian mammals of the Western Interior are thus based entirely on negative evidence since there are no Paleocene faunas known from there.

Paleoclimatic evidence also does not fully correspond to hypotheses involving a Central American origin for Clarkforkian and Wasatchian mammals of the Western Interior. There does not appear to have been a marked climatic cooling relating to a southward dispersal of Western Interior mammalian faunas to Central America at the beginning of the Tiffanian. If such a cooling event took place at all, it was probably not until somewhat later. Although climates appear to have ameliorated again in the Clarkforkian and Wasatchian (beginning in the late Tiffanian), there is no evidence that the beginnings of these two land-mammal ages are marked by a major influx of mammals from Central America. Instead, the warming climates appear to have opened up northern corridors for a major emigration of mammals from Asia and possibly Europe during the Clarkforkian and Wasatchian. Any dispersal between North America and Europe would have been facilitated by the formation of a subaerial route, the Thulean Bridge, along the proto-Icelandic "hot spot" (McKenna, 1975a).

Geophysical evidence is of little assistance in testing the Central American hypothesis since the tectonic history of Central America is very complicated and currently highly controversial; geophysics cannot be used independently and convincingly to support or refute any particular biogeographic hypothesis involving the early Tertiary of Central America.

Route 4. Southeastern United States to the Western Interior

Schiebout's (1979, 1981) hypothesis that the southeastern United States was a distinct biogeographic province during the Paleocene separated from the Western Interior by the Cannonball Sea, and that it potentially served as a source for early Eocene mammals of the Western Interior, is not corroborated by the available but admittedly sparse evidence from fossil mammals. Schiebout predicted that, when found, Paleocene and early Eocene faunas of the United States Southeast would have stronger affinities with those of Europe than with those of the Western Interior. However, of the three genera now known from the Paleocene of the southeastern United States, one is new and the other two are shared with the Western Interior. None of the three families that these genera belong to is known from the European Paleocene. This suggests therefore, that Paleocene mammals were capable of dispersing between the Western Interior and the eastern part of the continent during at least some, and possibly most or all, of the Paleocene and, therefore, that the southeastern United States did not serve as an important, isolated refugium in which mammals that later dispersed into the Western Interior in the early Eocene evolved. This is also supported by paleogeographic reconstruction of the Cannonball Sea that differs radically from that presented by Sloan (1969) and reproduced by Schiebout (1981). More recent evidence suggests that the Cannonball Sea extended southward from the Arctic Ocean (Marincovich and others, 1985) rather than northward from the Gulf of Mexico.

Route 5. Arctic Canada to the Western Interior

There currently is little or no evidence to support an Arctic Canadian origin of Clarkforkian and Wasatchian mammalian faunas of the Western Interior as proposed by Hickey and others (1983, 1984) simply because there is no independent evidence to support an earlier age assignment for the only faunas of relevance from there (Kent and others, 1984; Norris and Miall, 1984; Flynn and others, 1984; Spicer and others, 1987). The late Paleocene age assigned to the Ellesmere Island faunas is probably erroneous (it is, at the very least, not independently verified); there appears to be no reason at this time to believe that it is anything other than early and middle Eocene in age. The proposal of faunal heterochrony, though possible, is currently unsupported. Furthermore, a recent compilation of faunal resemblance indices by Flynn (1986, p. 335) indicates "that Ellesmere Island was part of a single, continuous European/North American faunal realm" during the early Eocene.

Route 6. Europe to the Western Interior via Greenland and Arctic Canada

Paleogeographic data indicate that one and, at times, two potential dispersal routes between Europe and North America were present during the late Paleocene and early Eocene. The Faunal Resemblance Indices in Table 2 indicate that there was increasing similarity between the faunas of Europe and North America through the Tiffanian, Clarkforkian, and Wasatchian Land-Mammal Ages. This increasing resemblance probably can be best explained by the progressively warm temperatures that allowed for the opening up of a high-latitude land corridor between North America and Europe during this time interval.

None of the above data, however, provide an indication of *directions* of dispersal of particular taxa between Europe and North America. Ideally, the ancestor of a taxon should be known from an earlier horizon in one biogeographic province (in this case, Europe) to be identified as an emigrant to the second (in this case, North America). Such determinations are encumbered by large gaps in the Paleogene fossil record, particularly in Europe, and by the absence of definitive evidence with which to *independently* verify bio- or geochronological correlations to the degree of precision required for determination of relative ages of particular faunas between North America and Europe.

A number of early Ypresian mammalian faunas from Europe have been assigned a Clarkforkian-equivalent age. Russell and others (1982b) referred to these faunas as Dormaal Reference Level faunas. Russell and others listed the Belgian faunas of Dormaal and Erquelinnes; the French faunas of Meudon, Pourcy, and Rians; and the English fauna of Kyson as belonging to this level. To these can be added the faunas from Palette, France (Godinot, 1984; Godinot and others, 1987), and Silveirinha, Portugal (Antunes and Russell, 1981). Their Clarkforkian age equivalence has been inferred from three lines of evidence: (1) the presence of tillodonts, rodents, and coryphodontids, which are known from Clarkforkian horizons in the Western Interior and are taken as heralds of the Clarkforkian; (2) the presence of *Plesiadapis russelli* (only at Meudon, and represented by only eight isolated teeth), which according to Gingerich (1976), is closely related to Clarkforkian *P. cookei* from the Western Interior; and (3) the presence of more primitive species of higher taxa (e.g., Artiodactyla, Perissodactyla, Omomyidae, Adapidae, Hyaenodontidae, Didelphini) that are also present in the Wa1 faunal zone of the Western Interior. The presence of tillodonts, rodents, and coryphodontids, however, cannot be taken as evidence of a Clarkforkian-equivalent age to the exclusion of a later age. These same taxa are also present in the Wasatchian. The close relationship between *Plesiadapis russelli* and *P. cookei* also does not necessarily indicate age equivalence (see also McKenna and others, 1977). The morphologic differences between the two species may reflect not only geographic but also temporal separation. Furthermore, there now appears to be some doubt about the identity of the specimens originally assigned to *P. russelli* (Dashzeveg, 1982; Lucas, 1984). Finally, the recent discovery in the Western Interior by Gingerich (1982a, 1986, 1987) of a Wasatchian fauna, designated Wa0, that is stratigraphically below and therefore older than the previously known Wa1 faunas further casts doubt on the Clarkforkian age assigned to the European faunas listed above. The Wa0 fauna occurs in a stratigraphic interval within what was previously identified as the Clarkforkian-Wasatchian boundary sandstone of Kraus (1980; see Gingerich, 1986). According to Gingerich (1986), however, species represented in the Wa0 fauna, at least those of Adapidae and Perissodactyla, are as primitive or more primitive than those in the so-called Clarkforkian-equivalent faunas of Europe.

A Clarkforkian-equivalent age therefore cannot be maintained for the Dormaal Reference Level faunas on present evidence; they are more likely earliest Wasatchian equivalents, a position also held by Dashzeveg (1982). Comparisons of Faunal Resemblance Indices indicate a much stronger affinity of Dormaal Reference Level faunas with composite Wasatchian faunas than with composite Clarkforkian faunas (Table 2). As stated by Flynn and Novacek (1984, p. 153), "closely similar faunas must be regarded as roughly contemporaneous unless there is clear, independent evidence to the contrary." Such evidence is lacking, and it does not seem possible, therefore, to maintain on the basis of the available data that artiodactyls, perissodactyls, adapids, omomyids, and hyaenodontids arrived significantly earlier in Europe than in North America (or Asia, for that matter). Nor, in our view, is there conclusive evidence to suggest that these taxa necessarily arrived earlier in North America than in Europe (or Asia). There is a large gap in the European record of fossil mammals between the Dormaal Reference Level faunas and those that Russell and others (1982b) and Savage and Russell (1983) regard as Cernaysian (see also Aubry, 1985). It is not inconceivable, furthermore, that an even earlier Wasatchian fauna than Wa0 has been erased before deposition of the Clarkforkian-Wasatchian boundary sandstone in the Clark's Fork Basin. Therefore, without contradictory geochronological or independent nonmammalian biochronological evidence, there is no reason to assume faunal heterochrony between the Dormaal Reference Level faunas and those from the earliest Wasatchian of the Western Interior. In conclusion, there is no unequivocal evidence that Perissodactyla, Artiodactyla, Adapidae, Omomyidae, or Hyaenodontidae or their immediate ancestors were present in earlier Clarkforkian faunas of North America, Cernaysian faunas of Europe, or Nongshanian faunas of Asia.

Route 7. Asia to the Western Interior via Beringia

In their review of Paleogene mammalian paleontology and stratigraphy of Asia, Russell and Zhai (1987) noted that the Asian Paleocene and early Eocene strata, although now fairly well sampled paleontologically, cannot be precisely correlated with the better known sequences in Europe and North America. In part this is because of the lack of correlation with marine biostratigraphic scales but also in part because of the relatively high degree of endemism of Asian mammals exhibited during the Paleocene and early Eocene. Flynn's (1986) recent assessment of faunal similarity among holarctic continents, for instance, led him to conclude (p. 335), "It is clear from the faunal similarity matrices . . ., at all taxonomic levels, that the now more completely known Early Eocene fauna of Asia was distinct and isolated in a separate faunal realm from the faunas of Europe and North America." This endemism was even more pronounced during the Paleocene (Table 2). Nonetheless, some taxa are shared between Asia and North America, and an analysis of the phylogenetic affinities of the first appearances of mammalian taxa in the Clarkforkian of the Western Interior suggests that they may represent emigrants from Asia. This is indicated particularly by rodents and tillodonts; the case is not clear for coryphodontids and miacids. Specifically, rodents appear to be derivable from eurymyloids, which are present in the Nongshanian and Shanghuanian of Asia. Tillodonts are known from these earlier Asian horizons as well but do not appear in North America until the beginning of the Clarkforkian. It seems likely that dispersal of these taxa into the Western Interior from Asia across the high-latitude land corridor of Beringia was permitted by the increasingly warm temperatures at this time.

Just as for the Ypresian of Europe, there is no evidence to indicate that the beginnings of the Bumbanian of Asia and the Wasatchian of North America were not isochronous (see also Dashzeveg, 1982). The only possible Wasatchian suprageneric taxon that first appears in the Western Interior fossil record after its appearance in Asia, and which is not known from elsewhere, is Didymoconidae. As noted above, however, the affinities of the only purported didymoconid from the Western Interior, *Wyolestes,* are not yet firmly established.

Route 8. Africa to the Western Interior via Asia and Beringia

Three Paleocene and early Eocene localities containing mammals have been discovered recently in Africa. They therefore have a direct bearing on previous hypotheses implicating Africa as the center of origin for certain supraspecific taxa that appear in the late Paleocene or early Eocene of the Western Interior. The new localities are: (1) Adrar Mgorn 1 (Morocco, late Paleocene—Cappetta and others, 1978, 1987; Gheerbrant, 1987), (2)

Chambi (central Tunisia, early Eocene—Hartenberger and others, 1985; Hartenberger, 1986; Crochet, 1986), and (3) El Kohol (Algeria, late early Eocene—Mahboubi and others, 1983, 1984a, 1984b, 1986; Crochet, 1984).

The late Paleocene Adrar Mgorn 1 fauna contains a palaeoryctid proteutherian (cf. *Cimolestes* sp.), a lipotyphlan insectivoran (*Afrodon chleuhi*), and not precisely identifiable remains of carnivores (Miacidae?), creodonts (Hyaenodontidae?), primates (Omomyidae?), and "condylarths." Although the material is scrappy, most of the mammalian taxa known suggest Laurasian affinities (Cappetta and others, 1987; Gheerbrant, 1987, 1988); there is little indication that the late Paleocene mammalian fauna of Africa, as represented by the few specimens from this locality, was evolving in isolation.

The Chambi locality is the oldest known Eocene locality in Africa (Harbenberger and others, 1985). The preliminary faunal list includes marsupials, hyracoids, macroscelideans, palaeoryctid proteutherians, chiropterans, anomalurid and ischyromyid rodents, and a questionable omomyid primate. As such, the fauna consists of both largely endemic elements (hyracoids, macroscelideans, anomalurid rodents) as well as those shared with Holarctica, the latter indicating the possibility of faunal exchange between Africa and Europe in the late Paleocene or earliest Eocene (Hartenberger and others, 1985). Interestingly, of the five major groups (Artiodactyla, Perissodactyla, Adapidae, Omomyidae, Hyaenodontidae) that dispersed throughout Holarctica at the beginning of Wasatchian or Wasatchian-equivalent time, only omomyids are possibly represented in the Chambi fauna (but this record has yet to be documented). It cannot be determined whether this record of an omomyid, if confirmed, is the result of dispersal from Europe or of in situ evolution in Africa (which, in turn, rests in part on the confirmation of the Adrar Mgorn record of a possible omomyid).

The mammalian fauna from El Kohol also is not well known; the preliminary list includes marsupials, lipotyphlans, creodonts, hyracoids, and proboscideans. At least two of these taxa (lipotyphlans and creodonts) suggest Holarctic affinities, although the more precise relationships of these forms have not been, or cannot be, determined because of the fragmentary nature of the material (Mahboubi and others, 1984b, 1986). *Garatherium mahboubii* was assigned to the Marsupialia (Mahboubi and others, 1983; Crochet, 1984) and was thought to be of Gondwanian origin (Jaeger and Martin, 1984; Mahboubi and others, 1986), but the identification of the single tooth assigned to this species as that of a marsupial is questionable (Bown and Simons, 1984). In any case, the presence of both hyracoids and proboscideans in the El Kohol fauna suggests some degree of isolation from Europe (Mahboubi and others, 1984b, 1986). As at Chambi, unquestionable representatives of the Artiodactyla, Perissodactyla, Adapidae, Omomyidae, and Hyaenodontidae have not been found at El Kohol, although an indeterminate creodont may be referrable to the Hyaenodontidae. On this basis, Mahboubi and others (1986) suggested that Africa was the biogeographic province in which creodonts may have differentiated from palaeoryctids. Like the record of a possible omomyid from Chambi, it cannot be determined if the creodont dispersed to Africa or originated there.

There is therefore little support for the hypothesis that Africa was the source for some of the mammalian immigrant taxa (at least Artiodactyla, Perissodactyla, and Adapidae) that appear at the beginning of the Wasatchian in the Western Interior. The Euramerican affinities of the Adrar Mgorn fauna suggest that Africa was not completely isolated faunally during the Paleocene (Mahboubi and others, 1984b; Hartenberger and others, 1985; Cappetta and others, 1987; Gheerbrant, 1987). On the other hand, perissodactyls, artiodactyls, adapids, omomyids, and hyaenodontids are not represented with certainty in any of the known late Paleocene or early Eocene mammalian faunas of Africa. The late Paleocene and early Eocene of Africa are admittedly still very poorly sampled, but if that continent was the biogeographic source for some or all of these mammalian taxa, their presence even in these small collections might have been expected.

Finally, if Africa was the source for some of the supraspecific taxa that ultimately found their way into North America in the early Eocene, it seems as likely, or more so, that they did so by way of Europe rather than by way of Arabia, Asia, and Beringia as suggested by Gingerich (1986). The Euramerican affinities of several of the taxa found in the three localities from the late Paleocene and early Eocene of Africa indicate that dispersal between Europe and Africa across the Tethys was, in fact, possible in the early Paleogene.

SUMMARY AND CONCLUSIONS

There are no known suprageneric mammalian taxa that appear at the beginning of the Tiffanian Land-Mammal Age in the Western Interior of North America. Although not all lineages are well studied, it appears that the genera that do appear at this time can be derived from late Torrejonian forms in the same region. As such, support is withdrawn from hypotheses implicating derivation of early Tiffanian first appearances from more northern faunas in association with a progressively more temperate climate. The only Tiffanian taxa that may have had their origins from outside of the Western Interior include the Palaeanodonta, Dinocerata, and Arctostylopidae, but these taxa arrived in the late Tiffanian and their source(s) remains unresolved.

The data presented here do not support an origin of Clarkforkian mammalian faunas of the Western Interior from anywhere else in the New World (South America, Central America, southeastern United States, or Arctic Canada), as previously hypothesized. New higher taxa that appeared at the beginning of the Clarkforkian were either derived from Tiffanian forms in the same region or, at least in the case of rodents and tillodonts, from Asia. Coryphodontids, which also appear at the beginning of the Clarkforkian, may have dispersed from or to the Western Interior; the available evidence does not permit a choice.

Determination of the biogeographic origin(s) of mammalian supraspecific taxa that appear at the beginning of the Wasatchian is more problematic than for those that appear at the beginnings of the Tiffanian or Clarkforkian. Some appear to have been derived from Clarkforkian forms of the Western Interior, some from South America (Didelphini), some from Europe (Meniscotheriidae), and some possibly from Asia (?Didymoconidae), but the origin of most (Artiodactyla, Perissodactyla, hyaenodontid Creodonta, omomyid and adapid Primates) probably was in some area other than those listed above. Despite a vast improvement over the last decade in the late Paleocene mammalian fossil record (particularly in central Asia), there is no compelling evidence that these taxa, or their direct ancestors (or at least more primitive sister taxa) were present in North America, Europe, or Asia before Wasatchian (or Wasatchian-equivalent) time. The late Paleocene fossil record on the northern continents is now sufficiently detailed that evidence of these taxa or their immediate forebears would be expected to have been recorded, if they had been present. By elimination, therefore, it seems likely that these taxa dispersed from some contiguous or nearby southern landmass.

In terms of large landmasses, the two most likely southern centers of origin for Artiodactyla, Perissodactyla, Hyaenodontidae, Omomyidae, and Adapidae are Africa and the Indian subcontinent (Krause and Maas, 1988). Unfortunately, there are no records of mammals from Africa between the Early Cretaceous (Jacobs and others, 1988) and the late Paleocene (Cappetta and others, 1987; Gheerbrant, 1987, 1988). The few taxa now known from the late Paleocene and early Eocene of Africa suggest that at least artiodactyls, perissodactyls, and adapids were not present; tentative identifications of omomyids and hyaenodontids remain to be confirmed by more diagnostic material. Furthermore, the Euramerican affinities of the only known late Paleocene fauna from Africa, which is from Adrar Mgorn 1 in Morocco, suggest that Africa was not completely isolated from Europe at this time.

The Indian subcontinent previously has not been seriously considered as a biogeographic source for mammalian higher taxa that appear in the Western Interior of North America at the beginning of the Wasatchian for several reasons: (1) during the last 20 or so years, other centers of origin, including Europe, central Asia, Central America, South America, Africa, as well as other regions of North America have been suggested for almost all groups; (2) large gaps existed in the Late Cretaceous through middle Eocene vertebrate fossil record of central Asia, and particularly in that of the Indian subcontinent; (3) dispersal of mammals after collision of the Indian subcontinent with central Asia generally was assumed to have been from the latter to the former (e.g., Sahni and Kumar, 1974; Lucas and Schoch, 1981; Sahni and others, 1981, 1983; Hartenberger, 1982; de Bonis and others, 1985); and (4) collision of the Indian subcontinent with central Asia was thought by most workers to have occurred in the middle Eocene (Sahni and Kumar, 1974) or later (Gansser, 1964; Raiverman, 1972; Athavale, 1973; Dewey and Burke, 1973; Molnar and Tapponnier, 1975; Blow and Hamilton, 1975; Sclater and others, 1976; Pierce, 1978; Barron and others, 1981; Parrish, 1987), that is, well after representatives of many of the mammalian higher taxa that appeared at the beginning of the Wasatchian were already well established and widely distributed on all three holarctic continents.

An assessment of recent paleontologic and geologic evidence suggests that the Indian subcontinent must now be given serious consideration as the biogeographic source for one or more of the following groups: Artiodactyla, Perissodactyla, Adapidae, Omomyidae, and Hyaenodontidae (and perhaps also Cetacea, but a discussion of this taxon falls outside the scope of this paper because its members do not appear at the beginning of the Wasatchian in the Western Interior). Three independent lines of evidence are discussed in the following paragraphs.

1. Recently discovered Late Cretaceous vertebrate faunas from the Indian subcontinent are very similar to those from Africa and Madagascar (e.g., Sahni and others, 1981, 1986; Sahni, 1984a, 1984b, 1986; Colbert, 1984; Mathur and Srivastava, 1987), although Indian vertebrates with Laurasian affinities are also known from close to the Cretaceous-Tertiary boundary (e.g., Sahni and others, 1982; de Bonis and others, 1985). This indicates that the Late Cretaceous vertebrate fauna of the Indian subcontinent was not highly endemic, as one would have suspected from the older geophysical evidence, which suggested that the subcontinent was an island since its separation from the rest of Gondwanaland in the Late Triassic (but see Chatterjee, 1984; and Chatterjee and Hotton, 1985, for an extreme opposite view). Srivastava (1983, p. 152) supported an Indian-African affinity for Late Cretaceous plate assemblages, stating that "new elements introduced in north Africa in the Late Maastrichtian appear simultaneously in the east coast sediments of India" (see also Hallam, 1981). In a summary of the vertebrate evidence, Colbert (1984, p. 35) concluded, "the fossils show that the ligation between India and Africa must have been maintained well into late Cretaceous time, after which rifting took place, and Peninsular India drifted rapidly to the northeast." Included in the Late Cretaceous vertebrate fauna is a recently described eutherian mammal, *Deccanolestes,* which documents that placental mammals had the opportunity to colonize the subcontinent at that time (Prasad and Sahni, 1988). Alternatively (but, in our opinion, less plausibly), mammals that ultimately gave rise to placentals may have existed on the Indian subcontinent prior to Late Cretaceous time. Symmetrodonts are known from the Lower Jurassic Kota Formation (Datta and others, 1978; Datta, 1981; Yadigiri, 1984, 1985), but it seems unlikely that these particular forms had anything to do directly with later placental evolution. Therefore, it is quite possible that one or more groups of placental mammals "boarded" the Indian subcontinent in Late Cretaceous time and then evolved in isolation and achieved an independent character during at least some of the Paleocene as the subcontinent drifted northeastward toward Asia.

2. The Indian subcontinent was adjacent to the eastern margin of Madagascar immediately prior to anomaly 33b time, that is, sometime in the Campanian age of the Late Cretaceous (Powell, 1979; Molnar and others, 1988; Tarling, 1988) and well

after Madagascar had assumed its present position relative to Africa (Rabinowitz and others, 1983; Coffin and Rabinowitz, 1987). Africa, which apparently was not subdivided by a trans-Saharan seaway at this time (Gee, 1988), was faunally connected with Madagascar during the Late Cretaceous (Taquet, 1982; Rage, 1988). Thereafter, the Indian subcontinent drifted rapidly northeastward. Recent paleomagnetic, stratigraphic, and sedimentologic evidence indicates that the Indian subcontinent may have collided with central Asia in approximately early Eocene time (Powell and Conaghan, 1973; Curray and others, 1982; Wells, 1983; Besse and others, 1984; Patriat and Achache, 1984; Klootwijk and others, 1985), considerably earlier than suggested previously (see above). Although the exact form, nature, and timing of this collision are problematic, it is relevant to note also that several blocks of continental crust that were adrift in the Indian Ocean north of the Indian subcontinent could well have served as stepping stones to central Asia as it was approached by the Indian subcontinent (e.g., Allègre and others, 1984). These blocks, as well as a large northern part of the subcontinent, have been accreted to, or subducted under, central Asia.

3. Unfortunately, Paleocene mammals are unknown from the Indian subcontinent. The earliest known, identifiable Tertiary mammals from there are from early Eocene horizons in the lower part of the Subathu Formation of India (Subathu and Dharampur localities), and in the lower part of the Kuldana Formation of Pakistan (the Barbora localities; Russell and Zhai, 1987). Although the collections are small, the known fauna consists of artiodactyls (Raoellidae and Dichobunidae), perissodactyls (Isectolophidae and Hyracodontidae), primates (Omomyidae), proboscideans (Anthracobunidae), cetaceans (indet.), sirenians (indet.), rodents (Paramyidae? and Ctenodactylidae), and carnivorans (indet.) (Russell and Zhai, 1987; Russell and Gingerich, 1987). While it is implicit in most studies of Asian mammalian paleobiogeography that mammals dispersed to (rather than from) the Indian subcontinent from central Asia after contact was made between the two landmasses, present evidence suggests that such a claim can be made with any degree of certainty only for rodents. Eurymyloids, from the late Paleocene of central Asia, appear to be involved in the ancestry of rodents (see above). Raoellid artiodactyls and anthracobunid proboscideans, however, are unknown from anywhere other than the Indian subcontinent (although anthracobunids may be related to the more derived Moeritheriidae from Africa), and neither group has any known, more primitive close relative (Kumar and Jolly, 1986; Thewissen and others, 1987; Wells and Gingerich, 1983; West, 1980, 1984). This suggests origins of raoellids and anthracobunids on the Indian subcontinent, together with a lengthy period of endemism during the Paleocene. West (1980) also believed that south Asia may have been the geographic source of tethytheres (but see Gingerich and Russell, 1981; and Domning and others, 1986), the group to which proboscideans are now allocated.

The family Dichobunidae is well represented in the early Eocene faunas of Laurasia but, with the exception of *Diacodexis,* all dichobunids from the early and middle Eocene of the Indian subcontinent are endemic to that landmass (Thewissen and others, 1987). This too appears to attest to a previous period of isolation. Furthermore, the species of *Diacodexis* known from the early Eocene of the Indian subcontinent, *D. pakistanensis,* has been suggested to be the most primitive known artiodactyl (Thewissen and others, 1983, 1987). For the reasons stated previously, this cannot be considered as strong biogeographic evidence, but nonetheless, Kumar and Jolly (1986) used it to suggest that the Indian subcontinent may have been the center of origin for artiodactyls.

Artiodactyls, perissodactyls, adapids, omomyids, and hyaenodontids are all represented, and in some considerable diversity, in the much better known middle Eocene (or late early Eocene; Gingerich and others, 1983) faunas of the Indian subcontinent (upper part of the Subathu Formation of India and upper part of the Kuldana Formation and the Domanda Formation of Pakistan; see Russell and Zhai, 1987; and Russell and Gingerich, 1987). Artiodactyls are represented there by four families (Raoellidae, Dichobunidae, Anthracotheriidae, and Helohyidae), perissodactyls by no fewer than six families (Brontotheriidae, Isectolophidae, Helaletidae, Lophialetidae, Hyracodontidae, and Deperetellidae), primates by both of the only known Eocene families, Omomydae and Adapidae (Gingerich and others, 1983; Russell and Zhai, 1987; Russell and Gingerich, 1987), and hyaenodontids by *Paratritemnodon* (Russell and Zhai, 1987). The presence and diversity of these higher taxa in the middle Eocene (or late early Eocene) of the Indian subcontinent is, at the very least, consistent with a hypothesis of origin there.

A chronology of events—including a Late Cretaceous boarding of the Indian subcontinent by basal stocks of placental mammals from eastern Africa via Madagascar, a period of isolation during the Paleocene when the subcontinent was adrift in the Indian Ocean and when the mammalian fauna attained an endemic character, and an early Eocene collision with central Asia—therefore makes it possible that the Indian subcontinent was the biogeographic source for one or more of Artiodactyla, Perissodactyla, Adapidae, Omomyidae, and Hyaendontidae (Fig. 7). These taxa appear to have dispersed across Holarctica instantaneously (at current levels of resolution), if we are correct, that is, in assuming that the beginnings of the Wasatchian in North America, the Sparnacian in Europe, and the Bumbanian in Asia represent an essentially isochronous horizon (see above and Dashzeveg, 1982; Dashzeveg and Russell, 1988). The uncertainties concerning intercontinental age correlations between North America, Europe, and Asia prevent anything other than this assumption at the present time. Among early Tertiary large landmasses, the Indian subcontinent is unique in its combination of having been in the right places at the right times to provide for the development and the subsequent disembarking of several new higher taxa of mammals and in having fossils that demonstrate the appropriate stages of evolution to be consistent with such a hypothesis. This hypothesis can best be tested by the discovery of Paleocene and/or more early Eocene mammals from the Indian subcontinent.

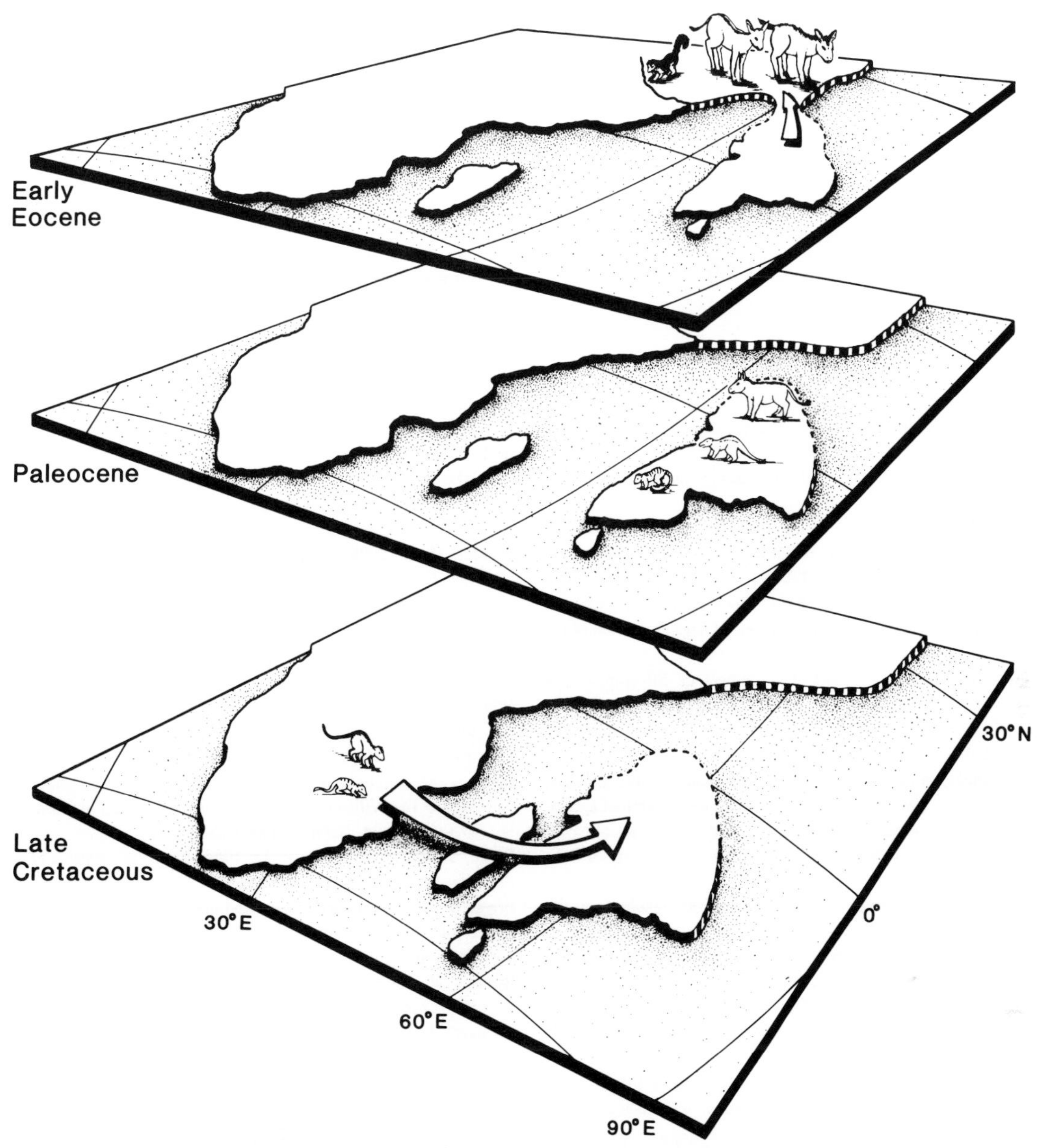

Figure 7. Diagram depicting hypothetical series of events leading to the origin of several placental higher taxa on the Indian subcontinent and their subsequent dispersal. Bottom: Dispersal of basal placental stocks from Africa to the Indian subcontinent via Madagascar in the Late Cretaceous. Middle: Evolution in isolation of one or more of Artiodactyla, Perissodactyla, Omomyidae, Adapidae, and/or Hyaenodontidae (and perhaps also Cetecea) on the Indian subcontinent during its northeastward progression toward central Asia in the Paleocene. Top: Collision of the Indian subcontinent with central Asia in approximately early Eocene time and the disembarking of these taxa, which subsequently dispersed throughout Holarctica.

ACKNOWLEDGMENTS

We are particularly grateful to Tom Bown and Ken Rose for the invitation to participate in the symposium and for the opportunity to contribute a chapter to this volume. We also thank J. D. Archibald, T. M. Bown, J. G. Fleagle, J. J. Flynn, P. D. Gingerich, M. Godinot, M. C. McKenna (on a muddy trek in the Wyoming wilderness in 1978), and K. D. Rose for helpful and stimulating discussions; R. L. Cifelli, M. Fischer, J. J. Flynn, J. Franzen, M. Godinot, F. E. Grine, J.-L. Hartenberger, J. Hooker, A. Jolly, S. G. Lucas, L. G. Marshall, M. C. McKenna, Jin Meng, P. Molnar, J. T. Parrish, K. D. Rose, D. E. Russell, A. Sahni, B. Sigé, R. Stucky, G. Upchurch, S. Wing, and J. Wolfe for correspondence and/or unpublished information and/or for making us aware of references that we had not seen; and R. C. Fox, M. C. McKenna, K. D. Rose, and D. E. Russell for their careful reviews of this paper. Access to Torrejonian and Tiffanian localities in the Crazy Mountains Basin, Montana, not to mention countless acts of assistance, were provided by members of the following families: Adams, Connelly, Cremer, Cooney, Donald, Ford, Glennie, Starr, and Tronrud. Their help is greatly appreciated. This study was supported by National Science Foundation grants BSR 84-06707 and BSR 87-22539 and grants from the National Geographic Society and the Center for Field Research/EARTHWATCH to DWK.

REFERENCES CITED

Adams, C. G., 1981, An outline of Tertiary palaeogeography, *in* Cocks, L.R.M., ed., The evolving Earth: Cambridge, Cambridge University Press, p. 221–235.

Allègre, C. J., and 34 others, 1984, Structure and evolution of the Himalaya-Tibet orogenic belt: Nature, v. 307, p. 17–22.

Anderson, T. H., and Schmidt, V. A., 1983, The evolution of Middle America and the Gulf of Mexico–Caribbean Sea region during Mesozoic time: Geological Society of America Bulletin, v. 94, p. 941–966.

Andrews, P., Lord, J. M., and Nesbit Evans, E. M., 1979, Patterns of ecological diversity in fossil and modern mammalian faunas: Biological Journal of the Linnean Society, v. 11, p. 177–205.

Antunes, M. T., and Russell, D. E., 1981, Le gisement de Silveirinha (Bas Mondego, Portugal); La plus ancienne faune de Vertébrés éocènes connue en Europe: Paris, Comptes Rendus Hebdomadaires des Seances de l'Academie des Sciences, v. 293, p. 1099–1102.

Archibald, J. D., Clemens, W. A., Gingerich, P. D., Krause, D. W., Lindsay, E. H., and Rose, K. D., 1987, First North American land mammal ages of the Cenozoic Era, *in* Woodburne, M. O., ed., Cenozoic mammals of North America; Geochronology and biostratigraphy: Berkeley, University of California Press, p. 24–76.

Athavale, R. N., 1973, Inferences from recent palaeomagnetic results from the northern margin of the Indian Plate and the tectonic evolution of India, *in* Tarling, D. H., and Runcorn, S. K., eds., Implications of continental drift to the earth sciences: New York, Academic Press, p. 117–130.

Aubry, M.-P., 1985, Northwestern European Paleogene magnetostratigraphy, biostratigraphy, and paleogeography; Calcareous nannofossil evidence: Geology, v. 13, p. 198–202.

Azzaroli, A., 1981, Cainozoic mammals and the biogeography of the Island of Sardinia, western Mediterranean: Palaeogeography, Palaeoclimatology, Palaeoecology, v. 36, p. 107–111.

Badamgarav, D., and Reshetov, V. Yu., 1985, Paleontologiya i stratigrafiya paleogena zaaltayskoy gobi (Paleontology and stratigraphy of the Paleogene of the Zaaltay Gobi), *in* Trofimov, B., ed., Sovmestnaya Sovetsko-Mongolskaya Paleontologicheskaya Ekspeditsiya, v. 25, 104 p. (in Russian with English summary).

Barron, E. J., Harrison, C.G.A., Sloan, J. L., and Hay, W. W., 1981, Paleogeography; 180 million years ago to the present: Eclogae Geologicae Helvetiae, v. 74, p. 443–470.

Berggren, W. A., and Hollister, C. D., 1977, Plate tectonics and paleocirculation; Commotion in the ocean: Tectonophysics, v. 38, p. 11–48.

Berggren, W. A., Kent, D. V., and Flynn, J. J., 1985, Jurassic to Paleogene; Part 2, Paleogene geochronology and chronostratigraphy, *in* Snelling, N. J., ed., Geochronology and the geological record: Geological Society of London Special Paper 10, p. 141–195.

Besse, J., Courtillot, V., Pozzi, J. P., Westphal, M., and Zhou, Y. X., 1984, Palaeomagnetic estimates of crustal shortening in the Himalayan thrusts and Zango suture: Nature, v. 311, p. 621–626.

Blow, R. A., and Hamilton, N., 1975, Palaeomagnetic evidence from DSDP cores of northward drift of India: Nature, v. 257, p. 570–572.

Boersma, A., 1984, Campanian through Paleocene paleotemperature and carbon isotope sequence and the Cretaceous-Tertiary boundary in the Atlantic Ocean, *in* Berggren, W. A., and Van Couvering, J. A., eds., Catastrophes and Earth history: Princeton, New Jersey, Princeton University Press, p. 247–277.

Bonaparte, J. F., 1986a, A new and unusual Late Cretaceous mammal from Patagonia: Journal of Vertebrate Paleontology, v. 6, p. 264–270.

——, 1986b, History of the terrestrial Cretaceous vertebrates of Gondwana, *in* 4th, Congreso Argentino de Paleontologia y Bioestratigraphia, Simposio "Evolucion de los vertebrados mesozoicos" (J. Bonaparte, Organizador): Mendoza, Argentina, Editorial Inca, v. 2, p. 63–95.

——, 1987, The Late Cretaceous fauna of Los Alamitos, Patagonia, Argentina; Part 8, The mammals: Revista del Museo Argentino de Ciencias Naturales "Bernardino Rivadavia," v. 3, p. 163–169.

Bonini, W. E., Hargraves, R. B., and Shagam, R., eds., 1984, The Caribbean–South American plate boundary and regional tectonics: Geological Society of America Memoir 162, 421 p.

de Bonis, L., Bouvrain, G., Buffetaut, E., Denys, C., Geraads, D., Jaeger, J.-J., Martin, M., Mazin, J.-M., and Rage, J.-C., 1985, Contribution des Vertébrés à l'histoire de la Téthys et des continents péritéthysiens: Societe Géologique de France Bulletin, v. 1, p. 781–786.

Boulter, M. C., 1984, Palaeobotanical evidence for land-surface temperature in the European Palaeogene, *in* Brenchley, P., ed., Fossils and climate: Chichester, John Wiley and Sons, Ltd., p. 35–47.

Bown, T. M., 1979, Geology and mammalian paleontology of the Sand Creek facies, lower Willwood Formation (lower Eocene), Washakie County, Wyoming: Wyoming Geological Survey Memoir 2, 152 p.

Bown, T. M., and Rose, K. D., 1987, Patterns of dental evolution in early Eocene anaptomorphine primates (Omomyidae) from the Bighorn Basin, Wyoming: Journal of Paleontology, v. 61, part 2, p. 1–162.

Bown, T. M., and Schankler, D., 1982, A review of the Proteutheria and Insectivora of the Willwood Formation (lower Eocene), Bighorn Basin, Wyoming: U.S. Geological Survey Bulletin 1523, 79 p.

Bown, T. M., and Simons, E. L., 1984, Reply *to* Jaeger and Martin 'African marsupials; Vicariance or dispersal?': Nature, v. 312, p. 379–380.

Briggs, J. C., 1987, Biogeography and plate tectonics; Developments in palaeontology and stratigraphy 10: Amsterdam, Elsevier Scientific Publishing Company, 204 p.

Bruns, T. R., 1983, Model for the origin of the Yakutat block, an accreting terrane in the northern Gulf of Alaska: Geology, v. 11, p. 718–721.

Buchardt, B., 1978, Oxygen isotope paleotemperatures from the Tertiary period in the North Sea area: Nature, v. 275, p. 121–123.

Burchfiel, B. C., and Royden, L., 1982, Carpathian Foreland fold and thrust belt and its relation to Pannonian and other basins: American Association of Petroleum Geologists, v. 66, p. 1179–1195.

Butler, R. F., Krause, D. W., and Gingerich, P. D., 1987, Magnetic polarity stratigraphy and biostratigraphy of middle-late Paleocene continental deposits of south-central Montana: Journal of Geology, v. 95, p. 647–657.

Cappetta, H., Jaeger, J.-J., Sabatier, M., Sigé, B., Sudre, J., and Vianey-Liaud, M., 1978, Découverte dans le Paléocène du Maroc des plus anciens mammifères euthériens d'Afrique: Géobios, v. 11, p. 257–262.

Cappetta, H., Jaeger, J.-J., Sigé, B., Sudre, J., and Vianey-Liaud, M., 1987, Compléments et précisions biostratigraphiques sur la faune Paléocène à Mammifères et Sélaciens du Bassin d'Ouarzazate (Maroc): Tertiary Research, v. 8, p. 147–157.

Cavelier, C., and Pomerol, C., 1986, Stratigraphy of the Paleogene: Societe Géologique de France Bulletin, v. 2, p. 255–265.

Chatterjee, S., 1984, The drift of India; A conflict in plate tectonics: Societe Géologique de France Memoires 147, p. 43–48.

Chatterjee, S., and Hotton, N., III, 1985, The paleoposition of India: Journal of Southeast Asian Earth Sciences, v. 1, p. 145–189.

Chow, M., and Wang, B.-Y., 1979, Relationship between the pantodonts and tillodonts and classification of the order Pantodonta: Vertebrata PalAsiatica, v. 17, p. 37–48.

Cifelli, R. L., 1983, The origin and affinities of the South American Condylarthra and early Tertiary Litopterna (Mammalia): American Museum Novitates, no. 2772, p. 1–49.

Cifelli, R. L., Schaff, C. R., and McKenna, M. C., 1987, The affinities of the Arctostylopidae (Mammalia): Journal of Vertebrate Paleontology, v. 7, p. 14A.

——, 1989, The relationships of the Arctostylopidae (Mammalia); New data and interpretation: Cambridge, Massachusetts, Harvard University Museum of Comparative Zoology Bulletin, v. 152, p. 1–44.

Clemens, W. A., 1979, Marsupialia, *in* Lillegraven, J. A., Kielan-Jaworowska, Z., and Clemens, W. A., eds., Mesozoic mammals; The first two-thirds of mammalian history: Berkeley, University of California Press, p. 192–220.

Coffin, M. F., and Rabinowitz, P. D., 1987, Reconstruction of Madagascar and Africa; Evidence from the Davie Fracture Zone and Western Somali Basin: Journal of Geophysical Research, v. 92, p. 9385–9406.

Colbert, E. H., 1984, Mesozoic reptiles, India and Gondwanaland: India Journal of Earth Sciences, v. 11, p. 25–37.

Cope, E. D., 1883, The ancestor of *Coryphodon:* American Naturalist, v. 17, p. 406–407.

Crochet, J.-Y., 1977, Les Didelphidae (Marsupicarnivora, Marsupialia) holarctiques tertiares: Paris, Comptes Rendus Hebdomadaires des Seances de l'Academie des Sciences, v. 284, p. 357–360.

——, 1979, Données nouvelles sur l'histoire paléogéographique des Didelphidae (Marsupialia): Paris, Comptes Rendus Hebdomadaires des Seances de l'Academie des Sciences, v. 288, p. 1457–1460.

——, 1980, Les Marsupiaux du Tertiare d'Europe: Paris, Education Fondation Singer-Polignac, v. 1, p. 1–279.

——, 1984, *Garatherium mahboubii* nov. gen., nov. sp., marsupial de l'Éocène inférieur d'El Kohol (Sud-Oranais, Algérie): Annales de Paleontologie, v. 70, p. 275–294.

——, 1986, *Kasserinotherium tunisiense* nov. gen., nov. sp., troisième marsupial découvert en Afrique (Eocène inférieur de Tunisie): Paris, Comptes Rendus Hebdomadaires des Seances de l'Academie des Sciences, v. 302, p. 923–926.

Crochet, J.-Y., and Sigé, B., 1983, Les Mammifères Montiens de Hainin (Paléocène moyen de Belgique); Part 3, Marsupiaux: Palaeovertebrata, v. 13, p. 51–64.

Crowley, T. J., Short, D. A., Mengel, J. G., and North, G. R., 1986, Role of seasonality in the evolution of climate during the last 100 million years: Science, v. 231, p. 579–584.

Curray, J. R., Emmel, F. J., Moore, D. G., and Raitt, R. W., 1982, Structure, tectonics, and geological history of the northeastern Indian Ocean, *in* Nairn, A.E.M., and Stehli, F. G., eds., The ocean basins and margins; v. 6, The Indian Ocean: New York, Plenum Press, p. 399–450.

Cvancara, A. V., 1966, Revision of the fauna of the Cannonball Formation (Paleocene) of North and South Dakota: Ann Arbor, University of Michigan Museum of Paleontology Contributions, v. 20, 97 p.

Dashzeveg, D., 1982, La faune de Mammifères du Paléogène inférieur de Naran-Bulak (Asie centrale) et ses corrélations avec l'Europe et l'Amerique du Nord: Societe Géologique de France Bulletin, v. 24, p. 275–281.

Dashzeveg, D., and Russell, D. E., 1988, Paleocene and Eocene Mixodontia (Mammalia, Glires) of Mongolia and China: Palaeontology, v. 31, p. 129–164.

Datta, P. M., 1981, The first Jurassic mammal from India: Zoological Journal of the Linnean Society, v. 73, p. 307–312.

Datta, P. M., Yadagiri, P., and Rao, B.R.J., 1978, Discovery of Early Jurassic micromammals from Upper Gondwana sequence of Pranhita-Godavari valley, India: Journal of the Geological Society of India, v. 19, p. 64–68.

Dawson, M. R., Li, C.-K., and Qi, T., 1984, Eocene ctenodactyloid rodents (Mammalia) of eastern and central Asia, *in* Mengel, R. M., ed., Papers in vertebrate paleontology honoring Robert Warren Wilson: Pittsburgh, Pennsylvania, Carnegie Museum of Natural History Special Publication 9, p. 138–150.

Dercourt, J., and 16 others, 1985, Présentation de 9 cartes paléogéographiques au 1/20,000,000^{e} s'etendant de l'Atlantique au Pamir pour la période du Lias à l'Actuel: Societe Géologique de France Bulletin, v. 1, p. 637–652.

Dewey, J. F., and Burke, K.C.A., 1973, Tibetan, Variscan, and Precambrian basement reactivation; Products of continental collision: Journal of Geology, v. 81, p. 683–692.

Domning, D. P., Ray, C. E., and McKenna, M. C., 1986, Two new Oligocene desmostylians and a discussion of tethytherian systematics: Smithsonian Contributions to Paleobiology, no. 59, p. 1–56.

Donnelly, T. W., 1985, Mesozoic and Cenozoic plate evolution in the Caribbean region, *in* Stehli, F. G., and Webb, S. D., eds., The great American biotic interchange: New York, Plenum Press, p. 89–121.

Dorf, E., 1964, The use of fossil plants in paleoclimatic interpretation, *in* Nairn, A.E.M., ed., Problems in paleoclimatology: London, Interscience, p. 13–31.

Eisenberg, J. F., 1981, The mammalian radiations; An analysis of trends in evolution, adaptation, and behavior: Chicago, University of Chicago Press, 610 p.

Emry, R. J., 1970, A North American Oligocene pangolin and other additions to the Pholidota: American Museum of Natural History Bulletin 142, p. 455–510.

Feldman, R. M., 1972, First report of *Herglossa ulrichi* (White, 1882) (Cephalopoda: Nautiloida) from the Cannonball Formation (Paleocene) of North Dakota, U.S.A.: Malacologia, v. 11, p. 407–413.

Ferrusquia-Villafranca, I., 1978, Distribution of Cenozoic vertebrate faunas in Middle America and problems of migration between North and South America: Universidad Nacional Autonoma de Mexico Instituto de Geologia Boletin, v. 101, p. 193–321.

Fischer, M., 1985, Hyracoidea should be included in the Perissodactyla Owen 1848 again: Fourth International Theriological Congress Abstract 0194.

——, 1986, Die Stellung der Schliefer (Hyracoidea) im phylogenetischen System der Eutheria: Courier Forschungsinstitut Senckenberg, v. 84, p. 1–132.

Fleming, T. H., 1973, Numbers of mammal species in North and Central American forest communities: Ecology, v. 54, p. 555–563.

Flessa, K. W., 1975, Area, continental drift, and mammalian diversity: Paleobiology, v. 1, p. 189–194.

Flynn, J. J., 1986, Faunal provinces and the Simpson Coefficient: Laramie, University of Wyoming Contributions to Geology Special Paper 3, p. 317–338.

Flynn, J. J., and Galiano, H., 1982, Phylogeny of early Tertiary Carnivora, with a description of a new species of *Protictis* from the middle Eocene of northwestern Wyoming: American Museum Novitates, no. 2725, p. 1–64.

Flynn, J. J., and Novacek, M. J., 1984, Early Eocene vertebrates from Baja California; Evidence for intracontinental age correlations: Science, v. 224, p. 151–153.

Flynn, J. J., MacFadden, B. J., and McKenna, M. C., 1984, Land-mammal ages, faunal heterochrony, and temporal resolution in Cenozoic terrestrial sequences: Journal of Geology, v. 92, p. 687–705.

Fox, R. C., 1983, Notes on the North American Tertiary marsupials *Herpetotherium* and *Peradectes:* Canadian Journal of Earth Sciences, v. 20, p. 1565–1578.

——, 1987, The succession of Paleocene mammals in western Canada: Geological Society of America Abstracts with Program, v. 19, p. 275–276.

Franzen, J. L., 1987, Ein neuer Primate aus dem Mitteleozän der Grube Messel (Deutschland, S-Hessen): Courier Forschungsinstitut Senckenberg, v. 91, p. 151–187.

Gansser, A., 1964, Geology of the Himalayas: New York, Wiley-Interscience, 289 p.

Gazin, C. L., 1959, Early Tertiary *Apheliscus* and *Phenacodaptes* as pantolestid insectivores: Smithsonian Miscellaneous Collections, v. 139, p. 1–7.

Gee, H., 1988, Cretaceous unity and diversity: Nature, v. 332, p. 487.

Gheerbrant, E., 1987, Les vertébrés continentaux de l'Adrar Mgorn (Maroc, Paléocène); une dispersion de mammifères transtéthysienne aux environs de la limite mésozoique/cénoique?: Paris, Geodinamica Acta, v. 1, p. 233–246.

——, 1988, *Afrodon chleuhi* nov. gen., nov. sp., "insectivore" (Mammalia, Eutheria) lipotyphlé (?) du Paléocène marocain; Données préliminaires: Paris, Comptes Rendus Hebdomadaires des Seances de l'Academic des

Sciences, v. 307, p. 1303–1309.
Gingerich, P. D., 1975, New North American Plesiadapidae (Mammalia, Primates) and a biostratigraphic zonation of the middle and upper Paleocene: Ann Arbor, University of Michigan Museum of Paleontology Contributions, v. 24, p. 135–148.
——, 1976, Cranial anatomy and evolution of early Tertiary Plesiadapidae (Mammalia, Primates): Ann Arbor, University of Michigan Papers on Paleontology, v 15, p. 1–141.
——, 1977a, Patterns of evolution in the mammalian fossil record, *in* Hallam, A., ed., Patterns of evolution: Amsterdam, Elsevier Scientific Publishing Company, p. 469–500.
——, 1977b, Radiation of Eocene Adapidae in Europe: Géobios, Mémoire Spécial, v. 1, p. 165–182.
——, 1978, New Condylarthra (Mammalia) from the Paleocene and early Eocene of North America: Ann Arbor, University of Michigan Museum of Paleontology Contributions, v. 25, p. 1–9.
——, 1980a, Evolutionary patterns in early Cenozoic mammals: Annual Review of Earth and Planetary Sciences, v. 8, p. 407–424.
——, 1980b, *Tytthaena parrisi,* oldest known oxyaenid (Mammalia, Creodonta) from the late Paleocene of western North America: Journal of Paleontology, v. 54, p. 570–576.
——, 1981a, Early Cenozoic Omomyidae and the evolutionary history of tarsiiform primates: Journal of Human Evolution, v. 10, p. 345–374.
——, 1981b, Radiation of early Cenozoic Didymoconidae (Condylarthra, Mesonychia) in Asia, with a new genus from the early Eocene of western North America: Journal of Mammalogy, v. 62, p. 526–538.
——, 1981c, Early Cenozoic Omomyidae and the evolutionary history of tarsiiform primates: Journal of Human Evolution, v. 10, p. 345–374.
——, 1982a, Paleocene *"Meniscotherium semicingulatum"* and the first appearance of Meniscotheriidae (Condylarthra) in North America: Journal of Mammalogy, v. 63, p. 488–491.
——, 1982b, Climate, continental relationships, sea level, and the geography of primate evolution: American Journal of Physical Anthropology, v. 57, p. 193.
——, 1982c, Studies on Paleocene and early Eocene Apatemyidae (Mammalia, Insectivora); 2, *Labidolemur* and *Apatemys* from the early Wasatchian of the Clark's Fork Basin, Wyoming: Ann Arbor, University of Michigan Museum of Paleontology Contributions, v. 26, p. 57–69.
——, 1982d, Second species of *Wyolestes* (Condylarthra, Mesonychia) from the early Eocene of western North America: Journal of Mammalogy, v. 63, p. 706–709.
——, 1983a, Paleocene-Eocene faunal zones and a preliminary analysis of Laramide structural deformation in the Clark's Fork Basin, Wyoming: Wyoming Geological Association 34th Guidebook, p. 185–195.
——, 1983b, Systematics of early Eocene Miacidae (Mammalia, Carnivora) in the Clark's Fork Basin, Wyoming: Ann Arbor, University of Michigan Museum of Paleontology Contributions, v. 26, p. 197–225.
——, 1985, South American mammals in the Paleocene of North America, *in* Stehli, F. G., and Webb, S. D., eds., The great American biotic interchange: New York, Plenum Press, p. 123–137.
——, 1986, Early Eocene *Cantius torresi;* Oldest primate of modern aspect from North America: Nature, v. 319, p. 319–321.
——, 1987, Origin of modern orders of mammals; Paleocene-Eocene faunal evolution in the Clark's Fork Basin, Wyoming: Geological Society of America Abstracts with Programs, v. 19, p. 278.
Gingerich, P. D., and Gunnell, G. F., 1979, Systematics and evolution of the genus *Esthonyx* (Mammalia, Tillodontia) in the early Eocene of North America: Ann Arbor, University of Michigan Museum of Paleontology Contributions, v. 25, p. 125–153.
Gingerich, P. D., and Rose, K. D., 1977, Preliminary report on the American Clark Fork mammal fauna, and its correlation with similar faunas in Europe and Asia: Géobios, Mémoire Spécial, v. 1, p. 39–45.
——, 1982, Studies on Paleocene and early Eocene Apatemyidae (Mammalia, Insectivora); 1, Dentition of Clarkforkian *Labidolemur kayi:* Ann Arbor, University of Michigan Museum of Paleontology Contributions, v. 26, p. 49–55.
Gingerich, P. D., and Russell, D. E., 1981, *Pakicetus inachus,* a new archaeocete (Mammalia, Cetacea) from the early-middle Eocene Kuldana Formation of Kohat (Pakistan): Ann Arbor, University of Michigan Museum of Paleontology Contributions, v. 25, p. 235–246.
Gingerich, P. D., Wells, N. A., Russell, D. E., and Ibrahim Shah, S. M., 1983, Origin of whales in epicontinental remnant seas; New evidence from the early Eocene of Pakistan: Science, v. 220, p. 403–406.
Glanz, W. E., 1982, Adaptive zones of Neotropical mammals; A comparison of some temperate and tropical patterns: Pittsburgh, Pennsylvania, University of Pittsburgh Pymatuning Laboratory of Ecology Special Publication 6, p. 95–110.
Godinot, M., 1978, Diagnoses de trois nouvelles espèces de mammifères du Sparnacien de Provence: Compte Rendu Sommaire des Seances de la Societe Géologique de France, fasc. 6, p. 286–288.
——, 1981, Les Mammifères de Rians (Eocene inferieur, Provence): Paleovertebrata, v. 10, p. 43–126.
——, 1982, Aspects nouveaux des échanges entre les faunes mammaliennes d'Europe et d'Amérique du Nord a la base de l'Eocène: Géobios, Mémoire Special, no. 6, p. 403–412.
——, 1984, Un nouveau genre de Paromomyidae (Primates) de l'Eocène Inférieur d'Europe: Folia Primatologica, v. 43, p. 84–96.
Godinot, M., Broin, F. de, Buffetaut, E., Rage, J.-C., and Russell, D. E., 1978, Dormaal; Une des plus anciennes faunes éocènes d'Europe: Paris, Comptes Rendus Hebdomadaires des Seances de l'Academie des Sciences, v. 287, p. 1273–1276.
Godinot, M., Crochet, J.-Y., Hartenberger, J.-L., Lange-Badre, B., Russell, D. E., and Sigé, B., 1987, Nouvelles données sur les mammifères de Palette (Eocène inférieur, Provence): Münchner Geowissenschaftliche Abhandlungen, v. 10, p. 273–288.
Graham, R. W., 1976, Late Wisconsin mammalian faunas and environmental gradients of the eastern United States: Paleobiology, v. 2, p. 343–350.
——, 1986, Response of mammalian communities to environmental changes during the late Quaternary, *in* Diamond, J., and Case, T. J., eds., Community ecology: New York, Harper and Row Publishers, p. 300–313.
Grayson, D. K., 1973, On the methodology of faunal analysis: American Antiquity, v. 38, p. 432–439.
——, 1978, Minimum numbers and sample size in vertebrate faunal analysis: American Antiquity, v. 43, p. 53–65.
Gunnell, G. F., 1985, Systematics of early Eocene Microsyopinae (Mammalia, Primates) in the Clark's Fork Basin, Wyoming: Ann Arbor, University of Michigan Museum of Paleontology Contributions, v. 27, p. 51–71.
Hallam, A., 1981, Biogeographic relations between the northern and southern continents during the Mesozoic and Cenozoic: Geologische Rundschau, v. 70, p. 583–595.
Haq, B. U., Premoli-Silva, I., and Lohmann, G. P., 1977, Calcareous plankton paleobiogeographical evidence for major climatic fluctuations in the early Cenozoic Atlantic Ocean: Journal of Geophysical Research, v. 82, p. 3861–3876.
Harper, C. W., Jr., 1975, Standing diversity of fossil groups in successive intervals of geologic time; A new measure: Journal of Paleontology, v. 49, p. 752–757.
Harbenberger, J.-L., 1977, A propos de l'origine des Rongeurs: Géobios, Mémoire Spécial, v. 1, p. 183–193.
——, 1980, Données et hypothèses sur la radiation initiale des Rongeurs: Palaeovertebrata, Mémoire Jubilee R. Lavocat, p. 285–301.
——, 1982, Vertebrate faunal exchanges between Indian subcontinent and central Asia in early Tertiary times: Bollettino della Società Paleontologica Italiana, v. 21, p. 283–288.
——, 1986, Hypothèse paléontologique sur l'origine des Macroscelidea (Mammalia): Paris, Comptes Rendus Hebdomadaires des Seances de l'Academie des Sciences, v. 302, p. 247–249.
Hartenberger, J.-L., Martinez, C., and Ben Said, A., 1985, Decouverte de Mam-

mifères d'âge Éocène inférieur en Tunisie Centrale: Paris, Comptes Rendus Hebdomadaires des Seances de l'Academie des Sciences, v. 301, p. 649–652.

Hickey, L., 1977, Stratigraphy and paleobotany of the Golden Valley Formation (early Tertiary) of western North Dakota: Geological Society of America Memoir 150, 181 p.

—— , 1980, Paleocene stratigraphy and flora of the Clark's Fork Basin, *in* Gingerich, P. D., ed., Early Cenozoic paleontology and stratigraphy of the Bighorn Basin, Wyoming: Ann Arbor, University of Michigan Papers in Paleontology, v. 24, p. 33–49.

Hickey, L., West, R. M., Dawson, M. R., and Choi, D. K., 1983, Arctic terrestrial biota; Paleomagnetic evidence of age disparity with mid-northern latitudes during the late Cretaceous and early Tertiary: Science, v. 221, p. 1153–1156.

—— , 1984, Reply *to* technical comments *by* Kent and others and Norris and Miall 'Arctic biostratigraphic heterochroneity': Science, v. 224, p. 175–176.

Hillhouse, J. W., and Gromme, C. S., 1982, Limits to northward drift of the Paleocene Cantwell Formation, central Alaska: Geology, v. 10, p. 552–556.

—— , 1983, Wrangellia in southern Alaska 50 million years ago: EOS Transactions of the American Geophysical Union, v. 64, p. 687.

Hoffstetter, R., 1972, Relationships, origins, and history of the ceboid monkeys and caviomorph rodents; A modern reinterpretation: Evolutionary Biology, v. 6, p. 323–347.

—— , 1974, Phylogeny and geographical deployment of the Primates: Journal of Human Evolution, v. 3, p. 327–350.

Hooker, J. J., 1979, Two new condylarths (Mammalia) from the early Eocene of southern England: British Museum of Natural History (Geology) Bulletin, v. 32, p. 43–56.

—— , 1980, The succession of *Hyracotherium* (Perissodactyla, Mammalia) in the English early Eocene: British Museum of Natural History (Geology) Bulletin, v. 33, p. 101–114.

Hooker, J. J., and Insole, A. N., 1980, The distribution of mammals in the English Palaeogene: Tertiary Research, v. 3, p. 31–45.

Huang Xueshi and Zheng Jiajian, 1987, A new pantodont-like mammal from the Paleocene of Chienshan Basin, Anhui: Vertebrata PalAsiatica, v. 1, p. 20–31.

Hurlbert, S. H., 1971, The nonconcept of species diversity; A critique and alternative parameters: Ecology, v. 52, p. 577–586.

Hutchinson, G. E., 1961, The paradox of the plankton: American Naturalist, v. 95, p. 137–145.

Jacobs, L. L., Congleston, J. D., Brunet, M., Dejax, J., Flynn, L. J., Hell, J. V., and Mouchelin, G., 1988, Mammal teeth from the Cretaceous of Africa: Nature, v. 336, p. 158–160.

Jaeger, J. J., and Martin, M., 1984, African marsupials; Vicariance or dispersal?: Nature, v. 312, p. 379.

Karr, J. R., 1971, Structure of avian communities in selected Panama and Illinois habitats: Ecological Monographs, v. 41, p. 207–233.

Kent, D. V., McKenna, M. C., Opdyke, N. D., Flynn, J. J., and MacFadden, B. J., 1984, Arctic biostratigraphic heterochroneity: Science, v. 224, p. 173–174.

Kihm, A. J., 1984, Early Eocene mammalian faunas of the Piceance Creek Basin, northwestern Colorado [Ph.D. thesis]: Boulder, University of Colorado, 390 p.

Klootwijk, C. T., Conaghan, P. J., and Powell, C. McA., 1985, The Himalayan Arc; Large-scale continental subduction, oroclinal bending and back-arc spreading: Earth and Planetary Science Letters, v. 75, p. 167–183.

Korth, W. W., 1984, Earliest Tertiary evolution and radiation of rodents in North America: Carnegie Museum of Natural History Bulletin, v. 24, p. 1–71.

Kraus, M. J., 1980, Genesis of a fluvial sheet sandstone, Willwood Formation, northwest Wyoming, *in* Gingerich, P. D., ed., Early Cenozoic paleontology and stratigraphy of the Bighorn Basin, Wyoming: Ann Arbor, University of Michigan Papers in Paleontology, v. 24, p. 87–94.

Krause, D. W., 1982, Multituberculates from the Wasatchian Land-Mammal Age, early Eocene, of western North America: Journal of Paleontology, v. 56, p. 271–294.

—— , 1986, Competitive exclusion and taxonomic displacement in the fossil record; The case of rodents and multituberculates in North America: Laramie, University of Wyoming Contributions to Geology Special Paper 3, p. 95–117.

Krause, D. W., and Gingerich, P. D., 1983, Mammalian fauna from Douglass Quarry, earliest Tiffanian (late Paleocene) of the eastern Crazy Mountain Basin, Montana: Ann Arbor, University of Michigan Museum of Paleontology Contributions, v. 26, p. 157–196.

Krause, D. W., and Maas, M. C., 1988, The biogeographic origins of Clarkforkian and Wasatchian mammalian immigrants to the Western Interior: Journal of Vertebrate Paleontology, v. 8, p. 19A.

Krishtalka, L., and Stucky, R. K., 1983a, Revision of the Wind River faunas, early Eocene of central Wyoming; Part 3, Marsupialia: Annals of Carnegie Museum, v. 52, p. 205–227.

—— , 1983b, Paleocene and Eocene marsupials of North America: Annals of Carnegie Museum, v. 52, p. 229–263.

Krishtalka, L., and 10 others, 1987, Eocene (Wasatchian through Duchesnean) biochronology of North America, *in* Woodburne, M. O., ed., Cenozoic mammals of North America; Geochronology and biostratigraphy: Berkeley, University of California Press, p. 77–117.

Krutzsch, W., 1967, Der Florenwechsel im Alttertiar Mitteleuropas auf Grund von Zonengliederung im Jungtertiar Mitteleuropas: Abhandlungen Des Zentralen Geologischen Institutes, v. 10, p. 83–98.

Kumar, K., and Jolly, A., 1986, Earliest artiodactyl (*Diacodexis,* Dichobunidae: Mammalia) from the Eocene of Kalakot, north-western Himalaya, India: I.S.G. Bulletin, v. 2, p. 20–30.

Kurten, B., 1966, Holarctic land connexions in the early Tertiary: Commentationes Biologicae Societas Scientarium Fennica, v. 29, p. 1–5.

Lasker, H., 1976, Effects of differential preservation on the measurement of taxonomic diversity: Paleobiology, v. 2, p. 84–93.

Lavocat, R., 1974, The interrelationships between the African and South American rodents and their bearing on the problem of the origin of South American monkeys: Journal of Human Evolution, v. 3, p. 323–326.

—— , 1980, The implications of rodent paleontology and biogeography to the geographical sources and origin of platyrrhine primates, *in* Ciochon, R. L., and Chiarelli, A. B., eds., Evolutionary biology of the New World monkeys and continental drift: New York, Plenum Press, p. 93–102.

Leffingwell, H. A., 1971, Palynology of the Lance (Late Cretaceous) and Fort Union (Paleocene) Formations in the Type Lance area, Wyoming, *in* Kosanke, R. M., and Cross, A. T., eds., Symposium on palynology of the Late Cretaceous and early Tertiary: Geological Society of America Special Paper 127, p. 1–64.

Lemke, R. W., 1960, Geology of the Sioux River area, North Dakota: U.S. Geological Survey Professional Paper 325, 138 p.

Li, C.-K., 1977, Paleocene eurymyloids (Anagalida, Mammalia) of Qianshan, Anhui: Vertebrate PalAsiatica, v. 15, p. 104–118.

Li, C.-K., and Ting, S.-Y., 1983, The Paleogene mammals of China: Carnegie Museum of Natural History Bulletin 21, p. 1–98.

—— , 1985, Possible phylogenetic relationship of eurymylids and rodents, with comments on mimotonids, *in* Luckett, W. P., and Hartenberger, J.-L., eds., Evolutionary relationships among rodents; A multidisciplinary analysis: New York, Plenum Press, p. 35–58.

Lowrie, W., and Alvarez, W., 1981, One hundred million years of geomagnetic polarity history: Geology, v. 9, p. 392–397.

Lucas, S. G., 1984, Systematics, biostratigraphy, and evolution of early Cenozoic *Coryphodon* (Mammalia, Pantodonta) [Ph.D. thesis]: New Haven, Connecticut, Yale University, 649 p.

Lucas, S. G., and Schoch, R. M., 1981, *Basalina,* a tillodont from the Eocene of Pakistan: Mitteilungen der Bayerischen Staatssammlung für Palaontologie und Historische Geologie, v. 21, p. 89–95.

Lucas, S. G., and Tong, Y., 1988, A new coryphodontid (Mammalia, Pantodonta) from the Eocene of China: Journal of Vertebrate Paleontology, v. 7, p. 362–372.

Luckett, W. P., and Harbenberger, J.-L., 1985, Evolutionary relationships among rodents; Comments and conclusions, *in* Luckett, W. P., and Hartenberger, J.-L., eds., Evolutionary relationships among rodents; A multidisciplinary analysis: New York, Plenum Press, p. 685–712.

MacArthur, R. H., 1960, On the relative abundance of species: American Naturalist, v. 44, p. 25–36.
—— , 1965, Patterns of species diversity: Biological Reviews, v. 40, p. 510–533.
MacFadden, B. J., 1988, Horses, the fossil record, and evolution; A current perspective, *in* Hecht, M. K., Wallace, B., and Prance, G. T., eds., Evolutionary biology: New York, Plenum Press, v. 22, p. 131–158.
Mahboubi, M., Ameur, R., Crochet, J.-Y., and Jaeger, J.-J., 1983, Première découverte d'un Marsupial en Afrique: Paris, Comptes Rendus Hebdomadaires des Seances de l'Academie des Sciences, v. 297, p. 691–694.
—— , 1984a, Earliest known proboscidean from early Eocene of north-west Africa: Nature, v. 308, p. 543–544.
—— , 1984b, Implications paléobiogéographiques de la découverte d'une nouvelle localité Éocène à vértebrés continentaux en Afrique Nord-occidentale: El Kohol (Sud-Oranais, Algérie): Géobios, no. 17, p. 625–629.
—— , 1986, El Kohol (Saharan Atlas, Algeria); A new Eocene mammal locality in northwestern Africa; Stratigraphical, phylogenetic, and paleobiogeographical data: Palaeontographica, Atb. A, v. 192, p. 15–49.
Marincovich, L., Jr., Brouwers, E. M., and Carter, D. L., 1985, Early Tertiary marine fossils from northern Alaska; Implications for Arctic Ocean paleogeography and faunal evolution: Geology, v. 13, p. 770–773.
Marshall, L. G., 1987, Systematics of Itaboraian (middle Paleocene) age opossum-like marsupials from the limestone quarry at Sao Jose de Itaborai, Brazil, *in* Archer, M., ed., Possums and opossums: Sydney, New South Wales, Royal Zoological Society, p. 91–160.
Marshall, L. G., and de Muizon, C., 1988, The dawn of the Age of Mammals in South America: National Geographic Research, v. 4, p. 23–55.
Mathur, U. B., and Srivastava, S., 1987, Dinosaur teeth from Lameta Group (Upper Cretaceous) of Kheda District, Gujarat: Journal Geological Society of India, v. 29, p. 554–566.
Matthew, W. D., 1915, Climate and evolution: New York Academy of Science Annals, v. 24, p. 171–138.
—— , 1928, The evolution of mammals in the Eocene: Zoological Society of London Proceedings, part 4, 1927, p. 947–985.
Matthew, W. D., and Granger, W., 1925, Fauna and correlation of the Gashato Formation of Mongolia: American Museum Novitates, no. 376, p. 1–12.
McCoy, E. D., and Connor, E. F., 1980, Latitudinal gradients in the species diversity of North American mammals: Evolution, v. 34, p. 193–203.
McKenna, M. C., 1967, Classification, range, and deployment of the prosimian primates: Colloques Internationaux du Centre National de la Recherche Scientifique, no. 163, p. 603–610.
—— , 1969, The origin and early differentiation of therian mammals: New York Academy of Sciences Annals, v. 167, p. 217–240.
—— , 1975a, Fossil mammals and early Eocene North Atlantic land continuity: Missouri Botanical Gardens Annals, v. 62, p. 335–353.
—— , 1975b, Toward a phylogenetic classification of the Mammalia, *in* Luckett, W. P., and Szalay, F. S., eds., Phylogeny of the primates; A multidisciplinary approach: New York, Plenum Press, p. 21–46.
—— , 1980, Early history and biogeography of South America's extinct land mammals, *in* Ciochon, R. L., and Chiarelli, A. B., eds., Evolutionary biology of the New World monkeys and continental drift: New York, Plenum Press, p. 43–77.
—— , 1983a, Holarctic landmass rearrangement, cosmic events, and Cenozoic terrestrial organisms: Missouri Botanical Gardens Annals, v. 70, p. 459–489.
—— , 1983b, Cenozoic paleogeography of North Atlantic land bridges, *in* Bott, M.H.P., Saxon, S., Talwani, M., and Thiede, J., eds., Structure and development of the Greenland–Scotland Ridge: New York, Plenum Press, p. 351–399.
McKenna, M. C., Engelmann, G. F., and Barghoorn, S. F., 1977, Review *of* 'Cranial anatomy and evolution of Early Tertiary Plesiadapidae (Mammalia, Primates)' by Philip D. Gingerich: Systematic Zoology, v. 26, p. 233–238.
Molnar, P., and Tapponier, P., 1975, Cenozoic tectonics of Asia; Effects of a continental collision: Science, v. 189, p. 419–426.
Molnar, P., Pardo-Casas, F., and Stock, J., 1988, The Cenozoic and Late Cretaceous evolution of the Indian Ocean Basin; Uncertainties in the reconstructed positions of the Indian, Arabian and Antarctic plates: Basin Research, vol. 1, p. 23–40.
Morris, W. J., 1966, Fossil mammals from Baja California; New evidence on early Tertiary migrations: Science, v. 153, p. 1376–1378.
—— , 1968, A new early Tertiary perissodactyl, *Hyracotherium seekinsi,* from Baja California: Los Angeles County Museum Contributions in Science, no. 151, p. 1–11.
de Muizon, C., and Marshall, L. G., 1987, Le plus ancien Pantodonte (Mammalia), du Cretacé supérieur de Bolivie: Paris, Comptes Rendus Hebdomadaires des Seances de l'Academie des Sciences, v. 304, Serie II, no. 5, p. 205–208.
Norris, G., and Miall, A. D., 1984, Arctic biostratigraphic heterochroneity: Science, v. 224, p. 174–175.
Novacek, M. J., 1985, Cranial evidence for rodent affinities, *in* Luckett, W. P., and Hartenberger, J.-L., eds., Evolutionary relationships among rodents; A multidisciplinary analysis: New York, Plenum Press, p. 59–81.
Novacek, M. J., Flynn, J. J., Ferrusquia-Villafranca, I., and Cipolletti, R. M., 1987, An early Eocene (Wasatchian) mammal fauna from Baja California: National Geographic Research, v. 3, p. 376–388.
Osman, R. W., and Whitlatch, E., 1978, Patterns of species diversity; Fact or artifact?: Paleobiology, v. 4, p. 41–54.
Parrish, J. T., 1987, Global palaeogeography and palaeoclimate of the Late Cretaceous and early Tertiary, *in* Friis, E. M., Chaloner, W. G., and Crane, P. R., eds., The origins of angiosperms and their biological consequences: Cambridge, Cambridge University Press, p. 51–73.
Patriat, P., and Achache, J., 1984, India-Eurasia collision chronology has implications for crustal shortening and driving mechanism of plates: Nature, v. 311, p. 615–621.
Patterson, B., 1934, A contribution to the osteology of *Titanoides* and the relationships of the Amblypoda: American Philosophical Society Proceedings, v. 73, p. 71–101.
—— , 1939, New Pantodonta and Dinocerata from the upper Paleocene of western Colorado: Field Museum of Natural History Geology Series, v. 6, p. 351–384.
—— , 1958, Affinities of the Patagonian fossil mammal *Necrolestes:* Breviora, no. 94, p. 1–14.
Patterson, B., and Pascual, R., 1972, The fossil mammal fauna of South America, *in* Keast, A., Erk, F. C., and Glass, B., eds., Evolution, mammals, and southern continents: Albany, State University of New York Press, p. 247–309.
Peet, R. K., 1974, The measurement of species diversity: Annual Review of Ecology and Systematics, v. 5, p. 285–307.
Pianka, E. R., 1983, Evolutionary ecology: New York, Harper and Row Publishers, 416 p.
Pierce, J. W., 1978, The northward motion of India since the Late Cretaceous: Geophysical Journal of the Royal Astronomical Society, v. 52, p. 277–311.
Plafker, G., 1984, Comment *on* Model for the origin of the Yakutat block, an accreting terrane in the northern Gulf of Alaska': Geology, v. 12, p. 563–564.
Powell, C. McA., 1979, A speculative tectonic history of Pakistan and surroundings; Some constraints from the Indian Ocean, *in* Farah, A., and DeJong, K. A., eds., Geodynamics of Pakistan: Quetta, Geological Survey of Pakistan, p. 5–24.
Powell, C. McA., and Conaghan, P. J., 1973, Plate tectonics and the Himalayas: Earth and Planetary Science Letters, v. 20, p. 1–12.
Prasad, G.V.R., and Sahni, A., 1988, First Cretaceous mammal from India: Nature, v. 332, p. 638–640.
Prothero, D., and Manning, E., 1987, Phylogeny of the ungulates: Journal of Vertebrate Paleontology, v. 7, p. 23A.
Rabinowitz, P. D., Coffin, M. F., and Falvey, D., 1983, The separation of Madagascar and Africa: Science, v. 220, p. 67–69.
Radinsky, L., 1966, The adaptive radiation of the phenacodontid condylarths and the origin of the Perissodactyla: Evolution, v. 20, p. 408–417.

Radinsky, L., and Ting, S.-Y., 1984, The skull of *Ernanodon,* an unusual fossil mammal: Journal of Mammalogy, v. 65, p. 155–158.

Rage, J.-C., 1988, Gondwana, Tethys, and terrestrial vertebrates during the Mesozoic and Cainozoic, *in* Audley-Charles, M. G., and Hallam, A., eds., Gondwana and Tethys: New York, Oxford University Press, p. 255–273.

Raiverman, V., 1972, Time series and stratigraphic correlation of Cenozoic sediments in foothills of Himachal Pradesh: Himalayan Geology, v. 2, p. 82–101.

Rogers, J. S., 1975, Species density and taxonomic diversity of Texas amphibians and reptiles: Systematic Zoology, v. 25, p. 26–40.

Rose, K. D., 1975a, The Carpolestidae, early Tertiary primates from North America: Cambridge, Massachusetts, Harvard University Bulletin of the Museum of Comparative Zoology, v. 147, p. 1–74.

——, 1975b, *Elpidophorus,* the earliest dermopteran (Dermoptera, Plagiomenidae): Journal of Mammalogy, v. 56, p. 676–679.

——, 1978, A new Paleocene epoicotheriid (Mammalia), with comments on the Palaeanodonta: Journal of Paleontology, v. 52, p. 658–674.

——, 1979, A new Paleocene palaeanodont and the origin of the Metacheiromyidae (Mammalia): Breviora, v. 455, p. 1–14.

——, 1980, Clarkforkian Land-Mammal Age; Revised definition, zonation, and tentative intercontinental correlations: Science, v. 208, p. 744–746.

——, 1981a, The Clarkforkian Land-Mammal Age and mammalian faunal composition across the Paleocene-Eocene boundary: Ann Arbor, University of Michigan Papers on Paleontology, v. 26, p. 1–197.

——, 1981b, Composition and species diversity in Paleocene and Eocene mammal assemblages; An empirical study: Journal of Vertebrate Paleontology, v. 1, p. 367–388.

——, 1982, Skeleton of *Diacodexis,* oldest known artiodactyl: Science, v. 216, p. 621–623.

——, 1985a, Skeletal anatomy in early ungulates and the origin of modern ungulate orders: 4th International Theriological Congress, Abstract no. 0537.

——, 1985b, Comparative osteology of North American dichobunid artiodactyls: Journal of Paleontology, v. 59, p. 1203–1226.

——, 1987a, Climbing adaptations in the early Eocene mammal *Chriacus* and the origin of Artiodactyla: Science, v. 236, p. 314–316.

——, 1987b, New skeletal remains of Eocene palaeanodonts: Journal of Vertebrate Paleontology, v. 7, p. 24A.

Russell, D. E., and Dashzeveg, D., 1986, Early Eocene insectivores (Mammalia) from the People's Republic of Mongolia: Palaeontology, v. 29, p. 269–291.

Russell, D. E., and Gingerich, P. D., 1987, Nouveaux Primates de l'Éocène du Pakistan: Paris, Comptes Rendus Hebdomadaires des Seances de l'Academie des Sciences, v. 304, p. 209–214.

Russell, D. E., and Zhai, R.-J., 1987, The Paleogene of Asia; Mammals and stratigraphy: Museum National d'Histoire Naturelle, Sciences de la Terre, Memoires, v. 52, p. 1–488.

Russell, D. E., chairman, and 32 others, 1982a, Tetrapods of the northwest European Tertiary basin: Geologisches Jahrbuch, v. A60, p. 5–74.

Russell, D. E., Hartenberger, J.-L., Pomerol, Ch., Sen, S., Schmidt-Kittler, N., and Vianey-Liaud, M., 1982b, Mammals and stratigraphy; The Paleogene of Europe: Palaeovertebrata, Mémoire Extraordinaire 1982, p. 1–77.

Sahni, A., 1984a, Upper Cretaceous–early Palaeogene palaeobiogeography of India based on terrestrial vertebrate faunas: Societe Géologique de France, Memoires, no. 147, p. 125–137.

——, 1984b, Cretaceous-Paleocene terrestrial faunas of India; Lack of endemism during drift of the Indian plate: Science, v. 226, p. 441–443.

——, 1986, Cretaceous-Paleocene microfossil assemblages from the Infra and Intertrappeans of peninsular India; Implications for the drift of the Indian plate: Bulletin of the Geological Mining and Metallurgical Society of India, v. 54, p. 91–100.

Sahni, A., and Kumar, K., 1974, Palaeogene palaeobiogeography of the Indian subcontinent: Palaeogeography, Palaeoclimatology and Palaeoecology, v. 15, p. 209–226.

Sahni, A., Bhatia, S. B., Hartenberger, J.-L., Jaeger, J. J., Kumar, K., Sudre, J., and Vianey-Liaud, M., 1981, Vertebrates from the type section of the Subathu Formation and comments on the palaeobiogeography of the Indian subcontinent during the early Palaeogene: Indian Geological Association Bulletin, v. 14, p. 89–100.

Sahni, A., Kumar, K., Hartenberger, J.-L., Jaeger, J.-J., Rage, J.-C., Sudre, J., and Vianey-Liaud, M., 1982, Microvertébrés nouveaux des Trapps du Deccan (Inde); Mise en evidence d'une voie de communication terrestre probable entre la Laurasie et l'Inde à la limite Crétacé-Tertiaire: Societe Géologique de France Bulletin, v. 24, p. 1093–1099.

Sahni, A., Bhatia, S. B., and Kumar, K., 1983, Faunal evidence for the withdrawal of the Tethys in the Lesser Himalaya, northwestern India: Bolletino della Società Paleontologica Italiana, v. 22, p. 77–86.

Sahni, A., Prasad, G.V.R., and Rana, R. S., 1986, New palaeontological evidence for the age and initiation of the Deccan Volcanics, Central Peninsular India: Gondwana Geology Magazine, v. 1, p. 13–24.

Savage, D. E., 1971, The Sparnacian-Wasatchian mammalian fauna, early Eocene, of Europe and North America: Abhandlungen Des Hessischen Landesamptes für Bodenforschung, v. 60, p. 154–158.

Savage, D. E., and Russell, D. E., 1983, Mammalian paleofaunas of the world: Reading, Pennsylvania, Addison-Wesley Publishing Company, 432 p.

Savin, S. M., 1977, The history of the earth's surface temperature during the past 100 million years, *in* Donath, F. A., Stehli, F. G., and Wetherill, G. W., eds., Annual Review of Earth and Planetary Sciences: Palo Alto, California, Annual Reviews, Inc., p. 319–355.

Schall, J. J., and Pianka, E. R., 1978, Geographical trends in numbers of species: Science, v. 201, p. 679–686.

Schiebout, J. A., 1979, An overview of the terrestrial early Tertiary of southern North America; Fossil sites and paleopedology: New Orleans, Louisiana, Tulane Studies in Geology and Paleontology, v. 15, p. 75–94.

——, 1981, Effects of sea level changes on the distribution and evolution of early Tertiary mammals: Gulf Coast Association of Geological Societies Transactions, v. 31, p. 383–387.

Schoch, R. M., 1985, Preliminary description of a new late Paleocene land-mammal fauna from South Carolina, U.S.A.: Postilla, no. 196, p. 1–13.

——, 1986, Systematics, functional morphology, and macroevolution of the extinct mammalian order Taeniodonta: New Haven, Connecticut, Yale University Peabody Museum of Natural History Bulletin 42, p. 1–307.

Schoch, R. M., and Lucas, S. G., 1985, The phylogeny and classification of the Dinocerata (Mammalia, Eutheria): University of Uppsala Geological Institute Bulletin, v. 11, p. 31–50.

Schum, M., 1984, Phenetic structure and species richness in North and Central American bat faunas: Ecology, v. 65, p. 1315–1324.

Scillato-Yané, G. J., 1986, Los Xenarthra fosiles de Argentina (Mammalia, Edentata), *in* Pascual, R., organizador, Simposia "Evolucion de los vertebrados cenozoicos de America del sur": v. 2, p. 151–155.

Scillato-Yané, G. J., and Pascual, R., 1984, Un peculiar Paratheria (Edentata, Mammalia) del Paleoceno de Patagonia: Primeras Jornadas Argentinas de Paleontologia de Vertebrdos (La Plata, 1984), Resúmenes, 15 p.

——, 1985, Un peculiar Xenarthra del Paleoceno medio de Patagonia (Argentina); Su imporantancia en la sistematica de los Paratheria: Ameghiniana, v. 21, p. 173–176.

Sclater, J. G., Luyendyk, B. P., and Meinke, L., 1976, Magnetic lineations in the southern part of the Central Indian Basin: Geological Society of America Bulletin, v. 87, p. 371–378.

Shackleton, N. J., 1984, Oxygen isotope evidence for Cenozoic climatic change, *in* Brenchley, P. J., ed., Fossils and climate: Chichester, John Wiley and Sons, p. 352.

Simons, E. L., 1960, The Paleocene Pantodonta: American Philosophical Society Transactions, v. 50, p. 1–99.

Simpson, E. H., 1949, Measurement of diversity: Nature, v. 163, p. 688.

Simpson, G. G., 1928, A new mammalian fauna from the Fort Union of southern Montana: American Museum Novitates, no. 297, 15 p.

——, 1932, A new Paleocene mammal from a deep well in Louisiana: U.S. National Museum Proceedings, v. 82, p. 1–4.

——, 1937, The Fort Union of the Crazy Mountain Field, Montana, and its mammalian faunas: U.S. National Museum Bulletin, v. 169, 287 p.
——, 1947, Holarctic mammalian faunas and continental relationships during the Cenozoic: Geological Society of America Bulletin, v. 58, p. 613–688.
——, 1951, History of the fauna of Latin America, *in* Baitsell, G. A., ed., Science in progress: New Haven, Connecticut, Yale University Press, p. 369–408.
——, 1964, Species diversity of North America Recent mammals: Systematic Zoology, v. 13, p. 57–73.
——, 1965, The geography of evolution: Philadelphia, Pennsylvania, Chilton Books, 249 p.
——, 1978, Early mammals in South America; Fact, controversy, and mystery: American Philosophical Society Proceedings, v. 122, p. 318–328.
——, 1980, Splendid isolation; The curious history of South American mammals: New Haven, Connecticut, Yale University Press, 266 p.
Sloan, R. E., 1969, Cretaceous and Paleocene terrestrial communities of western North America: North American Paleontological Convention Proceedings, v. 1(E), p. 427–453.
——, 1987, Paleocene and latest Cretaceous mammal ages, biozones, magnetozones, rates of sedimentation, and evolution, *in* Fassett, J. E., and Rigby, J. K., Jr., The Cretaceous-Tertiary boundary in the San Juan and Raton Basins, New Mexico and Colorado: Geological Society of America Special Paper 209, p. 165–200.
Spicer, R. A., Wolfe, J. A., and Nichols, D. J., 1987, Alaskan Cretaceous-Tertiary floras and Arctic origins: Paleobiology, v. 13, p. 73–83.
Srivastava, S. K., 1983, Cretaceous phytogeoprovinces and paleogeography of the Indian plate based on palynological data, *in* Proceedings of the Symposium Cretaceous of India; Palaeoecology, palaeogeography, and time boundaries: Indian Association of Palynostratigraphers, Lucknow, India, p. 141–157.
Stanley, E. A., 1965, Upper Cretaceous and Paleocene plant microfossils and Paleocene dinoflagellates and hystrichosphaerids from northwestern South Dakota: Bulletins of American Paleontology, v. 49, p. 175–384.
Stone, D. B., and Packer, D. R., 1979, Paleomagnetic data from the Alaska Peninsula: Geological Society of America Bulletin, Part 1, v. 90, p. 545–560.
Stone, D. B., Panuska, B. C., and Packer, D. R., 1982, Paleolatitudes versus time for southern Alaska: Journal of Geophysical Research, v. 87, p. 3697–3707.
Storch, G., 1978, *Eomanis waldi,* ein Schuppentier aus dem Mittel-Eozän der "Grube Messel" bei Darmstadt (Mammalia, Pholidota): Senckenbergiana Lethaea, v. 59: p. 503–529.
——, 1981, *Eurotamandua joresi,* ein Myrmecophagide aus dem Eozän der "Grube Messel" bei Darmstadt (Mammalia, Xenarthra): Senckenbergiana Lethaea, v. 61, p. 247–289.
Sudre, J., and Russell, D. E., 1982, Les Mammifères Montiens de Hainin (Paléoceñe moyen de Belgique); Part 2, Les Condylarthres: Palaeovertebrata, v. 12, p. 173–184.
Szalay, F. S., 1968, The beginnings of primates: Evolution, v. 1, p. 19–36.
Szalay, F. S., and Li, C.-K., 1986, Middle Paleocene euprimate from southern China and the distribution of primates in the Paleogene: Journal of Human Evolution, v. 15, p. 387–397.
Szalay, F. S., and McKenna, M. C., 1971, Beginning of the Age of Mammals in Asia; The late Paleocene Gashato fauna, Mongolia: American Museum of Natural History Bulletin, v. 144, p. 269–318.
Taquet, P., 1982, Une connexion continentale entre Afrique et Madagascar au Crétacé supérieur; Données géologiques et paléontologiques, *in* Buffetaut, E., Janvier, P., Rage, J.-C., and Tassy, P., eds., Phylogénie et paléobiogéographie; Livre jubilaire en l'honneur de R. Hoffstetter: Geobios, Mémoire Spécial, v. 6, p. 385–391.
Tarling, D. H., 1980, The geologic evolution of South America with special reference to the last 200 million years, *in* Ciochon, R. L., and Chiarelli, A. B., eds., Evolutionary biology of the New World monkeys and continental drift: New York, Plenum Press, p. 1–41.
——, 1988, Gondwanaland and the evolution of the Indian Ocean, *in* Audley-Charles, M. G., and Hallam, A., eds., Gondwana and Tethys: New York, Oxford University Press, p. 61–77.
Thewissen, J.G.M., Gingerich, P. D., and Russell, D. E., 1987, Artiodactyla and Perissodactyla (Mammalia) from the early-middle Eocene Kuldana Formation of Kohat (Pakistan): Ann Arbor, University of Michigan Museum of Paleontology Contributions, v. 27, p. 247–274.
Thewissen, J.G.M., Russell, D. E., Gingerich, P. D., and Hussain, S. T., 1983, A new dichobunid artiodactyl (Mammalia) from the Eocene of north-west Pakistan: Proceedings of the Koninklijke Nederlandse Akademie van Wetenschappen, Series B, v. 86, p. 153–180.
Ting, S.-Y., 1979, A new edentate from the Paleocene of Guangdong: Vertebrata PalAsiatica, v. 17, p. 57–64.
Upchurch, G. A., Jr., and Wolfe, J. A., 1987, Mid-Cretaceous to early Tertiary vegetation and climate; Evidence from fossil leaves and woods, *in* Friis, E. M., Chaloner, W. G., and Crane, P. R., eds., The origins of angiosperms and their biological consequences: New York, Cambridge University Press, p. 75–105.
Van Devender, T. R., 1986, Climatic cadences and the composition of Chihuahuan desert communities; The late Pleistocene packrat midden record, *in* Diamond, J., and Case, T. J., eds., Community ecology: New York, Harper and Row Publishers, p. 285–299.
Van Valen, L., 1963, The origin and status of the mammalian order Tillodontia: Journal of Mammalogy, v. 44, p. 364–373.
——, 1966, Deltatheridia, a new order of mammals: American Museum of Natural History Bulletin, v. 132, p. 1–126.
——, 1971, Adaptive zones and the orders of mammals: Evolution, v. 25, p. 420–428.
——, 1978, The beginning of the Age of Mammals: Evolutionary Theory, v. 4, p. 45–80.
——, 1988, Faunas of a southern world: Nature, v. 333, p. 113.
Van Valen, L., and Sloan, R. E., 1966, The extinction ofthe multituberculates: Systematic Zoology, v. 15, p. 261–278.
Vianey-Liaud, M., 1979, Les Mammifères Montiens de Hainin (Paléocène moyen de Belgique); Part 1, Multituberculés: Palaeovertebrata, v. 9, p. 117–131.
Webb, S. D., 1985, Main pathways of mammalian diversification in North America, *in* Stehli, F. G., and Webb, S. D., eds., The great American biotic interchange: New York, Plenum Press, p. 201–217.
Wells, N. A., 1983, Transient streams in sand-poor redbeds; Early-middle Eocene Kuldana Formation of northern Pakistan: Special Publications, International Association of Sedimentologists, v. 6, p. 393–403.
Wells, N. A., and Gingerich, P. D., 1983, Review of Eocene Anthracobunidae (Mammalia, Proboscidea) with a new genus and species, *Jozaria palustris,* from the Kuldana Formation of Kohat (Pakistan): Ann Arbor, University of Michigan Museum of Paleontology Contributions, v. 26, p. 117–139.
West, R. M., 1976, The North American Phenacodontidae (Mammalia, Condylarthra): Milwaukee Public Museum Contributions to Biology and Geology, v. 6, p. 1–78.
——, 1980, Middle Eocene large mammal assemblage with Tethyan affinities, Ganda Kas region, Pakistan: Journal of Paleontology, v. 54, p. 508–533.
——, 1984, A review of the South Asian middle Eocene Moeritheriidae (Mammalia: Tethytheria): Societe Géologique de France Memoires, no. 147, p. 183–190.
Whittaker, R. H., 1972, Evolution and measurement of species diversity: Taxon, v. 21, p. 213–251.
——, 1977, Evolution of species diversity in land communities: Evolutionary Biology, v. 10, p. 1–67.
Wible, J. R., 1986, Transformations in the extracranial course of the internal carotid artery in mammalian phylogeny: Journal of Vertebrate Paleontology, v. 6, p. 313–325.
Wilson, J. W., III, 1974, Analytical zoogeography of North American mammals: Evolution, v. 28, p. 124–140.
Wing, S. L., 1984, A new basis for recognizing the Paleocene/Eocene boundary in Western Interior North America: Science, v. 226, p. 439–441.
——, 1987, Eocene and Oligocene floras and vegetation of the Rocky Mountains: Missouri Botanical Gardens Annals, v. 74, p. 748–784.
Wolfe, J. A., 1978, A paleobotanical interpretation of Tertiary climates in the Northern Hemisphere: American Scientist, v. 66, p. 694–703.

——, 1980, Tertiary climates and floristic interpretations at high latitudes in the Northern Hemisphere: Palaeogeography, Palaeoclimatology, Palaeoecology, v. 30, p. 313–323.

——, 1985, Distribution of major vegetational types during the Tertiary, *in* Sundquist, E. T., and Broecker, W. S., eds., The carbon cycle and atmospheric CO_2; Natural variations Archean to present: Geophysical Monograph 32, p. 357–375.

——, 1987a, Late Cretaceous-Cenozoic history of deciduousness and the terminal Cretaceous event: Paleobiology, v. 13, p. 215–226.

——, 1987b, An overview of the origins of the modern vegetation and flora of the northern Rocky Mountains: Missouri Botanical Gardens Annals, v. 74, p. 785–803.

Wolfe, J. A., and Hopkins, D. M., 1967, Climatic changes recorded by Tertiary land floras in northwestern North America, *in* Hatai, K., ed., Tertiary correlation and climatic changes in the Pacific, 11th Pacific Scientific Congress, Symposium 25: Sendai, Sasaki Printing and Publishing, p. 67–76.

Wolfe, J. A., and McCoy, S., Jr., 1984, Comment *on* 'Model for the origin of the Yakutat block, an accreting terrane in the northern Gulf of Alaska': Geology, v. 12, p. 564–565.

Wolfe, J. A., and Poore, R. Z., 1982, Tertiary marine and nonmarine climatic trends, *in* Geophysics Study Committee, Geophysics Research Board, Commission on Physical Sciences, Mathematics and Resources, National Research Council, compilers, Climate in earth history: Washington, D.C., National Academy Press, p. 154–158.

Wolfe, J. A., and Upchurch, G. R., Jr., 1986, Vegetation, climatic, and floral changes at the Cretaceous-Tertiary boundary: Nature, v. 324, p. 148–152.

——, 1987, Maestrichtian-Paleocene vegetation and climate of the Powder River Basin, Wyoming and Montana: Geological Society of America Abstracts with Programs, v. 19, p. 897.

Wood, A. E., 1962, The early Tertiary rodents of the family Paramyidae: American Philosophical Society Transactions, v. 52, p. 1–261.

——, 1977, The Rodentia as clues to Cenozoic migrations between the Americas and Europe and Africa: Milwaukee Public Museum Special Publications in Biology and Geology, v. 2, p. 95–109.

Wood, H. E., Chaney, R. W., Clark, J., Colbert, E. H., Jepsen, G. L., Reeside, J. B., Jr., and Stock, C., 1941, Nomenclature and correlation of the North American continental Tertiary: Geological Society of America Bulletin, v. 52, p. 1–48.

Wood, R. C., 1967, A review of the Clark Fork vertebrate fauna: Breviora, v. 257, p. 1–30.

Yadagiri, P., 1984, New symmetrodonts from Kota Formation (Early Jurassic), India: Geological Society of India Journal, v. 25, p. 514–521.

——, 1985, An amphidontid symmetrodont from the Early Jurassic Kota Formation, India: Zoological Journal of the Linnean Society, v. 85, p. 411–417.

Zheng Jia-jian, 1979, The Paleocene notoungulates of Jiang-xi, *in* A Symposium on Cretaceous and early Tertiary Red Beds of South China: Beijing, Science Press, Academica Sinica, p. 387–394. (in Chinese with English summary).

Ziegler, A. M., Scotese, C. R., and Barrett, S. F., 1983, Mesozoic and Cenozoic paleogeographic maps, *in* Brosche, P., and Sundermann, J., eds., Tidal friction and the Earth's rotation, v. 2: Berlin, Springer-Verlag, p. 240–252.

MANUSCRIPT ACCEPTED BY THE SOCIETY JUNE 12, 1989

Printed in U.S.A.

Geological Society of America
Special Paper 243
1990

Postcranial skeletal remains and adaptations in early Eocene mammals from the Willwood Formation, Bighorn Basin, Wyoming

Kenneth D. Rose
Department of Cell Biology and Anatomy, Johns Hopkins University School of Medicine, Baltimore, Maryland 21205

ABSTRACT

The Bighorn Basin has produced the richest and most diverse early Eocene mammalian faunas in the world and is the principal source of our knowledge of skeletal anatomy in these mammals. Until recently, most of our information on postcranial anatomy in early Eocene mammals came from the works of Matthew and his contemporaries. Considerable new evidence has been unearthed in the last 25 years, but very little of it has yet been described or even reported in the literature. Since 1979, a USGS–Johns Hopkins project working in the Wasatchian part of the Willwood Formation has collected more than 150 skeletal associations (representing more than 25 genera in 20 families), varying from several bones to virtually complete, articulated skeletons. Among these are important new specimens—some of them the first or the most nearly complete skeletons known—of *Palaeanodon, Alocodontulum, Microsyops, Phenacolemur, Cantius, Chriacus, Anacodon, Oxyaena, Prototomus, Didymictis, Vulpavus, Miacis, Phenacodus, Hyracotherium, Homogalax, Wasatchia,* and *Diacodexis.*

Comparison of characters such as limb proportions, long bone and joint structure, and ungual shape with those in extant forms whose behavior is documented enables inferences of locomotor capabilities in extinct mammals. A wide range of terrestrial adaptations is apparent in Willwood mammals, which include fossorial palaeanodonts, a large digger/rooter (*Ectoganus*), ambulatory (*Oxyaena, Didymictis*) or graviportal forms (*Coryphodon*), incipient cursors (*Phenacodus, Pachyaena*), more specialized cursors (*Hyracotherium*), small cursorial/saltatorial types (*Diacodexis, Wasatchia*), and small saltatorial mammals (leptictid insectivores). Arboreal locomotion was of at least two types: quadrupedal climbing and leaping (adapid primates), and scansorial claw-climbing (small arctocyonids and miacid carnivorans) that involved extreme tarsal mobility. Some postcranial modifications are strikingly similar to those in extant relatives of these Eocene mammals, suggesting that modification of skeletal form occurred well in advance of dental evolution.

INTRODUCTION

The Willwood Formation of the Bighorn Basin, Wyoming, has been a major source of information on early Eocene mammals. Teeth and jaws are the most abundant fossils from these beds (now numbering in the tens of thousands), but much of our knowledge of the postcranial skeleton of early Eocene mammals comes from Willwood deposits as well. Until recently, our understanding of the postcranial skeletons of Willwood mammals was based primarily on the work of Matthew (1915a, b, 1918) or earlier accounts (various works by Cope, Osborn, Wortman, and others; specific references are cited below). Additional postcranial specimens were subsequently collected by several institutions (e.g., American Museum of Natural History, Carnegie Museum,

Rose, K. D., 1990, Postcranial skeletal remains and adaptations in early Eocene mammals from the Willwood Formation, Bighorn Basin, Wyoming, *in* Bown, T. M., and Rose, K. D., eds., Dawn of the Age of Mammals in the northern part of the Rocky Mountain Interior, North America: Boulder, Colorado, Geological Society of America, Special Paper 243.

National Museum, Harvard, Princeton, Yale, University of Michigan) but, with few exceptions, these remain unpublished (and, in most cases, unstudied).

Recent work by a joint U.S. Geological Survey–Johns Hopkins University project has resulted in discovery of more than 150 skeletal associations, ranging from several bones of a single individual to fragmentary skeletons and, in a few cases, nearly complete, articulated individuals. All are catalogued into the vertebrate paleontology collection of the U.S. Geological Survey (USGS), Denver. Although the majority of bones have been collected from the surface after weathering from the outcrop, additional elements or fragments were almost always recovered by excavating and screen-washing both weathered and unweathered sediment. Most of these specimens were unequivocally associated with dentitions allowing confirmation of their taxonomic identity; many others have been confidently identified by comparison with such material. Study of many of these fossils is underway, but only a few of the most significant specimens have been described so far. Most specimens require substantial preparation before study. The purpose of this report is to place on record the most important new postcranial fossils collected by the USGS–JHU project through the 1987 field season, and to summarize what is known about the skeletons of Willwood mammals, focussing primarily on this new material. It is, therefore, a status report intended not as an exhaustive survey, but rather as preliminary (and stimulus) to further study.

Significant mammalian postcranial material was recovered from Willwood outcrops as early as the first paleontological expedition there in 1880 (Wortman, 1899; Gingerich, 1980), and some of the most important skeletons from the Willwood were found by collectors working for the the American Museum of Natural History in the 1890s and early 1900s. Subsequently, as emphasis on dental evidence grew, there seems to have been a tendency to leave postcranial fossils behind unless they were very well preserved or complete. This practice overlooked the fact that occurrences of apparently useless bone fragments or poorly preserved bones often become useful, even exceptional, skeletal specimens if collected and prepared properly. Concerning a skeleton of *Oxyaena* collected by the expeditions of 1880 and 1891, Wortman (1899, p. 140–141) wrote:

> During this latter expedition [1891], a new method was employed for securing the missing parts which had been washed out and covered up again by the accumulating debris. Wherever possible the loose dirt containing the fragments was gathered up and transferred to the nearest stream where it was washed out after the manner of the placer miner. In this way, wherever conditions were favorable, all the fragments were recovered, but in the case of the present skeleton, after every possible exertion, a large number of pieces necessary to complete the skeleton remained missing. . . . it was discovered that the specimen of the Cope collection [the *Oxyaena* skeleton collected by Wortman in 1880] furnished the missing parts of one and the same individual of the Museum specimen collected in 1891. . .

Needless to say, this technique remains a very profitable one. Like Wortman, we regularly screen-wash sediment from sites of important skeletal specimens and closely inspect these sites in subsequent seasons, almost invariably recovering additional fragments.

Postcranial skeletal material is now known for more than 35 genera of Willwood Wasatchian mammals. At least 25 genera (more than 30 species) are represented by postcrania in the USGS collection (Table 1). No unequivocally associated skeletal specimens have previously been reported for 20 percent of them (*Alocodontulum,* leptictids, *Microsyops, Phenacolemur, Anacodon*), and for as many more, the USGS material is substantially more complete than anything previously known and provides considerable new information about their anatomy (*Cantius, Chriacus, Ectocion, Homogalax, Diacodexis*). Skeletal reconstructions of representative taxa are shown in Figure 1. These fossils demonstrate a diversity of adaptations to terrestrial and arboreal life, some closely approximating adaptations that exist in recent mammals (and interpretable by identification of similar functionally associated character complexes), but others unlike any living forms. Some reveal a degree of specialization for particular habits or locomotor behavior that seems remarkable in the early Eocene, in these oldest representatives of some higher level groups.

Curiously, the occurrence of postcranial elements and skeletal associations in the Willwood Formation does not reflect the relative abundance of taxa as indicated by much more abundant teeth and jaws. Although some elements (e.g., astragali) are not uncommon, skeletons of some of the more common (dental) members of the fauna are exceedingly rare or incomplete (e.g., primates, rodents, *Hyopsodus,* tillodonts, artiodactyls), while skeletons of certain dentally rare taxa are disproportionately common (e.g., *Chriacus* and *Anacodon*).

The following account summarizes significant postcranial characters of Willwood mammals, as listed in Table 1, based principally on the most important new postcranial remains in the USGS collection. Where possible, preliminary interpretations are offered.

SURVEY OF POSTCRANIAL REMAINS

Palaeanodonta

Palaeanodonts were early Tertiary edentate-like mammals adapted for fossorial habits. They were first discovered in the Bridger Formation in 1903 (*Metacheiromys;* see Wortman, 1903; Osborn, 1904); not long after, more primitive forms were found in the Willwood Formation of the Bighorn Basin. The Willwood specimens were described as the metacheiromyid *Palaeanodon* (*P. ignavus* from the Wasatchian and *P. parvulus* from the Clarkforkian of the Clark's Fork Basin; Matthew, 1918), and several of them included skeletal remains. No additional postcranial fossils of *Palaeanodon* have since been reported. Our project has recovered several partial skeletons (Table 1) and numerous isolated elements of *Palaeanodon.* They vary noticeably in size, probably reflecting sexual dimorphism and/or the presence of more than one species; but they are here referred to *P. ignavus* pending adequate study. From the upper Willwood comes a second palae-

TABLE 1. WASATCHIAN MAMMALIAN POSTCRANIA FROM THE WILLWOOD FORMATION COLLECTED BY THE USGS–JHU PROJECT

Palaeanodonta
Palaeanodon ignavus–4726, 5896, 6000, 7209, 16471, 16494, 16499
**Alocodontulum atopum*–7208

Insectivora
cf. *Prodiacodon* sp.–16492, 16493
(*Palaeosinopa* spp.: Matthew, 1918)
(*Creotarsus lepidus*: Matthew, 1918)
(?*Pontifactor* sp.: = "*Nyctitherium celatum*" of Matthew, 1918)

Taeniodonta
Ectoganus copei–3838, 16498
(*E. gliriformis*: Schoch, 1986)

Primates
**Microsyops*–12804 (*M. angustidens*), 16647 (*M. latidens*)
**Phenacolemur* sp.–17847
**Cantius trigonodus*–4724, 5900, 21832-21834, 21782 (*Cantius* sp. or *Pelycodus jarrovii*)

Creodonta
Oxyaena–264, 952, 4752, 5910, 5911, 6093, 7144, 7145, 7160, 16483-16487, 21863 (all *O. forcipata* or *O. intermedia*); 7186 (*O. gulo*)
(*Palaeonictis occidentalis*: Matthew, 1915a; Sinclair and Jepsen, 1929)
(*Arfia* sp.–YPM 36932)
Prototomus sp.–1824, 6111, 7157, 16475
Tritemnodon sp.–4727, 21837
(*Prolimnocyon atavus*: Matthew, 1915a; D. L. Gebo, personal communication, 1987)

Carnivora
Didymictis protenus–1721, 5024, 5908, 6087, 16472, 16489, 21835, 21836
Vulpavus, cf. *V. canavus*–5025, 7143, 16488
cf. *Miacis* sp.–7161
(*Miacis exiguus*: Matthew, 1915a)
?*Uintacyon* sp.–16490
(*Vassacyon promicrodon*: Matthew, 1915a)

Condylarthra
**Chriacus* sp.–2353, 6073, 15404, 16495
(*Thryptacodon* sp.–UW 7421; Matthew, 1915a)
**Anacodon ursidens*–5026, 5031, 5902, 6092, 16476, 21856, 21857
(*Pachyaena* spp.: Matthew, 1915a)
Hapalodectes leptognathus–5912
Ectocion osbornianus–16496
Phenacodus primaevus–5033, 7146, 7158, 7162, 16477, 21838
Phenacodus vortmani–5022, 7159, 21855, 21859
(*Apheliscus* sp.: Matthew, 1918)

Tillodontia
Esthonyx bisulcatus–7551, 21839
(*Megalesthonyx hopsoni*: Rose, 1972)

Pantodonta
Coryphodon sp.–5903, 7187, 7190, 21840-21842, 21875

Perissodactyla
Hyracotherium spp.–5901,5904-5907, 6110, 16479, 16482, 21858, 21860
**Homogalax*, cf. *H. protapirinus*–1562, 21843
Heptodon, cf. *H. calciculus*–21861

Artiodactyla
**Diacodexis metsiacus*–2352, 5895
Wasatchia, cf. *W. dorseyana*–16470

Rodentia
Paramyids, unidentified–unsorted specimens

Note: All numbers are U.S.G.S. (Denver) catalogue numbers. Genera known only from postcrania in other collections are shown, with published references, in parentheses.
*Designates most complete or only known postcranial material.

anodont, the epoicotheriid *Alocodontulum atopum,* represented in our collection by six dentaries and a specimen consisting of both dentaries with associated skeletal fragments (USGS 7208). The two genera are readily distinguished by their mandibular tooth configuration: postcanines of *Alocodontulum* tend to be larger and extend to the back of the ramus, whereas in *Palaeanodon* the back of the ramus is edentulous.

Palaeanodon. Partial skeletons in the USGS collection, representing *Palaeanodon ignavus* or a closely similar unnamed species, include the following elements: dentaries, most of the vertebral column except for the caudal series (which is relatively incomplete), parts of the sternum, fragments of ribs and ossified sternal ribs, scapula (glenoid part), humerus, proximal and distal radius and ulna, some metacarpals and phalanges, ilium, ischium,

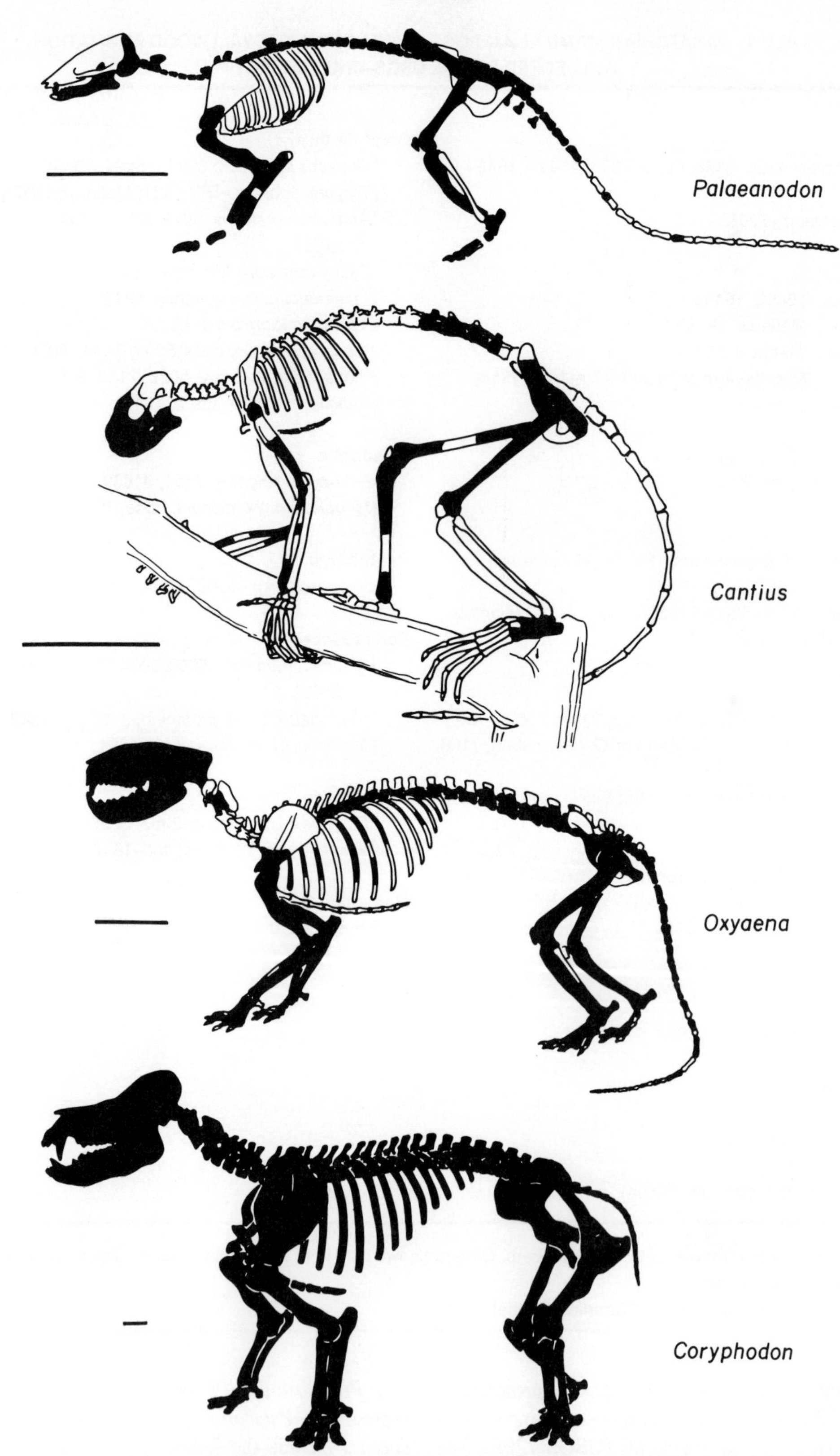

Figure 1. Skeletal reconstructions of some Wasatchian mammals from the Willwood Formation, based on specimens in Table 1 and previous published accounts. Shaded parts indicate either elements known in USGS specimens or all known elements, as specified. *Palaeanodon ignavus*—based largely on USGS 16471, but includes other material as well; *Cantius trigonodus*—based on USGS specimens, principally USGS 5900, superimposed on the skeleton of *Smilodectes* (adapted from Napier, 1970); *Oxyaena*

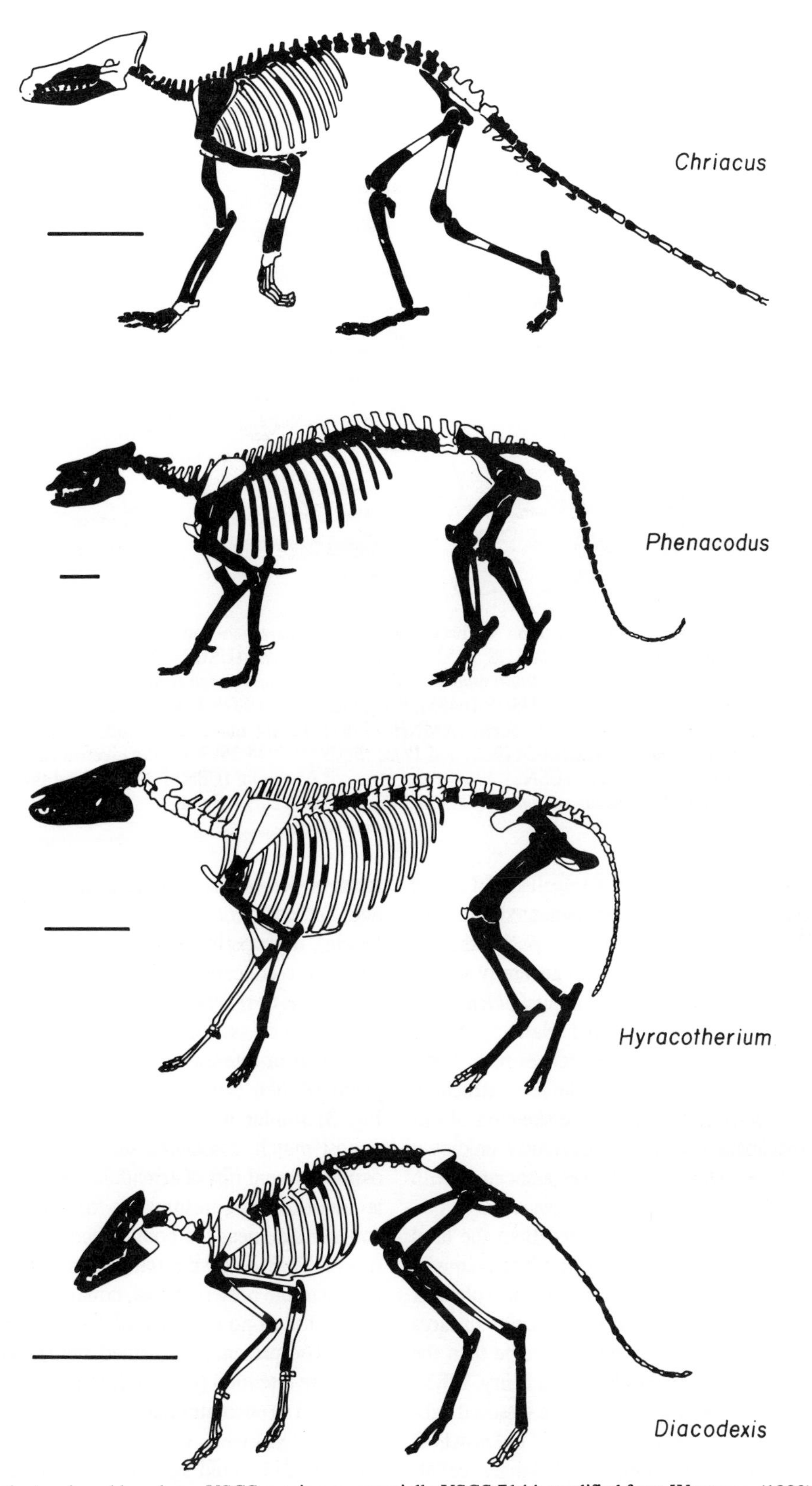

forcipata—based largely on USGS specimens, especially USGS 7144, modified from Wortman, (1899); *Coryphodon* sp.—all known parts, after Osborn (1898c); *Chriacus* sp.—USGS 2353, modified from Rose (1987); *Phenacodus primaevus*—all known parts, modified from Osborn (1898b); *Hyracotherium* sp.—USGS specimens, modified from Cope (1884); *Diacodexis metsiacus*—based mostly on USGS 2352, after Rose (1985). Scale bars = 10 cm.

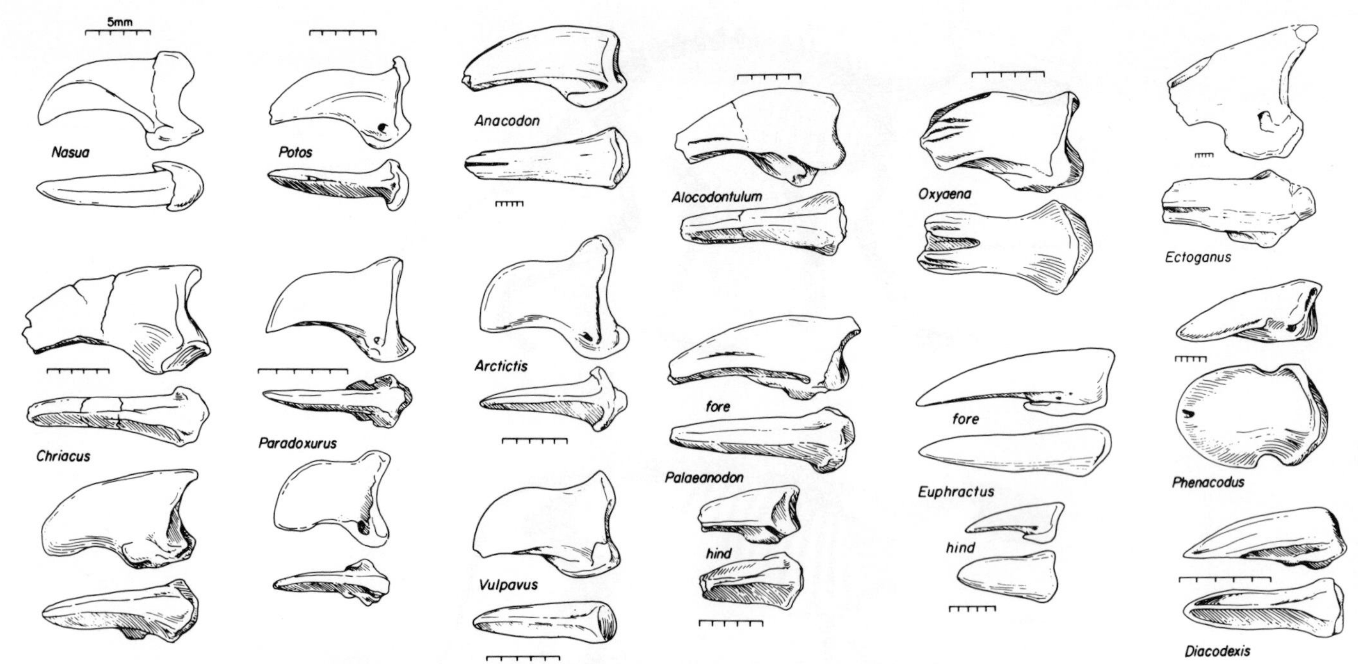

Figure 2. Ungual phalanges of some Willwood Wasatchian mammals and similar unguals of some extant mammals. *Chriacus* sp. (USGS 2353), *Anacodon ursidens* (USGS 5902), *Vulpavus,* cf. *V. canavus* (USGS 5025), *Alocodontulum atopum* (USGS 7208), *Palaeanodon ignavus* (USGS 6000 and 16471), *Oxyaena* cf. *O. forcipata* (USGS 16485), *Ectoganus copei* (USGS 16498), *Phenacodus primaevus* (USGS 7146), *Diacodexis metsiacus* (AMNH 27787). Extant mammals include the procyonid carnivores *Nasua narica* (USNM 244900) and *Potos flavus* (USNM 258316), the viverrid carnivores *Paradoxurus hermaphroditus* (USNM 49868) and *Arctictis binturong* (USNM 49642), and the dasypodid xenarthran *Euphractus sexcinctus* (USNM 241105). Scale bars = 5 mm.

femur, proximal and distal tibia, calcaneus, astragalus, and a hind ungual phalanx. These new skeletons, supplemented by Matthew's (1918) material, are the basis for the composite skeletal reconstruction of *Palaeanodon* in Figure 1. It is generally similar to Bridgerian *Metacheiromys* (Simpson, 1931), but *Palaeanodon* may have had relatively longer hind limbs and relatively shorter distal limb segments. These proportional differences may, however, be an artifact of the composite nature of the reconstruction.

Although Matthew described much of the skeleton of *Palaeanodon,* the new specimens reveal two previously unknown aspects of particular interest. Ungual phalanges associated with two specimens (Fig. 2) indicate that the foreclaws were larger, more curved, and more transversely compressed than the hind claws. Moreover, the ungual phalanx of digit III of the manus in USGS 6000 articulated with an unusual middle (second) phalanx bearing large bony excrescences lateral to the articular surfaces (Figs. 3 and 4), approximating but even more extreme than the condition in the Oligocene *Xenocranium* (Rose and Emry, 1983). This lateral expansion is undoubtedly related to the fossorial habits of *Palaeanodon,* although its precise function is unknown. It suggests the presence of particularly strong interphalangeal collateral ligaments, which may have immobilized the proximal interphalangeal joint, and may also be related to enhanced power of the flexor digitorum superficialis, whose tendons inserted there. No other specimens like this have been recognized, and specimens of the manus of other palaeanodonts (e.g., *Metacheiromys* and *Pentapassalus;* see Matthew, 1918; Simpson, 1931; Fig. 4 herein), if properly reconstructed, appear to have "normal" phalanges. It is conceivable that the middle phalanx of USGS 6000 is anomalous or represents a senescent condition, but these possibilities cannot be assessed without additional material.

Also of interest are three flattened, slightly curved, and tapered rib-like elements (included with skeleton USGS 16471: Fig. 3) similar in preservation to, and with broken edges that almost match, associated sternal fragments. They resemble the ossified sternal ribs of armadillos. Two of them bear small articular depressions (presumably for the next anterior rib) on the concave side near the pointed end, as in posterior ossified sternal ribs of *Euphractus;* the third lacks this articular surface and may represent the posteriormost, nonarticulating sternal rib. (Simpson [1931] found no evidence of ossified sternal ribs in *Metacheiromys.*) The presence of ossified sternal ribs is often considered a primitive retention (e.g., McKenna, 1975), but could be a derived condition in edentates based on the complex articulation with the sternum (Novacek, 1982; Ding, 1987). If so, a similar state in *Palaeanodon* would again raise the probability of relationship to edentates. Unfortunately, whether or not *Palaeanodon* was specifically similar to edentates in this regard cannot be determined from the specimens at hand. Palaeanodonts lack other specializations characteristic of xenarthrans, such as xenarthrous vertebrae,

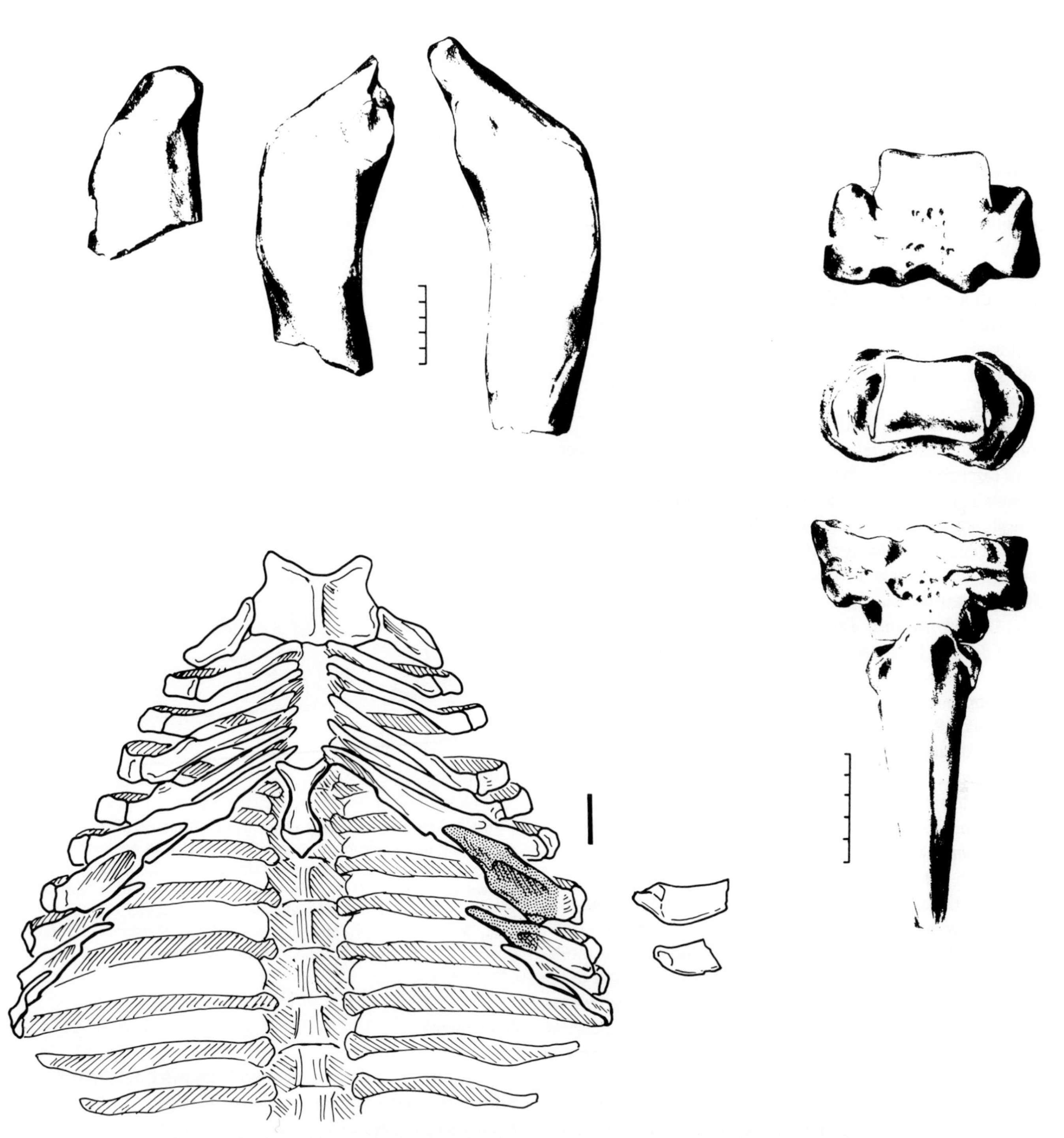

Figure 3. Recently discovered elements of *Palaeanodon ignavus.* Probable ossified sternal ribs shown above left (scale = 5 mm). Ventral view of rib cage in *Euphractus sexcinctus* (USNM 241105), below left, showing ossified sternal ribs; shaded ribs show inferred position of those of *Palaeanodon,* shown at same scale to the right (scale bar = 1 cm). Middle and ungual phalanges of third digit at right in ventral, distal, and dorsal views (top-bottom) (scale = 5 mm).

a true synsacrum, and dermal ossifications. Nonetheless, the rarity of ossified sternal ribs among eutherians suggests that their presence in a palaeanodont could be significant for questions of relationship.

Other distinctive skeletal characteristics of *Palaeanodon* include: relatively short and robust limb and foot elements; humerus (see Fig. 5) with a broad, distally extending deltopectoral shelf, a prominent medial epicondyle, and a flaring supinator crest; radial head (see Fig. 6) ovoid with a flat ulnar facet (similar to the condition in armadillos and indicating little or no supinatory capability), sacrum composed of two true sacrals and two pseudosacrals (proximal caudals); femur longer than the tibia (Fig. 7) and with a prominent third trochanter situated just proximal to midshaft; distal femur broader than deep and with a moderately shallow patellar groove; distal tibia and astragalar trochlea, broad and shallowly grooved; tibia and fibula occasionally fused distally.

Except for the third digit of the manus, *Palaeanodon* is postcranially less specialized than geologically younger palaeanodonts (especially epoicotheres). Compared to other early Eocene mammals (with the probable exception of pantolestids), however, it displays a host of fossorial specializations. It seems probable that it occupied a niche similar to that of armadillos or terrestrial anteaters and manids today, engaging in considerable forelimb digging or tearing (through termite mounds or rotting logs).

Alocodontulum. The holotype of *Alocodontulum atopum* (YRM 30790), from upper Graybullian strata (Yale locality 348, 400 m), is a maxillary dentition encased in a very hard nodule containing many presumably associated bone fragments (Rose and others, 1977). Among the latter are an astragalus and distal radius of distinctly palaeanodont morphology; apart from some rib fragments, other elements in the nodule are obscured by matrix or are too shattered to be of much use. USGS 7208, therefore, provides the first significant evidence of the postcranial anatomy of *Alcodontulum.* It represents a subadult individual and includes both dentaries, left ulna, distal radii, Mc II and III, and three ungual phalanges of the manus (Figs. 2 and 4). Although quite similar to bones of *Palaeanodon,* these elements differ slightly in ways approaching *Pentapassalus:* the ulna has a relatively longer olecranon and shorter distal shaft (i.e., distal to the semilunar notch), the metacarpals are shorter and more robust, the terminal phalanges are dorsoventrally deeper, more curved, and have larger subungual processes (Fig. 4). (The ratio of olecranon:distal ulna in *Metacheiromys* = 0.53 [unknown in *Palaeanodon* but probably equivalent or less], in *Alocodontulum* = 0.57 [for shaft lacking epiphyses], for *Pentapassalus pearcei* = 0.68.) These traits suggest more fossorial tendencies in early epoicotheres compared with metacheiromyids.

Alocodontulum appears to be closely allied with the epoicotheres *Pentapassalus* and perhaps *Tubulodon* (later Wasatchian of the Wind River and Washakie basins). New material at hand strengthens the probability of this relationship but is not adequate to determine if any of them are congeneric.

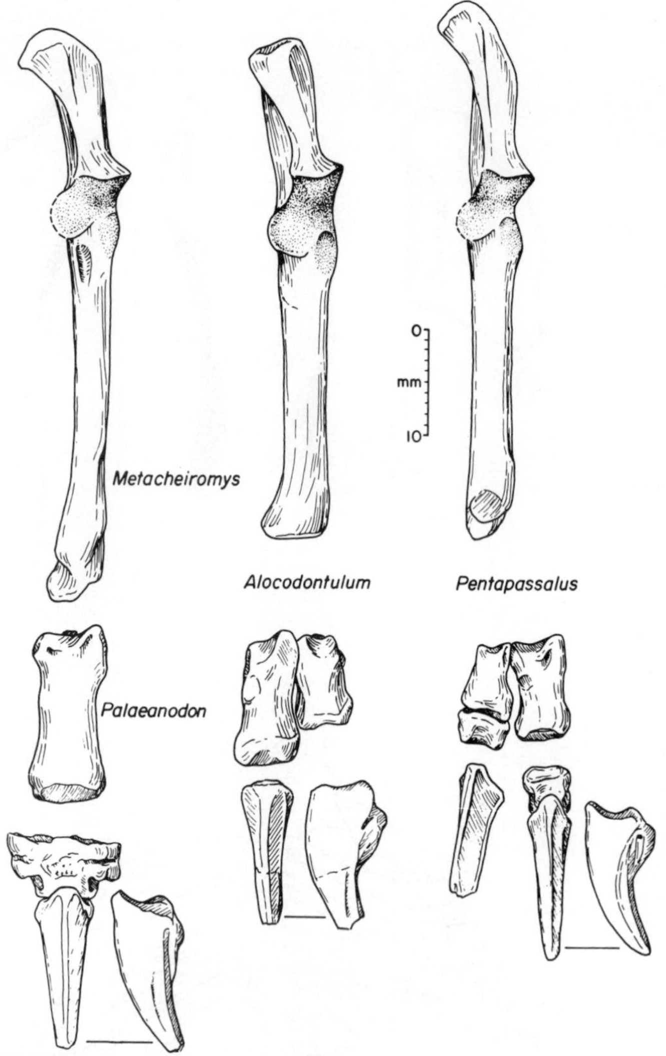

Figure 4. Left ulna and some elements of left manus of palaeanodonts, comparing metacheiromyid morphology (left) with epoicotheriid morphology. *Metacheiromys marshi* (USNM 26132; no complete ulna known for *Palaeanodon*), *Palaeanodon ignavus* (Mc III, USGS 16471; phalanges, USGS 6000), *Alocodontulum atopum* (USGS 7208, subadult, ulna, Mc III and IV, and ungual phalanx), *Pentapassalus pearcei* (USNM 20028, ulna, Mc II and III and phalanges).

Insectivora

Insectivore postcrania are poorly known from the Willwood Formation, being limited until now principally to fragments of *Palaeosinopa, Creotarsus,* and *Pontifactor* (= *Nyctitherium celatum,* possibly a bat) that were illustrated or mentioned by Matthew (1918). Matthew (1918, p. 603) reported "three long slender bones, suitable in size and proportions for the shafts of chiropteran fore limb bones" associated with jaws of *Nyctitherium celatum,* and suggested that they might belong to a bat. Gingerich (1987) accepted this view and described the first known teeth of Willwood Clarkforkian bats, thus affirming the existence of Chiroptera in the Willwood. Whether or not Matthew's specimen represents a bat remains controversial; hence, no unquestioned bats have yet been reported from the Wasatchian part of the Willwood Formation.

Two specimens in the USGS collection include postcrania of a leptictid (cf. *Prodiacodon* sp.). Skeletal remains of leptictids have not previously been reported from the Willwood and are probably scarce in existing collections because of their small size. USGS 16492 includes parts of the femur, pelvis, and other bones in a concretion associated with dentaries. More nearly complete is USGS 16493, a partial skeleton including jaws and parts (including articular surfaces) of humeri, ulna, femora, tibia, astragali, and calcaneus. Many critical details of this specimen remain obscured by very hard matrix; however, certain significant features can be discerned. The greater trochanter projects well above the head of the femur, and the distal femur is narrow and anteroposteriorly deep, with a narrow patellar trochlea. The fibula is distally synostosed to the rather slender tibia for nearly half its length, as in Oligocene *Leptictis,* the Eocene lipotyphlan *Macrocranion* (Maier, 1979), and various extant mammals adapted for jumping (e.g., *Tarsius, Jaculus, Rhynchocyon;* Barnett and Napier, 1953), but unlike *Prodiacodon* from the Paleocene, in which the two elements are joined only at the distal extremity (Matthew, 1918, 1937). The astragalus is elongate, and its lateral trochlear crest is conspicuously larger and higher than the medial crest, closely resembling that in *Prodiacodon* (Szalay, 1966). The astragalar trochlea is deeply grooved (and the lateral trochlear ridge more elevated than the medial), thereby largely restricting cruroastragalar mobility to flexion and extension. All of these traits are suggestive of specialization for saltatorial locomotion.

Taeniodonta

Taeniodonts are among the rarest elements of the Willwood Wasatchian fauna, being known from fewer than 20 specimens referable to two species of the genus *Ectoganus, E. gliriformis,* and *E. copei* (Schoch, 1986). In his comprehensive review of taeniodonts, Schoch reported only a small number of fragmentary postcranial skeletal specimens of the genus, of which only three come from the Willwood Formation. A fragmentary femur (PU 13173, referred to *E. gliriformis* by Schoch), a well-preserved humerus (YPM 27201, which Schoch did not assign to species), and two postcranial associations in our collection (USGS 3838, allocated by Schoch to *E. copei copei,* and USGS 16498, collected after Schoch's study) represent *Ectoganus.* The USGS specimens include vertebrae and fragments of the ulna, pelvis, femur, tibia, tarsals, metapodials, and phalanges. All elements are massive, with exaggerated crests and processes. The ungual phalanges (Fig. 2) are large, curved, and laterally compressed, with a very prominent subungual process. *Ectoganus* was one of the largest early Eocene mammals, equalled or surpassed in size only by *Coryphodon* and *Pachyaena.* Its osteological characteristics indicate a propensity for digging and tearing, particularly using the forelimbs (Coombs, 1983; Schoch, 1986).

Primates

Primates are relatively diverse in the Wasatchian part of the Willwood Formation; as many as 30 or more species in 19 genera (plesiadapiforms, adapids, and omomyids) are currently recognized. They typically account for 15 to 20 percent of samples (minimum number of individuals based on jaws and teeth) from Wasatchian sites (Rose, 1981). Despite this prevalence, postcranial bones are rare and are known in only three genera: two plesiadapiforms (*Microsyops* and *Phenacolemur*) and one euprimate (the adapid *Cantius*). Even these are relatively incomplete, and our knowledge remains very inadequate.

Microsyops. No postcranial elements unquestionably referable to this genus have yet been described. Two recently collected specimens are the oldest of the genus (and the first from the Bighorn Basin) to include postcranial remains. USGS 16647 (*Microsyops latidens*) includes maxillary dentitions associated with most of the humerus, proximal radius and ulna, vertebrae, and other bone fragments. A proximal humerus is associated with jaws (USGS 12804) of the somewhat older (upper "Graybullian") *M. angustidens.* These specimens are under study by K. C. Beard and will be described elsewhere.

Phenacolemur. USGS 17847, recovered from late Wasatchian strata in 1986 and 1987, is the most nearly complete of four Willwood specimens known that preserve postcrania associated with dentitions of *Phenacolemur.* Most limb elements are represented (parts of the scapula, humerus, radius, ulna, pelvis, femur, tibia, fibula, calcaneus, and phalanges), as well as vertebrae. The other specimens, all from the Clark's Fork Basin, are in the collections of the University of Michigan Museum of Paleontology (UM 66440, 86352) and the Harvard Museum of Comparative Zoology (MCZ 19449). These are the first known undoubted postcranials of the genus; they will also be described by K. C. Beard. Preliminary examination indicates that they are generally similar to comparable elements of *Plesiadapis* (Beard, personal communication, 1988).

Cantius. *Cantius* is the oldest lemur-like primate and one of the oldest known euprimates; hence, knowledge of its skeletal anatomy is potentially very important for understanding primate evolution. Matthew (1915b) illustrated the first known skeletal elements of this genus (then referred to *Pelycodus frugivorus,* but almost certainly representing *Cantius trigonodus* or *C. abditus*), tarsal bones from the Willwood Formation. Although species of *Cantius* are among the most common Willwood Wasatchian mammals, no significant additions were made to the postcranial anatomy of *Cantius* until the discovery in 1982 of a partial skeleton of *C. trigonodus* (USGS 5900), described by Rose and Walker (1985). It includes dentaries and a partial skull, five trunk vertebrae, a fragmentary distal humerus, the proximal ulna, the radius except for the head, the right ilium and ischium, proximal and distal right femur and a nearly complete left femur, proximal and distal tibia, the astragalus, and the proximal end of metatarsal I (Fig. 1). A few additional parts (e.g., the radial head, terminal phalanges, the navicular, the cuboid, and caudal vertebrae) are present in other specimens of the genus (e.g., USGS 4724 and 21832—*C. trigonodus;* USGS 18338, 18339—*C. mckennai;* USGS 21782—a large unnamed species of *Cantius* or possibly *Pelycodus jarrovii*). More than 50 isolated limb and foot elements of *Cantius* have also been identified in our collection by M. Dagosto and D. Gebo (e.g., Gebo, 1987, 1988).

In nearly all respects, the osteology of early Wasatchian (Graybullian) *Cantius* is very similar to that of its Bridgerian descendant *Notharctus* (Gregory, 1920). The skeleton of *Cantius* is distinctive in the following characters: the humerus is broad distally with a prominent brachialis flange; the proximal ulna is laterally compressed and convex posteriorly with a well-developed olecranon; the radius is slender and bowed, with an ovoid, almost round head (indicating substantial supinatory capability); the ischium is relatively much longer than in most extant lemurs; the hind limbs (see Fig. 7) are slender and much longer than the forelimbs, and the femur was probably longer than the tibia; the distal femur is particularly lemur-like: the patellar groove is narrow and elevated, with the lateral ridge rounded and more prominent; the tibial crest is prominent with a distally situated inferior tibial tuberosity; the distal tibia and talar trochlea are narrow and shallowly grooved; the neck of the talus is elongate, and its head is ovoid and almost spherical; metatarsal I has a large peroneal tubercle; and the terminal phalanges bore nails (Covert, 1988).

From what is now known about its skeleton, *Cantius* was an arboreal quadrupedal adapiform, capable of leaping, and probably resembled living lemurs in general appearance and locomotor behavior (Rose and Walker, 1985).

Creodonta

Oxyaena. Based on both dental and postcranial remains, *Oxyaena* is the most common larger carnivorous mammal in the Willwood fauna. Relatively complete skeletal remains of *Oxyaena* (principally *O. forcipata*; more fragmentary specimens of *O. intermedia* and *O. platypus*) have been known from the Bighorn Basin since the late 1800s or early 1900s (Osborn and Wortman, 1892; Wortman, 1894, 1899; Osborn, 1900; Matthew, 1915a). Denison (1938) provided a relatively detailed description of the skeleton.

A skeleton of the closely allied genus *Palaeonictis* (*P. occidentalis*) from the Willwood Formation was described by Sinclair and Jepsen (1929). (Matthew [1915a] earlier reported a hind foot skeleton and associated fragments that he referred questionably to *Palaeonictis.*) It closely resembles the skeleton of *Oxyaena.*

The USGS collection includes 16 partial skeletons of *Oxyaena* (Table 1), several of which are well preserved and relatively complete. Based on size, most specimens appear to represent *O. forcipata* or the slightly more primitive *O. intermedia,* but there is at least one individual of *O. gulo* (USGS 7186). Most nearly complete but not the best preserved (because it is partly covered by a thin hematite veneer and cracked, probably from postmortem subaerial exposure) is USGS 7144, a partly articulated skeleton including jaw fragments, most of the thoracic and lumbar vertebrae and numerous caudals, ribs, scapulae (glenoid), partial humeri and radii, ulnae, carpals and metacarpals, pelvic fragments, complete femora, tibiae, a distal fibula, the complete tarsus and metatarsus, and some phalanges. Particularly well preserved limb and foot elements are contained in certain other specimens (USGS 5910, 16485, 21863).

Compared to those of many other Willwood mammals, the skeleton of *Oxyaena* was relatively long bodied and short limbed. There were 20 trunk vertebrae, three (fused) sacrals, and a long, well-developed tail probably numbering more than 20 caudals (Denison, 1938).

Characteristics of the forelimb skeleton of *Oxyaena* may be summarized as follows: humerus (Fig. 5) with a large greater tuberosity that does not project noticeably above the head, a prominent deltopectoral crest that extends distally two-thirds the length of the shaft, and a moderately developed supinator crest; distal humerus broad, with a wide, cylindrical capitulum, a prominent medial epicondyle, and a large entepicondylar foramen; both ulna and radius are well developed, the radius conspicuously shorter than the humerus and with an ovoid head bearing a prominent capitular eminence (Fig. 6; limits supinatory capability; see Davis, 1964); ulna slightly bowed (convex posteriorly), and olecranon moderately large and slightly incurved; alternating carpus, with a centrale present (Osborn and Wortman, 1892); manus pentadactyl and spreading, metacarpals and phalanges not markedly elongate; unguals sharply curved dorsally (but less so on the plantar surface), deeply fissured, not markedly laterally compressed, and bearing a large subungual process.

Hind-limb characteristics of *Oxyaena* (Figs. 7 and 8) include: femur moderately long, with the greater trochanter lower than the head, a prominent lesser trochanter projecting slightly posteriad, and a weakly developed third trochanter situated on the proximal third of the shaft; distal femur almost as deep (anteroposteriorly) as broad, with a wide, slightly elevated and moderately deep patellar groove (deeper than in climbers such as *Chriacus* and *Miacis,* shallower than in runners such as *Hyracotherium* and *Diacodexis*); tibia shorter than femur, anteroposteriorly deep and laterally compressed proximally but almost round in cross section near the distal articulation; tibial crest not distinctly set off from the shaft, and the distal articulation of the tibia shallow; fibula strong; astragalus with a shallow trochlea, the fibular facet set at a right angle to the trochlea, the head broad and rounded but dorsoventrally compressed, and the astragalar foramen high (limiting plantarflexion); calcaneus short (especially distally), with a prominent peroneal process; cuboid with a large astragalar facet; metatarsals slightly longer than metacarpals; hallux (and pollex; Denison, 1938) somewhat divergent, pes (Fig. 8) otherwise similar to manus.

The skeleton of *Oxyaena* has proven difficult to interpret. The feet have been considered plantigrade (Cope, 1884), digitigrade (Wortman, 1899), subdigitigrade (Osborn, 1900), and again plantigrade (Denison, 1938). Based on conformation of the feet and astragalus, Sinclair and Jepsen (1929) considered the

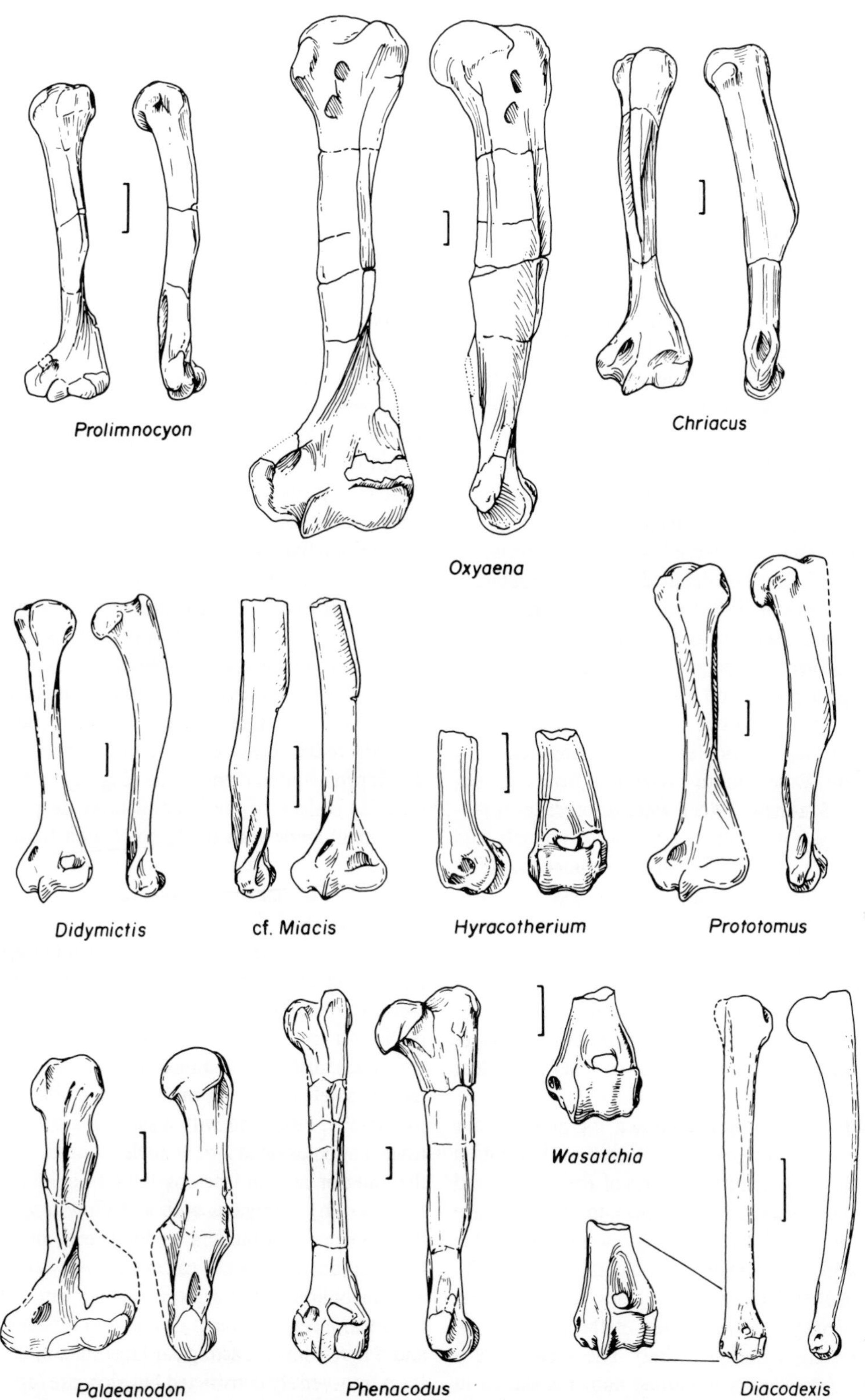

Figure 5. Left humeri of some Willwood Wasatchian mammals. *Prolimnocyon* sp. (DPC 5364), *Oxyaena*, cf. *O. forcipata* (USGS 5910), *Chriacus* sp. (USGS 2353), *Didymictis protenus* (USGS 21835, 21836), cf. *Miacis* sp. (USGS 7161), *Hyracotherium* sp. (USGS 5905, 6110), *Prototomus* sp. (USGS 7157, 16475), *Palaeanodon ignavus* (USGS 4726), *Phenacodus vortmani* (USGS 7159), *Wasatchia*, cf. *W. dorseyana* (USGS 16470), *Diacodexis metsiacus* (USGS 2352, 5895). Scale bars = 1 cm.

closely similar *Palaeonictis* to have been plantigrade. There is, in fact, substantial morphologic evidence in support of plantigrady (Ginsburg, 1961) in the skeleton of *Oxyaena:* the proportions of the limb and foot elements are similar to those in extant plantigrade carnivores (e.g., the femur is longer than the tibia, the metatarsals are short); the humerus has a strong deltopectoral crest and a well-developed entepicondyle; the ulna has a prominent incurved olecranon; the femur has a small third trochanter and a moderately broad patellar groove; the calcaneus is short and the sustentaculum tali very distal; the astragalus has a shallow trochlea and a broad, dorsoventrally compressed head; and the first and fifth metatarsals remain well developed. Gunnell (1988) also concluded that *Oxyaena* had plantigrade feet, although the smaller Clarkforkian–early Wasatchian oxyaenid *Dipsalidictis* may have been subdigitigrade.

Wortman suggested that *Patriofelis,* and by inference *Oxyaena,* was aquatic or semiaquatic, foreshadowing anatomical modifications that occur in pinnipeds. In contrast, Osborn (1900, p. 270) concluded that these taxa were "powerful terrestrial, or partly arboreal, animals." Based on detailed study of the skeleton in various creodonts, Denison (1938) rejected the interpretation that *Oxyaena* was aquatic, semiaquatic, or arboreal. He argued that certain anatomical features suggesting arboreal habits (e.g., the somewhat divergent pollex and hallux, the well-developed deltopectoral crest and entepicondyle of the humerus, the strong tail, and the shallow astragalar trochlea and associated relatively mobile crurotalar joint) were primitive retentions from an arboreal ancestry (see also Matthew, 1909), whereas other traits militate against arboreality (e.g., short, blunt, fissured unguals, and the limited capacity for supination). Denison concluded that *Oxyaena* was a terrestrial ambulatory form. Nonetheless, the possibility that *Oxyaena* was able to climb, even if infrequently, seems plausible. Indeed, Gunnell (1988) reported that *Dipsalidictis* had relatively mobile feet and was likely to have been scansorial. Further study is clearly needed, and the considerable new skeletal material now known for *Oxyaena* should contribute to a better understanding of locomotor capabilities in this prevalent creodont.

Hyaenodontids. Postcrania are known for at least four different hyaenodontid genera from the Willwood Formation (Table 1). Two of them (unidentified species of *Prototomus* and *Tritemnodon*) are represented in our collection. Fragmentary postcranial remains of early Wasatchian *Arfia* were found in the southern Bighorn Basin by a Yale expedition conducted by E. L. Simons, and additional skeletal remains of this genus have recently been collected in the northern part of the basin (P. D. Gingerich, 1989, personal communication). The skeleton of a fourth hyaenodontid (*Prolimnocyon atavus;* humerus shown in Fig. 5) is under study by D. L. Gebo and the author and is not discussed here. Willwood skeletons of hyaenodontids are relatively incomplete and have only been subject to preliminary study. The taxonomy of early hyaenodontids is in need of revision, and the allocations made here (based on dental morphology) must be considered tentative.

Four specimens in the USGS collection represent what has generally been called *Prototomus.* USGS 16475 is subadult, with unfused epiphyses but fully erupted third molars; the other specimens are adult. Included in one or more of them are dentaries, maxillae, vertebrae (thoracic, lumbar, and caudal), scapula (glenoid), humerus, proximal and distal ulna and radius, partial pelvis, nearly complete femur and tibia, tarsals (all except the mesocuneiform), fragmentary metatarsals, and phalanges.

Two specimens of *Tritemnodon* contain fragments of the dentary and maxilla, vertebrae (axis and most of thoracic and lumbar series and numerous caudals), scapula (glenoid), proximal and distal humeri, radii, and ulnae, pelvic fragments, proximal femur, calcaneus, entocuneiform, incomplete metapodials, and phalanges. There are no complete long bones among any of our specimens. Although complete (composite) elements can be restored in many cases, lengths are not certain. To the extent that these specimens can be compared, there are no significant differences except size between those assigned to different genera.

Distinctive characteristics of these hyaenodontid skeletons (some based on only one of the taxa represented) include: humeral head projecting slightly above the tuberosities; deltopectoral crest elevated, narrow, and extending distally just beyond midshaft; distal humerus broad, olecranon fossa rather shallow, supinator crest of moderate size, and entepicondyle and foramen large; olecranon process well developed but not markedly incurved; radial head ovoid and relatively flat, with a well-developed capitular eminence (Fig. 6); anterior margin of the articular surface of the femoral head steeply angled, forming a very small angle with the femoral shaft (indicative of a less abducted femoral posture and more closely approximating the condition in cursorial than ambulatory carnivores; see Jenkins and Camazine, 1977, Fig. 8); greater trochanter projecting above the femoral head; lesser trochanter directed posteriad; third trochanter distinct and more distal than in Bridgerian hyaenodontids (Matthew, 1909); femoral shaft distinctly bowed at the third trochanter (medial border of the femur noticeably concave, as in *Oxyaena*); distal femur almost as deep as broad, with an elevated medial patellar ridge (as in *Oxyaena*); tibia laterally compressed proximally, distal end with a shallow astragalar facet but a large tibial malleolus set at a right angle to the facet and a projection on the anterior margin (thereby limiting mobility especially during dorsiflexion); astragalus with a shallow trochlea, a fossa for the tibial malleolus, a moderately long neck, and a broad, dorsoventrally compressed head; calcaneus with broad astragalar facets and an oblique cuboid facet; cuboid with a somewhat helical calcaneal facet (but much shorter and broader than in *Chriacus*) and a much smaller astragalar facet than in *Oxyaena;* entocuneiform transversely constricted but elongate (approximating that in *Phenacodus* and unlike that of *Chriacus,* indicative of a more reduced and less divergent hallux than in *Oxyaena* or *Chriacus;* see also Denison, 1938); metatarsals slender and moderately elongate (longer than in *Oxyaena*); unguals more curved and laterally compressed than in *Oxyaena* and fissured at the tip.

Although Willwood hyaenodontids were generally similar

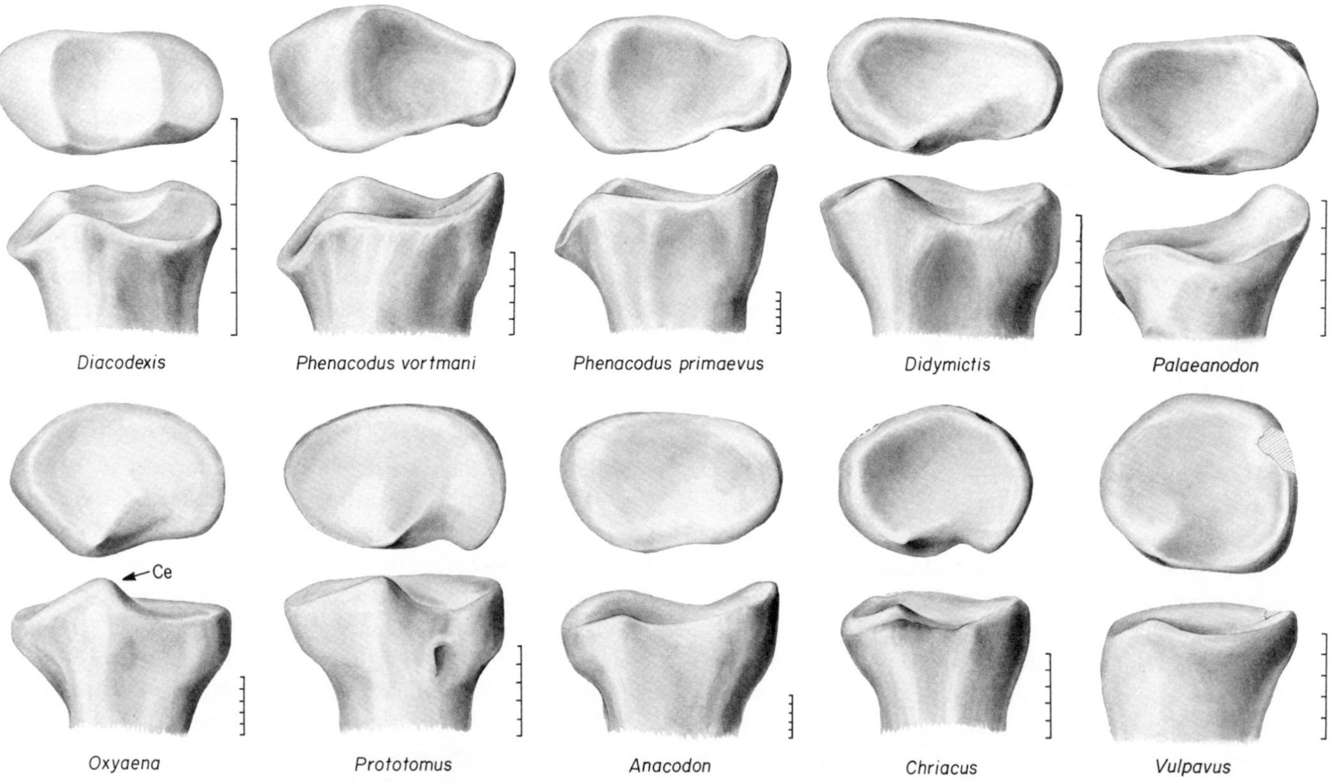

Figure 6. Left radial heads of some Willwood Wasatchian mammals. *Diacodexis metsiacus* (USGS 2352), *Phenacodus vortmani* (USGS 7159), *Phenacodus primaevus* (USGS 7146, 21838), *Didymictis protenus* (USGS 21836), *Palaeanodon ignavus* (USGS 16471), *Oxyaena*, cf. *O. forcipata* (USGS 5911, 16485), *Prototomus* sp. (USGS 7157), *Anacodon ursidens* (USGS 6092, 21857), *Chriacus* sp. (USGS 15404), *Vulpavus*, cf. *V. canavus* (USGS 5025); *Ce* indicates capitular eminence. Scales = 5 mm.

in size and in many anatomical features to the arctocyonid *Chriacus*, close comparison indicates that they were much more limited than *Chriacus* in their range of supinatory mobility of the radius, femoral abduction, tarsal flexibility, and hallucal divergence. Matthew (1909) and Denison (1938) considered hyaenodontids to be terrestrial ambulatory to subcursorial animals (some later members such as *Hyaenodon* were cursorial; Mellett, 1977), although neither dealt specifically with early Eocene members. (Matthew [1915a] illustrated some fragmentary postcrania attributed to *Sinopa hians* but did not discuss its adaptations.) According to Matthew (1909, p. 466), proviverrine hyaenodontids (the group to which all Willwood forms are assigned) "correspond more or less in adaptation with the civets." In several of the characters listed above, Willwood hyaenodontids are indeed more similar to terrestrial ambulatory civets such as *Viverra* than to arboreal types; but some aspects of their limb structure and joint anatomy suggest that they retained the ability to climb.

Carnivora

At least five genera of primitive Carnivora are known from skeletal material from the Willwood Formation; four of them, the viverravid *Didymictis* and the miacids *Vulpavus, Miacis,* and *Uintacyon,* are represented in our collection (Table 1; identified by associated dentitions). Most of these specimens, like the hyaenodontids, are fragmentary and require further preparation before they can be studied in detail; accordingly the following remarks are preliminary.

Matthew (1915a) reported skeletal material of *Didymictis, Miacis,* and *Vassacyon* from the Bighorn Basin Wasatchian, but he illustrated better material of Wasatchian *Didymictis* from the Wind River Basin and earlier had described important postcranial specimens of *Miacis, Vulpavus,* and *Uintacyon* from the Bridger Basin middle Eocene (Matthew, 1909). In comparable parts, the Wasatchian forms discussed here are very similar to their Bridgerian relatives.

Didymictis. Postcranial specimens of *Didymictis* in our collection vary somewhat in size and may belong to more than one species; however, pending revision of Wasatchian *Didymictis,* all are here referred to *D. protenus.* (They probably belong in the groups that have been called *D. protenus curtidens* and *D. protenus lysitensis;* see Matthew, 1915a; Simpson, 1937). Elements represented in one or more specimens include the skull, dentaries, vertebrae, sternebrae, scapula (glenoid), humerus, ulna, proximal and distal radius, pelvis, proximal and distal femur, tibia, distal fibula, tarsals (astragalus, calcaneus, navicular, and cuboid), and fragments of metapodials and phalanges.

The humerus (Fig. 5) is characterized by a prominent

greater tuberosity that projects above the head. Proximally the shaft is laterally compressed, but otherwise the deltopectoral crest is much less prominent than that in arctocyonids, creodonts, or miacids (such as *Vulpavus*); however, its form is very similar to that in extant viverrids and procyonids. Distally the humerus is broad, the entepicondyle and entepicondylar foramen large, the supinator crest moderate, and the olecranon fossa deep (and perforate?). The ulna (Fig. 9) is slightly sigmoid, its proximal two-thirds convex posteriorly and the distal third concave posteriorly, and the radial facet is relatively flat. The radial head (Fig. 6) is ovoid, with a moderately expressed capitular eminence (distinct but less prominent than in creodonts). In the femur, the greater trochanter is about even with the top of the femoral head, the lesser trochanter projects distinctly posteriad (rather than only mediad), and the third trochanter forms a moderate tubercle just distal to the level of the lesser trochanter. Distally the femur is slightly broader than deep and the patellar trochlea is broad and moderately grooved. The proximal half of the tibia is laterally compressed, and the cnemial crest extends almost to midshaft. The tibial malleolus is large and set at a right angle to the astragalar articulation. The astragalar trochlea and distal tibia are somewhat grooved, not almost flat as in miacids, and the astragalar head is deep and convex in a dorsoplantar plane (rather than transversely broad and convex in a mediolateral plane as in miacids; Matthew, 1909; Gingerich, 1983). There is a small astragalocuboid articulation. The pes is pentadactyl and generalized, with laterally compressed, slightly curved ungual phalanges (Matthew, 1915a).

Didymictis is generally considered to have been a terrestrial ambulatory form, in contrast to the more arboreal miacids (e.g., Matthew, 1909). Much of its skeleton, however, is primitive and suggests a scansorial or arboreal ancestry, and *Didymictis* itself may have retained some capability to climb. For example, the humerus, compared to that in extant carnivores, is morphologically intermediate between those of terrestrial and arboreal viverrids and procyonids. The ovoid radial head and flat radial facet on the ulna, however, indicate restriction in supinatory capability, and its tarsal morphology reflects a similar limitation of ankle flexibility compared to that of habitual climbers. In addition, the sinuous shape of the ulna (Fig. 9), the deep olecranon fossa (permitting greater antebrachial extension), and the posteriad projection of the lesser trochanter more closely resemble the conditions in extant terrestrial rather than arboreal carnivores.

Miacids. Five partial skeletons in our collection belong to the miacid genera *Vulpavus* (cf. *V. canavus,* three specimens), cf. *Miacis* sp., and ?*Uintacyon* sp. With one exception (USGS 7161, cf. *Miacis* sp.), these are fragmented skeletons that require considerable additional preparation. Apart from size and robustness (cf. *Miacis* is smaller and more gracile), they appear to be closely similar, as Matthew (1909) also observed. This applies equally to *Vassacyon* (Matthew, 1915a). USGS 7161 includes a dentary and very well-preserved distal humeri, a complete tibia, proximal and distal femora, and several vertebrae. The other specimens include a skull (USGS 16488), dentaries and/or maxillae, vertebrae, proximal and distal humerus and radius, proximal ulna, pelvis, femur, proximal and distal tibia, distal fibula, tarsals (calcaneus, astragalus, cuboid, navicular), metatarsals, and phalanges.

Preliminary comparisons of these specimens with those of *Didymictis* reveal several significant contrasts: in miacids (sensu stricto) the deltopectoral crest of the humerus is sharp and elevated (even more so in *Vulpavus* than in cf. *Miacis;* Fig. 5), the greater tuberosity does not project as high as the humeral head (Matthew, 1909), the supinator crest is more pronounced, the olecranon fossa is markedly shallower, the radial head (Fig. 6) is nearly round (even rounder than in *Chriacus* and comparable only to primates among Willwood mammals), and the corresponding radial facet on the ulna is more curved, the lesser trochanter projects mediad (not posteromediad), the patellar groove is shallower and the distal femur relatively broader (cf. *Miacis;* Fig. 7), the distal tibia and astragalar trochlea are markedly less grooved, the tibial malleolus is smaller and less sharply angled from the trochlear facet, the astragalar head is much broader and more dorsoventrally compressed, and the ungual phalanges (Fig. 2) are laterally compressed, shorter, deeper in dorsoplantar dimension, and more curved along their dorsal margins (closely resembling those of extant arboreal viverrids such as *Paradoxurus*).

Matthew (1909) considered the Viverravidae to be terrestrial (and *Didymictis* subdigitigrade; Matthew, 1937) and the Miacidae to be plantigrade and arboreal. He (Matthew, 1909) compared *Miacis* in size and proportions to the recent procyonid *Bassariscus* and considered *Vulpavus* particularly similar to recent *Potos* and *Arctictis* (Procyonidae and Viverridae, respectively). Among the most arboreal living carnivores, these genera display particular specializations for climbing (Trapp, 1972; Jenkins and McClearn, 1984) that are approximated in Willwood miacids. The osteological characters listed above by which miacids differ from *Didymictis* are resemblances (primitive?) to arctocyonids such as *Chriacus* and are indicative of or associated with enhanced climbing ability (e.g., habitually flexed elbows and knees, extensive supinatory capability of the radius, a very flexible tarsus perhaps permitting hind-foot reversibility). This evidence suggests that miacids (sensu stricto) were among the most arboreally adapted Willwood mammals.

Condylarthra

Condylarths of several families occur in the Willwood Wasatchian fauna and include some of the most prevalent forms. Matthew reported fragmentary postcrania of the ?hyopsodontid *Apheliscus* and of the mesonychid *Pachyaena* (Matthew, 1915a). Nearly complete skeletons are known for two species of *Phenacodus* (*P. primaevus* and *P. vortmani;* Cope, 1884; Osborn, 1898b). Curiously, however, good postcranial specimens of *Hyopsodus,* the most common mammal in most Willwood Wasatchian localities, are rare (although isolated elements and minor associations are present in our collection). The most important condylarth specimens collected by the USGS-JHU project are skeletal associations of the arctocyonids *Chriacus* sp. and *Ana-*

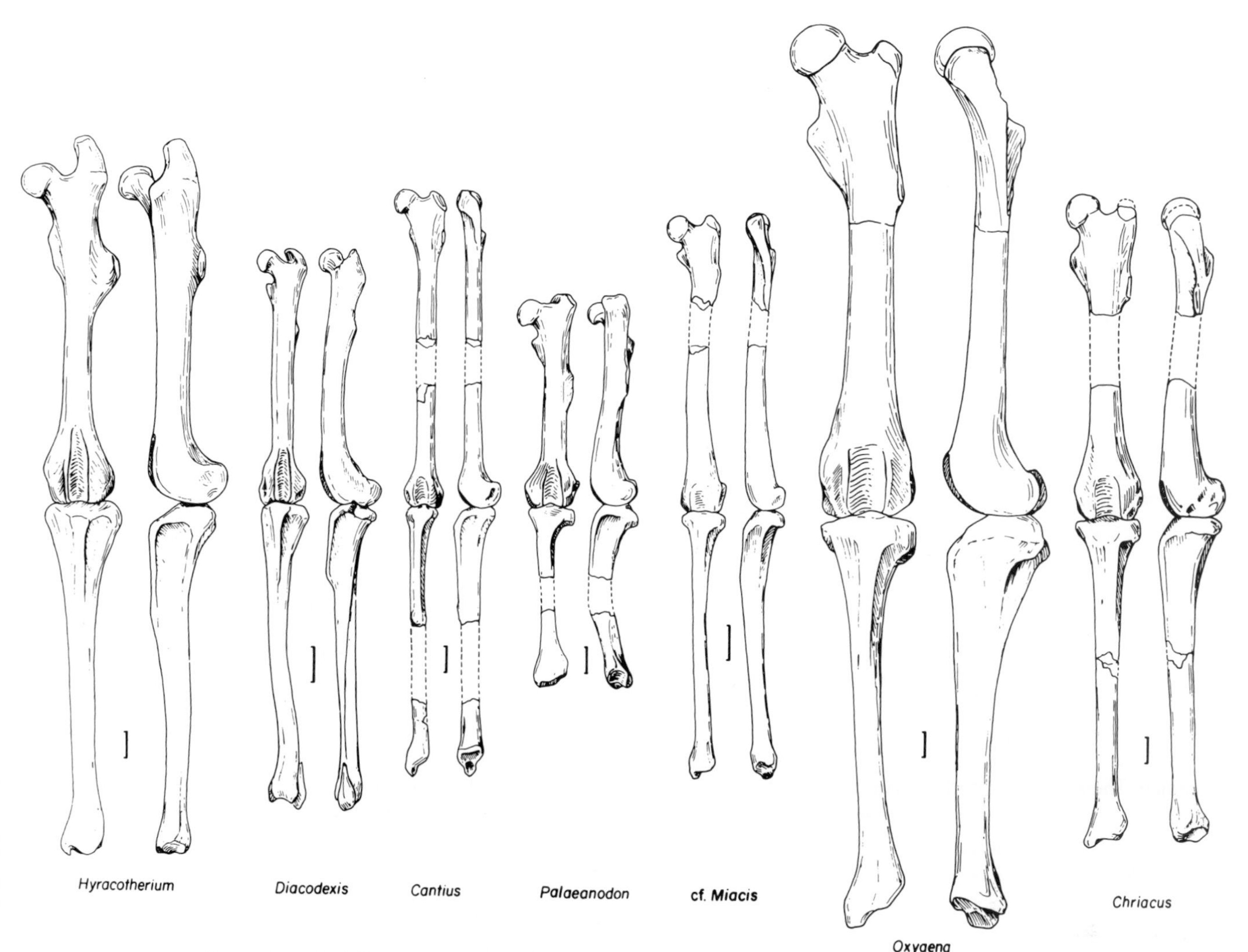

Figure 7. Left femur and tibia of some Willwood Wasatchian mammals. *Hyracotherium* sp. (USGS 5904, 5907, 21858), *Diacodexis metsiacus* (USGS 2352), *Cantius trigonodus* (USGS 5900), *Palaeanodon ignavus* (USGS 16471; proportions and tibia shape after Matthew, 1918), cf. *Miacis* sp. (USGS 7161), *Oxyaena*, cf. *O. forcipata* (USGS 7144, 21863), *Chriacus* sp. (USGS 2353). Scales = 1 cm.

codon ursidens and of the phenacodontids *Phenacodus primaevus, P. vortmani,* and *Ectocion osbornianus.*

Mesonychids. Relatively complete limb material is known in *Pachyaena ossifraga* (Cope, 1884; Matthew, 1915a) and *P. gigantea* (USNM 14915), and numerous characters indicate that *Pachyaena* was a cursorially adapted form (e.g., relatively long and slender limb elements; a greater tuberosity that projects above the humeral head; distal humerus transversely narrow, with a very deep and possibly perforate olecranon fossa; ulna concave posteriorly (see Fig. 9); radius positioned mostly anterior to the ulna; distal femur narrow and deep, with a well-defined trochlear groove). Indeed, the forelimb bones and the feet of *Pachyaena* are remarkably similar in general appearance (as well as size) to those of *Phenacodus primaevus* (e.g., deeply grooved astragalar trochlea; narrow tarsus; short, broad, dorsoventrally compressed phalanges; and fissured hooflike unguals). Compared to *Phenacodus,* however, *Pachyaena* has a more specialized astragalar head (slightly grooved rather than convex), a longer navicular, relatively longer metapodials, and a functionally tetradactyl, more paraxonic pes (in the sense that the first digit is vestigial)—all characters suggesting more progressive cursorial capability than in *Phenacodus.* Resemblances in postcranial osteology between *Pachyaena* and *Phenacodus* probably evolved in parallel from an early Paleocene (or older) arctocyonid common ancestor and, therefore, are not indicative of special relationship.

A very fragmentary specimen of the small mesonychid *Hapalodectes* (USGS 5912) includes jaws, cervical vertebrae, partial scapulae, and a distal humerus. Little can be detected before further preparation; but it may be observed that the distal humerus, as in *Pachyaena* and *Mesonyx,* is comparatively narrow, with a well-defined, narrow trochlea, deep olecranon fossa (or foramen?, obscured by matrix), and modest entepicondyle lacking a

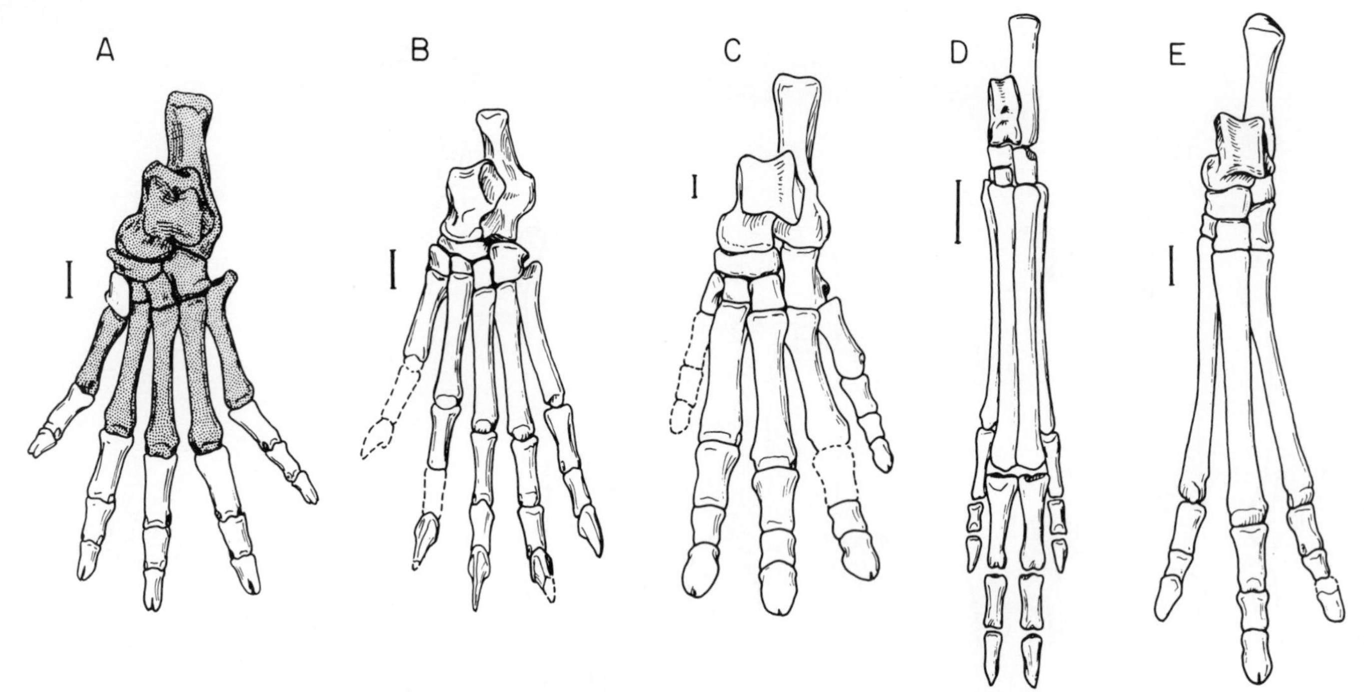

Figure 8. Left tarsus and pes of some Willwood Wasatchian mammals. A, *Oxyaena forcipata* (after Denison, 1938; shading indicates elements preserved in USGS 7144; B, *Chriacus* sp. (USGS 2353, after Rose, 1987); C, *Phenacodus primaevus* (USGS 7146); D, *Diacodexis metsiacus* (USGS 2352 and AMNH 27787, modified from Rose, 1985); E, *Hyracotherium* sp. (after Kitts, 1956). Scales = 1 cm.

foramen. These features are consistent with cursorial adaptations (restriction of lateral mobility but enhanced capability for extension at the elbow) seen in other mesonychids.

Chriacus. Several skeletal specimens of the arctocyonid *Chriacus* are known, the best of which is an extraordinarily complete, articulated skeleton (USGS 2353; Fig. 1) missing only the skull roof, parts of the sacrum and caudal series, the distal part of the left manus, and a few other bits. It is the most complete skeleton known for the genus and is one of the most intact early Eocene mammal skeletons known. In addition, parts of the appendicular skeleton (particularly the tarsus) are exquisitely preserved in a partial skeleton, USGS 15404. These and other specimens indicate an animal similar in size, proportions, and skeletal adaptations to the living coatimundi *Nasua* (Rose, 1987; see Fig. 10 herein).

Skeletal characteristics of *Chriacus* include: humerus with a prominent deltopectoral ridge and a large entepicondyle (Fig. 5); radial head ovoid to nearly round (Fig. 6), indicating substantial supinatory capability; olecranon process well developed and angled slightly anterior to long axis of ulna, indicating habitual elbow flexion; hip articulation indicating substantial hind limb abduction (see Jenkins and Camazine, 1977); distal femur with a broad, relatively shallow patellar groove; talocrural, subtalar, and midtarsal joints specialized to promote an exceptional range of motion, permitting hind-foot reversal associated with headfirst descent (judging from shape and extent of articular surfaces, flexibility was probably at least as great as in *Nasua* but less than in *Potos;* see Jenkins and McClearn, 1984); and pentadactyl, plantigrade feet with a somewhat divergent first digit (Fig. 8) and curved, laterally compressed unguals that closely resemble those of extant arboreal viverrid and procyonid carnivores (Fig. 2). The tail was long, and the caudal vertebrae are relatively broad, with flaring transverse processes and hemal processes—characters indicating that the tail was important for balance or was possibly semiprehensile. All these traits are typical of scansorial mammals, particularly those known to spend a significant amount of time in the trees. Like certain living viverrids and procyonids, *Chriacus* was an ambulatory form that clearly was also a proficient climber (see Fig. 10).

Thryptacodon. Matthew (1915a) illustrated skeletal fragments of two species of the arctocyonid *Thryptacodon* (*T. antiquus* and *T. olseni*) from the Willwood Formation, the only published postcranial specimens of the genus. A significant part of a skeleton of the small arctocyonid *Thryptacodon* sp. (UW 7421) comes from a lower Willwood (early Wasatchian) site in the Sand Creek facies, southeast of Worland. A second partial skeleton (YPMPU 14690) comes from lower Willwood strata in the central Bighorn Basin. These specimens differ in only minor respects from *Chriacus,* notably in the articular surfaces of the astragalus, which suggest slightly less flexibility at all three ankle joints. Nonetheless, it can be confidently inferred that *Thryptacodon* was a capable climber whose skeleton was structurally and adaptively similar to that of *Nasua* (Fig. 9).

Anacodon. The specimens reported here (Table 1) are the only known postcranial remains of this rare, moderately large, bearlike arctocyonid. With one exception, all specimens are asso-

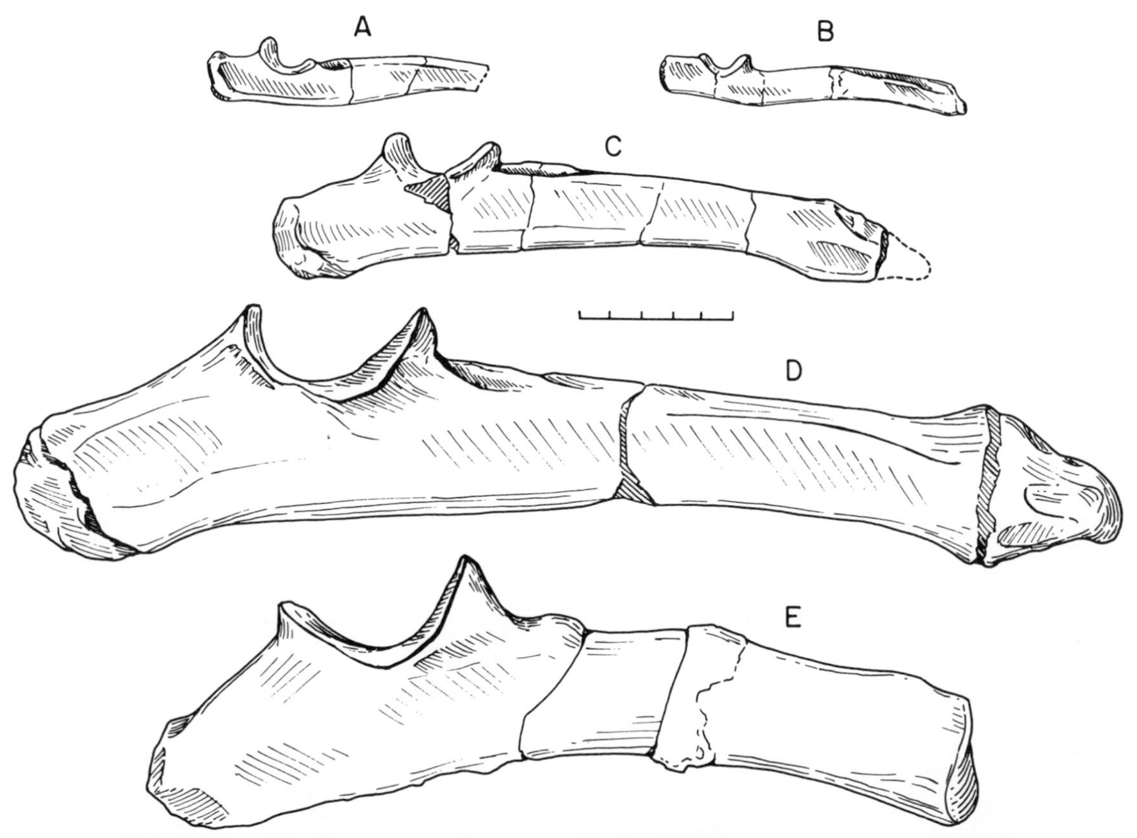

Figure 9. Left ulnae (medial view) of some Willwood Wasatchian mammals. A, *Phenacodus vortmani* (USGS 7159); B, *Didymictis protenus* (USGS 1721); C, *Phenacodus primaevus* (USGS 7146); D, *Pachyaena gigantea* (USNM 14915); E, *Coryphodon* sp. (USGS 7190, 21840). Scale = 5 cm.

ciated with teeth of *Anacodon ursidens,* so their taxonomic identity is unequivocal. The exception, USGS 21857 (associated incomplete left humerus, radius, and ulna), was first thought to pertain to the taeniodont *Ectoganus* because of its large size and robustness, but close comparison indicates that these bones belong to *Anacodon* (a large *A. ursidens* or perhaps *A. cultridens*). Most of the skeleton is now represented, including vertebrae, fragmentary scapula, humerus, radius, ulna, some carpals, pelvis, femur, tibia, fibula, most of the tarsus, metapodials, and phalanges.

The appendicular skeleton (see Fig. 11) of *Anacodon* is robust, the forelimbs apparently more so than the hind; the forelimbs are only slightly shorter than the hind limbs. (Some proportional details are uncertain because of distortion.) In many respects the skeleton is a larger, more robust version of *Chriacus* (Fig. 10); but there are also some notable and significant differences. The forelimb skeleton can be characterized as follows: the humerus is more robust than in *Chriacus* and bears a very prominent, elevated and shelflike deltopectoral crest; the distal humerus is broad, the medial epicondyle is large, and the supinator crest is well developed (relatively more so than in *Chriacus*); the radial head is much more elliptical, and the radioulnar facet less rounded than in *Chriacus* (Fig. 6), hence supinatory ability was much less than in *Chriacus;* the radius is expanded distally and has a prominent pronator ridge; the ulna is laterally compressed and remarkably deep (anteroposteriorly; relatively much deeper than in *Chriacus*).

In the hind limb, the femur is about the same length as the tibia or slightly longer; the fovea capitis femoris is very large, and the anterior margin of the articular surface of the head makes a wide angle with the shaft (about 50 degrees), suggesting an abducted hind limb posture (Jenkins and Camazine, 1977); the lesser trochanter is prominent and projects mediad, while the third trochanter is situated proximally on the shaft (opposite the lesser trochanter) and is continuous with the greater trochanter; the patellar groove is relatively slightly deeper than in *Chriacus;* the distal tibia and astragalar trochlea are shallow (but slightly more grooved than in *Chriacus*); the metapodials and phalanges are not elongated; and the ungual phalanges are curved, weakly fissured at the tip, and laterally compressed, with large subungual processes (Fig. 2). The unguals are more transversely compressed but otherwise similar to those of Paleocene *Arctocyon primaevus* (Russell, 1964).

The tarsal bones and ankle joints differ from those of *Chriacus* in several features that indicate a less flexible ankle in *Anacodon.* For instance, *Anacodon* has an interlocking spur on the tibial malleolus that fits into a depression on the astragalus, severely restricting mediolateral (especially medial) movement of

Figure 10. Life restoration of *Chriacus* sp., based on USGS 2353 and 15404. Drawn by Elaine Kasmer, under supervision of K. Rose. Reproduced in color on front cover.

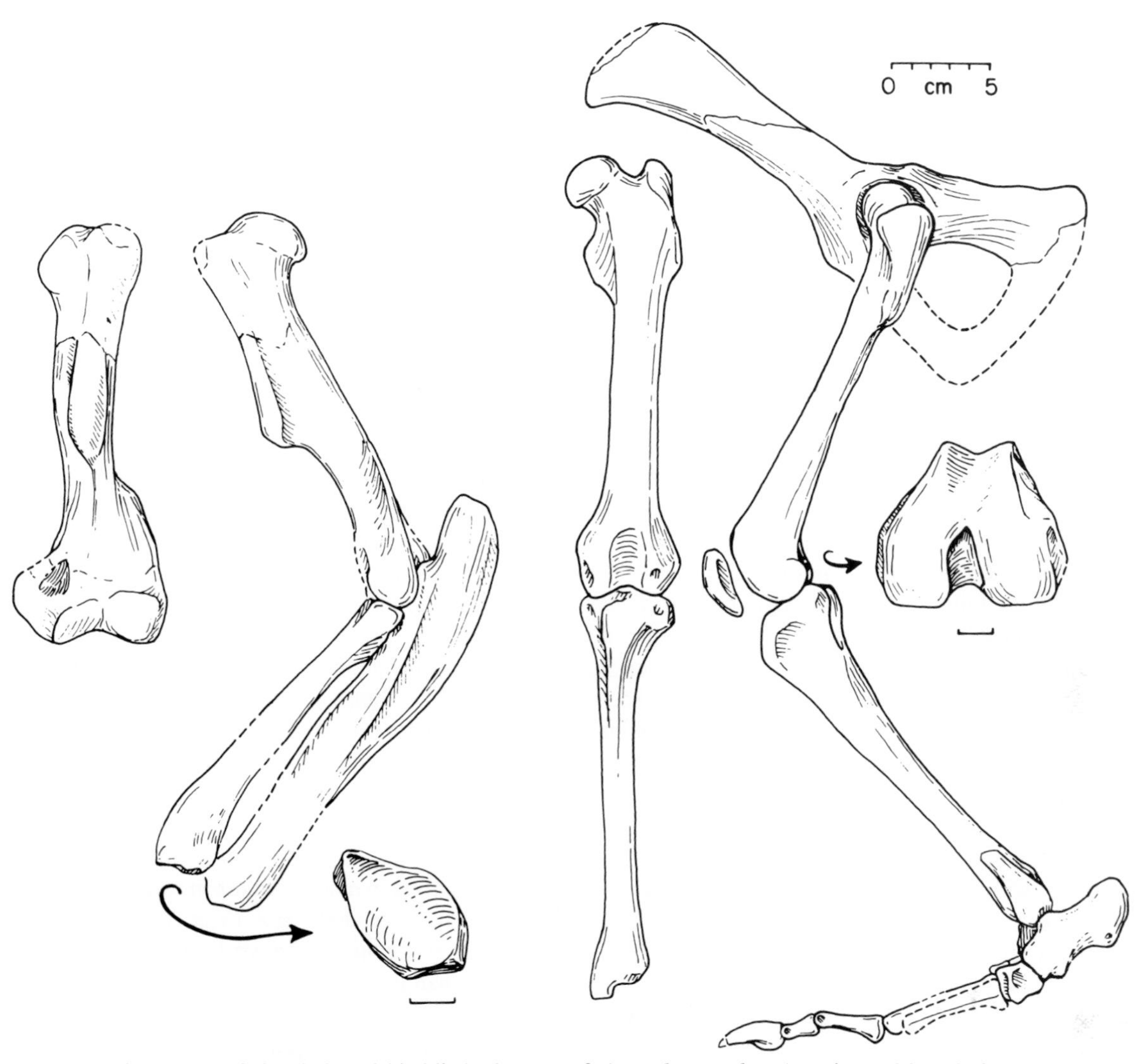

Figure 11. Left forelimb and hind-limb elements of *Anacodon ursidens* (anterior and lateral views) reconstructed from USGS 5026, 5031, 5902, 21856, and 21857. Scales on distal views of radius and femur = 1 cm.

the pes during dorsiflexion. A similar situation exists in the tibioastragalar joint of hyracoids (Fischer, 1986; see also Coombs, 1983). Other characters include a large convex fibular facet on the calcaneus and matching concave facet on the distal fibula (as in *Arctocyon ferox* [AMNH 3268, 16543] and *Arctocyonides mumak* [YPMPU 18703]), a large concave facet on the cuboid for the astragalar head (as in *Arctocyon primaevus* [Russell, 1964] and *Oxyaena*), and a less extensive posterior calcaneal facet. The calcaneus is distinctive; it is short and dorsoventrally deep, with the tuber calcanei expanded both transversely and dorsoventrally, the ventral projection forming a salient posterior plantar (heel) process (Fig. 11). A smaller anterior plantar process is also present, so that the plantar profile of the calcaneus is strongly arched. The giant anteater *Myrmecophaga tridactyla* and, to a lesser extent, certain hominoids (*Pongo, Homo*) and ursids (e.g., see Stains, 1973) approximate this unusual calcanean morphology, which is otherwise infrequent among eutherians. The heel process in these and some other mammals provides origin for the flexor digitorum superficialis (= flexor digitorum brevis in humans; Evans and Christensen, 1979) and for the plantar aponeurosis. Sarmiento (1983), who was principally interested in the function of the heel process in anthropoid primates, argued that the heel process is associated with grasping ability in slow climbers, and that its presence in terrestrial forms such as *Myrmecophaga* reflects ancestry from slow climbers. This hypothesis is weakened, at least for *Myrmecophaga,* by the absence of a heel process in arboreal edentates. In bears, the weak heel process is apparently associated with their plantigrade posture (Sarmiento, 1983). A similar relationship may apply to *Anacodon* as well, but, in view of the absence or weak development of a heel process in other arctocyonids (including even the larger *Arctocyonides mumak*), this is far from certain. Further investigation of the precise function of the heel process in *Anacodon* is clearly warranted.

Based on this preliminary examination of the skeleton, *Ana-*

codon appears to have been plantigrade, primarily terrestrial and ambulatory, and capable of powerful digging or tearing and also of climbing. This is consistent with Russell's (1964) interpretation of the somewhat similar *Arctocyon*. Bears may represent the closest living analogue of *Anacodon*.

Phenacodus. Cope (1884) described and illustrated nearly complete skeletons of two species of *Phenacodus, P. primaevus* (Fig. 1) and the much smaller *P. vortmani*. Postcranial remains of *Phenacodus* are not uncommon in the Willwood Formation; the USGS collection includes several specimens of each of these two species. Because of their overall similarity, they are described together below, with a few significant differences noted. Of *P. primaevus,* the most notable specimen is USGS 7146, which consists of the nearly complete left tarsus and pes (Fig. 8), distal fibula, left ulna (Fig. 9) and proximal radius, carpals, numerous vertebrae, and associated fragments, in excellent preservation. Well-preserved foot elements are also represented in USGS 7158, 7162 (which also includes distal tibias), and 16477. USGS 21838 preserves most of the left tibia and a distal radius. Of *P. vortmani,* USGS 7159 includes jaws and well-preserved forelimbs up to the carpus; USGS 5022 includes fragments of forelimb elements, pelvis, femur, tibia, calcaneus, and vertebrae; and USGS 21855 is a partial skeleton with vertebrae, parts of the scapula, proximal and distal humerus, pelvis, proximal and distal tibia, and articulated tarsus and metatarsals.

The appendicular skeleton displays numerous traits indicative of incipient cursorial capability. The humerus (Fig. 5) is moderately robust, with a prominent, elevated greater tuberosity projecting well above the humeral head, a long but weakly developed deltopectoral crest extending two-thirds the length of the shaft, and, notably, a relatively narrow distal end with a deep olecranon fossa and a large supratrochlear foramen. The medial epicondyle, though less projecting than in arctocyonids or creodonts, remains well developed, and an entepicondylar foramen is present. The ulna and the radius are primitively robust and separate (as are the tibia and the fibula). In *P. primaevus,* the ulna is slightly concave posteriorly, and the prominent olecranon is inclined sharply backward (about 30 degrees) from the axis of the shaft (Fig. 9; cf. Van Valkenburgh, 1987). In *P. vortmani,* the proximal end of the ulnar shaft is slightly convex posteriorly, the distal end is slightly concave, and the olecranon shows only the slightest backward inclination. The oval radial head (Fig. 6), which is positioned largely anterior to the ulna, is twice as broad in mediolateral diameter as in anteroposterior diameter and is marked by an elevated ridge on its articular surface that matches the deep humeral trochlea. The construction of the elbow region thus severely limits supination while maximizing antebrachial extension, despite the presence of a long olecranon. The carpus is relatively little changed from the condition in middle Paleocene phenacodontids: it is broad and essentially serial (*P. primaevus;* the arrangement in *P. vortmani* is slightly alternating; Radinsky, 1966).

Characteristics of the hind-limb skeleton include prominent femoral trochanters (the third trochanter is located just proximal to midshaft), distal femur deeper anteroposteriorly than broad and with a well-defined patellar groove (but not as deep as in *Hyracotherium*), tibia with a weaker cnemial crest than in Paleocene phenacodontids, tarsus relatively broad, astragalus with a deeply grooved trochlea (but not as deep as in *Hyracotherium*) and a rounded head (not grooved as in primitive perissodactyls and artiodactyls), navicular deeply concave, and proximal cuboid surface dorsoplantarly convex. The structure of the tarsal joints indicates that limited inversion and eversion were possible, occurring largely at the subtalar and astragalonavicular joints (Radinsky, 1966); this capability was probably limited by the calcaneocuboid joint, which is configured more for flexion-extension. The tarsal joints had less flexibility than in the arctocyonids discussed above, but more than in true ungulates.

The phalanges in both manus and pes are broad and not especially elongate, and the unguals clearly bore hooves (Figs. 2, 8). Although pentadactyl, *Phenacodus* was functionally tridactyl (digits I and V are conspicuously reduced in both manus and pes) and incipiently mesaxonic.

The two species differ significantly in size—*Phenacodus primaevus* being about twice the size, more robust, and having a body weight perhaps an order of magnitude greater than *P. vortmani*—but otherwise their structure is quite similar. As noted above, a conspicuous distinction may be seen in the form of the ulna, particularly the olecranon. In general, the ulnar type that occurs in *P. primaevus* would suggest that this species was more specialized for cursorial locomotion than was *P. vortmani;* but similar variation in ulnar structure in carnivores tends to be exaggerated by larger body size (Van Valkenburgh, 1987). The smaller *P. vortmani* has proportionally longer distal hind limb segments than in *P. primaevus* (Table 2), which, contrary to the evidence of the ulna, suggests that the smaller species was a slightly better runner (or jumper). Further study will be required to evaluate more precisely the locomotor capabilities of different species of *Phenacodus*.

The anatomy of the limbs and joints indicates that *Phenacodus* was a terrestrial ambulatory form that, compared to its middle Paleocene progenitors, had begun to specialize for running (Radinsky, 1966). Indeed, no other Willwood mammals, except mesonychids and the much more specialized true ungulates, show as many modifications toward cursorial habits.

Ectocion. Based on dental remains, the small phenacodontid *Ectocion osbornianus* was the commonest mammal in the Clarkforkian of the Bighorn Basin, and it remained moderately common in the early Wasatchian (Rose, 1981). Nonetheless, few postcranial remains are known and none has ever been described (although Granger [1915] briefly alluded to a few skeletal fragments). We have recovered only a single postcranial association—a snout, dentaries, and partial skeleton, including at least parts of the radius, femur, tibia, and tarsus, encased in hematitic matrix (USGS 16496). Although little can be clearly discerned without substantial preparation, it is possible to detect that the radial head is ovoid, with a relatively flat ulnar facet, the third trochanter is well developed, and the astragalar trochlea and distal tibial articulation are deeply grooved—all, not surprisingly, features closely resembling *Phenacodus.* Thus *Ectocion* was also a terrestrial quadruped showing incipient cursorial adaptations.

Tillodontia

Esthonyx. Although jaws and teeth of the early Eocene tillodont *Esthonyx* are not uncommon in the Willwood Formation, the postcranial skeleton in members of this genus remains poorly known, represented only by a few very fragmentary specimens (Cope, 1884; Gazin, 1953) that reveal little about its adaptations. Two specimens of *E. bisulcatus* in our collection (USGS 7551 and 21839, bones associated with dentitions) add to this meager sample. They have not yet been studied but are noted here for completeness. USGS 21839 includes the axis, lumbar, and sacral vertebrae, as well as fragments of the femur and tibia. USGS 7551 includes phalanges, fragmentary metapodials, and a distal humerus. The latter is broad, with a very large entepicondyle and a deep olecranon fossa (although perforate in the specimen, this is probably an artifact). The supinator crest is not particularly prominent, in contrast to the condition in *Trogosus* (Gazin, 1953). An associated terminal phalanx is laterally compressed and appears to resemble closely that of *Trogosus* and, even more so, those of *Chriacus* and *Thryptacodon.* The other phalanges also resemble those of *Chriacus,* and the distal humerus differs but little from *Chriacus.* These similarities are sufficient to raise doubt about the association with *Esthonyx* teeth; alternatively, they may simply indicate close correspondence in these parts between *Esthonyx* and small arctocyonids.

Megalesthonyx. Several bones of the manus are associated with the holotype specimen of *Megalesthonyx hopsoni* (YPM 18767), from the upper Willwood Formation ("Lostcabinian" or upper *Heptodon* Range Zone of Schankler, 1980). They are structurally similar to those of *Esthonyx* and *Trogosus* and roughly intermediate in proportions (Rose, 1972). The terminal phalanges are curved and laterally compressed, as in other tillodonts. These fragments add little to the inadequate understanding of tillodont adaptation, but they are not inconsistent with scansorial habits or, more likely, forelimb digging or tearing, as suggested by Coombs (1983).

Pantodonta

The pantodont *Coryphodon* is the most common large mammalian herbivore in the Willwood fauna; remains, usually consisting of teeth or fragments of teeth but often including bones, are ubiquitous. Although *Coryphodon* is not among the four or five most common Wasatchian mammals, its relative abundance is surely underestimated in collections because of the tendency to leave behind fragmentary (but identifiable) teeth. Species-level taxonomy of *Coryphodon* has long been in disarray (at least 35 species have been proposed; Lucas, 1982), and a formal revision of North American species has not been published in nearly a century; consequently it is not possible to assign our material confidently to species.

Partial skeletons of *Coryphodon* have been known from the Bighorn Basin since the late nineteenth century (Cope, 1884; Osborn, 1898a, c; see also Gilmore, 1936; Patterson, 1939; Simons, 1960). Together these specimens represent all the skeletal components (Fig. 1). Concentrations of complete skeletons have since been discovered elsewhere (e.g., Lucas, 1981). The material listed in Table 1 is fragmentary, but all long bones are represented, as well as tarsals, metapodials, phalanges, and vertebrae.

Coryphodon was a short-limbed, heavily built animal, with a relatively large head equipped with prominent long canines reminiscent of those of hippopotamuses. The larger species attained the size of a steer. The humerus is robust, with a prominent deltoid ridge extending distally more than half the length of the bone, and large epicondyles. The radius and the ulna are stout, the radius conspicuously shorter than the humerus, and the ulna with a large olecranon. The ulna (Fig. 9) is bowed, concave posteriorly as in *Phenacodus primaevus* and *Pachyaena,* but it is relatively shorter and more robust and its humeral articular surface is much broader. The radial head is much broader in transverse diameter than in anteroposterior diameter (Table 2), and the radioulnar articulation is only slightly curved; hence, little supination was possible. The femur, though robust, was described by Osborn (1898a) as "long and rather slender." It is about 50 percent longer than the tibia, which, however, is remarkably short and robust. The greater trochanter is lower than the femoral head, and the third trochanter is prominent and situated at about midshaft. The feet are broad and pentadactyl, with short, robust metapodials and phalanges and unfissured hooflike unguals. Simons (1960) described the feet as "much more elephantine" (p. 70) and "much more graviportal . . . than in other pantodonts" (p. 12). The stance was probably plantigrade, possibly showing a slight tendency toward a digitigrade condition (Osborn, 1898a). *Coryphodon* is considered to have been semiaquatic, probably analogous to the hippopotamus in general habits (Simons, 1960).

Perissodactyla

Skeletal remains of at least three genera are included in the collection (Table 1): *Hyracotherium, Homogalax,* and *Heptodon.* Although they represent three different families (Equidae, Isectolophidae, Helaletidae, respectively) and two suborders (Hippomorpha and Ceratomorpha), at this very early stage in perissodactyl evolution they are, with minor exceptions, quite similar.

Perissodactyls have been considered a derivative of phenacodontid condylarths (e.g., Radinsky, 1966). Although this view is losing popularity, close resemblances in what are believed to be derived characters do exist in many parts of the skeleton. Whatever its relationships, *Phenacodus* appears to have been incipiently cursorial and is not very different osteologically from what would be expected in a precursor of perissodactyls; hence, some of the description that follows is given with reference to *Phenacodus.*

Hyracotherium. Although the skeletal anatomy of *Hyracotherium* is relatively well known, good skeletons are rare and skeletal remains of any sort are much less common than might be expected from their dental abundance. In 1904, Loomis collected a substantial part of a skeleton of *Hyracotherium* for Amherst College (Gingerich, 1980)—perhaps the first from the Bighorn Basin. Kitts (1956) reported three relatively complete skeletons from the Bighorn Basin (AMNH 15428, 15820, and an unspecified skeleton in the collection of the California Institute of Technology). The USGS collection includes about a dozen partial skeletons probably belonging to more than one species (Table 1); the proper identification and validity of species represented, however, must await an up-to-date systematic revision. One specimen (USGS 5901) is a substantial part of an articulated skeleton with cervical, lumbar, and thoracic vertebrae, scapula (glenoid part), proximal humerus, nearly complete radius and ulna, complete carpus and much of the manus, middle and distal femur, and nearly complete and articulated tibia, fibula, tarsus, metatarsus, and proximal phalanges. Much of the latter specimen is encased in hard calcareous matrix so that few details can be observed without further preparation. Additional elements represented in other USGS material include part of the sacrum, the distal humerus, the pelvis, and complete femora and tibiae.

The following characteristics of the postcranial skeleton of *Hyracotherium* are based on our specimens, supplemented by the descriptions by Kitts (1956) and Radinsky (1966). Noteworthy in the vertebral column are the nearly vertical lumbar neural spines (suggesting that the spine in these early perissodactyls was less flexible than in *Phenacodus*), a sacrum composed of five fused vertebrae (three in *Phenacodus*), and the lack of known caudal vertebrae (negative evidence, but perhaps suggestive that the tail was shorter and more gracile than in *Phenacodus*). Compared to most other Willwood mammals, the humerus has a high greater tuberosity (projecting above the humeral head or even with it), a weak deltopectoral crest, and a narrow distal end, with reduced epicondyles, no entepicondylar foramen, and a large supratrochlear foramen. The lateral part of the distal articulation forms a transversely constricted intercondylar ridge (flanked laterally by a narrow shelf), rather than a rounded capitulum as in *Phenacodus.* The ulna and the radius are somewhat bowed (concave posteriorly), the olecranon bends slightly backward, and the radial head is ovoid and similar to that in *Phenacodus* but with the condylar facet more constricted and more nearly equal in breadth to the trochlear facet. Thus the structure of the elbow joints prohibited supination and limited motion primarily to a sagittal plane. The carpus is transversely narrow, and its bony elements are strongly alternating (interlocking) in arrangement (Radinsky, 1966), which promoted stability and restricted lateral or rotatory mobility. There are four digits in the manus (digit I is lost), but it is already mesaxonic, with digits II and IV slightly shorter than digit III and digit V more reduced. The metapodials are elongate and the ungual phalanges bore hooves.

The pelvis is characterized by a broad, dorsally flaring iliac blade, which provided a more dorsal origin for the gluteus medius muscle (Kitts, 1956; Hussain, 1975), a powerful extensor (and abductor) of the thigh. Such a dorsal expansion is typical of cursorial ungulates. Importance of the gluteus medius is also reflected in the very high greater trochanter of the femur (its insertion), projecting well above the head (Fig. 7). As in *Phenacodus,* the femur also has well-developed lesser and third trochanters (the latter relatively more proximal than in *Phenacodus*) and is narrow and deep distally, with a well-defined patellar trochlea. The tibia is comparatively slender, with a shorter cnemial crest than in *Phenacodus.* The distal articular surface is deeply grooved to accommodate the astragalus. The fibula is very gracile but remains separate. Like the forefoot, the tarsus and metatarsus are narrow. The deeply grooved astragalar trochlea and shallowly grooved head are recognizably similar to those in later perissodactyls. These features, together with the narrow calcaneus with its interlocking (rather than convex) posterior astragalocalcanear articulation and the saddle-shaped astragalonavicular and calcaneocuboid joints, strengthened the ankle joints while restricting motion to a sagittal plane. The pes is three-toed and mesaxonic, with the metatarsals (Mt) arranged proximally in an arch such that Mt II and IV are situated on the medial and lateral plantar surfaces of Mt III; the lateral metapodials are reduced. The limbs are relatively elongated distally, particularly the metapodials (Table 2). The digits were hoofed (Fig. 8). In combination, these features are clear evidence for a "nearly unguligrade" stance (Kitts, 1956). The overwhelming majority of these traits are modifications for speed rather than power, and for joint stability and sagittal mobility, adaptations associated with cursorial habits in *Hyracotherium.* Apart from the dichobunid artiodactyls, the early perissodactyls were the most cursorially adapted Willwood mammals.

Homogalax. No postcrania have previously been reported for this primitive perissodactyl, usually considered the oldest known tapiroid. Hooker (1984) suggested that Isectolophidae (including *Homogalax*) are even more primitive than true tapiroids and are the sister group of both Ancylopoda and Ceratomorpha. At least two USGS specimens represent *Homogalax,* cf. *H. protapirinus.* The better of the two, USGS 21843, includes jaw fragments associated with the glenoid part of the scapula, proximal and distal humerus, proximal ulna and radius, proximal and distal femur, distal tibia, astragalus, calcaneus, metapodial fragments, and phalanges. Preliminary comparisons indicate that nearly all elements are very similar to their counterparts in *Hyracotherium.* The distal humerus, however, is conspicuously more robust—specifically in having a much larger medial epicondyle—than in *Hyracotherium* and the primitive tapiroid *Heptodon.* It is superficially more similar to that of *Phenacodus vortmani,* but it has a slightly larger ectepicondyle and a more constricted capitulum (intercondylar ridge) with a small shelf lateral to it, and it lacks an entepicondylar foramen. Kitts (1956) stated that the astragalus and calcaneus of *Homogalax* are "more condylarth-like" than in *Hyracotherium. Homogalax* was unquestionably one of the more specialized runners of its day, but its humeral morphology suggests that it may have been somewhat more primitive than *Hyracotherium* in this regard, thus supporting Hooker's hypothesis.

Heptodon. Skeletal remains of this early helaletid tapir are poorly known from the Bighorn Basin; however, Radinsky (1965) described a good skeleton of *Heptodon posticus* from the Wind River Basin. The best USGS specimen (21861, Table 1), a subadult individual probably representing the smaller species *H. calciculus,* includes longbone fragments, separated epiphyses, tarsal and pedal elements, and vertebrae; however, it is too fragmentary and distorted to yield much information. Both Kitts (1956) and Radinsky (1965) remarked on the extraordinary postcranial similarity among early Eocene perissodactyls. In comparable parts, USGS 21861 shows no significant departure from either *Heptodon* (Radinsky, 1965) or *Hyracotherium.*

Artiodactyla

Diacodexis. Except for a few fragmentary bones from the Bighorn Basin described under the name *Pantolestes brachystomus* (Cope, 1884), the postcranial skeleton of this oldest known artiodactyl remained virtually unknown until a few years ago. A second, very incomplete specimen was collected by Troxell in 1936 (but was unrecognized until much later), and two much better specimens were found over the last decade by USGS–Johns Hopkins field parties (Table 1). Based on these specimens, which include a nearly complete skeleton (USGS 2352), Rose (1982, 1985) reconstructed and described the skeleton of *Diacodexis metsiacus* (Fig. 1). It is remarkably specialized for an early Eocene mammal, even markedly surpassing the cursorial modifications of the early perissodactyls.

Characteristics of the axial skeleton include a relatively short trunk, tapered sacrum comprising three fused vertebrae, and moderately long but gracile tail. The neural spines of the lumbar vertebrae are strongly inclined craniad, and the vertebral column was probably conspicuously arched, suggesting the capability to increase stride by flexing and extending the vertebral column (see also Franzen, 1981, Fig. 11), in contrast to the inferred manner of locomotion in early perissodactyls.

The limb elements are comparatively long and slender, particularly distally, and the forelimbs were evidently much shorter than the hind limbs (although complete forelimb elements, except for the humerus, are unknown). A small bone in the pectoral girdle has been interpreted as a possible vestigial clavicle. Clavicles are not known to exist in any other genus of artiodactyl.

Forelimb traits include a slender humerus (Fig. 5) with the greater tuberosity higher than the head, and a very weak deltoid crest (marked by a line rather than a raised crest). The humerus is narrow distally, with a narrow trochlea, a constricted intercondylar ridge (i.e., modified capitulum), and a deep, perforate olecranon fossa. The entepicondyle is moderately developed but without a foramen. The ulna and the radius are primitively separate; the ulna has a prominent but relatively short olecranon; the radial head is ovoid with a depression for the intercondylar ridge (Fig. 6) and is oriented fully anterior to the ulna (thus preventing supination). The manus is unknown.

The iliac blade is expanded mediodorsally, reflecting the increased importance of the gluteus medius as a rapid extensor of the thigh. Other hind-limb characteristics (Fig. 7) are: moderately developed greater and lesser trochanters and a rudimentary third trochanter (the latter absent in most later artiodactyls); distal femur much deeper (anteroposteriorly) than broad, with posteriorly projecting condyles and an elevated, sharply defined patellar trochlea; tibia slender, with a short tibial crest; distal tibia deeply grooved for the astragalar head; fibula reduced to a very slender rod sometimes fused to the tibia distally; tarsus narrow and specialized; astragalus with a deeply grooved trochlea, sharp trochlear ridges, and a grooved head that articulates with the navicular and the cuboid; cuboid separate, ecto- and mesocuneiform fused; pes essentially four toed (vestigial first metatarsal present) and paraxonic; metatarsals (Mt) slender and very elongate and arranged in an arch in coronal section, Mt III and IV longer and greater in diameter than Mt II and V; phalanges somewhat elongate, unguals somewhat hooflike but rather constricted mediolaterally (Figs. 2, 8). The phalangeal articulations, together with several of the traits listed above, suggest that *Diacodexis* was unguligrade.

Almost all the osteological features and modifications listed above are specializations for a cursorial-saltatorial pattern of locomotion, such as is seen in rabbits or chevrotains—that is, frequent but usually brief episodes of running and/or jumping. At small body size, cursorial and saltatorial modes may intergrade considerably. In postcranial characters, *Diacodexis* is the most derived cursorial-saltatorial mammal in the Willwood fauna. Nearly every bone bears the hallmark of this adaptation and is, consequently, immediately recognizable as belonging to an artiodactyl.

Wasatchia. The taxonomy of larger Wasatchian dichobunids, *Wasatchia* and *Bunophorus,* is poorly understood, the most recent revision being that of Sinclair (1914). Most recent workers have tended to follow Van Valen (1971) in synonymizing the two genera, but they are probably distinct.

A single very incomplete skeletal specimen of a large dichobunid (USGS 16470, probably *Wasatchia dorseyana*) has been recognized in our collection, consisting of lumbar and caudal vertebrae, a distal humerus (Fig. 4), proximal and distal femur and tibia, calcaneus, astragalus, fragmentary metapodials, and other fragments associated with a dentary with M_3. Dichobunids are postcranially uniform, differing significantly only in size (Rose, 1985); *Wasatchia* is larger and correspondingly more robust than *Diacodexis,* but otherwise very similar. Except for a few isolated elements, no other postcranial specimens of *Wasatchia* or *Bunophorus* are known from the Bighorn Basin. Comparable parts of the USGS specimen are 30 to 50 percent larger than geologically younger specimens of *Bunophorus* from the Wind River and Huerfano basins (Rose, 1985).

SUMMARY AND DISCUSSION

Postcrania are known for more than 35 genera of mammals from the Wasatchian part of the Willwood Formation—an impressive figure for the early Eocene, representing about half of the genera recorded in the Willwood assemblage. About one-third of those known from postcrania are represented by relatively complete skeletons, sufficient to allow reliable reconstructions. Although much remains to be learned—only one genus in six is known well enough to restore its skeleton—these data provide considerable information (unavailable from other sources, such as the abundant dental remains) about the anatomical characteristics and habits of early Eocene mammals of the Bighorn Basin.

These fossils demonstrate that Willwood Wasatchian mammals were as diversified in their postcranial skeletons as in their skulls and dentitions. There were arboreal forms (adapted in at least two distinctly different ways), terrestrial types (from graviportal to generalized ambulatory to cursorial), and fossorial forms. Swimmers and flyers probably existed, too, but their bones either have not been found or have not been recognized (with the possible exception of bats, see Insectivora above). Although there were several archaic mammals that have no close living analogues, many Willwood mammals had postcranial skeletons similar—even strikingly so—to extant forms (which are often but not always their relatives), thus permitting more reliable inferences regarding adaptations. In some of these groups (e.g., euprimates, artiodactyls, perissodactyls), certain characteristic postcranial specializations were achieved very early in their history (see also Schaeffer, 1948). Subsequent modifications occurred, but diagnostic postcranial traits were already established in these Wasatchian forms. For the most part, we know little about the origins of these specializations, but some of them may have arisen by selection for "excessive constructions"—characters related to uncommon but important (often extreme) aspects of behavior (Gans, 1979).

Comparison of various simple limb indices and ratios of joint dimensions (Table 2) provides some insight into functional capabilities of Willwood mammals. For example, high crural and femorometatarsal indices (in excess of 100 and 40, respectively) suggest cursorial abilities. In cursors, the distal femur is usually deeper (anteroposteriorly) than broad; diggers and climbers tend to have a mediolaterally broad distal femur. The radial head tends to be very elliptical in cursors and other forms in which it is advantageous to limit movement at the proximal radioulnar joint (ratio of diameters >160); a rounder radial head (lower ratio, nearer to 100) is typical of mammals that supinate freely (e.g., climbers).

Willwood mammals display at least two different adaptations to arboreality: typical scansorial claw-climbing, and friction grip climbing (Cartmill, 1985). The first is exemplified by the arctocyonids *Chriacus* and *Thryptacodon* and the miacid carnivores (e.g., *Vulpavus*), which approximated living viverrids and procyonids in postcranial anatomy and inferred locomotor capabilities. They employed sharp claws, probable hind-foot reversibility, and a strong balancing (?semiprehensile) tail to maneuver among branches. The second type, exemplified by adapid primates, were quadrupedal climbers and leapers with relatively long hind limbs, nails rather than claws, and opposable first digits that enhanced their grasp. Both of these groups had the capacity for extensive supination of the forearm. Other Willwood mammals (e.g., *Anacodon*, hyaenodontids, *Didymictis*) probably possessed the ability to climb, but anatomical modifications indicating restricted joint mobility suggest that they spent considerably more time on the ground.

A broad range of terrestrial locomotor behavior may be seen among Willwood mammals. The most specialized for speed, the ancient artiodactyls *Diacodexis* and *Wasatchia*, were similar in postcranial anatomy and presumed locomotor habits to recent chevrotains. Because of their small size, they were not strictly cursorial but probably coupled running and jumping when necessary, like living *Hyemoschus* (Dubost, 1975). The still-smaller leptictid insectivores appear to have been more strictly saltatorial. Early perissodactyls were also cursorially adapted but were somewhat larger than the early artiodactyls. Their postcranial skeletons resemble those of extant tapirs, except for being smaller, more gracile, and having a more flexible vertebral column (Radinsky, 1965; Janis, 1984). This suggests that *Hyracotherium*, *Homogalax*, and *Heptodon* may have been a little more agile than recent *Tapirus* but otherwise similar in locomotor capabilities. Less specialized for running, but nevertheless possessing unequivocal anatomical modifications toward cursorial habits, were the phenacodontids and mesonychids, which include the largest early Eocene mammals to show specializations for speed.

The largest herbivore in the fauna, *Coryphodon*, appears to have been an ambulatory, graviportal form, perhaps somewhat amphibious, without particular modifications for any other specialized locomotor behavior. Nearly the same size was the taeniodont *Ectoganus*, whose limb bones are more robust and suggest the tendency to use the forelimbs for powerful digging or tearing; the aardvark may provide the closest living analogue (Schoch, 1986). Smaller, but no less proficient diggers, were the palaeanodonts *Palaeanodon* and *Alocodontulum*, which already in the Wasatchian had achieved impressive fossorial specializations. Analogous to armadillos and manids, they must have been capable of efficient burrowing or ripping into termite mounds or rotting logs in search of food.

Oxyaenid creodonts were apparently also terrestrial ambulatory mammals, unusual for their comparatively short limbs and large heads. Matthew (1909, p. 412) considered them "to correspond in the Eocene fauna to the Mustelinae among modern Carnivora," a comparison that may be valid. Reevaluation of oxyaenid postcranial anatomy is overdue and should provide new insights.

From the foregoing it is obvious that our understanding of early Eocene mammals need not be restricted to the dentition. The Willwood Formation of the Bighorn Basin has yielded and

TABLE 2. SELECTED LIMB INDICES AND RATIOS OF JOINT-SURFACE DIMENSIONS IN SOME WILLWOOD MAMMALS

Taxon	Crural	Fem-Mt	Fem Cond	Brachial	Hum-Mc	Radial Head
Palaeanodon ignavus (AMNH 15137; USGS 4726, 16471)	84	....	83, 87	....	26*	162
Cantius trigonodus (USGS 4724, 5900)	74-93*	....	106, 107	....	....	114
Oxyaena spp. (USGS 4752, 5910, 5911, 7144, 7145, 7186, 16485)	84	25	99, 100	61*	....	$\overline{X}$ = 157, n = 5 (OR = 145-170)
Tritemnodon sp. (USGS 4727)	....	....	....	....	....	156
Prototomus sp. (USGS 6111, 7157)	....	....	93	....	....	161
Didymictis protenus (USGS 5024, 21836)	....	....	....	....	....	161, 166
cf. *Miacis* sp. (USGS 7161)	....	....	88	....	....	....
?*Uintacyon* sp. (USGS 16490)	....	....	81	....	....	....
Vulpavus, cf. *V. canavus* (USGS 5025)	....	....	88	....	....	117
Chriacus sp. (USGS 2353, 15404)	98*	30-33*	84, 87	84	30	131
Thryptacodon sp. (UW 7421)	....	....	....	....	....	132
Anacodon ursidens (USGS 5026, 5902, 6092, 21856, 21857)	90, 95	....	84, 87	....	....	171, 174
Phenacodus primaevus (Cope, 1884; Gregory, 1912; USGS 7146, 21838)	85, 95	32	117	84, 87	42	194, 200
Phenacodus vortmani (Cope, 1884; Gregory, 1912; USGS 5022, 7159	100, 108	36, 38	116, 136	80, 83	34	169, 188
Coryphodon sp. (Osborn, 1898c; USGS 21875)	60, 61	12, 14	....	65, 68	20	168, 171
Hyracotherium sp. (Cope, 1884; USGS 16479, 21858)	104	41	$\overline{X}$=116, N = 4 (OR = 109-129)	97	....	200
Diacodexis metsiacus (USGS 2352)	117	64	130	....	....	165

Notes: Crural index = (tibia length/femur length) x 100; Fem-Mt (Femorometatarsal index) = (Mt III length/femur length) x 100; Fem Cond = ratio of maximum anteroposterior depth to maximum mediolateral breadth of femoral condyles x 100; Brachial index = (radius length/humerus length) x 100; Hum-Mc (Humerometacarpal index) = (Mc III length/humerus length) x 100; Radial Head = ratio of maximum and minimum diameters x 100.

*Estimated

continues to provide substantial information on the anatomy of early Eocene mammals. Detailed studies have so far been undertaken on only a few fossil skeletons, and significant advances in our knowledge may be expected as more thorough analyses are completed. Such studies are a prerequisite to a more accurate understanding of ancient vertebrate communities.

ACKNOWLEDGMENTS

Nearly all the fossils reported here were collected on joint USGS–Johns Hopkins field parties, conducted with T. M. Bown, since 1979. We owe these discoveries to the diligence and dedication of the many participants in this field work, in particular, K. C. Beard, D. L. Gebo, M. J. Kraus, D. T. Rasmussen, J. J. Rose, and E. L. Simons.

For access to fossil and recent skeletal specimens I thank T. M. Bown, M. Carleton, R. J. Emry, J. A. Lillegraven, M. C. McKenna, J. H. Ostrom, E. L. Simons, R. W. Thorington, Jr., and M. Turner. I am especially grateful to E. Kasmer, who prepared all of the illustrations. I also thank T. Urquhart for photography, K. C. Beard, T. M. Bown, M. Dagosto, J. G. Fleagle, D. L. Gebo, F. A. Jenkins, Jr., B. Van Valkenburgh, and A. C. Walker for informative discussion, and T. M. Bown, J. L. Franzen, and F. A. Jenkins, Jr. for helpful review of the manuscript.

Field work and research have been supported in part by NSF grants BSR-8215099 and BSR-8500732, National Geographic Society grant 2366-81, The J. J. Hopkins Fund, and the USGS.

REFERENCES CITED

Barnett, C. H., and Napier, J. R., 1953, The rotatory mobility of the fibula in eutherian mammals: London, Journal of Anatomy, v. 87, p. 11–21.

Cartmill, M., 1985, Climbing, *in* Hildebrand, M., Bramble, D. M., Liem, K. F., and Wake, D. B., eds., Functional vertebrate morphology: Cambridge, Belknap Press, p. 73–88.

Coombs, M. C., 1983, Large mammalian clawed herbivores; A comparative study: Transactions of the American Philosophical Society, v. 73, part 7, p. 1–96.

Cope, E. D., 1884, The Vertebrata of the Tertiary formations of the West: Report of the U.S. Geological Survey of the Territories, v. 3, p. 1–1009.

Covert, H. H., 1988, Ankle and foot morphology of *Cantius mckennai;* Adaptations and phylogenetic implications: Journal of Human Evolution, v. 17, p. 57–70.

Davis, D. D., 1964, The giant panda; A morphological study of evolutionary mechanisms: Fieldiana, Zoology Memoirs, v. 3, p. 1–339.

Denison, R. H., 1938, The broad-skulled Pseudocreodi: Annals of the New York Academy of Sciences, v. 37, p. 163–256.

Ding Su-yin, 1987, A Paleocene edentate from Nanxiong Basin, Guangdong: Palaeontologia Sinica, no. 173, p. 1–118.

Dubost, G., 1975, Le comportement du chevrotain africain, *Hyemoschus aquaticus* Ogilby: Zeitschrift für Tierpsychologie, v. 37, p. 403–501.

Evans, H. E., and Christensen, G. C., 1979, Miller's anatomy of the dog, 2nd ed.: Philadelphia, W. B. Saunders Company, 1181 p.

Fischer, M. S., 1986, Die Stellung der Schliefer (Hyracoidea) im phylogenetischen System der Eutheria: Courier Forschungsinstitut Senckenberg, no. 84, p. 1–132.

Franzen, J. L., 1981, Das erste Skelett eines Dichobuniden (Mammalia, Artiodactyla), geborgen aus mitteleozänen Ölschiefern der "Grube Messel" bei Darmstadt (Deutschland, S-Hessen); Senckenbergiana lethaea, v. 61, p. 299–353.

Gans, C., 1979, Momentarily excessive construction as the basis for protoadaptation: Evolution, v. 33, p. 227–233.

Gazin, C. L., 1953, The Tillodontia; An early Tertiary order of mammals: Smithsonian Miscellaneous Collections, v. 121, no. 10, p. 1–110.

Gebo, D. L., 1987, Humeral morphology of *Cantius,* an early Eocene adapid: Folia Primatologica, v. 49, p. 52–56.

—— , 1988, Foot morphology and locomotor adaptation in Eocene primates: Folia Primatologica, v. 50, p. 3–41.

Gilmore, C. W., 1936, Fossil hunting in Montana and Wyoming: Explorations and Field-Work of the Smithsonian Institution, 1935, p. 1–4.

Gingerich, P. D., 1980, History of early Cenozoic vertebrate paleontology in the Bighorn Basin: Ann Arbor, University of Michigan Papers on Paleontology, no. 24, p. 7–24.

—— , 1983, Systematics of early Eocene Miacidae (Mammalia, Carnivora) in the Clark's Fork Basin, Wyoming: Ann Arbor, University of Michigan Contributions from the Museum of Paleontology, v. 26, p. 197–225.

—— , 1987, Early Eocene bats (Mammalia, Chiroptera) and other vertebrates in freshwater limestones of the Willwood Formation, Clark's Fork Basin, Wyoming: Ann Arbor, University of Michigan Contributions from the Museum of Paleontology, v. 27, p. 275–320.

Ginsburg, L., 1961, Plantigradie et digitigradie chez les carnivores fissipèdes: Mammalia, v. 25, p. 1–21.

Granger, W., 1915, Part 3, Order Condylarthra; Families Phenacodontidae and Meniscotheriidae, *in* Matthew, W. D., and Granger, W., eds., A revision of the lower Eocene Wasatch and Wind River faunas: Bulletin of the American Museum of Natural History, v. 34, p. 329–360.

Gregory, W. K., 1912, Notes on the principles of quadrupedal locomotion and on the mechanism of the limbs in hoofed animals: Annals of the New York Academy of Sciences, v. 22, p. 267–294.

—— , 1920, On the structure and relations of *Notharctus,* an American Eocene primate: Memoirs of the American Museum of Natural History, new series, v. 3, p. 49–243.

Gunnell, G. F., 1988, Locomotor adaptations in early Eocene Oxyaenidae (Mammalia, Creodonta): Journal of Vertebrate Paleontology, v. 8, p. 17A.

Hooker, J. J., 1984, A primitive ceratomorph (Perissodactyla, Mammalia) from the early Tertiary of Europe: Zoological Journal of the Linnean Society, v. 82, p. 229–244.

Hussain, S. T., 1975, Evolutionary and functional anatomy of the pelvic limb in fossil and recent Equidae (Perissodactyla, Mammalia): Anatomy, Histology, and Embryology, v. 4, p. 179–222.

Janis, C., 1984, Tapirs as living fossils, *in* Eldredge, N., and Stanley, S. M., eds., Living fossils: New York, Springer, p. 80–86.

Jenkins, F. A., Jr., and Camazine, S. M., 1977, Hip structure and locomotion in ambulatory and cursorial carnivores: London, Journal of Zoology, v. 181, p. 351–370.

Jenkins, F. A., Jr., and McClearn, D., 1984, Mechanisms of hind foot reversal in climbing mammals: Journal of Morphology, v. 182, p. 197–219.

Kitts, D. B., 1956, American *Hyracotherium* (Perissodactyla, Equidae): Bulletin of the America Museum of Natural History, v. 110, p. 1–60.

Lucas, S. G., 1981, Drought and vertebrate fossil assemblages; A *Coryphodon* (Mammalia; Pantodonta) quarry from the early Eocene of New Mexico: Geological Society of America Abstracts with Programs, v. 13, p. 202.

—— , 1982, The phylogeny and composition of the order Pantodonta (Mammalia, Eutheria): Third North American Paleontological Convention Proceedings, v. 2, p. 337–342.

Maier, W., 1979, *Macrocranion tupaiodon,* an adapisoricid (?) insectivore from the Eocene of "Grube Messel" (Western Germany): Paläontologische Zeitschrift, v. 53, p. 38–62.

Matthew, W. D., 1909, The Carnivora and Insectivora of the Bridger Basin, middle Eocene: Memoirs of the American Museum of Natural History, v. 9, p. 289–567.

—— , 1915a, Part 1, Order Ferae (Carnivora); Suborder Creodonta, *in* Matthew, W. D., and Granger, W., eds., A revision of the lower Eocene Wasatch and Wind River faunas: Bulletin of the American Museum of Natural History, v. 34, p. 4–103.

—— , 1915b, Part 4, Entelonychia, Primates, Insectivora (part), *in* Matthew, W. D., and Granger, W., eds., A revision of the lower Eocene Wasatch and Wind River faunas: Bulletin of the America Museum of Natural History, v. 34, p. 429–483.

—— , 1918, Part 5, Insectivora (continued), Glires, Edentata, *in* Matthew, W. D., and Granger, W., eds., A revision of the lower Eocene Wasatch and Wind River faunas: Bulletin of the American Museum of Natural History, v. 38, p. 565–657.

—— , 1937, Paleocene faunas of the San Juan Basin, New Mexico: Transactions of the American Philosophical Society, new series, v. 30, p. 1–510.

McKenna, M. C., 1975, Toward a phylogenetic classification of the Mammalia, *in* Luckett, W. P., and Szalay, F. S., eds., Phylogeny of the Primates: New York, Plenum, p. 21–46.

Mellett, J. S., 1977, Paleobiology of North American *Hyaenodon* (Mammalia, Creodonta): Contributions to Vertebrate Evolution, v. 1, p. 1–134.

Napier, J., 1970, The roots of mankind: Washington, Smithsonian Institution Press, 240 p.

Novacek, M. J., 1982, Information for molecular studies from anatomical and fossil evidence on higher eutherian phylogeny, *in* Goodman, M., ed., Macromolecular sequences in systematic and evolutionary biology: New York, Plenum, p. 3–41.

Osborn, H. F., 1898a, A complete skeleton of *Coryphodon radians;* Notes upon the locomotion of this animal: Bulletin of the American Museum of Natural History, v. 10, p. 81–91.

—— , 1898b, Remounted skeleton of *Phenacodus primaevus;* Comparison with *Euprotogonia:* Bulletin of the American Museum of Natural History, v. 10, p. 159–164.

—— , 1898c, Evolution of the Amblypoda; Part 1, Taligrada and Pantodonta: Bulletin of the American Museum of Natural History, v. 10, p. 169–218.

—— , 1900, *Oxyaena* and *Patriofelis* restudied as terrestrial creodonts: Bulletin of the American Museum of Natural History, v. 13, p. 269–279.

—— , 1904, An armadillo from the middle Eocene (Bridger) of North America: Bulletin of the American Museum of Natural History, v. 20, p. 163–165.

Osborn, H. F., and Wortman, J. L., 1892, Fossil mammals of the Wahsatch and Wind River beds; Collection of 1891: Bulletin of the American Museum of Natural History, v. 4, p. 81–147.

Patterson, B., 1939, A skeleton of *Coryphodon:* Proceedings of the New England Zoological Club, v. 17, p. 97–110.

Radinsky, L. B., 1965, Evolution of the tapiroid skeleton from *Heptodon* to *Tapirus:* Bulletin of the Museum of Comparative Zoology, v. 134, p. 69–106.

—— , 1966, The adaptive radiation of the phenacodontid condylarths and the origin of the Perissodactyla: Evolution, v. 20, p. 408–417.

Rose, K. D., 1972, A new tillodont from the Eocene upper Willwood Formation of Wyoming: Postilla, no. 155, p. 1–13.

—— , 1981, The Clarkforkian Land-Mammal Age and mammalian faunal composition across the Paleocene-Eocene boundary: Ann Arbor, University of Michigan Papers on Paleontology, no. 26, p. 1–197.

—— , 1982, Skeleton of *Diacodexis,* oldest known artiodactyl: Science, v. 216, p. 621–623.

—— , 1985, Comparative osteology of North American dichobunid artiodactyls: Journal of Paleontology, v. 59, p. 1203–1226.

—— , 1987, Climbing adaptations in the early Eocene mammal *Chriacus* and the origin of Artiodactyla: Science, v. 236, p. 314–316.

Rose, K. D., and Emry, R. J., 1983, Extraordinary fossorial adaptations in the Oligocene palaeanodonts *Epoicotherium* and *Xenocranium* (Mammalia): Journal of Morphology, v. 175, p. 33–56.

Rose, K. D., and Walker, A., 1985, The skeleton of early Eocene *Cantius,* oldest lemuriform primate: American Journal of Physical Anthropology, v. 66, p. 73–89.

Rose, K. D., Bown, T. M., and Simons, E. L., 1977, An unusual new mammal from the early Eocene of Wyoming: Postilla, no. 172, p. 1–10.

Russell, D. E., 1964, Les mammifères paléocènes d'Europe: Mémoires du Muséum National d'Histoire Naturelle, série C, v. 13, p. 1–324.

Sarmiento, E. E., 1983, The significance of the heel process in anthropoids: International Journal of Primatology, v. 4, p. 127–152.

Schaeffer, B., 1948, The origin of a mammalian ordinal character: Evolution, v. 2, p. 164–175.

Schankler, D. M., 1980, Faunal zonation of the Willwood Formation in the central Bighorn Basin, Wyoming, *in* Gingerich, P. D., ed., Early Cenozoic paleontology and stratigraphy of the Bighorn Basin, Wyoming: Ann Arbor, University of Michigan Papers on Paleontology, no. 24, p. 99–114.

Schoch, R. M., 1986, Systematics, functional morphology, and macroevolution of the extinct mammalian order Taeniodonta: New Haven, Connecticut, Yale University Peabody Museum of Natural History Bulletin 42, p. 1–307.

Simons, E. L., 1960, The Paleocene Pantodonta: Transactions of the American Philosophical Society, new series, v. 50, part 6, p. 1–99.

Simpson, G. G., 1931, *Metacheiromys* and the Edentata: Bulletin of the American Museum of Natural History, v. 59, p. 295–381.

—— , 1937, Notes on the Clark Fork, upper Paleocene, fauna: American Museum Novitates, no. 954, p. 1–24.

Sinclair, W. J., 1914, A revision of the bunodont Artiodactyla of the middle and lower Eocene of North America: Bulletin of the American Museum of Natural History, v. 33, p. 267–295.

Sinclair, W. J., and Jepsen, G. L., 1929, A mounted skeleton of *Palaeonictis:* Proceedings of the American Philosophical Society, v. 68, p. 163–173.

Stains, H. J., 1973, Comparative study of the calcanea of members of the Ursidae and Procyonidae: Bulletin of the Southern California Academy of Sciences, v. 72, p. 137–148.

Szalay, F. S., 1966, The tarsus of the Paleocene leptictid *Prodiacodon* (Insectivora, Mammalia): American Museum Novitates, no. 2267, p. 1–13.

Trapp, G. R., 1972, Some anatomical and behavioral adaptations of ringtails, *Bassariscus astutus:* Journal of Mammalogy, v. 53, p. 549–557.

Van Valen, L., 1971, Toward the origin of artiodactyls: Evolution, v. 25, p. 523–529.

Van Valkenburgh, B., 1987, Skeletal indicators of locomotor behavior in living and extinct carnivores: Journal of Vertebrate Paleontology, v. 7, p. 162–182.

Wortman, J. L., 1984, Osteology of *Patriofelis,* a middle Eocene creodont: Bulletin of the American Museum of Natural History, v. 6, p. 129–164.

—— , 1899, Restoration of *Oxyaena lupina* Cope, with descriptions of certain new species of Eocene creodonts: Bulletin of the American Museum of Natural History, v. 12, p. 139–148.

—— , 1903, Studies of Eocene Mammalia in the Marsh collection, Peabody Museum; Part 2, Primates; Suborder Cheiromyoidea: American Journal of Science, v. 16, p. 345–368.

MANUSCRIPT ACCEPTED BY THE SOCIETY JUNE 12, 1989

Printed in U.S.A.

Geological Society of America
Special Paper 243
1990

Systematic lateral variation in the distribution of fossil mammals in alluvial paleosols, lower Eocene Willwood Formation, Wyoming

Thomas M. Bown
Paleontology and Stratigraphy Branch, U.S. Geological Survey, Box 25046, Denver Federal Center, Denver, Colorado 80225
K. Christopher Beard
Department of Cell Biology and Anatomy, Johns Hopkins University School of Medicine, Baltimore, Maryland 21205

ABSTRACT

Willwood Formation paleosols are ranked on a scale of 0 to 5 on the basis of their relative maturity (= relative time required to form). In the lateral dimension, the least mature soils were developed more proximal to ancient channel belts, whereas the more mature paleosols formed in areas more distant to channel belts. Quantitative study shows that both mammalian taxonomic composition and taphonomic completeness vary systematically with the maturity of these paleosols.

Species-level differences in taxonomic composition are identified for pedofacies sequences located at the 442-m and 546-m levels of the Willwood Formation. At 442 m, *Cantius frugivorus* and *Hyopsodus* sp., cf. *H. minor* account for practically all of the adapiform primate and hyopsodontid condylarth faunas in stage 3 to 4 paleosols (which are distally located with respect to the ancient channel belt). Laterally adjacent and stratigraphically equivalent stage 1 to 2 paleosols (proximally located with respect to the ancient channel belt), are instead dominated by *Cantius* sp. nov. and *Hyopsodus* sp., cf. *H. miticulus*. Intermediate proportions of these taxa occur at localities in paleosols of intermediate maturity (stage 2 to 3 paleosols) at the 442-m level. At 546 m, the otherwise relatively rare species *Hyopsodus powellianus* makes up nearly 50 percent of the hyopsodontid fauna at some localities developed in stage 1 paleosols; elsewhere in this pedofacies the species *Hyopsodus minor* and *H. lysitensis* make up the overwhelming majority of the *Hyopsodus*. Also at 546 m, the adapiform primates *Cantius abditus* and *"Copelemur" feretutus* exhibit reversals in relative abundance from proximal to distal localities across the pedofacies; *Cantius* is more abundant in proximal localities and *"Copelemur"* is dominant in distal localities.

Ordinal-level differences in taxonomic composition were detected at localities in two distinct pedofacies lying at or slightly above Biohorizon C (= "Graybullian-Lysitean" boundary). There, Condylarthra and Artiodactyla are more common in immature (stages 1 to 2) than mature (stage 4) paleosols, whereas the reverse is true for Primates, Carnivora, Rodentia, and Perissodactyla.

Lateral controls on completeness of skeletal elements, as related to lateral variation in sedimentation rate, are also evident. Proportions of less complete skeletal elements are considerably higher at localities developed in mature paleosols, where sedimentation rates were low. These findings underscore the inherent relatedness of geographic distribution of taxa, taphonomy, and sedimentology and suggest that intrabasinal differences in microhabitat had a significant effect on the local taxonomic composition of the Willwood mammalian fauna.

Bown, T. M., and Beard, K. C., 1990, Systematic lateral variation in the distribution of fossil mammals in alluvial paleosols, lower Eocene Willwood Formation, Wyoming, *in* Bown, T. M., and Rose, K. D., eds., Dawn of the Age of Mammals in the northern part of the Rocky Mountain Interior, North America: Boulder, Colorado, Geological Society of America, Special Paper 243.

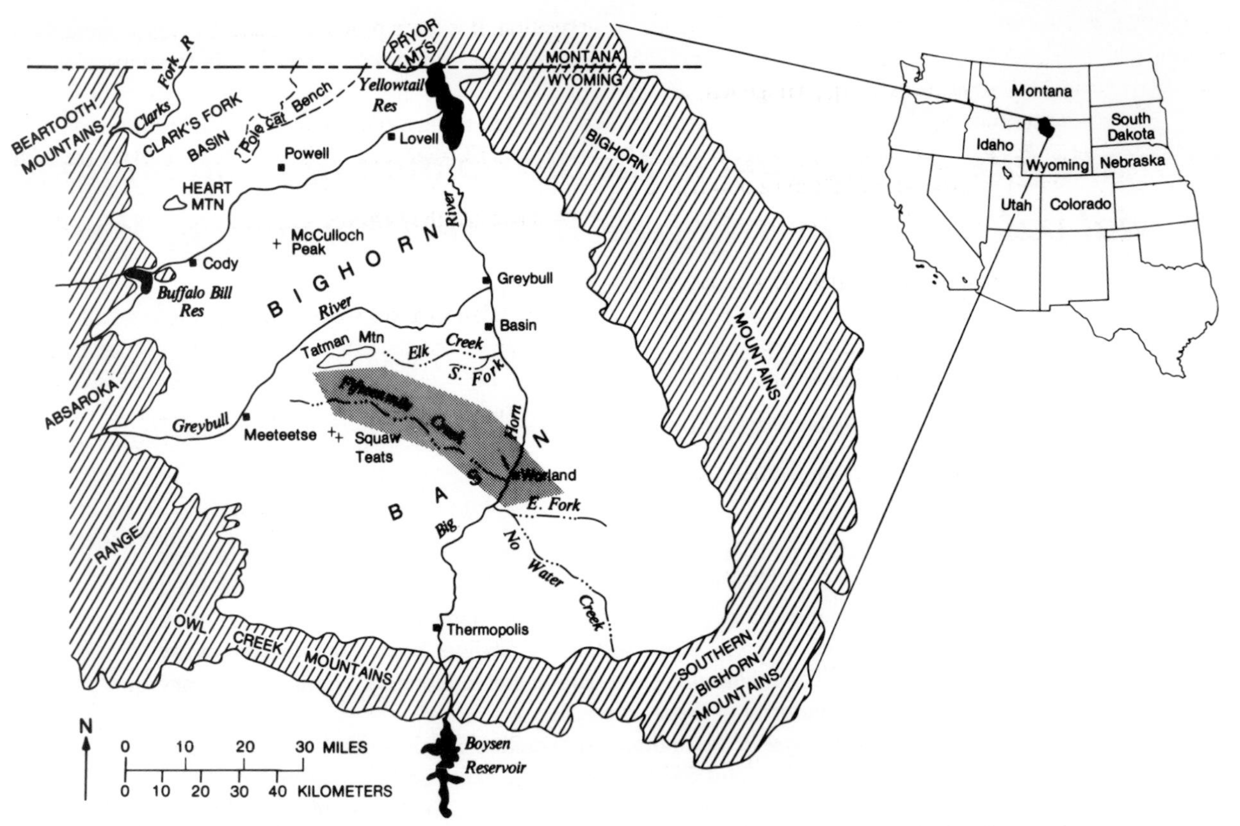

Figure 1. Location of area of study in Bighorn Basin (shaded), with respect to other physical features in northwest Wyoming.

INTRODUCTION

This chapter describes the initial results of empirical studies of lateral variation in mammalian faunal composition and fossil preservation at two richly fossiliferous and extensively exposed stratigraphic levels of the lower Eocene Willwood Formation, Bighorn Basin, Wyoming (Fig. 1). Lateral variation in these parameters are assessed in the context of their relations to sedimentological and paleopedological controls (Bown, 1987).

The Willwood Formation consists of approximately 770 m of fluvial sandstones and variegated mudstones. Previous work has demonstrated that the coloration of Willwood mudstones resulted from varying degrees of early Eocene soil (hereafter termed paleosol) development (Neasham and Vondra, 1972; Bown, 1979; Bown and Kraus, 1981a, 1987; Kraus, 1985, 1987; Kraus and Bown, 1986). Fossil vertebrates are abundant in Willwood rocks, especially the mudstones, in which they were preferentially concentrated by passive mechanisms in the upper levels (generally A horizons) of the developing paleosols (Bown, 1979, 1985); Bown and Kraus, 1981b). The amount of time represented by each of these paleosol accumulations of fossil vertebrates is believed to be the same as that required for development of the paleosols themselves; that is, in analogy with structurally similar modern soils, about 1,000 to 30,000 years (Bown, 1985; Kraus and Bown, 1986; Bown and Kraus, 1987).

Due to the exceptional abundance of fossil vertebrates in the Willwood Formation, their unusually dense stratigraphic distribution (Gingerich, 1980; Rose and Bown, 1986; Bown and Rose, 1987), and the widespread exposure of Willwood rocks, the Willwood mammal succession has been the most important fossil mammal succession in the world for evolutionary studies (e.g., Gingerich, 1974, 1976, 1979; Gingerich and Simons, 1977; Gingerich and Gunnell, 1979; Bown, 1979; Rose, 1981a; Bown and Rose, 1987). This chapter is the first concerning any rocks and fossils that examines aspects of the lateral distribution and preservation of fossil mammals in coeval, and relatable, sedimentologic and paleopedologic contexts.

In 1981, the senior author began a program of section measuring, paleosol studies, and field correlation of the then 250-odd U.S. Geological Survey (USGS) fossil vertebrate localities in the Willwood Formation of the central and southern Bighorn Basin. This section was correlated with the University of Wyoming (UW) and Yale Peabody Museum (YPM) localities in earlier Willwood sections published by Bown (1979) and Schankler (1980), respectively. As of this writing, 860 Willwood fossil vertebrate localities are now tied to this composite section (Bown and Rose, 1990).

Some comparative reference is made to work by others in which a much thinner Willwood section was utilized in biostratigraphic studies of *Hyopsodus, Cantius,* and *"Copelemur"* (e.g.,

Gingerich, 1974, 1976; Gingerich and Simons, 1977). That section is about 549 m (1,800 ft) thick, and was measured in 1965 by G. E. Meyer and L. Radinsky (see, e.g., Gingerich, 1976, p. 6–9). It is generally acknowledged to be erroneously thin. Van Houten (1944), Neasham and Vondra (1972), and Schankler (1980) reported Willwood thicknesses of 762 m (2,500 ft), 701 m (2,300 ft) and 773 m (2,535 ft), respectively. The section used here is a combination of sections measured by the senior author (up to the base of the *Lambdotherium* Range Zone: 599 m = 1,965 ft in Bown's section), correlated with that of Schankler (1980) at the 410-m and 465-m levels (see Bown and Rose, 1987, Figs. 2 to 3 for location of certain of these lines of section).

Measuring the central Bighorn Basin Willwood section and correlating localities to it entailed tracing Willwood rocks laterally, commonly for considerable distances, at numerous stratigraphic levels. In the course of these endeavors, it was found that the maturation sequence of paleosols established for the Willwood Formation (arbitrary stages 0 to 5 of maturity, where maturity is a function of time required to form; see Bown, 1985; Bown and Kraus, 1987; Kraus, 1987) is also developed more or less sequentially in the lateral dimension (Bown, 1985; Bown and Kraus, 1987). That is, several relatively immature stage 1 paleosols that are distributed on thick alluvial ridge sediments are equivalent in time to a relatively more mature stage 4 or 5 paleosol formed coevally on less thick sediments on more distal parts of the flood basin. Several intervening stage 2 to 4 paleosols are geographically distributed more or less sequentially from proximal to distal parts of the flood basin. The prism of rock defined by this system, thick (and containing many immature paleosols) on the alluvial ridge and thinner (with fewer, more mature paleosols) on the distal flood basin, is termed the pedofacies (Bown and Kraus, 1987; see Fig. 2, this paper). Because Willwood fossil mammals are abundant in most paleosols of any stage of maturity, and because the relative maturation stage of the paleosols can be easily established in the field, a means of controlling the coeval lateral distributions of fossils for given stratigraphic intervals of the Willwood Formation is established. This control is inherently related not only to paleosol maturity, but also to paleogeographic position with respect to local ancient stream systems (Fig. 2).

No complete pedofacies (i.e., paleosol complexes that are laterally continuous from stage 0 through stage 5) are known for the Willwood Formation, due to the combined complications of structural dip, truncation of sequences by younger channel complexes, and areas obscured by vegetation. Nonetheless, innumerable partial pedofacies are well preserved, and these supply ample evidence for the lateral relations of the paleosol maturation stages. The lateral relations of Willwood paleosols appear to have been controlled by their geographic positions (with respect to the alluvial ridge and the distal flood plain) and by local depositional histories (during which continued paleosol maturation was arrested by deposition). Therefore, the formation of the paleosols (and the accumulations of vertebrate fossils in them) is directly related to sedimentary processes. Changes in drainage and paleosol type laterally were doubtless accompanied by differences in vegetation. It is these probable lateral differences in flora and the known lateral differences in small-scale geomorphology that we believe influenced mammalian distributions as well.

To test this hypothesis, fossils from localities in two partial pedofacies, situated at the 442- and 546-m levels of the Willwood Formation, were examined. These particular pedofacies were chosen because they are well exposed, they have extensive lateral stratigraphic control, and their several fossil vertebrate localities are easily distinguished and richly fossiliferous. The exceptional fossil abundances allow some of the most reliable information on differential taxonomic composition to be afforded by sympatric species of the condylarth *Hyopsodus* and the adapiform primates *Cantius* and *"Copelemur"*. In addition, differential relative abundances of fossil mammals at the ordinal level were examined for stage 1 and stage 4 paleosols at the 470- to 481-m interval of the Willwood Formation. Finally, the differential distribution of fossil vertebrate remains in terms of their relative completeness was examined at the 442- and 546-m levels.

442-meter level

In the central Bighorn Basin, the 442-m level of the Willwood Formation is well exposed south and east of Red Butte, in the extensive badlands between Elk and Dorsey Creeks (north of the Elk Creek Rim), and in the middle part of the drainage of Dry Cottonwood Creek, where rocks at this stratigraphic level are exposed in a broad dome (see Bown and Rose, 1987, Fig. 3). Although it is possible that parts of two distinct pedofacies are developed in the area studied, there is a general increase in paleosol maturity from northwest to southeast at this level, across a distance of some 18 km.

At the west margin of this sequence, USGS fossil vertebrate localities D-1588, D-1657, D-1660, D-1682, and D-1688 are developed in immature paleosols, ranging from stage 1 paleosols at the most remote locality (D-1660), through stage 1–2 paleosols at locality D-1588 (see Bown and Kraus, 1987, Fig. 3, to visualize different paleosol maturities). These are variously termed proximal and very proximal localities in Figures 2 and 3. Intermediate paleosols (stages 2 to 3) are developed at localities D-1311 and D-1310, respectively, some 13 km to the southeast. Distal paleosols are represented by stages 3 to 4 paleosols southeast of Red Butte, at localities D-1203, D-1204, D-1208, and D-1693. These localities lie approximately 5 km farther southeast than intermediate localities D-1311 and D-1310, and have been directly correlated with the latter by tracing the pedofacies laterally.

The complex of localities at and near locality D-1588 is more extraneous and cannot be directly correlated with the more southeasterly localities cited above because of the local westerly structural dip. However, D-1588, D-1657, D-1660, D-1682, and D-1688 lie at the same stratigraphic level in directly correlated sections. Moreover, the suite of localities at the 442-m level listed above are the only known sites yielding the new species of *Cantius* discussed below and, with the exception of locality D-1699 at the 463-m level, are the only localities in this part of the basin that yield the rare omomyid primate *Steinius vespertinus* (Bown and Rose, 1984, 1987). The latter species is, as far as is known, confined to the 438- to 463-m interval of the Willwood Formation (Bown and Rose, 1987). Therefore, the correlation of the proximal northwestern localities with the intermediate and distal localities farther southeast is virtually certain.

Hyopsodus

Two species of *Hyopsodus* are present at the 442-m level of the Willwood Formation. In 1974, Gingerich identified the smaller of these two species as *Hyopsodus latidens* and the larger as *H. miticulus*; however, in 1976 he revised these identifications to *Hyopsodus minor* and *H. miticulus,* respectively. These names are tentatively accepted here, pending a general morphologic review of the numerous Wasatchian species attributed to this genus.

In addition to its larger overall tooth size, specimens referred to *Hyopsodus* sp., cf. *H. miticulus* (hereafter, *H. miticulus*) at this stratigraphic level differ from those of *H.* sp., cf. *H. minor* (hereafter *H. minor*) in a few relatively minor details of morphology. *Hyopsodus minor* tends to have generally more bunodont cusps, especially the paracones and metacones of the first two upper

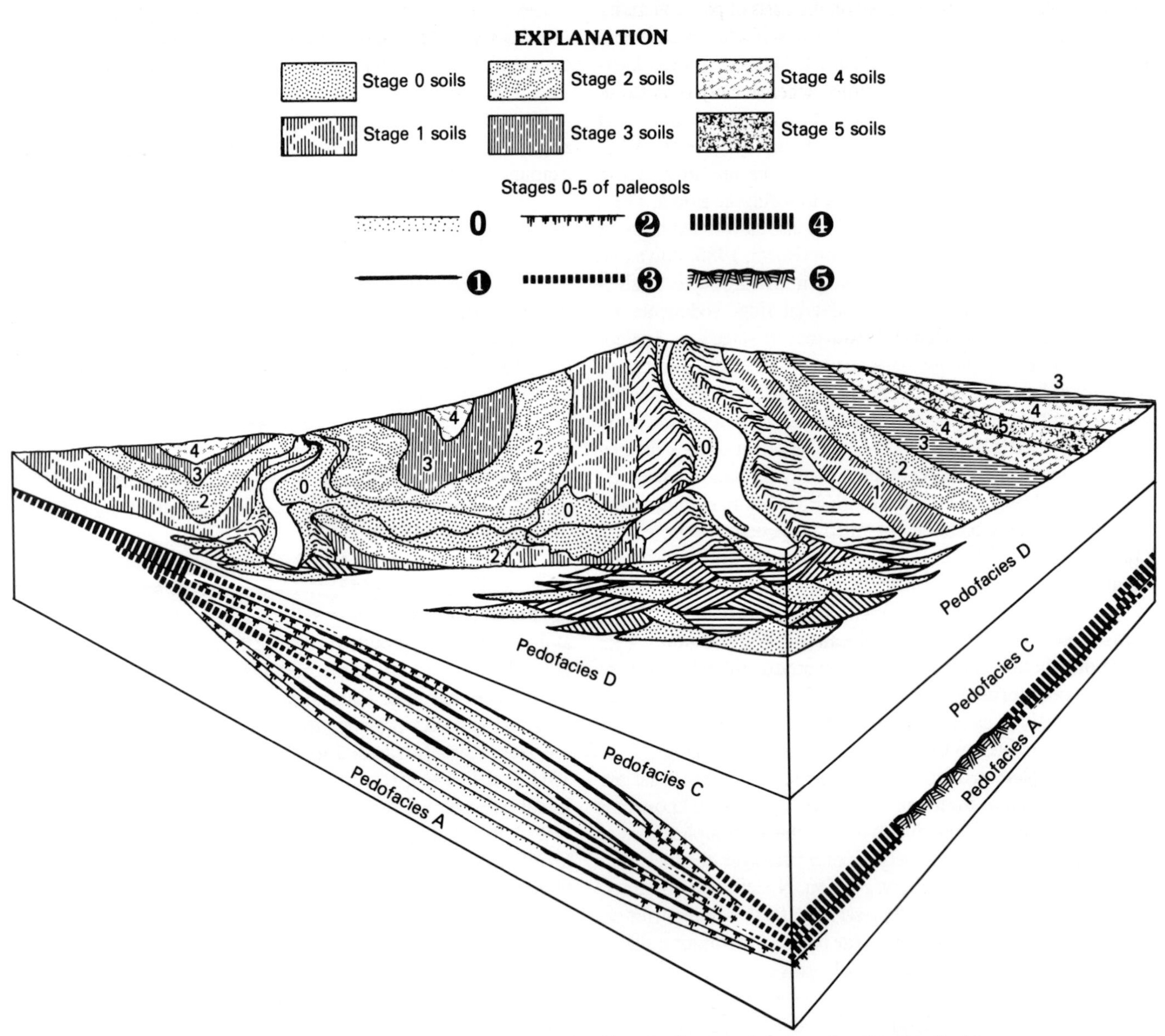

Figure 2. Block diagram showing, top, lateral relationships of soil maturation stages 0 to 5 to the alluvial ridge and distal flood plain; bottom, lateral relations of buried paleosols, with maturation stages 0 to 5 depicted in cross section. Four subsurface pedofacies (Bown and Kraus, 1987) are depicted (pedofacies A, B, C, D). For clarity, only pedofacies B is diagrammed. Taken together, pedofacies A, B, C, and D comprise part of a pedofacies sequence (Kraus, 1987). Not to scale; height of alluvial ridge rarely exceeds 20 m, and distance from alluvial ridge to distal flood plain is commonly as much as 15–20 km.

molars, which are slightly to markedly less compressed buccolingually. In *H. miticulus* these cusps are more crescentic. In contrast to *Hyopsodus minor, H. miticulus* tends to have a less pronounced paraconid and a more shelf-like or crested (less cuspidate) hypoconulid on the first two lower molars. Specimens referred to *H. miticulus* also commonly possess a distinct buccal cingulid beneath the protoconid and hypoflexid of the first two lower molars, a structure seen much less frequently in the smaller *H. minor.*

Figure 3 depicts the differential relative abundances of *Hyopsodus minor* and *H. miticulus* at the 442-m level of the Willwood Formation, at localities identified as distal, intermediate, proximal, and very proximal on the basis of paleosol maturities. Species identifications were made using both biometric and morphological information. As Figure 3 illustrates, only *Hyopsodus minor* is present in distal localities, whereas only *H. miticulus* occurs at the very proximal locality. One specimen at the intermediate locdality D-1311 could not be positively identified with either species (but it is apparently one or the other), and *Hyopsodus miticulus* appears to be slightly more abundant than *H. minor* at the proximal localities. The two species were obviously sympatric geographically as well as coeval in time within the same basin and, on a much smaller geographic scale, were sympatric at the same localities over part of their geographic range. Statistical analysis of the data illustrated in Figure 3 allows us to reject the null hypothesis that the differences in lateral abundance of these two species are due to random sampling error (chi-square test, $p < 0.01$). The pattern of differential relative abundance of these two species at the 442-m level of the Willwood Formation suggests instead that *Hyopsodus miticulus* may have preferred the conditions that prevailed closer to streams, with poorly drained,

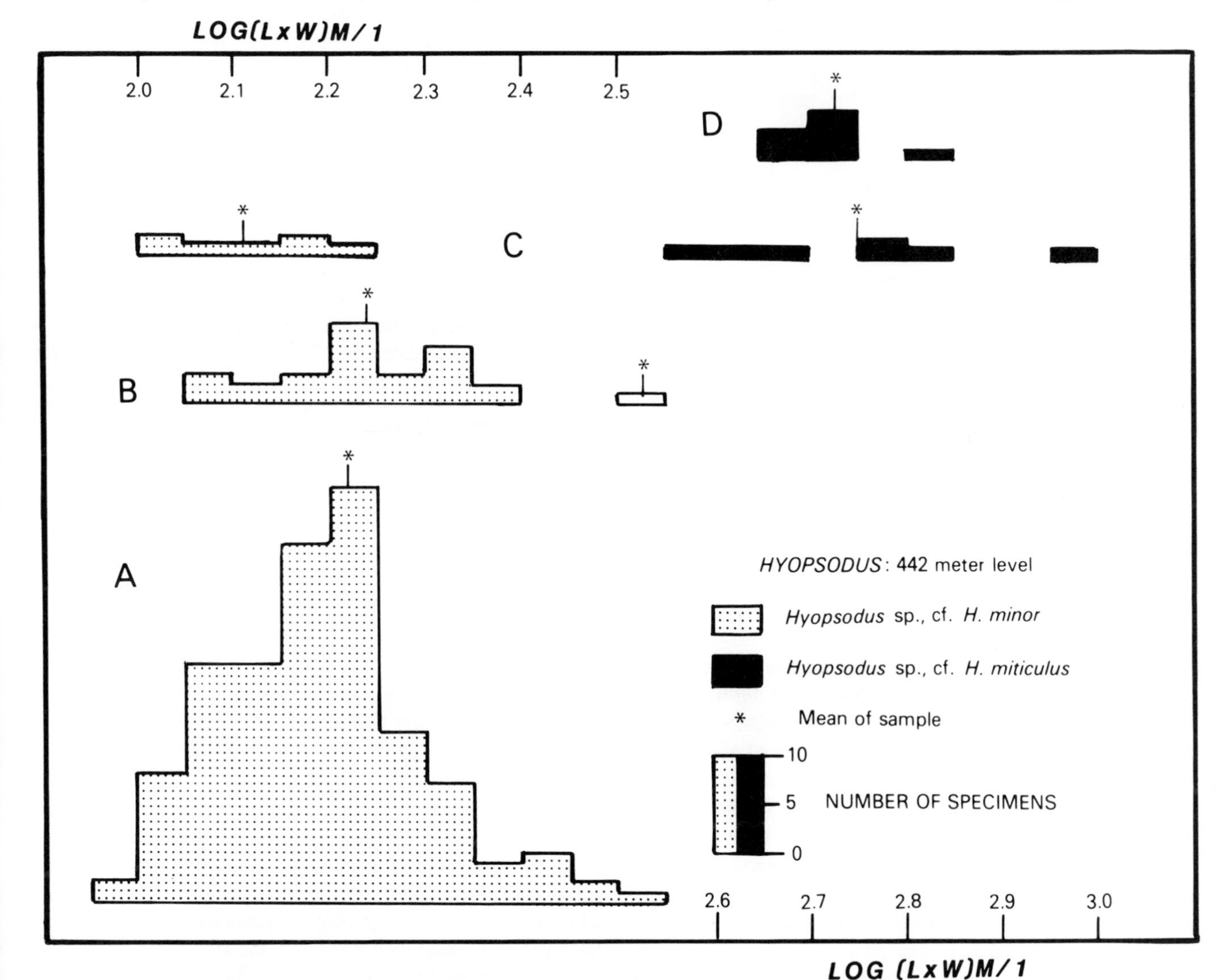

Figure 3. Histograms of Log (L×W) of the lower first molar (L = length, W = width) in two species of *Hyopsodus* at the 442-m level of the Willwood Formation. A = USGS localities D-1203, D-1204, D-1208, D-1693 (stage 3+ paleosols, distal); B = USGS localities D-1310, D-1311 (stage 2 to 3 paleosols, intermediate); C = USGS locality D-1588 (stages 1 to 2 paleosols, proximal); D = USGS locality D-1660 (stage 1 paleosol, very proximal). Data through the 1986 field season was utilized.

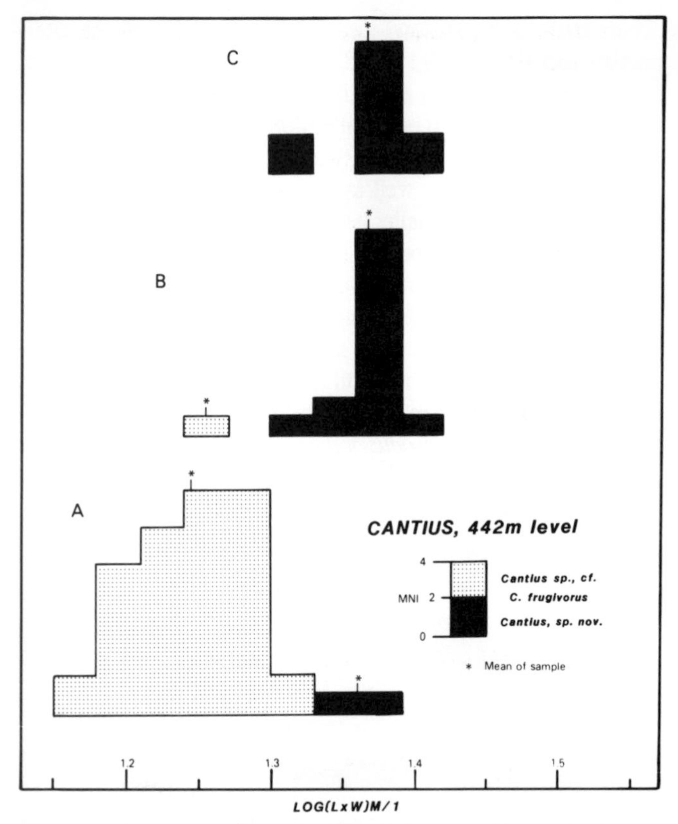

Figure 4. Histograms of Log (L×W) of the lower first molar in two species of *Cantius* at the 442-m level of the Willwood Formation. A = USGS localities D-1204, D-1693 (stage 3+ paleosols, distal); B = USGS localities D-1310, D-1311 (stage 2 to 3 paleosols, intermediate); C = USGS localities D-1588, D-1660, D-1682, D-1688 (stage 0 to 2 paleosols, proximal). Data through 1986 field season was utilized.

less mature paleosols, while *H. minor* appears to have been somewhat more eurytopic.

Cantius

Two species of the primitive adapiform primate genus *Cantius* also coexisted in the southern Bighorn Basin during the time interval represented by the 442-m level of the Willwood Formation. Like the two species of *Hyopsodus* discussed above, these two species of *Cantius* are also distinguished from each other on the basis of both size (Fig. 4) and morphology. The species with the smaller cheek teeth is close in morphology to *Cantius frugivorus* from the San Juan Basin of New Mexico (and is here hesitantly referred to that species), but may instead represent the Bighorn Basin species *C. trigonodus,* which it also closely resembles (Beard, 1988, believes *C. frugivorus* to be a senior synonym of the commonly recognized *Cantius "trigonodus"*). Whatever the proper name to apply to these specimens, they represent the end member of the anagenetic Bighorn Basin lineage *Cantius ralstoni—C. mckennai-C. "trigonodus"* proposed by Gingerich and Simons (1977; see also Gingerich and Haskin, 1981).

The larger *Cantius* species at the 442-m level (here *Cantius,* sp. nov.) is undescribed and was probably an immigrant into the Bighorn Basin at about the time represented by this level. The species is clearly more closely related to later Willwood species of *Cantius* (e.g., *C. abditus*) than is *C. frugivorus.* The new, large species has much broader lower molars (especially at the talonid) than *C. frugivorus,* and the paraconid on the first lower molar is situated more distolingually than is the case in the latter species. It may be distinguished from the similar sized species *Pelycodus jarrovii* (Gingerich and Haskin, 1981; Rose and Bown, 1984) by its somewhat smaller and narrower molars, presence of a mesostyle, and more anteriorly situated paraconid on the first lower molar.

Specimens of *Cantius* species are much less abundant than those of *Hyopsodus* species. Therefore, for the study of *Cantius,* localities at the 442-m level are separated into only three categories of relative proximity to stream channels: proximal, intermediate, and distal. As in the data presented for *Hyopsodus* above, the differences in lateral distribution for the two species of *Cantius* at the 442-m level are inconsistent with the null hypothesis of equal representation of the two species across the pedofacies (chi-square test, $p < 0.001$). Though not very abundant at any single locality, *Cantius* sp. nov. appears to have been more cosmopolitan in its habitat preference, being present throughout the preserved pedofacies. *Cantius frugivorus,* on the other hand, overwhelmingly preferred conditions in the distal parts of flood plains, at least at this time. Only one specimen of this species is known from intermediate localities; none are known from the proximal localities.

It is interesting that the solitary specimen of *C. frugivorus* known from the intermediate localities comes from D-1310, a stage 3 paleosol, rather than from D-1311, a less mature stage 2 paleosol. This finding lends additional weight to the apparent preference of *C. frugivorus* for habitats of more mature paleosols characteristic of the more distal flood plain.

546-meter level

The 546-m level of the Willwood Formation is well exposed over a large area in the upper parts of the drainages of Timber and Fifteenmile Creeks and in Bobcat Draw, marginal to Tatman Mountain and the Squaw Teats Divide in the southern Bighorn Basin (see Bown and Rose, 1987, Fig. 3). At and near the confluence of Timber Creek and Bobcat Draw, a major multistorey, multilateral fluvial sandstone and its associated levee and proximal flood-plain deposits are well exposed. These rocks, especially the levee mudstones, are exceptionally rich in fossil vertebrates, and lateral stratigraphic and pedofacies control is superb for distances of more than 2 km from the sand body.

Eighteen fossil vertebrate localities have been established in this area. These localities have been segregated, as above, into proximal, intermediate, and distal categories, reflecting their relative lateral positions in the local pedofacies. The terms proximal, intermediate, and distal are not tied to soil stages 0 to 5, but are

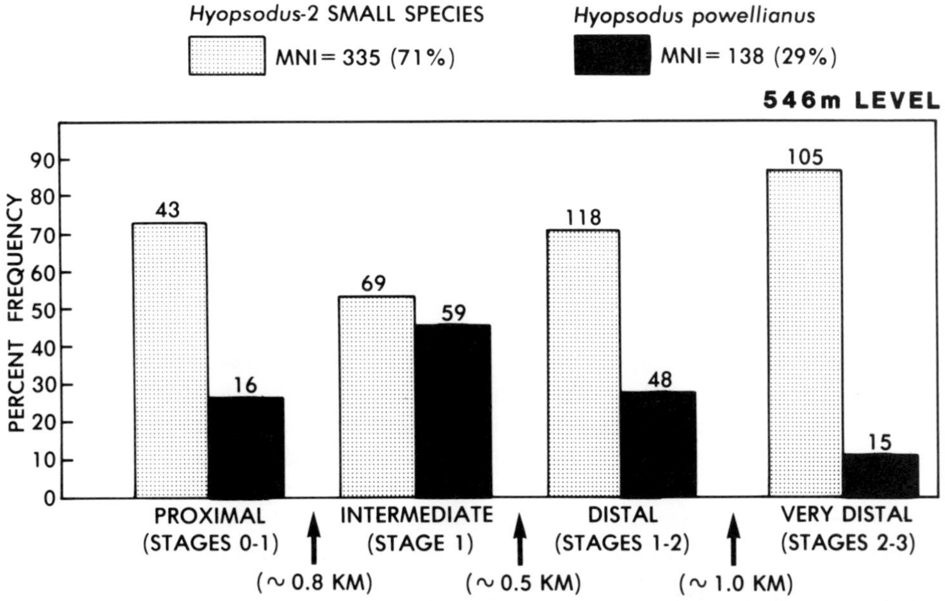

Figure 5. Percent frequency distributions of *Hyopsodus* at the 546-m level of the Willwood Formation with respect to lateral position and paleosol maturity. Data through the 1986 field season was utilized.

used only to group the fossils by their relative proximity to ancient alluvial ridges. Therefore, at the 546-m level, the proximal localities are: USGS D-1464, D-1467, D-1581, and Duke University Primate Center (DPC) 15, all of which are developed on stage 0 to 1 paleosols. Intermediate localities are developed on stage 1 paleosols and include localities: USGS D-1256, D-1643, D-1582, Yale Peabody Museum (YPM) 190, YPM 192 (part), YPM 315, and DPC 16. Distal localities, developed in stage 1 to 2 paleosols, are: USGS D-1558, D-1574, D-1575, D-1576, YPM 181, YPM 192 (part), and YPM 193. A single very distal locality, USGS D-1583, represents stage 2 to 3 paleosols at this meter level. YPM 193 (distal) and USGS D-1583 are two of the richest fossil vertebrate localities in the Bighorn Basin. In species of both the condylarth *Hyopsodus* and the adapiform primates *Cantius* and *"Copelemur,"* all of which are abundant at this stratigraphic level, lateral constraints on distribution are evident and similar to those seen at the 442-m level.

Hyopsodus

Three species of *Hyopsodus* coexisted in the southern Bighorn Basin during the time interval represented by the 546-m level of the Willwood Formation. The largest is clearly *Hyopsodus powellianus.* The species with the smallest molars is the same species as the highest stratigraphic records of what Gingerich (1974, 1976) called *Hyopsodus minor.* This species is exceedingly rare, being known from only seven specimens (minimum number of individuals) [MNI] = 3), all from proximal to intermediate localities. In order to simplify the visual presentation of data, these specimens were combined with the much more abundant specimens of a species of *Hyopsodus* of intermediate size (*H. lysitensis* of Gingerich, 1974, 1976) in Figures 5 and 6. This grouping in no way affects the results of our analysis of lateral variation in the distribution of *Hyopsodus* species at the 546-m level. These three species are readily separable on the basis of molar size alone, a character in which there is virtually no overlap at this stratigraphic level (recorded as 1,350 ft [411 m] by Gingerich, 1976, Fig. 5).

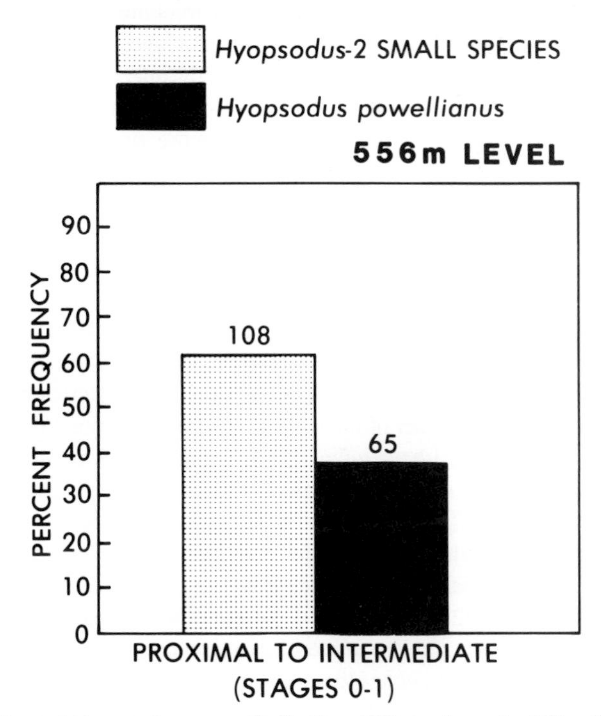

Figure 6. Percent frequency distribution of *Hyopsodus* at the 556-m level of the Willwood Formation with respect to lateral position and paleosol maturity. Numbers at tops of bars are minimum numbers of individuals (MNI). Compare with Figure 5. Data through the 1986 field season was utilized.

The two small species (overwhelmingly represented by *Hyopsodus lysitensis*) are more abundant than *H. powellianus* throughout the pedofacies. However, their relative abundances with respect to lateral position in the pedofacies show an interesting pattern (Fig. 5). Whereas the two small species of *Hyopsodus* are about 2.5 times as abundant as *H. powellianus* at proximal and distal localities (stages 0 to 1 and 1 to 2 paleosols, respectively), they are seven times as abundant as the latter at the very distal locality, D-1583 (stage 2 to 3 paleosols), yet only 17 percent more abundant at intermediate localities (stage 1 paleosols). Stated differently, the two small *Hyopsodus* species seem to have preferred habitats characteristic of the proximal alluvial ridge and in distal to very distal flood-plain localities, whereas the much larger *H. powellianus* appears to have preferentially inhabited intermediate (distal alluvial ridge) environments developed on stage 1 paleosols.

Statistical analysis of these data demonstrate that the observed differences in species distributions across the pedofacies cannot be attributed to random sampling error (chi-square test, $p < 0.025$). The very large numbers of specimens involved, the highly significant MNI, and the steady arithmetic increase in proportion of the two small species versus *H. powellianus* across the pedofacies from stage 1 through stage 3 paleosols, make the *Hyopsodus* data at 546 m good evidence of systematic lateral variation in abundance. To the authors, it seems conclusive that some sort of environmental factors controlled this pattern of *Hyopsodus* distribution. However, just what these factors were is conjectural. Keeping in mind that counts of minimum numbers of individuals (MNI) tend to overestimate abundances of rare species, whereas specimen counts tend to overestimate abundances of common species, it is likely that the *Hyopsodus* distribution at 546 m is, if anything, understated and more significant than depicted.

One environmental parameter that may have been important is paleosol wetness. Yellow and orange mudstones are more prominent in paleosols at intermediate localities than at others. This coloration is an indication of hydrated ferric iron compounds in the paleosols (Bown and Kraus, 1981a, 1987) and probably indicates damper paleosol conditions. Therefore, it is likely that the wettest paleosols were situated in intermediate localities, at the contact of the alluvial ridge and the flood plain. This wetness would have influenced the composition of vegetation and, by inference, mammalian distributions (e.g., that of *Hyopsodus powellianus*; see Fig. 5). However, any such general relationship should also obtain for pedofacies lower in the section, although such relationships are not seen in the *Hyopsodus* data from the 442-m level. Such a pattern of relative abundance is similarly lacking in the numerically more constrained adapiform primate data at either the 442- or 546-m levels. Nevertheless, there is no reason to suppose that the environmental factor, whatever it was, operated equally on ecologically different *Hyopsodus* and *Cantius*-*"Copelemur,"* or even on distinct species of *Hyopsodus*. Moreover, *Hyopsodus powellianus* is not present at the 442-m level.

Ten meters above the pedofacies examined at the 546-m level lies USGS D-1473, another of the richest vertebrate fossil localities in the Willwood Formation. Because D-1473 is at a similar stratigraphic position and beccause the locality has yielded abundant remains of *Hyopsodus* (MNI = 173), the relative abundance of *Hyopsodus* species at that locality is of interest as a check on that of the equivalent pedofacies position at the 546-m level.

Locality D-1473 is developed in paleosols of stages 0 to 1, like those of localities D-1464, D-1467, D-1581, and DPC 15 at the 546-m level. Figure 6 illustrates the relative abundances of large and small species of *Hyopsodus* at D-1473 (556-m level), which are approximately intermediate between those seen in proximal and intermediate localities 10 m lower in the section (compare with proximal and intermediate distributions in Fig. 5). That is, the large species *Hyopsodus powellianus* is more abundant than expected at proximal localities, yet less abundant than represented at intermediate localities at the 546-m level (compare Figs. 4 and 5). This suggests that paleosols at D-1473 are nearer to stage 1 than to stage 0, which indeed (and obviating circularity of the argument), is the relation seen in the field (D-1473 is a levee sequence with several stage 1 paleosols and fewer stage 0 paleosols). Therefore, as was the case with the *Cantius* data at 442 m (localities D-1310 and D-1311), greater resolution of paleosol maturity appears to predict greater resolution of patterns of mammalian species abundance, and vice versa.

Unfortunately, this greater resolution is not always possible, being constrained by the ability to subdivide paleosol stages consistently and meaningfully. Such subdivision is easier in levee deposits than in distal flood-plain deposits because the prism of rock is thicker and possesses several immature paleosols, some of stage 0 and others of stage 1. Because fossils come from all parts of this sequence, the proportion of stage 0 and stage 1 paleosols provides further resolution of paleosol maturity. Levees are made up of alternating fine and relatively coarse sediment. In the Willwood Formation, levee deposits generally weather to form small cliffs or ridges, and the fossils from all parts of these deposits collect on adjacent valley floors. Thus, in most areas, segregation of collections from individual paleosols in levee deposits is impossible.

Cantius and *"Copelemur"*

Two species of adapiform primates, *Cantius abditus* and *"Copelemur" feretutus*, occur at the 546-m level of the Willwood Formation (Gingerich and Simons, 1977; Beard and others, 1986). Beard (1988) has demonstrated that Willwood species attributed by Gingerich and Simons (1977) to the genus *Copelemur* do not belong in that taxon. Pending formal revision of Willwood adapiform primates, we refer to these species here as *"Copelemur,"* in order to minimize future taxonomic confusion. *"Copelemur" feretutus* differs from sympatric *Cantius abditus* in being smaller and in possessing more acute molar cusps, especially on the lower molar talonids (Fig. 7). Although the lower molars of most specimens of *"Copelemur" feretutus* also differ

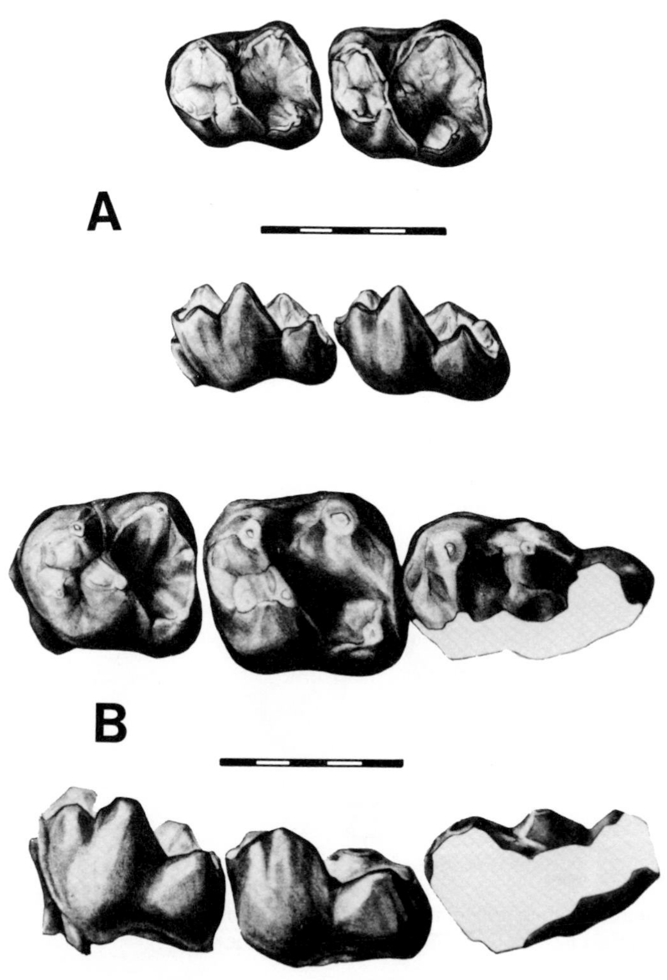

Figure 7. Lower molars of *"Copelemur"* and *Cantius* from the 546-m level of the Willwood Formation. A, *"Copelemur" feretutus,* occlusal (top) and lingual (bottom), views of USGS 10124, right M1 and 2; B, *Cantius abditus,* occlusal (top) and lingual (bottom) views of USGS 10116, right m1 to 3. Scale is 5 mm.

from those of *Cantius abditus* in possessing a distinct notch on the distolingual corner of the talonid (= "entoconid notch" of Gingerich and Simons, 1977; see Fig. 7), we have found that this structure is not typical of all specimens of the former species. Indeed, in some specimens of *"Copelemur" feretutus* (e.g., USGS 9571), the presence or absence of an entoconid notch varies along the molar series in a single dentary. The entoconid notch morphology might well be a transient character (Bown and Rose, 1987). The upper molars of *"Copelemur" feretutus,* especially M1, are distinguished from those of *Cantius abditus* in being relatively short and broad (more transverse) and in possessing less prominent mesostyles and pseudohypocones (compare Figs. 8A and B).

Although adapiform primates are collectively much less abundant than are species of *Hyopsodus* at all levels of the Willwood Formation, *Cantius* is the fourth most common Willwood genus (following *Hyopsodus, Hyracotherium,* and *Diacodexis;* see Bown, 1979; Rose, 1981b) at most localities. Unfortunately, our sample of adapiforms at the 546-m level, particularly at the proximal localities, is not yet large enough to render the interesting distributional trends illustrated in Figure 9 statistically significant. Nevertheless, the data show that *Cantius* decreases with respect to *"Copelemur"* from 60 to 36 percent to only 25 percent of the adapiform fauna from proximal to very distal localities in this pedofacies. Correspondingly, *"Copelemur"* increases from 40 to 64 percent to 75 percent. Where *"Copelemur"* does occur in the upper Willwood fauna, it is generally quite rare with respect to *Cantius* (Gingerich and Simons, 1977). It is obvious from independent field observation that less mature paleosols predominate in this part of the Willwood section. Therefore, the lateral relative abundances of these two species in paleosols at the 546-m level offer explanations for both the overall rarity of *"Copelemur"* in the Willwood Formation in general, and its extraordinary representation at this stratigraphic level. This knowledge allows us to predict that the immediate ancestors of *"Copelemur" feretutus* will probably also be most abundant in relatively mature paleosols.

PEDOFACIES CONTROL OF ORDINAL REPRESENTATION

From the foregoing, lateral variation in the relative abundances of at least some mammalian species appears to be directly related to systematic lateral changes in paleosol maturity. But what of other taxa, faunas at different stratigraphic levels, and in different and higher taxonomic categories? Obviously, these types of analyses demand a high level of systematic resolution involving numerous specimens from many localities. It is impossible to evaluate the possible interrelations without a clear idea of the systematics and morphologic variability of the mammalian groups studied. These studies also hinge on physical and other factors that cannot always be controlled or eliminated. For example, sufficient specimens are not yet known from most levels of the Willwood Formation for such analyses to produce meaningful results, and pedofacies are not sufficiently exposed at other levels.

Nonetheless, it is of interest to explore the lateral control on compositional differences at all possible taxonomic levels to gain an appreciation of the influence of the ancient lithotope on faunal composition. In the course of ongoing taphonomic studies of the Willwood Formation, faunal composition from species through ordinal levels and fossil bone element abundances were calculated for the 13 richest fossil vertebrate localities in the southern Bighorn Basin, utilizing all specimens collected through 1984 (see also Bown and Kraus, 1981b). These are: University of Wyoming V-73037 (34 m, stage 5; see Bown and Kraus, 1987: Fig. 3F); V-73125 (180 m, stage 5); USGS D-1389 (264 m, stage 3-4); D-1454 (409 m, stage 3); D-1326 (425 m, stage 2-3); D-1204 (442 m, stage 3+); D-1198 (=YPM 45 complex, 470 m, stage 1-2); D-1177 (481 m, stage 4); D-1162 (481 m, stage 1-2); D-1510 (490 m, stage 1); D-1256 (546 m, stage 1); D-1583 (546

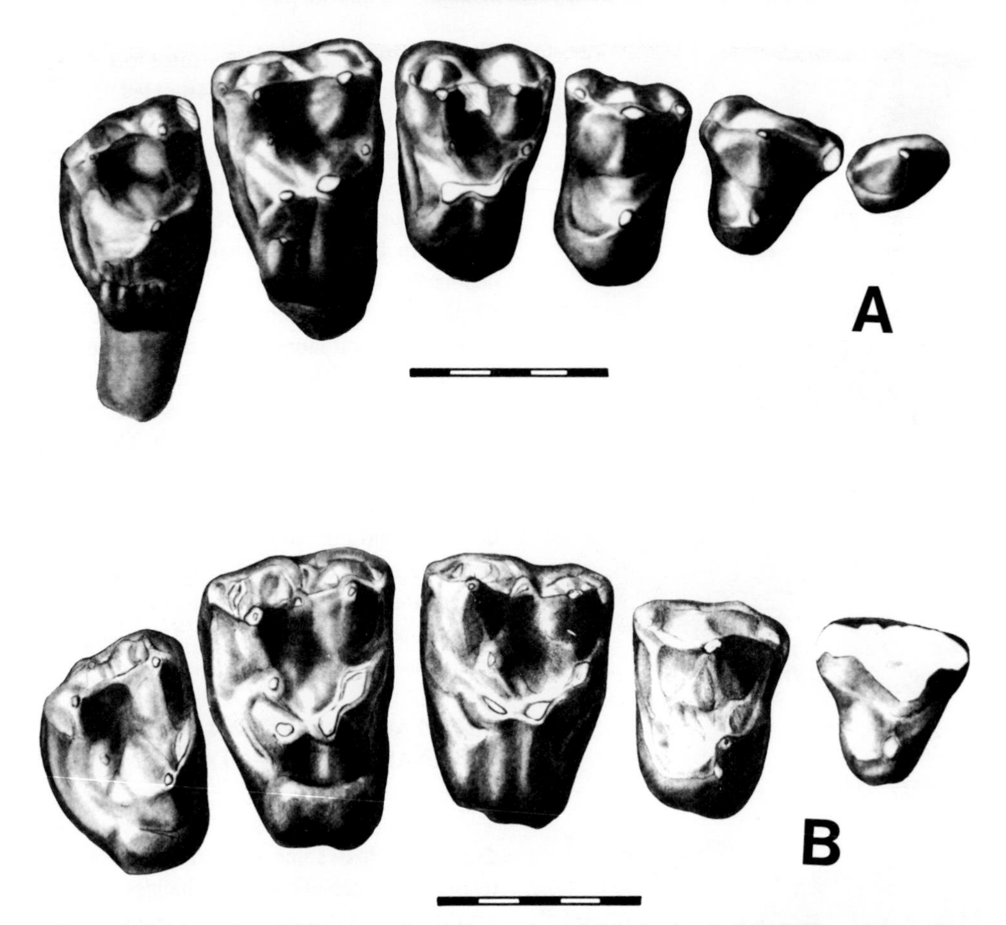

Figure 8. Upper molars of *"Copelemur"* and *Cantius* from the 546-m level of the Willwood Formation. A, *"Copelemur" feretutus,* occlusal view of USGS 7481, right P2-M3; B, *Cantius abditus,* occlusal view of USGS 9904, right P3 to M3. Scale is 5 mm.

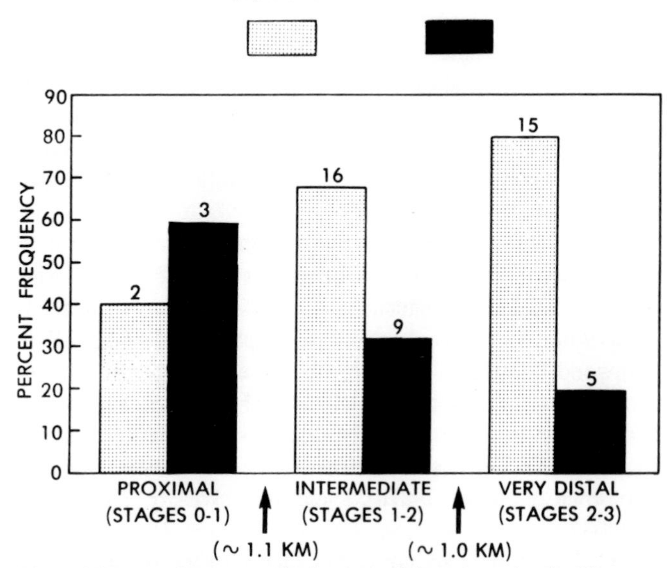

Figure 9. Percent frequency distributions of the sympatric adapid genera *"Copelemur"* and *Cantius* at the 546-m level of the Willwood Formation with respect to lateral position and paleosol maturity. Numbers at tops of bars are minimum numbers of individuals (MNI). Data through the 1986 field season was utilized.

m, stage 2 to 3); and D-1473 (556 m, stage 0 to 1). In all, more than 26,000 specimens were examined.

In analyzing paleofaunas for systematic lateral variation in mammalian composition, it is desirable to study the fauna in a single pedofacies. Barring this possibility due to constraints of exposure (no known continuously exposed pedofacies encompass more than 3+ stages), it is of utmost importance to temporally limit controlled compositional differences. This condition can be approximated by utilizing only localities for which approximate temporal equivalency can be easily determined, yet which represent two distinctly different paleosol maturities. The two localities that best meet these criteria, and for which adequate data are available, are D-1198 at 470 m (stages 1 to 2 but dominated by stage 1 paleosols) and D-1177 at 481 m (a single stage 4 paleosol). Although they form parts of two different pedofacies, these localities are additionally advantageous in that they are only 1.3 km apart, both are rich in vertebrate fossils, and both lie at or near the base of Biohorizon C (the old "Graybullian"-"Lysitean" boundary; Schankler, 1980). They both contain rich and diverse assemblages of the earliest middle Wasatchian mammals.

In Figure 10, mammalian composition at the ordinal level is depicted for both localities. Ignoring the rarer forms (Pantodonta,

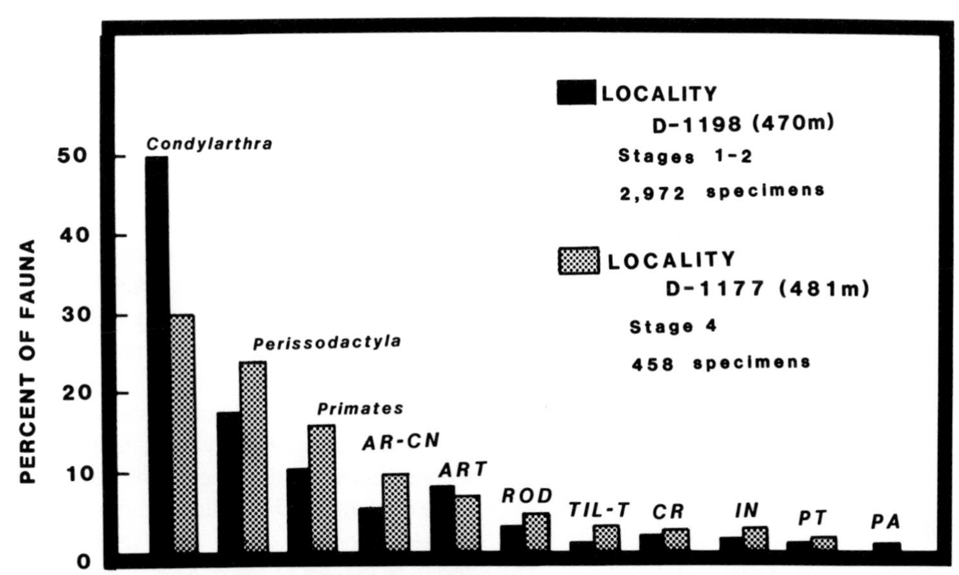

Figure 10. Relative representation of fossil mammals by order in relatively immature (stages 1 to 2) and relatively mature (stage 4) paleosols at the 470- and 481-m levels of the Willwood Formation. Representation is depicted as percent frequency of fauna at each locality. Data through the 1986 field season was utilized. AR-CN, Arctocyonia and Carnivora; ART, Artiodactyla; ROD, Rodentia; TIL-T, Tillodontia and Taeniodonta; CR, Creodonta; IN, "insectivorous" mammals (Leptictida, Pantolesta, Apatotheria, Lipotyphla); PT, Pantodonta; PA, Palaeanodonta.

Palaeanodonta), groups that appear more or less equivalent in distribution (Creodonta), and groups that are not only rare but artificial, certainly representing more than two orders (Proteutheria-Insectivora), several interesting differences are evident. Perhaps the most striking difference in distribution is that the Condylarthra and Artiodactyla are the only orders with greater representation in immature paleosols; the latter group by only 17 percent (not significant), but the condylarths by an impressive 67 percent (chi-square test, $p < 0.005$). Considering that the condylarths are the most abundant of all the mammals anywhere (represented by 1,471 and 137 specimens, 49 percent and 30 percent of all specimens at D-1198 and D-1177, respectively), their disproportionate numbers in less mature paleosols are highly significant.

Three other groups, Perissodactyla, Primates, and the combined orders Arctocyonia (which are very rare) and Carnivora, are all more abundant in the stage 4 paleosol at D-1177 and include enough specimens that their appreciably different relative abundances are also of interest. Perissodactyls are 37 percent more abundant (not significant), Primates are 56 percent more abundant (not significant), and combined Arctocyonia and Carnivora are 78 percent more abundant (chi-square test, $p < 0.05$) than in the more immature paleosols at D-1198. Differences in distribution of the Rodentia are difficult to assess because many species are represented by only a few specimens. The tillodont-taeniodont distribution is probably significant but is based on very few specimens (only two taeniodont specimens; the tillodonts belong to a single species of *Esthonyx*). Likewise, a single species of *Palaeanodon* is present only at D-1198, from where there are only four specimens. Although this distribution might at first appear to be significant, one *Palaeanodon* specimen is known for every 742 specimens at D-1198. The genus is so rare that, were it equally common at D-1198 and D-1177, it might require recovery of at least an additional 284 specimens at D-1177 to document its presence.

In sum, the differential ordinal representation of mammals between stages 1 to 2 and stage 4 paleosols at the 470- and 481-m levels of the Willwood Formation is suggestive of lateral controls on distribution at an appreciably higher taxonomic grade. The Condylarthra and the Artiodactyla are the sole orders with better representation in immature paleosols. The record of the Perissodactyla (37 percent more abundant in stage 4 paleosols) is also important because the perissodactyls are very common at all localities. It would be interesting to compare these results with those from stages 1 to 2 and stage 4 paleosols from other pedofacies from equivalent, higher, and lower stratigraphic levels of the Willwood Formation. Unfortunately, none are currently known. The 442-m level (see above) spans 0 to 3+ paleosol stages and might be an interesting comparison. However, the 442-m level lies below Biohorizon C, and the known biostratigraphically controlled faunal differences on either side of that boundary are great enough to render such a comparison almost meaningless.

LATERAL CONTROLS ON REPRESENTATION OF SKELETAL REMAINS

If at least some aspects of the taxonomic composition of the Willwood mammalian fauna are controlled in the lateral dimen-

TABLE 1. MEAN NUMBER OF TEETH PER GNATHIC SPECIMEN*

Position	Localities	Teeth/Specimen	N
442-m Level			
Proximal	D-1588, D-1657, D-1660, D-1682, D-1688	4.89	18
Intermediate	D-1310, D-1311	2.05	39
Distal	D-1204, D-1693	2.16	147
546-m Level			
Proximal	D-1464, D-1467, D-1581, DPC 15	2.76	21
Intermediate	D-1256, D-1463, D-1582, YPM 190 YPM 192 (part), YPM 315, DPC 16	2.31	104
Distal	D-1558, D-1574, D-1575, D-1576, YPM 181, YPM 192 (part), YPM 193	2.25	65

*For specimens of the adapid primates *"Copelemur"* and *Cantius* from localities at the 442- and 546-m levels of the Willwood Formation, all with respect to position of the localities on the ancient flood basins (proximal, intermediate, distal).

sion by paleosol maturity and the position of the fossils with respect to ancient streams, what about skeletal element composition (i.e., jaws, teeth, postcranial bones) of the assemblages? As observed above, by far the overwhelming number of Willwood fossil mammals occur in paleosols developed on flood-plain alluvium (Bown and Kraus, 1981b), and were passively accumulated in them by attrition at the former ground surfaces and subsequent sporadic burial (Bown, 1979; Bown and Kraus, 1981b). Because all of these fossils are in paleosols and presumably underwent little or no fluvial transport or breakage, any fundamental lateral differences in their representation at different localities across the pedofacies should be related to the different maturation stages (i.e., the amount of time they spent at the surface prior to burial).

According to the pedofacies model (Bown and Kraus, 1987), localities closest to the channel belt experienced a relatively greater rate of sediment accumulation than localities more distant from the channel. Today, these proximal localities are typified by relatively thick sequences of sediment upon which were developed a relatively large number of relatively immature paleosols. Conversely, localities more distant from channel belts are typified by less sediment and fewer paleosols, but paleosols of greater maturity. Therefore, proximal localities (immature paleosols) should yield a higher proportion of more complete specimens than distal localities (mature paleosols) because, in general, the remains in proximal positions were, on the average, buried more rapidly and suffered less from postmortem weathering, breakage, and scavenging.

To test this hypothesis, the intact teeth per specimen of *Cantius* at the 442-m level and of *Cantius* and *"Copelemur"* at the 546-m level were counted from proximal, intermediate, and distal localities in their respective pedofacies. Isolated teeth were recorded as one tooth per specimen, and dentaries and maxillae preserving two or more teeth were treated accordingly. These data are presented in Table 1. Although there are, on average, fewer teeth per specimen at intermediate localities than at distal localities at the 442-m level, available evidence suggests more teeth per specimen at proximal localities than at either intermediate or distal localities at both the 442-m (chi-square test, $p < 0.001$) and 546-m (n.s.) levels.

Unlike taxonomic composition, the composition of mammalian remains does not vary at different biostratigraphic levels. Consequently, trends in the relative numbers and completeness of these elements can be evaluated through a complete, though composite, pedofacies drawn from several distinct biostratigraphic levels. In Figures 11 through 14, several categories of mammalian remains, representing all of the fossils known from six of the richest localities through the 1984 field season, are depicted across a composite pedofacies reconstructed from localities in very fossiliferous stage 1 to 5 paleosols at six levels of the Willwood Formation. These are: V-73037 (34 m, stage 5); D-1177 (481 m, stage 4); D-1204 (442 m, stage 3+); D-1454 (409 m, stage 3); D-1198 (470 m, stage 1-2); and D-1256 (546 m, stage 1). In each of these figures, the abundance of more complete remains is compared with that of less complete remains, expressed as percent frequencies (Figs. 11 and 14), or with those of all other elements known from the localities, also expressed as percent frequencies (Figs. 12 and 13).

Figure 11 compares the abundance of jaws with two or more teeth with that of edentulous jaws. Little change in percent frequency is observed until stage 4, at which point the percent frequency of edentulous jaws rises and that for jaws with two or more teeth declines. Sample sizes are large enough to allow us to reject the null hypothesis of random sampling bias (chi-square test, $p < 0.005$). The pattern conforms well with that predicted by the taphonomic model suggested above for the pedofacies.

Figures 12 and 13 compare the abundances of even more complete remains with those of all other identified elements in the faunas. Jaws with four or more teeth (Fig. 12) show the same proportion with respect to all other faunal elements until stage 3+ paleosols, where they drop from 5.47 to 4.70 percent of the total known elements. In stage 4 and 5 paleosols, they decline further to 3.75 and 2.51 percent of the total known elements, respectively. Figure 13 compares the relative abundances of all of the most complete mammalian remains of any kind (skulls, palates, mandibles, complete long bones, and associations of two or more elements) with those of all other known elements. The proportion of the most complete remains increases from stage 1 through stage 3 paleosols to a high of 7.79 percent, then declines sharply to 5.01, 4.93, and 2.91 percent in stages 3+, 4, and 5 paleosols, respectively.

Finally, the abundance of isolated teeth was compared with

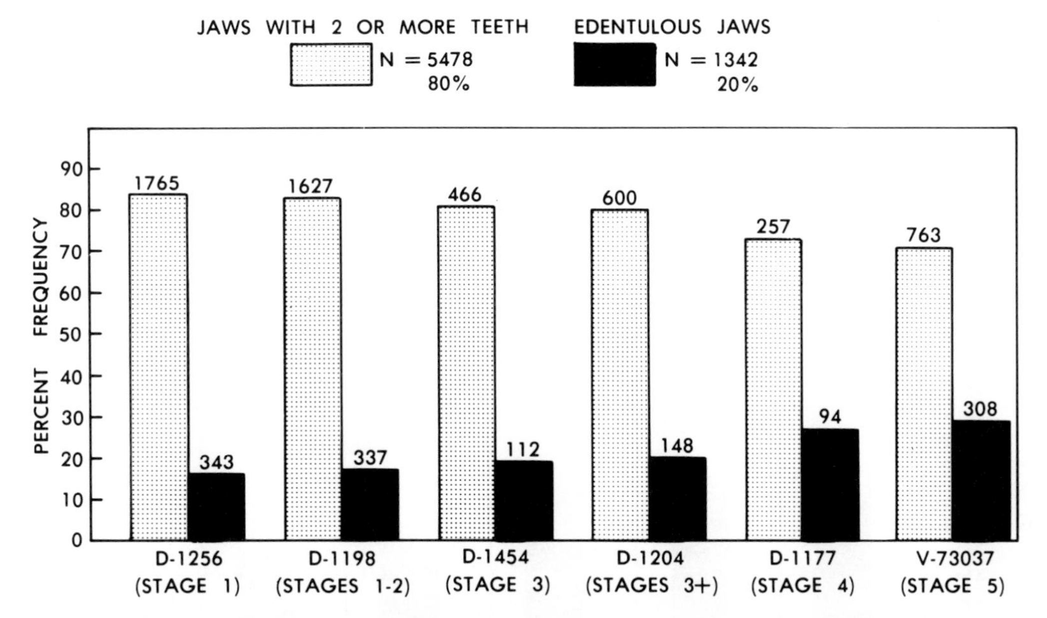

Figure 11. Percent frequency distributions of jaws (maxillae and mandibles) with two or more teeth, and edentulous jaws at six localities representing pedofacies soil stages 1 to 5 of the Willwood Formation. Data through the 1984 field season was utilized.

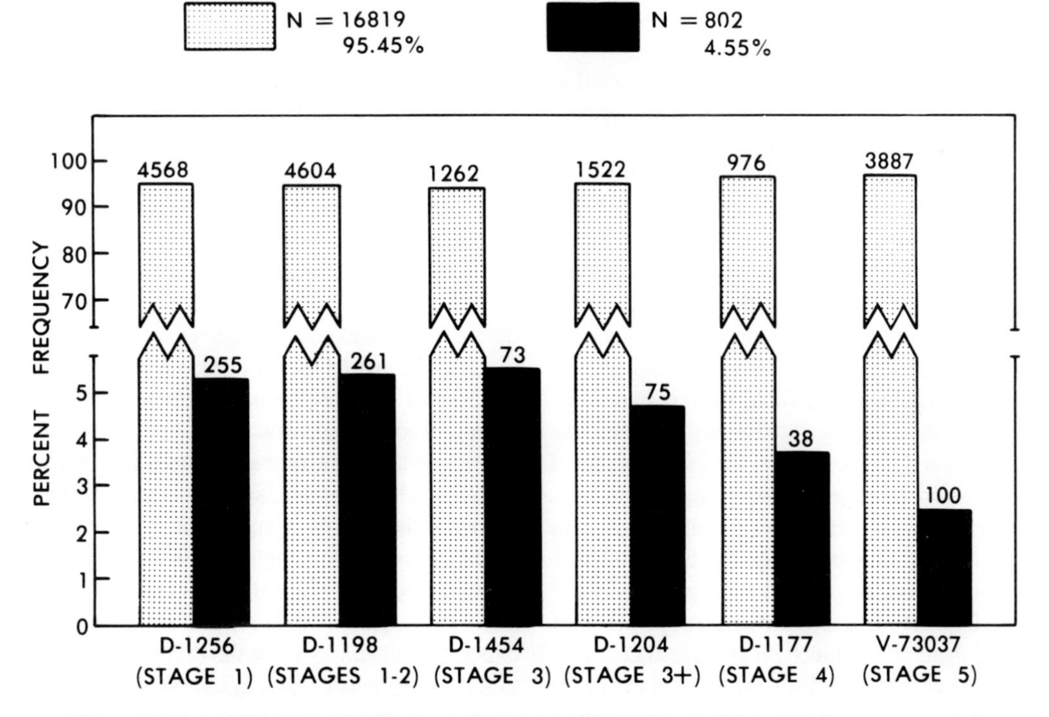

Figure 12. Percent frequency distributions of jaws (maxillae and mandibles) with four or more teeth, and all other vertebrate fossil elements at six localities representing pedofacies soil stages 1 through 5 of the Willwood Formation. Data through the 1984 field season was utilized.

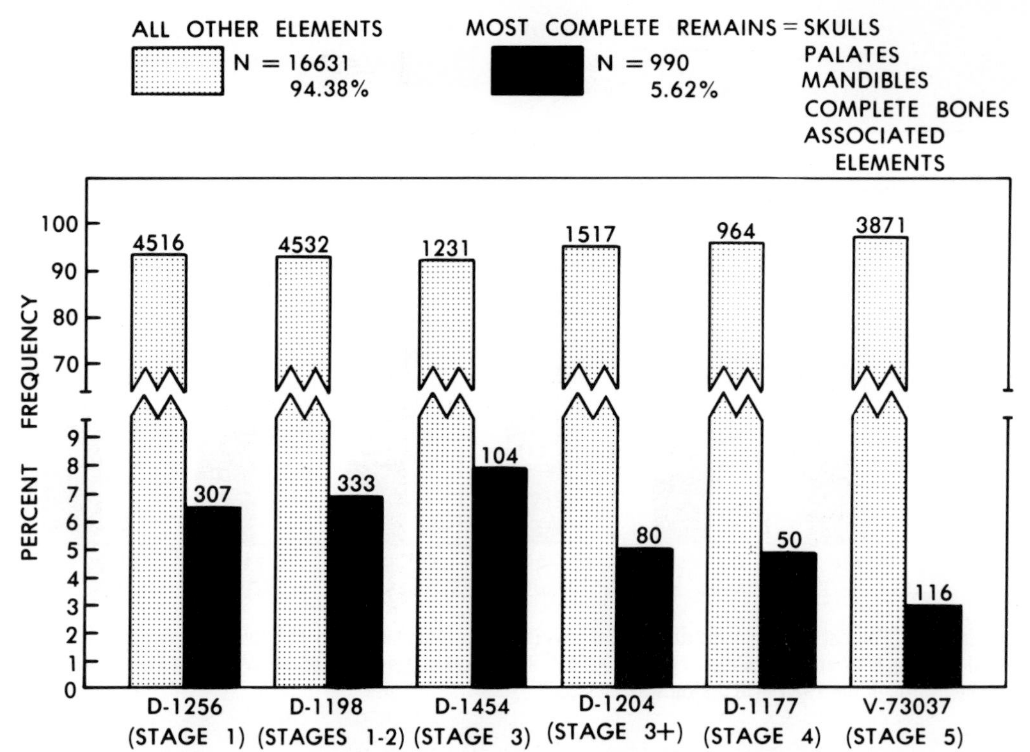

Figure 13. Percent frequency distributions of the most complete remains and all other vertebrate fossil elements at six localities representing pedofacies soil stages 1 to 5 of the Willwood Formation. Data through the 1984 field season was utilized.

that of all other gnathic remains (excluding edentulous jaws) as an indication of gnathic completeness (Fig. 14). From stage 1 through stage 3+ paleosols, the relative abundance of isolated teeth ranges from 44.3 to 49.9 percent. However, isolated teeth are more abundant than all other gnathic remains combined in stage 4 and 5 paleosols, first rising slightly to 52.9 percent (stage 4), then markedly to 69.3 percent (stage 5). Again, the pattern conforms with the pedofacies model, and the data are significantly different from those expected from random sampling error (chi-square test, $p < 0.001$).

The taphonomic data presented in Table 1 and in Figures 11 to 14 consistently indicate a strong relationship between the completeness of remains and the position of the localities lateral to ancient channel belts and, therefore, by position in the pedofacies maturation sequence. In general, the least complete remains show an increase in relative abundance distally; that is, in progressively more mature paleosols—exactly as predicted by the pedofacies taphonomic model. It is interesting, however, that for most of these fossils, there is no direct progressive increase in relative abundance from proximal to distal throughout the entire pedofacies. Such a situation occurs only in the case of edentulous jaws (Fig. 11), which increase progressively from 16.3 percent (stage 1), to 17.2 percent (stages 1 to 2), to 19.4 percent (stage 3), to 19.8 percent (stage 3+), to 26.8 percent (stage 4), to 28.8 percent (stage 5). Curiously, this progressive increase is not matched by the data for the complement to edentulous jaws (i.e., isolated teeth; Fig. 14). In the remainder of the data, there is little systematic change in relative abundance until the break between stage 3 and stage 3+ paleosols (Figs. 12 to 13), and that separating stage 3+ and stage 4 paleosols (Fig. 14). In Figure 11, it is seen that the most appreciable increase in relative abundance of incomplete remains (edentulous jaws) is also between stage 3+ (19.8 percent) and stage 4 (26.8 percent). To evaluate the reason for this peculiar (yet persistent) inconsistency, recourse to the nature of the paleosols themselves is necessary.

As defined by Bown (1985) and Bown and Kraus (1987), stage 2 and 3 paleosols are of one type (probably alfisols), whereas stage 4 and 5 paleosols are of another type (probably spodosols). It is therefore probably significant that the identified soil types also change appreciably in character from stage 3 to stage 4, just as do the records of mammalian remains in them. Regardless of the soil taxonomy applied to these paleosols, all five paleosol stages formed upon fluvial parent materials under equivalent climatic conditions. Therefore, their differing morphologic characteristics result simply from the differing relative amounts of time to form them (their maturity). It would therefore appear that the amount of time to transform a stage 3 paleosol into a stage 4 paleosol (and a stage 4 to a stage 5) is much greater than that necessary to change paleosols from stages 0 to 1, 1 to 2, or 2 to 3 (see Bown and Kraus, 1987: Fig. 3D to F with respect to rather marked stage 3 to 4 and 4 to 5 profile differences). It is probable that soil-forming processes slowed considerably through time

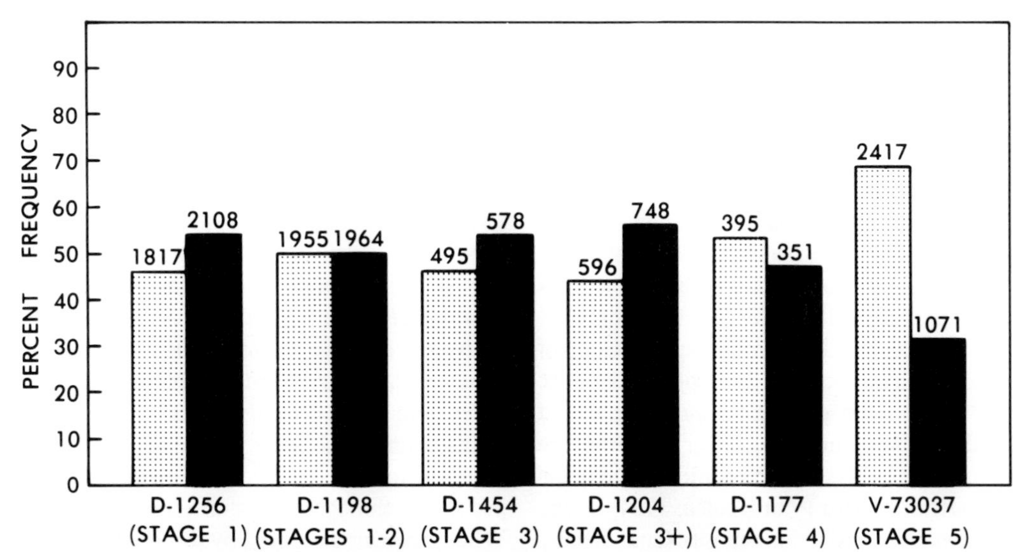

Figure 14. Percent frequency distributions of isolated teeth and all other gnathic remains (edentulous and dentulous maxillae and mandibles) at six localities representing pedofacies soil stages 1 to 5 of the Willwood Formation. Data through the 1984 field season was utilized.

simply because most clastic and soil plasma materials were translocated rather early in the paleosol maturation sequence (by stage 3; Bown and Kraus, 1987). Additional maturity (including increase in profile thickness—the hallmark of paleosol stages 4 and 5) required upward incremental addition of new sediment, downward cannibalization of previously buried paleosol profiles, and considerable additional time.

SUMMARY AND CONCLUSIONS

Pedofacies relations of alluvial paleosols in the Willwood Formation provide excellent control for evaluating coeval and approximately coeval assemblages of fossil mammals in the lateral dimension, so much so that they should be kept in mind by all workers examining faunal composition and taphonomic controls for assemblages derived from flood plain deposits. In the Willwood Formation, species of *Hyopsodus* (Condylarthra) and adapiform primates vary in relative abundance across the pedofacies from proximal to distal fluvial paleoenvironments and, consequently, from less mature to more mature paleosols. Mammalian faunal composition at the ordinal level also exhibits some striking differences between faunas from stage 1 to 2 paleosols and those from stage 4 paleosols. Taphonomically, the most complete fossil remains occur in greater proportion in relatively immature paleosols, proximal to alluvial ridges, where rates of sediment accumulation (and therefore burial) were relatively high. The profound causal interaction between the fossil fauna, its remains, paleosol maturity, and relative position on the ancient flood basin is strongly emphasized by these results.

There is little that is particularly surprising about these conclusions. Documentation of coeval distributional differences in modern mammalian taxa is both abundant and prosaic, and a similar relationship is only to be expected for fossil mammals. However, the taphonomic information presented here (new, and based on more than 18,000 specimens) is a major advance upon and departure from earlier studies by Shotwell (1955, 1963) and Clark and others (1967), to name a few.

The advantage of the pedofacies is that it: (1) provides an empirical, objective, and testable method for grouping collections from coeval localities in the lateral dimension (it even defines the temporal equivalence of these localities); (2) relates the fossil accumulations to paleosols, which also vary laterally in a systematic way and can be evaluated independently for paleoclimatic and temporal information; and (3) relates the accumulations to the coeval alluvial record, in terms of both its sedimentology and its local paleogeography. Moreover, because most ancient alluvial sequences contain numerous paleosols (Kraus and Bown, 1986), and these should possess similar pedofacies relations (though perhaps in different soil types), the pedofacies model and similar studies should be applicable to all ancient alluvial sequences that are both well exposed and rich in fossils. Possibly even more important, the systematic lateral variability of paleosols is related directly or indirectly to systematic small-scale variability in paleoenvironment. Therefore, future study of fossil accumulations grouped according to paleosol maturities will provide a way to approximate more closely true taxonomic diversity and its distribution and variation across Willwood flood basins at confined stratigraphic levels. As things stand now, we can no longer be even reasonably certain that a sample from a single locality (however abundant the fossils) affords anything close to an accurate estimate of species richness at one time over any significant geographic area. We must therefore also question the meaning (if any) of comparisons of species composition and richness at different individual localities, whether they are proximal or distal to one another, and whether or not they occur in the same structural basin. Discoveries in 1988 suggest that many new taxa might be found in paleosols with maturities unusual for the sequence or stratigraphic levels in which they occur.

For example, both *Hyposodus* sp., cf. *H. minor* and *Cantius frugivorus* preferred distal fluvial paleoenvironments at the 442-m level. Some parts of the Willwood section are locally or regionally dominated by proximal alluvial deposits with immature paleosols, whereas other sections exhibit greater representation of more mature paleosols developed on more distal alluvial deposits. In addition, because the axis of the southern Bighorn Basin shifted gradually westward through time, and because (in this part of the basin) the higher parts of the Willwood section are exposed in the west and the lower parts in the east (Van Houten, 1944; Bown, 1980), the density of proximal alluvial deposits (and thereby immature paleosols) increases both westward and upward in the section. Consequently, mammals that preferred more distal paleoenvironments are almost certainly underrepresented in the upper Willwood fauna and overrepresented in the fauna of the lower Willwood Formation. Add to this scenario a backdrop of gradually drying conditions from at least Biohorizon C upward, and the possible variables controlling Willwood mammal composition (even at specific localities) become both numerous and extremely complex. Therefore, it is obvious that to achieve a realistic, workable picture of lateral diversity in faunal composition, it is necessary to be aware of such formidable biases that are inherent in nearly every alluvial sequence.

Willwood mammals studied with respect to their pedofacies contexts thus far are rather eurytopic in distribution; however, it is quite clear that their proportions vary substantially in the lateral dimension. The distributions of some of the more stenotopic mammals (e.g., the records of the dermopteran *Plagiomene* and the perissodactyl *Homogalax;* Bown, 1979, 1980) are known to vary on practically microscopic and basin-wide scales, respectively. The distributions for paleosols (both maturity and sequencing) are also generally consistent vertically (pedofacies sequences of Kraus, 1987; Kraus and Bown, 1988); however, the faunal component is naturally more complicated in the vertical (temporal) dimension because of immigration, morphologic evolution and pseudoextinction, faunal turnover, and true extinction. Nonetheless, the possibilities for integrating temporal and lateral faunal data related to the pedofacies are exciting. As an additional example, the adapiform primate *"Copelemur" feretutus* appears to have preferred more distal fluvial paleoenvironments than its close, coeval relative, *Cantius abditus.*

But what about the ancestors and descendents (if any) of these taxa? Were they always so distributed? At the scale of individual drainage basins on the Willwood lithotope, what do the terms "allopatric" and "sympatric" really mean, how do they differ, and what do different distributions of these mammals through time and space contribute to the evolution of these groups? How are carnivorous mammals distributed across the pedofacies through time with respect to possible prey species? What problems are posed by dry soil/wet soil faunas? How are different paleosol suites distributed temporally? How are these related to character evolution? Some answers to these and other interesting questions now only await a workable systematic base for several groups of early Eocene mammals.

ACKNOWLEDGMENTS

We are grateful to Herbert H. Covert, Mary J. Kraus, Kenneth D. Rose, and Pat Shipman for reviewing earlier drafts of this paper, and for much helpful discussion. Figure 2 was skillfully constructed by Art Isom, and figures 6 and 7 were drawn by Elaine Kasmer. Collections of fossil mammals used in this study were obtained through the joint field research efforts of Bown at the U.S. Geological Survey and Kenneth D. Rose at The Johns Hopkins University. Financial support was provided in part by a G.K. Gilbert Professional Fellowship to T.M. Bown, and in part by National Science Foundation grants BSR-8500732 and BSR-8215099 to K.D. Rose.

REFERENCES CITED

Beard, K. C., 1988, New notharctine primate fossils from the early Eocene of New Mexico and southern Wyoming, and the phylogeny of Notharctinae: American Journal of Physical Anthropology, v. 75, p. 439–469.

Beard, K. C., Rose, K. D., and Bown, T. M., 1986, Dental variation in the early Eocene Adapidae *Cantius* and *Copelemur,* and some paleoecological implications [abs.]: American Journal of Physical Anthropology, v. 69, p. 174.

Bown, T. M., 1979, Geology and mammalian paleontology of the Sand Creek facies, lower Willwood Formation (lower Eocene), Washakie County, Wyoming: Geological Survey of Wyoming Memoir, v. 2, p. 1–151.

—— , 1980, The Willwood Formation (lower Eocene) of the southern Bighorn Basin, Wyoming, and its mammalian fauna, *in* Gingerich, P. D., ed., Early Cenozoic paleontology and stratigraphy of the Bighorn Basin, Wyoming: Ann Arbor, University of Michigan Papers on Paleontology, v. 24, p. 127–138.

—— , 1985, Maturation sequences in lower Eocene alluvial paleosols, Willwood Formation, *in* Flores, R. M., and Harvey, M., eds., Field guide to modern and ancient fluvial systems in the United States: Third International Fluvial Sedimentology Conference Guidebook, p. 20–26.

—— , 1987, The mammalian pedofacies; A novel approach to detailed verticolateral and temporal controls in vertebrate facies distribution and mammalian evolution [abs.], *in* Bown, T. M., and Rose, K. D., eds., Symposium, Dawn of the age of mammals in the northern part of the Rocky Mountain Interior, U.S.A.: Geological Society of America Abstracts with Programs, v. 19, p. 262.

Bown, T. M., and Kraus, M. J., 1981a, Lower Eocene alluvial paleosols (Willwood Formation, northwest Wyoming, U.S.A.), and their significance for paleoecology, paleoclimatology, and basin analysis: Paleogeography, Palaeoclimatology, Palaeoecology, v. 34, p. 1–30.

——, 1981b, Vertebrate fossil-bearing paleosol units (Willwood Formation, lower Eocene, northwest Wyoming, U.S.A.); Implications for taphonomy, biostratigraphy, and assemblage analysis: Palaeogeography, Palaeoclimatology, Palaeoecology, v. 34, p. 31–56.

——, 1987, Integration of channel and floodplain suites; 1, Developmental sequence and lateral relations of alluvial paleosols: Journal of Sedimentary Petrology, v. 57, p. 587–601.

Bown, T. M., and Rose, K. D., 1984, Reassessment of some early Eocene Omomyidae, with description of a new genus and three new species: Folia Primatologica, v. 43, p. 97–112.

——, 1987, Patterns of dental evolution in early Eocene anaptomorphine primates (Omomyidae) from the Bighorn Basin, Wyoming: Paleontological Society Memoir, v. 23, p. 1–162.

Bown, T. M., and Rose, K. D., 1990, Distribution and stratigraphic relationships of fossil vertebrate and plant localities in the Fort Union, Willwood, and Talman Formations (upper Paleocene to lower Eocene), central and southern Bighorn Basin, Wyoming: U.S. Geological Survey Bulletin (in press).

Clark, J., Beerbower, J. R., and Kietzke, K. K., 1967, Oligocene sedimentation, stratigraphy, and paleoecology and paleoclimatology in the Big Badlands of South Dakota: Fieldiana (Geology), v. 5, p. 1–158.

Gingerich, P. D., 1974, Stratigraphic record of early Eocene *Hyopsodus* and the geometry of mammalian phylogeny: Nature, v. 248, p. 107–109.

——, 1976, Paleontology and phylogeny; Patterns of evolution at the species level in early Tertiary mammals: American Journal of Science, v. 276, p. 1–28.

——, 1979, The stratophenetic approach to phylogeny reconstruction in vertebrate paleontology, *in* Cracraft, J., and Eldredge, N., eds., Phylogenetic analysis and paleontology: New York, Columbia University Press, p. 41–77.

Gingerich, P. D., 1980, The Bighorn Basin; Why is it so important?, *in* Gingerich, P. D., ed., Early Cenozoic paleontology and stratigraphy of the Bighorn Basin, Wyoming: Ann Arbor, University of Michigan Papers on Paleontology, v. 24, p. 1–5.

Gingerich, P. D., and Gunnell, G. F., 1979, Systematics and evolution of the genus *Esthonyx* (Mammalia, Tillodontia) in the early Eocene of North America: Ann Arbor, University of Michigan Contributions from the Museum of Paleontology, v. 25, p. 125–153.

Gingerich, P. D., and Haskin, R. A., 1981, Dentition of early Eocene *Pelycodus jarrovii* (Mammalia, Primates) and the generic attribution of species formerly referred to *Pelycodus:* Ann Arbor, University of Michigan Contributions from the Museum of Paleontology, v. 25, p. 327–337.

Gingerich, P. D., and Simons, E. L., 1977, Systematics, phylogeny, and evolution of early Eocene Adapidae (Mammalia, Primates) in North America: Ann Arbor, University of Michigan Contributions from the Museum of Paleontology, v. 24, p. 245–279.

Kraus, M. J., 1985, Sedimentology of early Tertiary rocks, northern Bighorn Basin, *in* Flores, R. M., and Harvey, M., eds., Field guide to modern and ancient fluvial systems in the United States: Third International Fluvial Sedimentology Conference Guidebook, p. 26–33.

——, 1987, Integration of channel and floodplain suites; 2, Vertical relations of alluvial paleosols: Journal of Sedimentary Petrology, v. 57, p. 602–612.

Kraus, M. J., and Bown, T. M., 1986, Paleosols and time resolution in alluvial stratigraphy, *in* Wright, V. P., ed., Paleosols; Their origin, classification, and interpretation: London, Blackwell, p. 180–207.

——, 1988, Pedofacies analysis; A new approach to reconstructing ancient fluvial sequences, *in* Reinhardt, J., and Siglio, W. R., eds., Paleosols and weathering through geologic time: Geological Society of America Special Paper 216, p. 143–152.

Neasham, J. W., and Vondra, C. F., 1972, Stratigraphy and petrology of the lower Eocene Willwood Formation, Bighorn Basin, Wyoming: Geological Society of America Bulletin, v. 83, p. 2167–2180.

Rose, K. D., 1981a, The Clarkforkian Land-Mammal Age and mammalian faunal composition across the Paleocene–Eocene boundary: Ann Arbor, University of Michigan Papers on Paleontology, no. 26, p. 1–197.

——, 1981b, Composition and species diversity in Paleocene and Eocene mammal assemblages; An empirical study: Journal of Vertebrate Paleontology, v. 1, p. 367–388.

Rose, K. D., and Bown, T. M., 1984, Early Eocene *Pelycodus jarrovii* (Primates, Adapidae) from Wyoming; Phylogenetic and biostratigraphic implications: Journal of Paleontology, v. 58, p. 1532–1535.

——, 1986, Gradual evolution and species discrimination in the fossil record, *in* Flanagan, K. M., and Lillegraven, J. A., eds., Vertebrates, phylogeny, and philosophy: Contributions to Geology Special Paper 3, p. 119–130.

Schankler, D. M., 1980, Faunal zonation of the Willwood Formation in the central Bighorn Basin, Wyoming, *in* Gingerich, P. D., ed., Early Cenozoic paleontology and stratigraphy of the Bighorn Basin, Wyoming: Ann Arbor, University of Michigan Papers on Paleontology, v. 24, p. 99–114.

Shotwell, J. A., 1955, An approach to the paleoecology of mammals: Ecology, v. 36, p. 327–337.

——, 1963, The Juntura Basin; Studies in earth history and paleoecology: Transactions of the American Philosophical Society, New Series, v. 53, p. 3–77.

Van Houten, F. B., 1944, Stratigraphy of the Willwood and Tatman Formations in northwestern Wyoming: Geological Society of America Bulletin, v. 55, p. 165–210.

MANUSCRIPT ACCEPTED BY THE SOCIETY JUNE 12, 1989

Printed in U.S.A.

Geological Society of America
Special Paper 243
1990

A statistical assessment of last appearances in the Eocene record of mammals

Catherine Badgley
Museum of Paleontology, University of Michigan, Ann Arbor, Michigan 48109

ABSTRACT

Changes in sample size may confound interpretation of faunal change in the fossil record. The record of early Eocene mammals from the Wasatchian Land-Mammal Age in the Clark's Fork Basin, Wyoming, shows a high correlation between the square root of sample size and species richness. This correlation suggests that Schankler's (1980) Biohorizon A, a faunal turnover composed mainly of disappearances, is largely an artifact of sampling fluctuation. Within the interval of Biohorizon A, sample size drops from record high to record low values.

Monte Carlo simulation of the drop in sample size across Biohorizon A demonstrates the role of sampling variation in producing artifacts of faunal change. The distribution of missing species resulting from the simulations provides a reliable estimate of the number of species that are likely to be missing at a specified sample size, even if they were present in the original population. Results of the simulations indicate that most of the 16 disappearances observed in the 200-m interval above Biohorizon A can be explained by low sample size alone. For each species that disappeared in the actual record, the frequency of absence in the simulations is a basis for comparing the likelihoods of two hypotheses: (1) that the species was present but not represented by fossils, and (2) that the species was absent. Evaluation of the likelihood ratio for these hypotheses indicates that *Arctodontomys wilsoni-A. nuptus, Phenacodus vortmani,* and *Homogalax* n. sp.-*H. semihians* are the lineages most likely to have disappeared over the interval of low sample size. Reconsideration of the biostratigraphic correlation between the central Bighorn Basin and the Clark's Fork Basin based on the first appearances of *Homogalax protapirinus* and *Tetonius matthewi/steini* supports a higher stratigraphic position for the interval corresponding to Biohorizon A in the Clark's Fork Basin than its original placement.

INTRODUCTION

The thick, fossiliferous sequences of alluvial sediments in the Bighorn Basin, Wyoming (Fig. 1), contain abundant paleontological data bearing on evolution within vertebrate lineages and on the changing composition of the preserved biota. The systematics and evolution of many lineages have received considerable attention (e.g., Gingerich, 1983b, 1985, 1986; Gunnell, 1985; Rose and Bown, 1984; Bartels, 1983; Bown and Rose, 1987). Studies of faunal change through time have been largely concerned with biostratigraphic zonation (e.g., Gingerich, 1983a; Gingerich and others, 1980; Rose, 1980, 1981a; Schankler, 1980; Archibald and others, 1987; Krishtalka and others, 1987). Ecological and taphonomic aspects of mammalian faunal composition have also been studied. Rose (1981b) evaluated changes in taxonomic composition and measures of mammalian diversity from middle Paleocene to early Eocene localities in terms of climatic changes and ecological interactions among faunal components. Schankler (1981) interpreted temporal gaps in the early Eocene record of phenacodontids from the Central Bighorn Basin as local extinc-

Badgley, C., 1990, A statistical assessment of last appearances in the Eocene record of mammals, *in* Bown, T. M., and Rose, K. D., eds., Dawn of the Age of Mammals in the northern part of the Rocky Mountain Interior, North America: Boulder, Colorado, Geological Society of America, Special Paper 243.

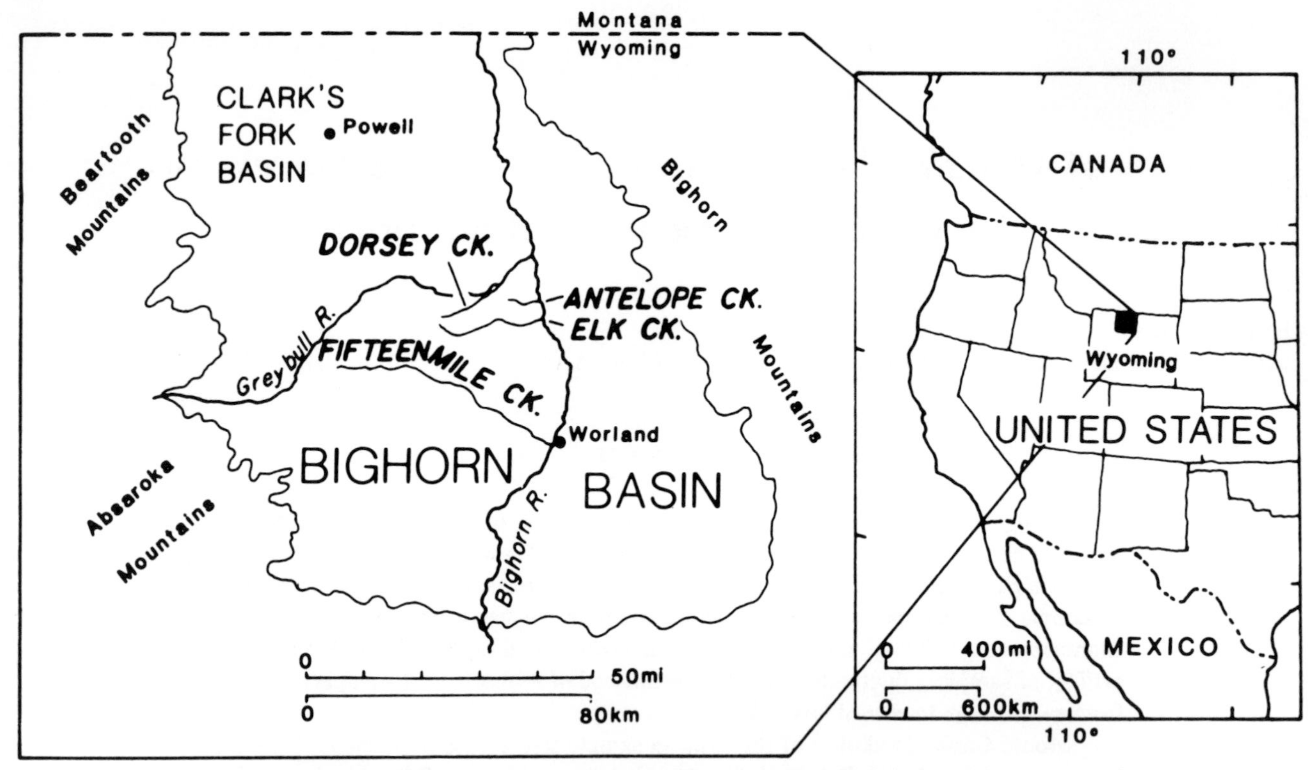

Figure 1. The Bighorn Basin in northwest Wyoming. The Clark's Fork Basin is the northwest corner of the Bighorn Basin. The study area of Schankler (1980) is around Elk Creek in the central Bighorn Basin. Reproduced from Badgley and Gingerich (1988).

tions and subsequent reappearances resulting from changes in vegetation and competition among phenacodontids. Bown and Kraus (1981) documented taphonomic characteristics of vertebrate assemblages in relation to facies and interpreted the assemblages as attritional accumulations on land surfaces, eventually incorporated into soil horizons. Winkler (1983) demonstrated differences in the composition of fossil assemblages in relation to paleosol characteristics and method of collection. Bown and Beard (this volume) document a correlation between faunal composition and paleosol maturity in laterally adjacent mammalian assemblages.

Here, I focus on the role of sampling variation in the record of faunal change in the Clark's Fork Basin. Badgley and Gingerich (1988) found that large fluctuations in sample size coincided with episodes of faunal turnover in the early Eocene mammalian record of the Clark's Fork Basin; they suggested that at least one of the faunal turnovers, Biohorizon A, was largely an artifact of sampling. The present study describes simulations designed to mimic the record of Biohorizon A in the Clark's Fork Basin. Results indicate that sampling variation can account for most of the observed species disappearances over an interval of about 200 m in which aggregate sample size is relatively low. The statistical likelihood ratio of species absence to species presence indicates which of the 16 species recorded as disappearing are most likely to have disappeared. Reevaluation of biostratigraphic correlation between the Clark's Fork Basin and the central Bighorn Basin supports a higher stratigraphic position for Biohorizon A than its original placement in the Clark's Fork Basin.

BACKGROUND

Considerable overlap in biostratigraphic records occurs between the Clark's Fork Basin and the central Bighorn Basin, some 100 km apart (Fig. 1). The record of faunal change known as Biohorizon A is an early Wasatchian species turnover, first documented by Schankler (1980), in the biostratigraphic record of the central Bighorn Basin. Biohorizon A separates the lower and upper units of the *Haplomylus-Ectocion* Range Zone in the biostratigraphic zonation of Schankler. In the Clark's Fork Basin, Biohorizon A has been interpreted to separate the Sandcouleean from Greybullian subages of the Wasatchian (Gingerich, 1983a; but see below). In both areas, Biohorizon A comprises a major pulse of disappearances and a minor pulse of appearances (Fig. 2 depicts the pattern originally recorded by Schankler). Also in both areas, some of the species that disappear at Biohorizon A reappear significantly higher in the stratigraphic section (Schankler, 1980, 1981; Badgley and Gingerich, 1988).

Biohorizon A in the Central Bighorn Basin was originally correlated with the interval from 1,750 to 1,775 m in the stratigraphic section of the Clark's Fork Basin, based on correlated

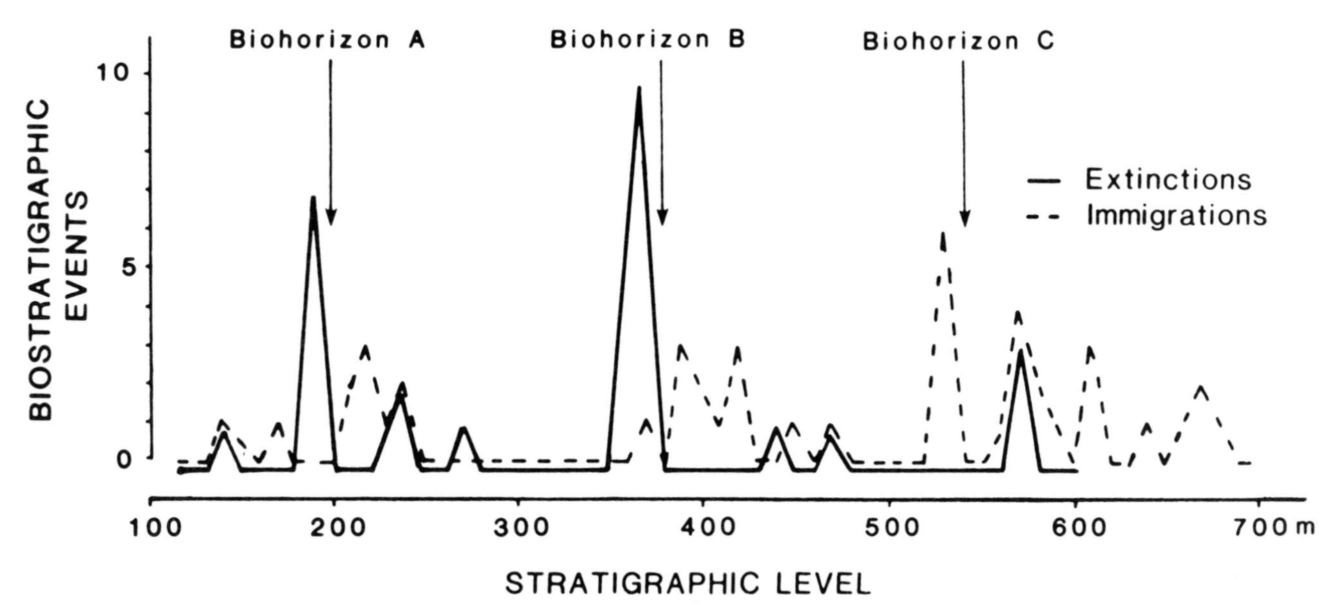

Figure 2. Schankler's episodes of faunal turnover in early Eocene mammals as indicated by the frequency of "extinctions" (disappearances, this paper) and "immigrations" (appearances, this paper) versus stratigraphic level in the central Bighorn Basin. The Clarkforkian-Wasatchian boundary lies below the lowest level depicted here. Reproduced from Badgley and Gingerich (1988).

changes in three mammalian lineages (Schankler, 1980). Badgley and Gingerich (1988) extended this interval up to 1,785 m to include the pulse of disappearances taken to characterize Biohorizon A. Figure 3 illustrates the biostratigraphic ranges of early Wasatchian mammals in the Clark's Fork Basin. In the interval 1,750 to 1,785 m, three species appear for the first time, seven species disappear from the local section (terminal disappearances), and five species begin gaps of 200 m or more (temporary disappearances—reappearing later in strata between 1,970 and 2,100 m). In addition, two species are known only in this interval, and two species are recorded by one or two specimens with gaps of 200 m or more, above or below this interval. These changes in faunal composition are summarized in Table 1. Changes in lineages through anagenesis are counted neither as appearances nor disappearances. (Interpretations of anagenesis follow Gingerich and Gunnell, 1979; Gingerich, 1980, 1983b, 1985, 1986; Gunnell, personal communication, 1986).

In the record of the Clark's Fork Basin, changes in the frequency of appearances and disappearances over time are correlated with sample size, as represented by the number of catalogued specimens in the University of Michigan Museum of Paleontology. Peaks of appearances coincide with peaks in sample size, and peaks of disappearances (including temporary disappearances) coincide with drops in sample size (Fig. 4). The coefficient of determination (r^2, based on Pearson's product moment correlation) between species richness and square root of sample size is 0.90 for the Wasatchian record of the Clark's Fork Basin (Badgley and Gingerich, 1988). The richest peak of the early Wasatchian (1,720 to 1,775 m) precedes a precipitous drop in sample size (1,780 m). The disappearances of Biohorizon A occur where sample size plunges from more than 1,100 specimens to fewer than 30 specimens in successive 20-m intervals. Above Biohorizon A, sample size remains relatively low with only minor increases. All of the disappearing taxa exhibit low frequencies in the stratigraphic intervals in which they are present, even when sample sizes are relatively high. Many of the disappearing species are carnivorous mammals that would have been rare in the original community because of their position in the trophic hierarchy.

These data suggest that Biohorizon A in the Clark's Fork Basin is largely an artifact of sampling. The covariation between species richness and sample size raises the question of whether any species would disappear if sample size could be increased substantially in the intervals where it is low. While this question is not directly amenable to investigation (the low sample sizes above Biohorizon A occur partly because of a sharp decrease in the availability and accessibility of exposures; Badgley and Gingerich, 1988), simulation of the drop in sample size from the rich intervals between 1,720 and 1,775 m addresses a significant component of this question.

DATA AND METHODS

Paleontological data

Data for this study come from the catalogued record of Wasatchian mammals from the Clark's Fork Basin at the University of Michigan Museum of Paleontology. Mammalian fossils collected and catalogued from 1975 through 1987 by University of Michigan field parties form the basis for Figure 3. This sample

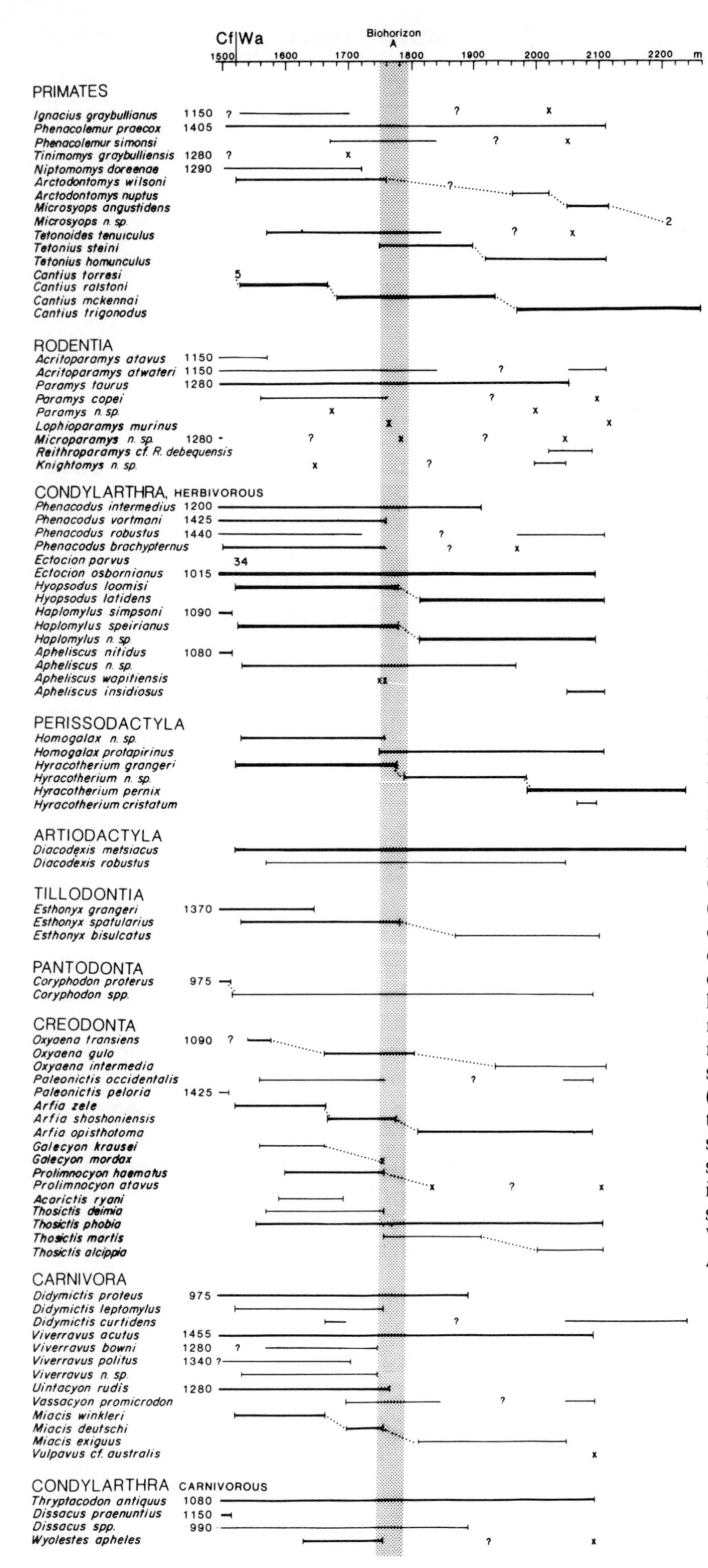

Figure 3. Biostratigraphic range chart of early Wasatchian mammals from the Clark's Fork Basin. Mammals are listed by order, with Condylarthra divided into herbivorous and carnivorous subunits. Excluded are mammalian orders comprising predominantly small taxa (less than 100 g estimated body weight), because remains of such taxa are generally too rare to be informative of the timing of faunal change. Species names, ranges, and phyletic transitions are taken principally from the University of Michigan Museum of Paleontology catalogue records as of 1985, with additional information from Gingerich and Gunnell (1979), Gingerich (1983b), Gunnell (1985), Ivy (1982), and Gingerich and Deutsch (1989). The quality of the record for each species is indicated by the width of the line. Heavy lines indicate records of good quality. Lines of medium width indicate records of medium quality. Thin lines indicate "spotty" records. Gaps of 200 m or greater are represented as blank areas with a question mark. Single specimens with large gaps below or above are marked as an X. For species represented by lines, small horizontal ticks at the base (respectively top) indicate the first (respectively last) records in the Clark's Fork Basin. The absence of a tick at the left (respectively right) indicates that the record continues below (respectively above) the interval represented. Numbers at the left indicate the stratigraphic position (in meters) at which the species is first recorded, if lower than the Clarkforkian-Wasatchian boundary. The stippled band marks Biohorizon A. Modified from Badgley and Gingerich (1988).

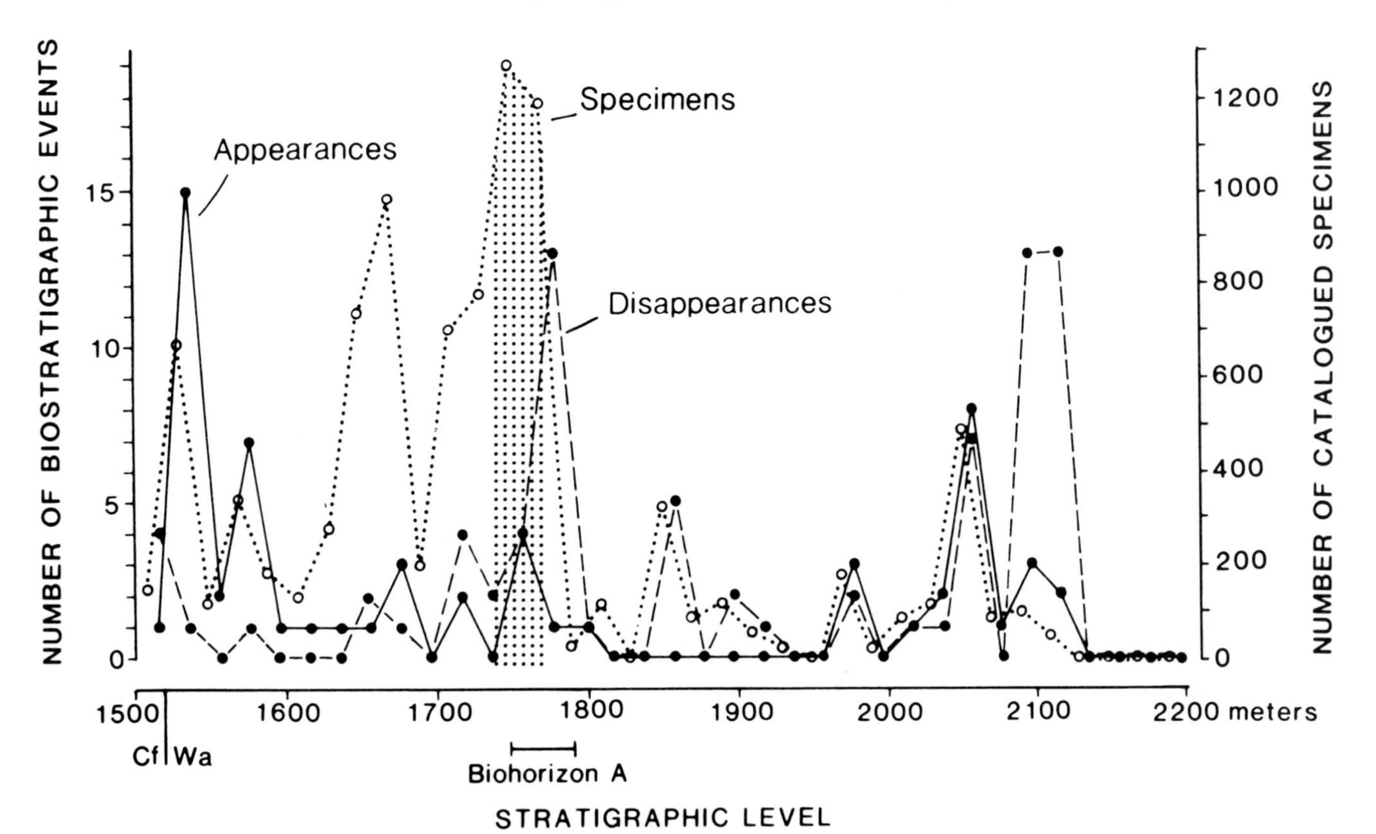

Figure 4. The number of appearances and disappearances, with gaps of 200 m or greater counting as temporary disappearances and later reappearances, plotted with sample size (number of catalogued specimens over 20-m interval). Data points representing appearances or disappearances are plotted at the interval boundaries. Points representing number of catalogued specimens are plotted in the middle of each interval. Stippled band marks the two intervals compiled for the faunal population of the simulations. Modified from Badgley and Gingerich (1988).

represents a selective culling, particularly of jaws and teeth, of surface fossil assemblages from restricted stratigraphic intervals. (Mammalian orders comprising predominantly rare taxa, such as Insectivora and Multituberculata, were omitted from consideration altogether). Further details of collecting and cataloguing procedures are given in Badgley and Gingerich (1988). The Wasatchian record spans 700 m of section in the Clark's Fork Basin. Individual localities are measured to the nearest 5 m; data for this study have been pooled over 20-m intervals.

The record of change in sample size and faunal composition in the vicinity of Biohorizon A (see Fig. 4) is the basis for the faunal population and sample sizes of the simulation models. A large peak in sample size, as represented by the number of catalogued specimens, occurs from 1,740 through 1,775 m, an interval that includes the early part of Biohorizon A. Above 1,780 m, sample size is persistently low, with minor increases around 1,840 m and around 2,040 m. The cumulative sample size for intervals 1,780 m and above does not reach values comparable to that of a single 20-m interval between 1,740 and 1,775 m until the interval up to 2,040 to 2,055 m. Since it is over this thick sequence of low sample size that the disappearances of Biohorizon A occur, aggregate sample sizes from this part of the record are most relevant to the simulations.

The disappearances recognized at Biohorizon A in the Clark's Fork Basin (Table 1) include both: (1) taxa not encountered again (terminal disappearances and taxa known only at this interval), and (2) taxa encountered again after a considerable gap (temporary disappearances and taxa represented by one specimen and large gaps). These disappearances are the same as those reported in Badgley and Gingerich (1988) with one exception. A revised stratigraphic range of *Acarictis ryani* (formerly hyaenodontid, n. gen. B, n. sp.) terminates now below Biohorizon A in the Clark's Fork Basin (Gingerich and Deutsch, 1989). Consideration of taxa that apparently disappear and then reappear recognizes the possibility that some species expanded and contracted their geographic ranges over time in response to changes in habitat (Schankler, 1981; Badgley and Gingerich, 1988; Bown and Beard, this volume). (Of course, such changes could occur over shorter time periods than those considered here, but it strains credibility to posit actual absence for every small interval in which a species is not recorded.) Of the 16 species that disappear at Biohorizon A, seven are recorded again at higher stratigraphic

TABLE 1. TAXA INVOLVED IN THE FAUNAL TURNOVER AT BIOHORIZON A IN THE CLARK'S FORK BASIN*

Appearances: 3
Tetonius steini
Homogalax protapirinus
Thosictis martis

Terminal Disappearances: 7
Phenacodus vortmani
Homogalax n. sp.
Thosictis deimia
Didymictis leptomylus
Viverravus bowni
Viverravus n. sp.
Uintacyon rudis

Temporary Disappearances: 5
Arctodontomys lineage (1,960 to 1,975 m)
Paramys copei (2,100 to 2,115 m)
Phenacodus brachypternus (1,960 to 1,975 m)
Paleonictis occidentalis (2,040 to 2,055 m)
Wyolestes apheles (2,080 to 2,095 m)

Taxa known only at this interval: 2
Apheliscus wapitiensis
Galecyon mordax

Taxa represented by one specimen and large gaps above/below: 2
Lophioparamys murinus (2,120 to 2,135 m)
Microparamys n. sp. (2,040 to 2,055 m)

*For taxa that reappear, the interval of reappearance is given in parentheses. Based on Figure 3 and modified from Badgley and Gingerich (1988).

intervals (indicated in Table 1). The *Arctodontomys* lineage and *Phenacodus brachypternus* reappear in the interval 1,960 to 1,975 m, and *Paleonictis occidentalis* and *Microparamys* n. sp. reappear in the interval 2,040 to 2,055 m. The three remaining reappearances occur at higher stratigraphic intervals (Fig. 3). Notably, almost all of the "appearances" above 1,900 m in Figure 4 are reappearances, following gaps of at least 200 m. The pattern of disappearances and reappearances over the interval 1,780 to 2,055 m is the basis for the sample sizes of the simulations.

Simulation models

The interval 1,740 to 1,755 m contains the faunal "population" from which samples are drawn (Fig. 4). This interval contains 45 species of mammals from the biostratigraphic range chart (Fig. 3). Abundances for these species are based on the number of catalogued specimens (N = 1,945) for this interval. This pooled 40-m interval is the best available representation of faunal composition for the simulations because of the large sample size and stratigraphic proximity to the long interval of low sample size. While the two component 20-m intervals do not contain identical species (Table 2), the differences involve mainly species that are quite rare, together making up less than 1 percent of the faunal sample in either subinterval. These differences in composition are themselves plausibly interpreted as arising from sampling fluctuation from a larger parent population of fossilized remains (Koch, 1987).

The sample sizes of the simulations are obtained from three long intervals over which disappearances and reappearances occur. From 1,780 to 1,955 m, 16 disappearances occur at an aggregate sample size of 552 specimens. By 1,975 m, three reappearances are recorded, for a total of 13 disappearances from 1,780 to 1,975 m at an aggregate sample size of 689 specimens. By 2,055 m, two additional reappearances occur, for a total of 11 remaining disappearances from 1,780 to 2,055 m at an aggregate sample size of 1,290 specimens. The purpose of using these sample sizes is to see how closely the simulations can reproduce the record of disappearances and reappearances.

From one faunal population, three sets of samples were drawn (Simulations I, II, and III), each at a different sample size (Table 3). The sampling procedure in the simulations was random sampling without replacement. This procedure accords an equal probability of selection to each individual (here, specimen) remaining in the population. Hence, the abundance of each species in a sample is a function of its abundance in the population. Simulations that produce sampling distributions of estimates, like this one, are Monte Carlo studies (Snedecor and Cochran, 1980). The simulation of subsamples was accomplished by converting the frequency distribution of Table 2 into a cumulative distribution so that each specimen corresponded to a unique integer. Then sets of random numbers—ranging in value between 1 and 1,945—were generated at the desired sample size and sorted into numerical ranges corresponding to species. For each set of simulations, 500 runs were performed. From each set of simulations, I determined the average and modal values of missing species, the sampling distribution of missing species, the average frequency of each species in the runs, and the frequency of species absences in 500 runs.

Paleontological hypotheses and statistical decisions

The results of the simulations facilitate evaluation of paleontological hypotheses about Biohorizon A. The broad paleontological issue is how much of the record of Biohorizon A in the Clark's Fork Basin could be an artifact of sampling fluctuation. More specifically: (1) How many missing species are expected on the basis of sampling variation alone? (2) Does the observed number of disappearances lie within the 95 percent confidence limits of the sampling distribution of missing species from the simulations? (3) If some species *did* disappear (at a specified sample size), which of the recorded species disappearances are most probable in light of the simulation results?

The sampling distribution of missing species indicates how many absences are expected in subsamples of three sizes. For a

TABLE 2. SPECIES ABUNDANCES FROM TWO ADJACENT INTERVALS OF LARGE SAMPLE SIZE*

Species	1,740 to 1,755 m	1,760 to 1,775 m	Combined
Phenacolemur praecox	17	17	34
Phenacolemur simonsi	2	1	3
Arctodontomys lineage	13	4	17
Tetonoides tenuiculus	12	24	36
Tetonius steini	8	47	55
Cantius mckennai	113	126	239
Acritoparamys atwateri	1	1	2
Paramys taurus	5	5	10
Paramys copei	0	1	1
Lophioparamys murinus	1	0	1
Microparamys n. sp.	0	1	1
Phenacodus intermedius	5	8	13
Phenacodus vortmani	4	2	6
Phenacodus brachypternus	7	4	11
Ectocion osbornianus	26	4	30
Hyopsodus loomisi	297	305	602
Haplomylus speirianus	147	167	314
Apheliscus n. sp.	7	5	12
Apheliscus wapitiensis	1	1	2
Homogalax n. sp.	9	10	19
Homogalax protapirinus	7	14	21
Hyracotherium grangeri	145	70	215
Diacodexis metsiacus	58	54	112
Diacodexis robustus	2	0	2
Esthonyx spatularius	20	7	27
Coryphodon spp.	19	9	28
Oxyaena gulo	9	2	11
Paleonictis occidentalis	1	1	2
Arfia shoshoniensis	20	13	33
Galecyon mordax	0	1	1
Prolimnocyon haematus	3	3	6
Thosictis deimia	1	4	5
Thosictis phobia	8	7	15
Thosictis martis	1	3	4
Didymictis proteus	0	2	2
Didymictis leptomylus	0	1	1
Viverravus acutus	3	5	8
Viverravus bowni	4	0	4
Viverravus politus	1	0	1
Viverravus n. sp.	2	0	2
Uintacyon rudis	3	3	6
Miacis deutschi	8	7	15
Thryptacodon antiquus	5	6	11
Dissacus spp.	1	1	2
Wyolestes apheles	1	2	3
Total	997	948	1945

*Abundances based on number of catalogued specimens at University of Michigan Museum of Paleontology.
Combined sample is faunal population for simulation.

TABLE 3. SAMPLE SIZE AND OBSERVED NUMBER OF SPECIES DISAPPEARANCES FOR STRATIGRAPHIC INTERVAL REPRESENTED IN EACH OF THREE SIMULATIONS

Simulation	I	II	III
Stratigraphic interval represented	1,780 to 1,955 m	1,780 to 1,975 m	1,780 to 2,055 m
Sample size	552	689	1,290
Observed number of disappearances	16	13	11

one-tailed test, the analogue of a t-value for the observed number of disappearances is determined from each sampling distribution by summing the number of missing species greater than or equal to the observed value and dividing by 500. When the observed number of disappearances is outside the confidence interval, then I turn to the relative likelihood of species absence versus presence to determine which of the recorded species disappearances are most plausible.

The likelihood of a hypothesis H, given data D and a statistical model, is proportional to P(D|H), the probability of the data under the hypothesis (Edwards, 1972). In the present situation, two hypotheses are mutually exclusive and exhaustive: H_a, the species is absent; and H_p, the species is present and unrepresented by fossils. The likelihood ratio for H_a relative to H_p is given by the expression $P(D|H_a)/P(D|H_p)$. (This expression is the portion of Bayes' Theorem representing the influence of data on alternate hypotheses. If the a priori probabilities of the hypotheses were brought into the analysis, then Bayes' Theorem would be the appropriate hypothesis-testing approach.)

Likelihood ratios were determined for the relevant species in two ways. The probability of observing 0 fossils under H_a is always 1. The probability of observing 0 fossils under H_p depends on the frequency of the species in the population and the size of the sample. The relative frequency of a species' absence in the simulation results is an empirical measure of $P(D|H_p)$. These frequencies in the simulation results are not strictly independent of each other, but the frequencies of the particular species under consideration are so small that their mutual dependence is negligible and does not confound the results of this approach. $P(D|H_p)$ can also be calculated using a Poisson model for the number of fossils found, given a specified frequency of the species and a specified sample size. The Poisson model is appropriate when the sample size is large but the event of interest is rare (Snedecor and Cochran, 1980). Under this model, the probability of observing 0 fossils is given by the expression

$$P(0 \text{ fossils}) = e^{-np},$$

where e is the base of natural logarithms, n is the sample size of the particular simulation, and p is the species' frequency in the collection sampled. These two approaches to evaluating H_p should give similar results.

RESULTS

The simulation results include the distribution of missing species for each set of 500 runs (Fig. 5), the average frequency of each species in each simulation (Table 4), and the relative frequency of species absence in each simulation (Table 5). Likelihood ratios (Table 6) determined from simulation results are based on the relative frequency of species absence (Table 5). The ranking of likelihood ratios (Table 7) indicates which species are most likely to have disappeared.

Simulation results

Figure 5 displays the distribution of missing species for Simulations I, II, and III. In the simulated samples, all missing species (simulated disappearances) are artifacts of sampling variation, since no absences occur in the original faunal population sampled. Hence, the distributions in Figure 5 illustrate effects of sample size alone on apparent changes in faunal composition. As sample size increases from Simulation I to Simulation III (Fig. 5a to c), the entire distribution shifts to the left, corresponding to greater average species representation (fewer species missing) in samples of larger size.

The number of missing species for Simulation I (Fig. 5a) ranges from four to 15 with a mode of 10. The observed value 16—species missing in the record of the Clark's Fork Basin from 1,780 to 1,955 m—falls not only outside the 95 percent confidence interval but outside the distribution altogether (see arrow). As many as 13 missing species lie within the 95 percent confidence interval, but at least three of the 16 species missing cannot be attributed to sampling variation at n = 552. The three taxa most likely to have disappeared over this stratigraphic interval are indicated below by evaluation of likelihood of species absence versus species presence.

The number of missing species in Simulation II (Fig. 5b) ranges from two to 14 with a mode of eight. The observed value 13—from 1,980 to 1,975 m—falls within the 5 percent rejection region of the distribution. Eleven disappearances lie within the 95 percent confidence interval of the distribution, but at least two of the 13 disappearances cannot be attributed to sampling variation at n = 689. The two species most likely to have disappeared are indicated below by determination of likelihood ratios.

The number of species missing in Simulation III (Fig. 5c) ranges from zero to seven with a mode of three. The sample size of 1,290, representing the stratigraphic interval 1,780 to 2,055 m, is approximately two-thirds the size of the original population sampled (n = 1,945) in the simulation. At n = 1,290, a few of the simulated runs contain all the species in the original population. The 11 observed disappearances up to 2,055 m fall well outside the distribution, but as many as five disappearances lie within the 95 percent confidence interval. Hence, at least six disappearances cannot be attributed to sampling variation in this simulation.

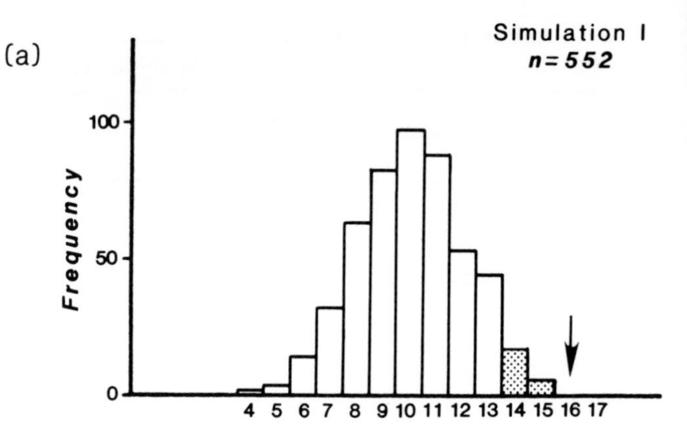

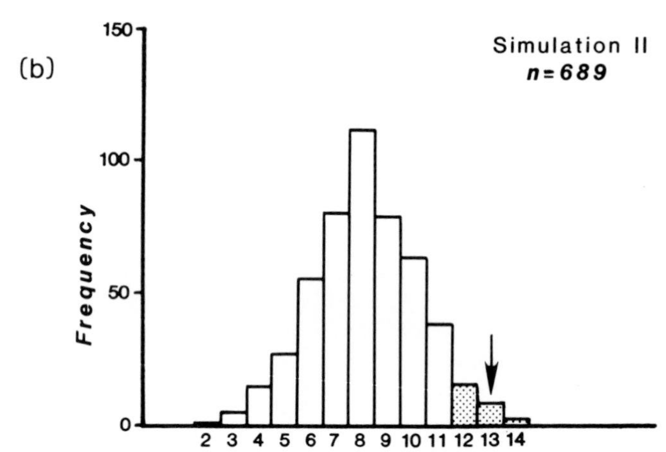

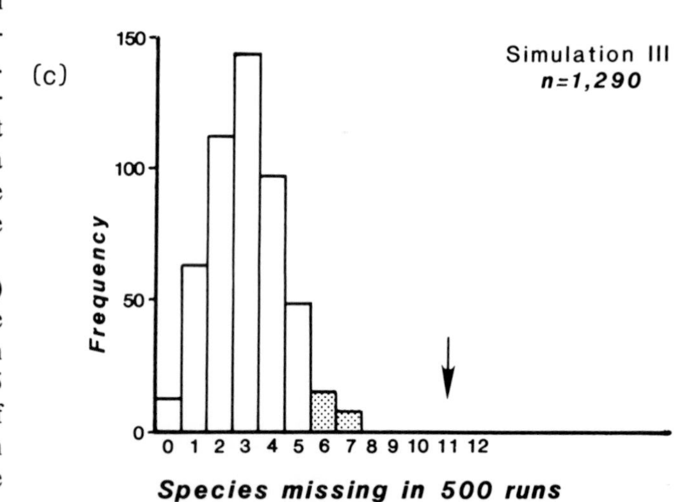

Figure 5. Sampling distributions of missing species for three simulations of 500 runs each (area under histograms = 500). (a) Results of Simulation I, mean = 10.1, mode = 10. (b) Results of Simulation II, mean = 8.1, mode = 8. (c) Results of Simulation III, mean = 3.0, mode = 3.

These simulations do not indicate what must have happened only what is probable under each of the simulation models. As the stratigraphic interval and corresponding time span under consideration become larger, it is more likely that real changes in species composition of the fauna occurred. The applicability of the simulation models, in which only sample size changes from one simulation to another, becomes less credible. When the stratigraphic record contains relatively small aggregate samples over large intervals, then information about faunal change is quite limited. The simulations simply reveal how many disappearances and reappearances would result from changes in sample size alone.

Table 4 contains the average frequency of each species in Simulations I, II, and III. These frequencies reflect the abundance of each species in the original population sampled. In Table 4, no species are absent altogether, even in results of Simulation I, but many species are represented by an average of less than one individual. The number of species with average frequencies less than 1.0 are 15 in Simulation I, 13 in Simulation II, and six in Simulation III. These species were absent from many individual runs. The average frequency of each species increases from Simulation I to III because of the increase in sample size. Nine of the 16 species observed to disappear at Biohorizon A have average frequencies of less than 1 (in 552) in Simulation I.

The relative frequency of species absences (average frequency of absence/sample size of simulation) indicates the percentage of runs in which each species did not occur (Table 5). For each species, these frequencies are inversely proportional to those of Table 4; that is, a species with a high relative frequency of absence in 500 runs has a low average frequency over all runs. A relative frequency of absence of zero means that the species was present in all runs. The number of such species is 15 in results of Simulation I, 20 in Simulation II, and 27 in Simulation III. The increase from Simulation I to III reflects the corresponding increase in sample size. Seven of the 16 species observed to disappear at Biohorizon A have relative frequencies of absence of more than 50 percent.

Likelihood ratios

The value of the likelihood ratio, H_a/H_p, depends on the frequency of the species under consideration, whether the ratio is determined empirically from the simulation results or calculated according to the Poisson model. The relative frequency of species absences (Table 5) is an empirical measure of $P(D|H_p)$, the probability of observing 0 fossils even when the species is present. For example, if a species was never absent in the simulated runs, we would not expect its observed absence in the record to be due to chance. The reciprocal of the relative frqeuency of absence is the value of the likelihood ratio favoring H_a. Table 6 displays likelihood ratios favoring species disappearance based on simulation results, for Simulations I to III, and based on Poisson calculations which depend on species frequencies in the faunal population sampled. Likelihood ratios range from indefinitely large (1/0) to 1, the smallest value possible in this context. For each species, the likelihood ratio increases from Simulation I to III, reflecting the increase in species frequency at larger sample sizes (Table 4). The frequency of tied ranks among the Poisson calculations obscures the general concordance between the simulation results and the Poisson calculations for a given simulation set.

TABLE 4. AVERAGE FREQUENCY OF EACH SPECIES FROM 500 RUNS IN EACH OF THREE SIMULATIONS

Species	Simulation I	Simultion II	Simulation III
Phenacolemur praecox	9.7	12.0	22.9
Phenacolemur simonsi	0.8	1.1	2.0
Arctodontomys lineage	4.8	6.1	11.3
Tetonoides tenuiculus	10.2	12.7	24.1
Tetonius steini	15.8	19.4	36.8
Cantius mckennai	68.5	85.9	159.5
Acritoparamys atwateri	0.6	0.7	1.3
Paramys taurus	2.7	3.5	6.7
Paramys copei	0.3	0.4	0.7
Lophioparamys murinus	0.3	0.4	0.7
Microparamys n. sp.	0.3	0.4	0.7
Phenacodus intermedius	3.7	4.5	8.7
Phenacodus vortmani	1.7	2.2	4.0
Phenacodus brachypternus	3.2	3.7	7.4
Ectocion osbornianus	8.5	10.7	19.8
Hyopsodus loomisi	172.1	215.4	401.6
Haplomylus speirianus	90.5	111.9	209.3
Apheliscus n. sp.	3.4	4.3	8.0
Apheliscus wapitiensis	0.5	0.7	1.3
Homogalax n. sp.	5.4	6.8	12.6
Homogalax protapirinus	5.7	7.6	14.0
Hyracotherium grangeri	61.4	76.6	143.3
Diacodexis metsiacus	30.5	38.0	72.0
Diacodexis robustus	0.6	0.7	1.3
Esthonyx spatularius	7.2	9.1	17.1
Coryphodon spp.	7.4	9.5	18.0
Oxyaena gulo	2.9	3.7	7.1
Paleonictis occidentalis	0.6	0.7	1.3
Arfia shoshoniensis	9.2	11.2	21.3
Galecyon mordax	0.3	0.3	0.6
Prolimnocyon haematus	1.7	2.0	3.9
Thosictis deimia	1.3	1.7	3.2
Thosictis phobia	4.1	5.2	9.6
Thosictis martis	1.1	1.4	2.5
Didymictis proteus	0.6	0.7	1.3
Didymictis leptomylus	0.3	0.4	0.7
Viverravus acutus	2.1	2.7	5.0
Viverravus bowni	1.1	1.4	2.6
Viverravus politus	0.3	0.3	0.6
Viverravus n. sp.	0.5	0.7	1.3
Uintacyon rudis	1.6	2.0	3.8
Miacis deutschi	4.0	5.0	9.7
Thryptacodon antiquus	3.0	3.6	7.1
Dissacus spp.	0.5	0.7	1.3
Wyolestes apheles	0.8	1.0	1.9

Ranking of likelihood ratios (Table 7) for species with a record of disappearance indicates which species are most likely to have disappeared. Under Simulation I, at least four disappearances are likely to have occurred (Fig. 5a). The three lineages with the highest likelihood ratios under Simulation I (Tables 6

TABLE 5. RELATIVE FREQUENCY OF EACH SPECIES' ABSENCE IN 500 RUNS FROM THREE SIMULATIONS

Species	Simulation I	Simulation II	Simulation III
Phenacolemur praecox	0	0	0
Phenacolemur simonsi	0.38	0.27	0.03
Arctodontomys lineage	0	0	0
Tetonoides tenuiculus	0	0	0
Tetonius steini	0	0	0
Cantius mckennai	0	0	0
Acritoparamys atwateri	0.50	0.43	0.10
Paramys taurus	0.04	0.02	0
Paramys copei	0.71	0.60	0.34
Lophioparamys murinus	0.68	0.64	0.29
Microparamys n. sp.	0.72	0.65	0.32
Phenacodus intermedius	0.01	0	0
Phenacodus vortmani	0.14	0.07	0
Phenacodus brachypternus	0.03	0.01	0
Ectocion osbornianus	0	0	0
Hyopsodus loomisi	0	0	0
Haplomylus speirianus	0	0	0
Apheliscus n. sp.	0.01	0	0
Apheliscus wapitiensis	0.53	0.39	0.13
Homogalax n. sp.	0.01	0	0
Homogalax protapirinus	0	0	0
Hyracotherium grangeri	0	0	0
Diacodexis metsiacus	0	0	0
Diacodexis robustus	0.53	0.43	0.14
Esthonyx spatularius	0	0	0
Coryphodon spp.	0	0	0
Oxyaena gulo	0.02	0	0
Paleonictis occidentalis	0.50	0.41	0.12
Arfia shoshoniensis	0	0	0
Galecyon mordax	0.71	0.65	0.35
Prolimnocyon haematus	0.14	0.10	0
Thosictis deimia	0.25	0.13	0.01
Thosictis phobia	0.01	0	0
Thosictis martis	0.26	0.17	0.02
Didymictis proteus	0.51	0.42	0.15
Didymictis leptomylus	0.71	0.62	0.33
Viverravus acutus	0.08	0.05	0
Viverravus bowni	0.29	0.20	0.02
Viverravus politus	0.72	0.66	0.35
Viverravus n. sp.	0.54	0.41	0.11
Uintacyon rudis	0.15	0.08	0
Miacis deutschi	0	0	0
Thryptacodon antiquus	0.02	0.01	0
Dissacus spp.	0.51	0.42	0.11
Wyolestes apheles	0.38	0.29	0.05

and 7) are *Arctodontomys wilsoni-A. nuptus, Phenacodus vortmani,* and *Homogalax* n. sp.-*H. semihians* (Table 8). The *Arctodontomys* lineage actually reappears between 1,960 and 1,975 m; hence, this analysis supports the interpretation that this lineage disappeared and later reappeared. Under Simulation II, at least two disappearances are likely to have occurred (Fig. 5b). The two species with the highest likelihood ratios under Simulation II are *Phenacodus vortmani* and *Homogalax* n. sp. Under Simulation III, at least six species disappearances are likely to have occurred (Fig. 5c). The six species with the highest likelihood ratios under Simulation III are *Phenacodus vortmani, Homogalax* n. sp., *Uintacyon rudis, Thosictis deimia, Viverravus bowni,* and *Wyolestes apheles.* Each simulation is an alternative view of the sampling problem that characterizes the Wasatchian record above 1,780 m. It is to be expected that some species show up as likely real disappearances under each simulation, since the likelihood ratios all reflect the same species abundances in the original population sampled.

DISCUSSION

Species disappearances

The simulation results indicate that a substantial drop in sample size may create disappearances of species in a biostratigraphic record even if no disappearances have actually occurred. The number of apparent losses and their species identities depend on the nature of the species abundance distribution of the population subsampled. A population with few common and many uncommon species, such as the faunal population utilized for these simulations, will appear to lose more species than will a population with many common and few uncommon species. Since many species-abundance distributions in nature are concave, with few common and many uncommon species, the demonstrated potential for apparent species losses at small sample sizes is broadly relevant. The relationship between sample size and species richness is well documented in the ecological literature (e.g., May, 1975; Simberloff, 1979) and is familiar in paleontology (e.g., Raup, 1975; Koch, 1987; Koch and Morgan, 1988). Analyses of biostratigraphic data, however, often do not include information about sample size and are thus vulnerable to erroneous interpretations of faunal change.

In the Wasatchian record of the Clark's Fork Basin, a significant number of species disappearances observed at Biohorizon A could have arisen from sampling variation, according to all three sets of simulations. The present sample sizes above 1,780 m are not large enough to establish confidence that many of the apparent disappearances are real. Among the species missing above 1,780 m, the most likely species to have disappeared are *Arctodontomys* lineage, *Homogalax* n. sp., *Phenacodus vortmani,* and *Phenacodus brachypternus,* according to the likelihood ratios in Tables 6 and 7. These likelihoods reflect the recorded frequencies of these species below 1,780 m. Species with relatively higher original frequencies are less susceptible to absence through sampling variation than species with lower original frequencies. These results do not indicate where between 1,780 and 1,955 m these disappearances occurred. As noted earlier, the *Arctodontomys* lineage (*A. nuptus*) and *Phenacodus brachypternus* reappear between 1,960 and 1,975 m (Table 1, Fig. 3). If the

TABLE 6. LIKELIHOOD RATIOS FAVORING HYPOTHESIS OF SPECIES ABSENCE, FROM SIMULATION RESULTS AND POISSON CALCULATIONS

Species	Simulation I	Simulation II	Simulation III	Poisson
Arctodontomys lineage	1/0	reappear---------------------------		1×10^5
Paramys copei	1.41	1.67	2.94	3.63
Lophioparamys murinus	1.47	1.56	3.45	3.63
Microparamys n. sp.	1.39	1.54	reappear	3.63
Phenacodus vortmani	100	1/0	1/0	47.94
Phenacodus brachypternus	33.33	reappear--------------------------		2×10^3
Apheliscus wapitiensis	1.89	2.56	7.69	3.63
Homogalax n. sp.	100	1/0	1/0	4×10^5
Paleonictis occidentalis	2.00	2.44	reappear	3.63
Galecyon mordax	1.41	1.54	2.86	3.63
Thosictis deimia	4.00	7.69	100	47.94
Didymictis leptomylus	1.41	1.61	3.03	3.63
Viverravus bowni	3.45	5.00	50.00	13.20
Viverravus n. sp.	1.85	2.44	9.09	3.63
Uintacyon rudis	6.67	12.50	1/0	47.94
Wyolestes apheles	2.63	3.45	20.00	13.20

TABLE 7. RANK ORDER OF SPECIES BY LIKELIHOOD RATIOS FAVORING SPECIES ABSENCE, FROM SIMULATION RESULTS AND POISSON CALCULATIONS

Species	Simulation I	Simulation II	Simulation III	Poisson
Arctodontomys lineage	1	--	--	2
Homogalax n. sp.	2	1	1	1
Phenacodus vortmani	2	1	1	4
Phenacodus brachypternus	3	--	--	3
Uintacyon rudis	4	2	1	4
Thosictis deimia	5	3	2	4
Viverravus bowni	6	4	3	5
Wyolestes apheles	7	5	4	5
Paleonictis occidentalis	8	7	--	6
Apheliscus wapitiensis	9	6	6	6
Viverravus n. sp.	10	7	5	6
Lophioparamys murinus	11	10	7	6
Paramys copei	12	8	9	6
Didymictis leptomylus	12	9	8	6
Galecyon mordax	12	11	10	6
Microparamys n. sp.	13	11	--	6

distribution of the faunal population is valid, then these lineages probably disappeared and reappeared—a pattern that characterizes *P. brachypternus* in the Elk Creek section (Schankler, 1980). Comparable taxonomic and biostratigraphic information is not available for *Arctodontomys* in the Elk Creek section.

Critique of the approach

In assessing the relevance of the simulation results, it is appropriate to review the validity of the assumptions underlying the model. A primary assumption is that the faunal population from the interval 1,740 to 1,775 m is a valid representation of faunal composition before the peak of disappearances at 1,780 m. The combined abundances of the 45 species in the faunal population (Table 2) mask differences in composition and abundance between the two component intervals. While most taxa have similar abundances in the two intervals, some taxa exhibit significant increases or decreases. For example, *Tetonoides tenuiculus* and *Tetonius steini* increase in abundance from the lower to upper subinterval; and the *Arctodontomys* lineage and *Hyracotherium grangeri* decrease significantly in abundance over the same subintervals. Thus, treating the combined faunal population as representative is not strictly correct. But does it make a difference in the results? Among the taxa recorded as disappearing (the 16 taxa of Table 1), three exhibit significant decreases from the lower to upper subinterval: *Arctodontomys* lineage, *Phenacodus brachypternus,* and *Viverravus bowni.* The first two are among the most likely real disappearances according to likelihood ratios of absence versus presence (Tables 6 and 7). The decrease from four to zero specimens of *Viverravus bowni* from the lower to upper subinterval suggests that this species has a higher likelihood of real disappearance than represented by the simulation results. *Thosictis deimia* exhibits an increase from one to four specimens over the two adjacent intervals of the faunal population. This increase suggests that the combined value of abundance is an underestimate for this species. If so, its likelihood ratio is also an underestimate. Thus, the minor trends in faunal composition within the component intervals are not greatly at odds with the results.

A second assumption to consider is that random sampling without replacement is the appropriate sampling plan. Under this procedure, each individual in the population can appear only once in a sample. This sampling plan corresponds to the belief that the population of fossils from which samples have been collected is not infinitely large. Since the purpose of the simulations is to discover how subsamples represent the composition of large but finite samples, sampling without replacement is more appropriate than sampling with replacement. Sampling with replacement results in even more missing species in the samples; since the frequencies of common species are never diminished, they dominate samples to a greater extent than they do under sampling without replacement.

Other aspects of the early Wasatchian record strengthen confidence in the approach represented by the simulations. Appearances generally occur in intervals of relatively large sample size (Fig. 4). Disappearances do not become a substantial part of the record until sample size drops and remains low over a long interval. Among the eight appearances that occur during the interval 2,040 to 2,055 m, seven are reappearances of species that have gaps of 200 m or more. These patterns suggest that sampling variation has strongly influenced the Wasatchian record in the Clark's Fork Basin.

Schankler's (1980, 1981) studies of faunal change in the Elk Creek section employed minimum numbers of individuals as the basis of species frequencies rather than number of catalogued specimens, as employed in Badgley and Gingerich (1988) and here. Estimating species frequencies by the minimum number of individuals tends to overestimate the abundances of rare taxa, as demonstrated in simulation experiments of Gilbert and others (1981). Different approaches to quantifying species abundances are appropriate in different taphonomic situations (Badgley, 1986). Schankler's use of the minimum number of individuals was a reasonable choice. I have employed the number of catalogued specimens primarily because some information about minimum number of individuals present at the time of collection (e.g., associated upper and lower jaws) was incorporated into the catalogue. Determining minimum numbers after cataloguing is risky, since for a given fossil locality, information about the precise provenance of specimens is rarely recovered. Unless the association of specimens is maintained at the time of collection, the a priori association of specimens that underlies the minimum-numbers approach is variable. The difference in methods of determining species abundances could have a minor effect on species frequencies used in the simulations but would not cause a major difference in the simulation results.

A final reflection on the approach taken here is to consider what kind of prior information would favor extending the use of likelihood ratios to a Bayesian approach for assessing hypotheses of presence and absence for a species. Briefly, a Bayesian approach to the comparison of hypotheses H_a and H_p would take the form of the likelihood ratios, as demonstrated here, modified by the prior probabilities of the hypotheses to determine the relative posterior probabilities of the hypotheses (e.g., McGee, 1971; Pilbeam and Vaisnys, 1975), according to the expression: $P(H_a|D)/P(H_p|D) = P(H_a)/P(H_p) \times P(D|H_a)/P(D|H_p)$. The posterior probability is the product of the prior probability of the hypothesis and the probability of the data given the hypothesis to be true. The ratio of prior probabilities must be greater than the reciprocal of the likelihood ratio in order to outweigh the effect of the likelihood ratio in determining the relative posterior probabilities.

Information incorporated in prior probabilities is usually empirical. Hypothetically, a correlated early Wasatchian section with similar facies characteristics, similar faunal composition, and consistently high sample sizes could provide plausible prior expectations of species frequencies over the interval correlated with 1,780 to 2,055 m. The record of Biohorizon A in the central Bighorn Basin meets the first two of these characteristics, but there also, sample size falls across Biohorizon A and remains relatively low up to the level of Biohorizon B (Schankler, personal communication, 1986). If in this hypothetical section, some of the species that disappeared at Biohorizon A in the Clark's Fork Basin were present over the correlated interval, then prior probabilities could be adjusted in favor of H_p; if species that disappeared in the Clark's Fork Basin were also absent in the correlated section, prior probabilities could be adjusted in favor of H_a. Alternatively, the appearance of a new kind of predator or evidence for a change in habitat (from paleobotanical or isotopic information) in the section under study could be the basis for modifying prior probabilities. How these prior probabilities would be adjusted is beyond the scope of this paper. The important point is that such information is potentially available.

Factors that affect sample size

Factors that potentially affect fossil productivity of a stratigraphic interval include area of outcrop, distribution of facies, and changes in the composition of the fauna (Badgley and Gingerich, 1988). Decrease in area of exposure at the level of 1,780 m and above in the Clark's Fork Basin has probably contributed significantly to declining sample size. A facies change also occurs around this level; multistoried sandstones over 30 m thick cap localities in the interval of Biohorizon A. This facies is typically unproductive of fossils in the Willwood Formation (Schankler, 1980; Bown and Kraus, 1981; Winkler, 1983). Ecological changes within the biota represent another possible mechanism of change in sample size—for example, through an increase in scavenging species—but there is no clear evidence in support of this factor.

Biohorizon A reconsidered

The record of Biohorizon A in the Clark's Fork Basin offers little certainty about the stratigraphic position or composition of faunal turnover because of the extreme changes in sample size from 1,740 to 1,795 m (Fig. 4). In the central Bighorn Basin, sample size and species richness also rise and fall in concert (Schankler, personal communication, 1988), raising the issue of whether some of the disappearances recorded in the Elk Creek section may also be produced by falling sample size. Appearances and disappearances of species seem not to be as highly correlated with changes in sample size as in the Clark's Fork Basin; this matter remains to be quantified.

The uncertainty of Biohorizon A in the Clark's Fork Basin indicates that a reconsideration of early Wasatchian biostratigraphic correlation between the central Bighorn Basin and the Clark's Fork Basin is in order. In the Elk Creek section, the faunal turnover named Biohorizon A defines the boundary between the Lower and Upper parts of the *Haplomylus-Ectocion* range zone (Schankler, 1980). The *Haplomylus-Ectocion* range zone itself is defined by the concurrent ranges of 13 species. The boundary between the Upper and Lower units is based on the disappearances of eight species and the appearances of seven species. A

TABLE 8. COMPARISON OF BIOSTRATIGRAPHIC RECORDS IN ELK CREEK SECTION (CENTRAL BIGHORN BASIN) AND CLARK'S FORK BASIN, FOR SPECIES THAT PARTICIPATE IN BIOHORIZON A AS ORIGINALLY DEFINED BY SCHANKLER (1980)

Elk Creek Section		Clark's Fork Basin
Extinctions at Biohorizon A:		
Oxyaena gulo		anagenetic change
Viverravus sp. 3		? (*Viverravus bowni* n. sp.)
Esthonyx spatularius		anagenetic change
Coryphodon sp. 1		?
Phenacodus vortmani	R	last appearance ~1,760 m
Phenacodus brachypternus	R	disappears ~1,760 to 1,960 m
Hyopsodus loomisi/minor	R	anagenetic change
Hyracotherium sp. 1	R	anagenetic change
Immigrations at Biohorizon A:		
Anacodon ursidens		not recorded
Didymictis curtidens/lysitensis		appears 1,665 m
Tritemnodon strenuus		? (*Thosictis* sp.)
Vulpavus australis		appears 2,095 m
Hyopsodus latidens/miticulus		anagenetic change
Esthonyx bisulcatus		anagenetic change
Diacodexis robustus		appears 1,570 m

R = Reappearance higher in section

comparison, species by species, of the records of these 15 species in the Elk Creek section and the Clark's Fork Basin reveals significant disparities (Table 8). Six of the 15 species interpreted by Schankler as local extinction or immigration are recorded as anagenetic changes within lineages for those species in the Clark's Fork Basin (Fig. 3; Gingerich and Gunnell, 1979; Gingerich, 1980, 1983b, 1985, 1986; Gunnell, personal communcation, 1986). The placement of some of these anagenetic transitions at Biohorizon A in the Clark's Fork Basin was dependent in part on the supposition of a faunal turnover; these changes within lineages do not provide independent support for the position of faunal change. For *Viverravus* and *Coryphodon,* there is insufficient taxonomic resolution in both records for effective comparison. *Phenacodus vortmani* disappears at Biohorizon A in the Elk Creek section; the present study indicates that *P. vortmani* probably disappeared in the Clark's Fork Basin somewhere between 1,760 and 1,955 m. This species reappears higher in the Elk Creek section but not higher in the Clark's Fork Basin (equivalent strata are missing). *Phenacodus brachypternus* disappears at 1,760 m and reappears (one specimen) at 1,960 m in the Clark's Fork Basin; this species also has a high likelihood of being absent during some part of the interval 1,760 to 1,955 m, according to the present analysis. *Anacodon ursidens* is not recorded in the Clark's Fork Basin. *Didymictis curtidens* and *Diacodexis robustus* appear low in the Wasatchian of the Clark's Fork Basin (Fig. 3); *Vulpavus* cf. *australis* appears at 2,095 m. Specimens once ascribed to the hyaenodontid *Tritemnodon* in the Clark's Fork Basin have been reassigned to the new genus *Thosictis* (Gingerich and Deutsch, 1989; personal communication, 1988). *Thosictis martia* appears at 1,750 m in the Clark's Fork Basin. If material assigned to *Tritemnodon strenuus* in Schankler (1980) actually corresponds to *Thosictis martia,* then the first appearance of this species would support the present correlation between the two areas. In total, the two records of *Phenacodus vortmani* and *P. brachypternus* may be similar in the relative timing of disappearance, and those of *Viverravus* n. sp. and *Thosictis* (= ?*Tritemnodon*) may become similar with further taxonomic resolution. But there is little agreement between the two records for the other 11 species. Thus, it remains to be seen whether Biohorizon A, as originally defined, will have regional biostratigraphic significance, as Woodburne (1987) claims for the broader aspects of Schankler's (1980) zonation.

Another approach to correlation between the two areas is to use the first appearances of *Homogalax protapirinus* and of *Tetonius matthewi/steini.* These taxa appear at 1,750 m in the Clark's Fork Basin (Fig. 3) and both are common after they appear (Table 2). In the Elk Creek section, the first appearances of *H. protapirinus* and *Tetonius matthewi/steini* occur in the vicinity of 100 m, considerably below Biohorizon A (Fig. 3, Schankler, 1980; Fig. 1, Stanley, 1982). The recent, comprehensive study of anaptomorphine primates in the Bighorn Basin supports this position for the first appearance of *Tetonius* in the central Bighorn Basin (Bown and Rose, 1987). Parallel patterns of morphological change within the lineage comprising species of *Tetonius* and *Pseudotetonius* in both the central Bighorn Basin and the Clark's Fork Basin suggest that the stratigraphic position of Schankler's Biohorizon A (200 m) in the Elk Creek section corresponds to approximately 1,860 m in the Clark's Fork Basin. This correlation places Biohorizon A in the long interval of low sample size in the Clark's Fork Basin. Correlation based on these common taxa is more reliable than one based on uncommon taxa. If this revised correlation is correct, then the details of Biohorizon A will be hard to elucidate until sample sizes increase substantially.

The likelihood of relocating this turnover higher in the section of the Clark's Fork Basin addresses a disparity in sediment accumulation rates posed by the original correlation of the Clarkforkian-Wasatchian boundary, Biohorizon A, and Biohorizon B between the central Bighorn Basin and the Clark's Fork Basin (Schankler, 1980, and personal communication, 1988). This correlation results in significantly higher sediment accumulation rates in the Clark's Fork Basin between 1,780 and 2,100 m relative to the earlier Wasatchian interval from 1,520 to 1,775 m. While plausible geological and paleontological mechanisms could result in such a change in sediment accumulation rate, a revised correlation may eliminate the disparity.

CONCLUSION

This simulation of disappearances in the Eocene record of mammals illustrates the potential effect of changing sample size on the apparent record of faunal change. The implications for biostratigraphic and paleoecological studies of faunal change through time are that common taxa should be used to define and characterize biostratigraphic zonation and that sample size should be monitoried in tandem with faunal change. A species turnover comprising either mainly disappearances or mainly appearances should be suspect as being a product of changes in sample size through the stratigraphic section. Carnivorous species, typically rare in terrestrial systems because of their position high in the trophic hierarchy, are unreliable indicators of faunal change unless they are persistently common within their documented ranges.

Large changes in sample size through time are likely to produce apparent faunal turnovers. Increases in sample size should produce first appearances or reappearances of taxa, and decreases in sample size should produce disappearances or notable gaps. The taxa most severely affected will be those that are uncommon even when sample sizes are large. This sampling principle explains why so many of the apparent disappearances at Biohorizon A involve carnivores of the Creodonta and Carnivora. Species abundance distributions of natural communities indicate that most species in nature are uncommon, so apparent faunal turnovers are potentially common. They are easily recognized if sample size is monitored along with the record of faunal change. Recognition of sampling effects permits analysis of real rather than apparent patterns of faunal change.

ACKNOWLEDGMENTS

Several people made valuable contributions to this study. Philip Gingerich provided information resulting from many years of work in the Clark's Fork Basin. Fred Bookstein gave substantial advice and insights regarding statistical analysis. David Schankler gave a thoughtful review of the manuscript and made available unpublished information. David Schultz contributed advice and help in programming. Fred Bookstein, Philip Gingerich, David Schankler, and Gerald Smith read and improved the manuscript with their comments. The figures were drawn primarily by Karen Klitz, with later modifications by Shayne Davidson and Bonnie Miljour. This research was supported in part by the National Science Foundation under grant EAR 83-05931.

REFERENCES CITED

Archibald, J. D., Gingerich, P. D., Lindsay, E. H., Clemens, W. A., Krause, D. W., and Rose, K. D., 1987, First North American land mammal ages of the Cenozoic Era, *in* Woodburne, M. O., ed., Cenozoic mammals of North America: Berkeley, University of California Press, p. 24–76.

Badgley, C., 1986, Counting individuals in mammalian fossil assemblages from fluvial environments: Palaios, v. 1, p. 328–338.

Badgley, C., and Gingerich, P. D., 1988, Sampling and faunal turnover in early Eocene mammals: Palaeogeography, Palaeoclimatology, Palaeoecology, v. 63, p. 141–157.

Bartels, W. S., 1983, A Paleocene-Eocene reptile fauna from the Bighorn Basin, Wyoming: Herpetologica, v. 39, p. 359–374.

Bown, T. M., and Kraus, M. J., 1981, Vertebrate fossil-bearing units (Willwood Formation, lower Eocene, northwest Wyoming, USA); Implications for taphonomy, biostratigraphy, and assemblage analysis: Palaeogeography, Palaeoclimatology, Palaeoecology, v. 34, p. 31–56.

Bown, T. M., and Rose, K. D., 1987, Patterns of dental evolution in early Eocene anaptomorphine primates (Omomyidae) from the Bighorn Basin, Wyoming: Journal of Paleontology, v. 61, Supplement to no. 5, p. 1–162.

Edwards, A.W.F., 1972, Likelihood; An account of the statistical concept of likelihood and its application to scientific inference: Cambridge, Cambridge University Press, 235 p.

Gilbert, A. S., Singer, B. H., and Perkins, D., Jr., 1981, Quantification experiments on computer-simulated faunal collections: Ossa, v. 8, p. 79–94.

Gingerich, P. D., 1980, Evolutionary patterns in early Cenozoic mammals: Annual Review of Earth and Planetary Sciences, v. 8, p. 407–424.

—— , 1983a, Paleocene–Eocene faunal zones and a preliminary analysis of Laramide structural deformation in the Clark's Fork Basin, Wyoming: Wyoming Geological Association 34th Annual Field Conference Guidebook, p. 185–195.

—— , 1983b, Systematics of early Eocene Miacidae (Mammalia, Carnivora) in the Clark's Fork Basin, Wyoming: Ann Arbor, University of Michigan Contributions from the Museum of Paleontology, v. 26, p. 197–225.

—— , 1985, Species in the fossil record; Concepts, trends, and transition: Paleobiology, v. 11, p. 27–41.

—— , 1986, Early Eocene *Cantius torresi;* Oldest primate of modern aspect from North America: Nature, v. 320, p. 319–321.

Gingerich, P. D., and Deutsch, H., 1989, Systematics and evolution of early Eocene Hyaenodontidae (Mammalia, Creodonta) in the Clark's Fork Basin, Wyoming: Ann Arbor, University of Michigan Contributions from the Museum of Paleontology (in press).

Gingerich, P. D., and Gunnell, G. F., 1979, Systematics and evolution of the genus *Esthonyx* (Mammalia, Tillodontia) in the early Eocene of North America: Ann Arbor, University of Michigan Contributions from the Museum of Paleontology, v. 25, p. 125–153.

Gingerich, P. D., Rose, K. D., and Krause, D. W., 1980, Early Cenozoic mammalian faunas of the Clark's Fork Basin–Polecat Bench area, northwestern Wyoming, *in* Gingerich, P. D., ed., Early Cenozoic paleontology and stratigraphy of the Bighorn Basin, Wyoming: Ann Arbor, University of Michigan Paper on Paleontology, no. 24, p. 51–64.

Gunnell, G. F., 1985, Systematics of early Eocene Microsyopinae (Mammalia, Primates) in the Clark's Fork Basin, Wyoming: Ann Arbor, University of Michigan Contributions from the Museum of Paleontology, v. 27, p. 51–71.

Ivy, L. D., 1982, Systematics and biostratigraphy of the earliest North American Rodentia (Mammalia), latest Paleocene and early Eocene of the Clark's Fork Basin, Wyoming [M.S. thesis]: Ann Arbor, University of Michigan, 135 p.

Koch, C. F., 1987, Prediction of sample size effects on the measured temporal and geographic distribution patterns of species: Paleobiology, v. 13, p. 100–107.

Koch, C. F., and Morgan, J. P., 1988, On the expected distribution of species' ranges: Paleobiology, v. 14, p. 126–138.

Krishtalka, L., and 10 others, 1987, Eocene (Wasatchian through Duchesnean) biochronology of North America, *in* Woodburne, M. O., ed., Cenozoic mammals of North America: Berkeley, University of California Press, p. 77–117.

May, R., 1975, Patterns of species abundance and diversity, *in* Cody, M. L., and Diamond, J. M., eds., Ecology and evolution of communities: Cambridge, Massachusetts, Belknap Press, p. 81–120.

McGee, V. E., 1971, Principles of statistics: New York, Appleton-Century Crofts, 373 p.

Pilbeam, D., and Vaisnys, J. R., 1975, Hypothesis testing in paleoanthropology, *in* Tuttle, R. H., ed., Paleoanthropology, morphology, and paleoecology: The Hague, p. 3–13.

Raup, D. M., 1975, Taxonomic diversity estimation using rarefaction: Paleobiology, v. 1, p. 333–342.

Rose, K. D., 1980, Clarkforkian Land-Mammal age; Revised definition, zonation, and tentative intercontinental correlations: Science, v. 208, p. 744–746.

—— , 1981a, The Clarkforkian Land-Mammal Age and mammalian faunal composition across the Paleocene–Eocene Boundary: Ann Arbor, University of Michigan Papers on Paleontology, no. 26, 197 p.

—— , 1981b, Composition and species diversity in Paleocene and Eocene mammal assemblages; An empirical study: Journal of Vertebrate Paleontology, v. 1, p. 367–388.

Rose, K. D., and Bown, T. M., 1984, Gradual phyletic evolution at the generic level in early Eocene omomyid primates: Nature, v. 309, p. 250–252.

Schankler, D. M., 1980, Faunal zonation of the Willwood Formation in the Central Bighorn Basin, Wyoming, *in* Gingerich, P. D., ed., Early Cenozoic paleontology and stratigraphy of the Bighorn Basin, Wyoming: Ann Arbor, University of Michigan Paper on Paleontology, no. 24, p. 99–110.

—— , 1981, Local extinction and ecological re-entry of early Eocene mammals: Nature, v. 293, p. 135–138.

Simberloff, D., 1979, Rarefaction as a distribution-free method of expressing and estimating diversity, *in* Grassle, J. F., Patil, G. P., Smith, W. K., and Taillie, C., eds., Ecological diversity in theory and practice: Fairland, Maryland, International Cooperative Publishing House, p. 159–176.

Snedecor, G. W., and Cochran, W. G., 1980, Statistical methods, 7th ed.: Ames, Iowa State University Press, 507 p.

Stanley, S. M., 1982, Macroevolution and the fossil record: Evolution, v. 36, p. 460–473.

Winkler, D. A., 1983, Paleoecology of an early Eocene mammalian fauna from paleosols in the Clark's Fork Basin, northwestern Wyoming (USA): Palaeogeography, Palaeoclimatology, Palaeoecology, v. 43, p. 261–298.

Woodburne, M. O., 1987, Mammal ages, stages, and zones, *in* Woodburne, M. O., ed., Cenozoic mammals of North America: Berkeley, University of California Press, p. 18–23.

Manuscript Accepted by the Society June 12, 1989

Printed in U.S.A.

Geological Society of America
Special Paper 243
1990

Geology, vertebrate fauna, and paleoecology of the Buck Spring Quarries (early Eocene, Wind River Formation), Wyoming

Richard K. Stucky*, Leonard Krishtalka, and Andrew D. Redline
Section of Vertebrate Fossils, Carnegie Museum of Natural History, 4400 Forbes Avenue, Pittsburgh, Pennsylvania 15218

ABSTRACT

The Buck Spring Quarries, located in the southern part of the type area of the Lost Cabin Member of the Wind River Formation, Wind River Basin, Wyoming, provide one of the richest assemblages of fossil vertebrates known from the latest Wasatchian Land-Mammal Age (ca. 50.5 Ma, Lostcabinian, early Eocene) of North America. More than 100 species of mammals, reptiles, birds, amphibians, and fishes are known. The quarries uniquely preserve associated skeletal remains, and complete skulls and dentitions of a large percentage of the vertebrates. The fossils come from a 2-m-thick sequence, which is composed primarily of mudstones, bioturbated limestone lenses, and laminated limestone/mudstone couplets. These sediments were deposited in a well-drained swamp or ponded area between 250 and 600 m away from a low-sinuosity stream. The fossils are especially common in limestones, where they accumulated as a result of (1) natural death (articulated specimens), (2) predator activity (coprolites and kill sites), and (3) very limited hydraulic transport of smaller bones.

Rarefaction estimates from surface and quarry collections suggest similar patterns of species richness, which are among the highest known for the Paleogene and compare favorably with penecontemporaneous Lostcabinian assemblages and slightly younger Gardnerbuttean (early Bridgerian) ones. The mammalian assemblage is dominated by small species of mammals, and has a body-size distribution and species diversity similar to modern tropical communities. The abundance of arboreal mammals indicates that a multistoried woodland habitat was in close proximity to the quarry area.

INTRODUCTION

Since 1975, paleontologists from the Carnegie Museum of Natural History have conducted an intensive survey for fossil vertebrates in the Wind River Formation of the northeastern Wind River Basin, Wyoming. During 1984, a very rich fossiliferous horizon, the "B-2 horizon," of late early Eocene age (latest Wasatchian Land-Mammal Age, Lostcabinian Subage), was discovered near Buck Spring in the type area of the Lost Cabin Member of the formation at locality K-6 (CM loc. 1040).

Since the discovery, extensive quarrying operations have been conducted at two sites in the B-2 horizon, Quarry-1 and Quarry-6 (Q-1, Q-6), which have produced an exceptional number of well-preserved early Eocene vertebrates. These quarry sites and other areas will continue to be highly productive in years to come. More than 100 species of mammals, reptiles, birds, amphibians, and fishes are now known from the B-2 horizon. Many of the species from the quarries are represented by their first known skulls and partial skeletons. Importantly, the smaller mammals and lizards, which are uncommon or rare elsewhere in

*Present address: Department of Earth Sciences, Denver Museum of Natural History, 2001 Colorado Boulevard, Denver, Colorado 80205.

Stucky, R. K., Krishtalka, L., and Redline, A. D., 1990, Geology, vertebrate fauna, and paleoecology of the Buck Spring Quarries (early Eocene, Wind River Formation), Wyoming, *in* Bown, T. M., and Rose, K. D., eds., Dawn of the Age of Mammals in the northern part of the Rocky Mountain Interior, North America: Boulder, Colorado, Geological Society of America, Special Paper 243.

the Wind River Formation, are abundant in the quarry assemblage. The fossiliferous rocks of the B-2 horizon represent a unique depositional environment that is dominated by otherwise scarce limestone units.

We are currently in the process of a complete analysis of the Wind River faunas, geology, and paleoecology (Stucky and Krishtalka, 1982, 1983; Korth, 1982; Krishtalka and Stucky, 1983, 1985; Stucky, 1984a, 1984b; Dawson and others, 1986; Hirsch and others, 1987; Stucky and others, 1987). The fossil materials from the Buck Spring Quarries will be central to the systematic revisions of all fossil vertebrates from the Wind River Formation. Much new anatomical information will be added to knowledge of early Eocene vertebrates.

This chapter provides the geological and paleobiological background on the Buck Spring Quarries, including: (1) a synopsis of the geology of the region, (2) an analysis of the depositional environments that account for the fossil accumulation, and (3) a preliminary summary of the mammalian assemblage and its paleoecology.

GEOLOGIC AGE OF THE ASSEMBLAGE

Stucky and Krishtalka (Stucky and Krishtalka, 1983; Stucky, 1984c; Krishtalka and others, 1987) have discussed in detail the age and relations of the fauna and sediments of the Lost Cabin Member in the Buck Spring area. The fauna from here, along with those from other areas in the Wind River Basin, is used to define the *Lambdotherium* Range Zone in the Wind River Formation (Stucky, 1984a). This zone is based on the first stratigraphic occurrence of *Lambdotherium popoagicum* at its lower boundary and the first appearance of *Eotitanops borealis, Hyrachyus* sp., and *Trogosus* sp. at its upper boundary; the latter also defines the lower boundary of the *Eotitanops* (= *Palaeosyops* in Stucky, 1984a) *borealis* Assemblage Zone. The *Lambdotherium* Range Zone is used to define the Lostcabinian Land-Mammal Subage, which represents the latest part of the Wasatchian Land-Mammal Age. The Lostcabinian is approximately latest early Eocene in age (Wood and others, 1941) and ranges from circa 50.5 to 51.0 Ma (Krishtalka and others, 1987). Faunas of Lostcabinian age also occur in the Green River and Bighorn basins of Wyoming, the Uinta basin of Utah, and the Piceance Creek and Huerfano basins of Colorado (Stucky, 1984c).

Recent stratigraphic studies of Lostcabinian faunas in the Wind River Basin allow further division into early and late parts that are defined by the presence of *Niptomomys* and *Loveina* in early Lostcabinian strata and by the presence of *Uintasorex* and *Shoshonius* in late Lostcabinian strata of the Wind River Formation. The presence of the latter two genera and the occurrence of mammals of middle Eocene aspect (*Eotitanops, Hyrachyus*) above the Buck Spring Quarries horizon (see Stucky, 1984a; Krishtalka and others, 1987) suggest that the assemblage is late Lostcabinian in age and very near the early/middle Eocene boundary.

Regional geology—Wind River Formation, northeastern Wind River Basin

The early Tertiary geology and paleontology of the northeastern Wind River Basin have been studied for the past century (for historical review see Stucky and Krishtalka, 1982; Stucky, 1984a). Thick, fossiliferous sediments of Eocene age, including the Wind River and Wagon Bed Formations, have provided rich assemblages of fossil vertebrates that serve as current standards of biostratigraphic comparison for much of the latter two-thirds of the Eocene (Krishtalka and others, 1987). In addition, rich vertebrate assemblages of late Paleocene age (Krishtalka and others, 1975) and middle Oligocene age (Setoguchi, 1978) have been collected. Figure 1 is a geologic map of the areas discussed in this chapter.

The Wind River Formation is the most extensively exposed Paleogene formation in this part of the Wind River Basin. Wind River sediments were deposited from the middle early to early middle Eocene (ca. 52 to 50 Ma) as the Big Horn Mountains were thrust to the south and the Owl Creek Mountains were uplifted during Laramide orogeny (Keefer, 1965, 1970). As a result of this tectonic activity, sediments of Mesozoic, Paleozoic, and Precambrian age were eroded sequentially from mountain highlands and deposited in the basin interior, accumulating more than 2,000 m of Wind River sediments along the basin axis. A composite section of approximately 500 m is exposed in the northeastern Wind River Basin.

The Wind River Formation unconformably overlies the Indian Meadows (early Eocene) and Fort Union (late Paleocene) Formations in the Badwater Creek area (Love, 1978), and the Indian Meadows Formation and Mesozoic strata in the Deadman Butte–Casper Arch area near Arminto (Stucky, 1984a; See Keefer, 1965). Along the northern border of the region, the Wind River Formation is in fault contact with sediments of the Wagon Bed Formation (middle and late Eocene) and undivided Oligo-Miocene rocks. In addition to Quaternary sediments, the formation is overlain by a middle Eocene volcaniclastic paraconglomerate that may be related to sediments in the lower part of the Wagon Bed Formation along Beaver Divide and at the base of Lysite Mountain (Stucky, 1984a; unpublished data).

In general, two major facies of the Wind River Formation can be recognized (Keefer, 1965): basinward and mountainward. Mountainward facies of the formation occur only around the periphery of the basin in the vicinity of present-day outcrops of pre-Tertiary sediments and are dominated by clastic debris deposited as alluvial fans and braidplains. Basinward facies occur throughout the interior of the basin and represent sediments deposited on flood plains and in channels of low-sinuosity, meandering streams as well as in paludal and lacustrine environments (Seeland, 1978a). Each of the two members of the Wind River Formation in the northeastern part of the basin, the Lysite and Lost Cabin Members (Sinclair and Granger, 1911; also see Tourtelot, 1948; Korth, 1982; Stucky and others, 1987), shows gradation and interfingering from one facies into the other (Fig.

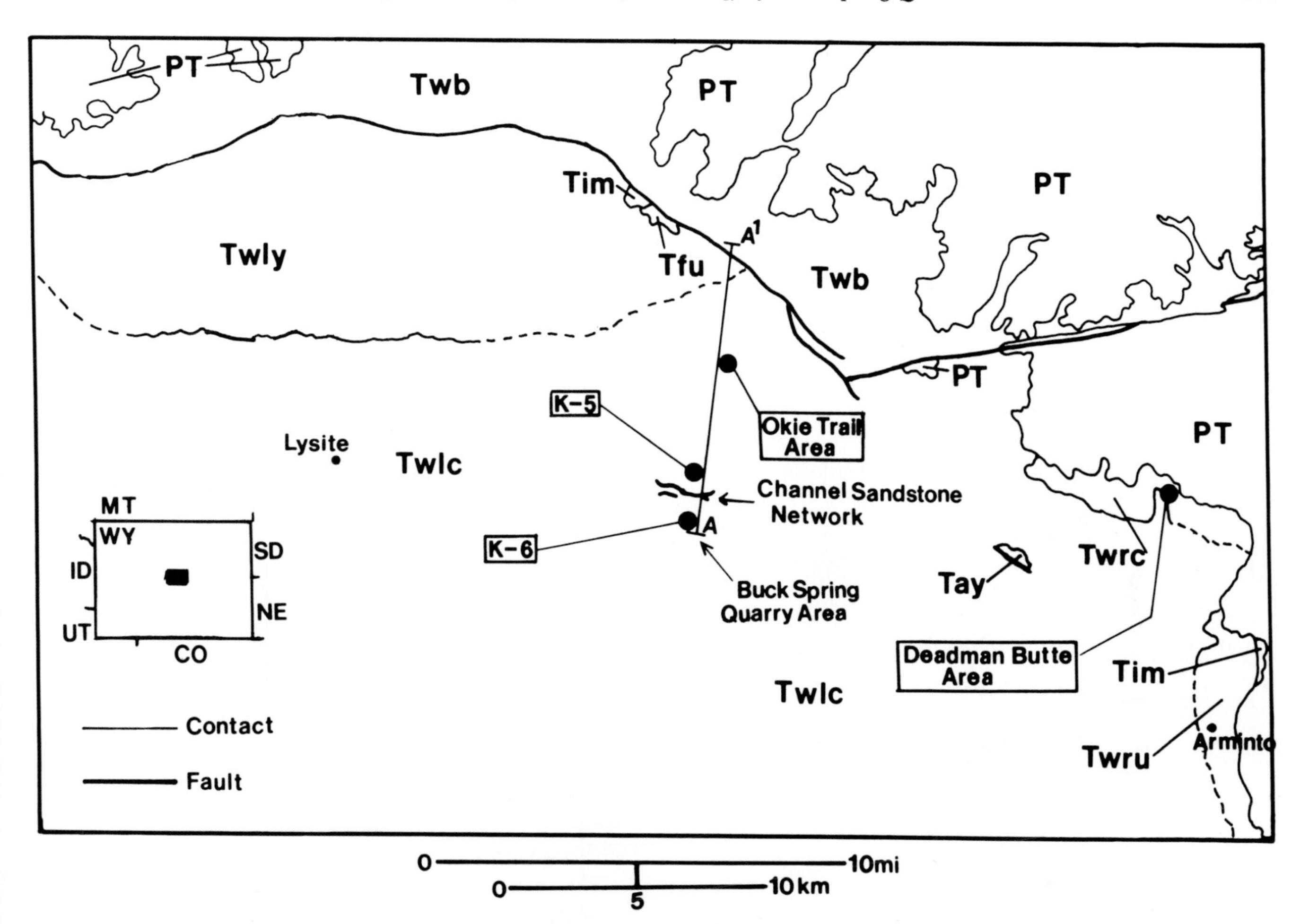

Figure 1. Geology of the northeastern Wind River Basin, Wyoming, showing inset map of location of study area and the areas discussed in the text. Stratigraphic sections shown in Figure 2 are along the transect line, A to A^1. Key to geologic units: PT, pre-Tertiary rocks; Tay, lower Wagon Bed Formation (middle Eocene); Tfu, Fort Union Formation (Paleocene); Tim, Indian Meadows Formation (early Eocene); Twb, Wagon Bed Formation (middle and late Eocene); Twly, Lysite Member, Wind River Formation (early Eocene); Twlc, Lost Cabin Member, Wind River Formation (early to middle Eocene); Twru, Wind River Formation undivided.

2). Lateral changes between these facies are best expressed in the Lost Cabin Member in the central portion of T.39N.,R.87W. to R.89W. Only the geology of the Lost Cabin Member is discussed here.

The Buck Spring Quarries lie approximately 12 km away from the Big Horn Mountain front in Lost Cabin sediments included in the basinward facies. The quarry sites occur almost directly above the east-west–trending axis of the Wind River Basin. The lateral changes between the two facies of the Lost Cabin Member are important for reconstructing the depositional setting and paleoenvironments that account for the unique preservation of the vertebrate fossils from the quarry sites.

Mountainward facies. Near the basin periphery in the Okie Trail area, the Lost Cabin Member is dominated by poorly sorted coarse-grained sandstones, cobble and boulder conglomerates (derived from Precambrian and Paleozoic rocks), and variegated sandy mudstones (alternating bands of red and gray color). The sandstones in this area are tabular sheets that vary in thickness from 0.6 to 5.0 m. Shallow channel troughs, no more than 2 m deep and from 5 to 150 m wide, occur at the base of many of the sandstones and usually contain clast-supported cobble conglomerates. Clast size in both conglomerates and sandstones decreases away from the mountain front into the basin interior. Cobble conglomerates all but disappear some 4 to 6 km into the basin interior. Indications of paleosol development in some sandstones and mudstones in the Okie Trail area include red coloration, color mottling, rhizoliths and burrows, and concentrations of calcareous glaebules and calcite cement below the top of the units. These paleosol horizons average 1 m in thickness and differ from those in the basinward facies because of their development on coarser parent materials. Fossil vertebrates are rare in the sandstones and mudstones of this area, except in one paleosol. The lithologic patterns of the mountainward facies persist across the front of the Big Horn Mountains and continue westward

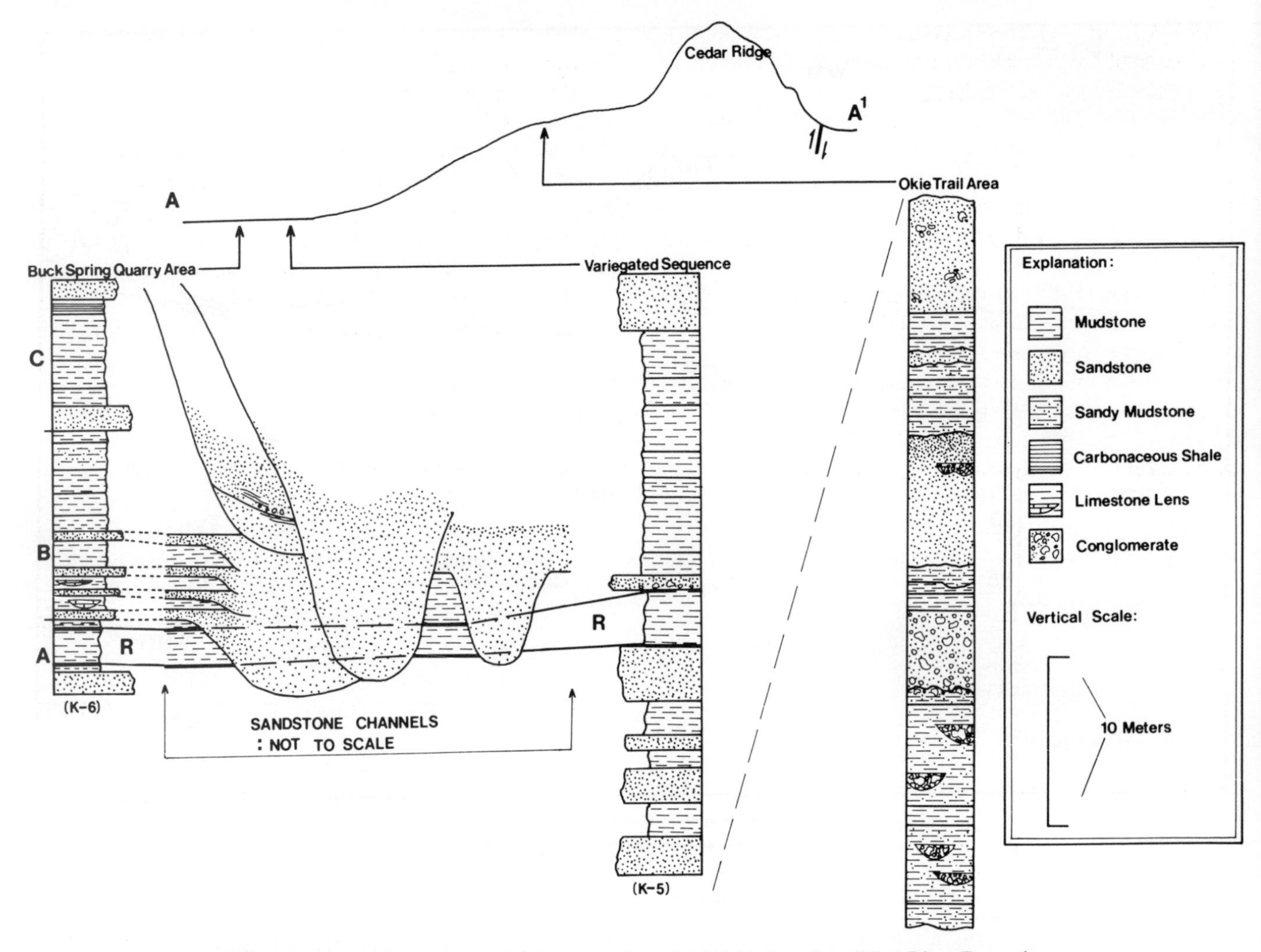

Figure 2. Stratigraphic relations of sections in the Lost Cabin Member of the Wind River Formation, measured along the transect, A to A^1, shown in Figure 1. Sediments of the mountainward facies are coarser grained in the Okie Trail area than are the sediments of the basinward facies in the Buck Spring area at K-5 and K-6. The horizon R at the base of the sequence in both areas is the spectrally mapped "dark red stratum" that occurs throughout the Buck Spring area.

along Cedar Ridge into exposures of the Lysite Member at the eastern end of the Owl Creek Mountains.

Basinward facies. Toward the basin interior, in the Buck Spring area, the lithology of the Lost Cabin Member shifts from the coarser deposits of the Okie Trail area to finer grained sediments dominated by variegated silty mudstones, gray mudstones and claystones, and predominantly medium to fine-grained sandstones. Sandstone body geometry changes from tabular to apron-channel form (see Bown, 1979) with deep channel incisions into underlying mudstones (Seeland, 1978a; Stucky, 1984a). Some sandstones contain intraformational conglomerates (gray mudstone pebbles, abraded bone fragments and isolated teeth) at their base. Allochthonous clasts were derived from Precambrian rocks and rarely exceed pebble size. Mudstones contain very few quartz sand particles. Paleosol horizons are similar to those in more mountainward areas but are developed more commonly on silty mudstones. Horizonation and color mottling are also more apparent in basinward paleosols (Stucky, 1984a, 1984b), and calcareous glaebules are generally smaller. Rarely, carbonaceous shales and limestones occur lateral to channel sandbodies. Carbonaceous shales are also present at the base of some channel sandbodies. Fossil vertebrates are often extremely common in red and gray mudstones.

Approximately 10 km to the south and west of the Buck Spring area, from Moneta to Pavilion Butte, large-scale multistoried sheet sandstones dominate basinward sediments (see Kraus and Middleton, 1987).

Geology of Buck Spring Quarries—Associated strata

The Buck Spring Quarries are located in a series of badland exposures within the type area of the Lost Cabin Member of the Wind River Formation (Sinclair and Granger, 1911; Tourtelot,

1948; Stucky, 1984a). These exposures consist of between 45 and 65 m of strata that are laterally continuous for 6 km north into the Okie Trail area and for 15 km west, slightly southwest of the town of Lysite. The Buck Spring area encompasses the eastern edge of this area and is divided into three subareas for the purposes of this paper (transect line A to A1, Figs. 1 and 2): from north to south, the K-5 area (CM locality 1039); the channel sandstone network; and the K-6 and quarry area (CM locality 1040). All quarry excavations have been conducted within a 6-hectare area (K-6) that lies between 250 and 600 m to the south of the channel sandstone network.

Channel sandstone network. The channel sandstone network includes four distinct channel sandbodies that have been mapped by conventional techniques and by remote-sensing images based on Thermal Infrared Multispectral Scanner data (TIMS; Plate I; Stucky and others, 1987). Three of the four channel sandbodies are distinct ribbons in exposures in the western part of the Buck Spring area. These sandbodies converge, however, approximately 1.0 km northwest of the quarry excavations where they are multistoried and multilateral, forming a "sheet-like" sandbody. From this point they continue along an east-west band for 1.3 km where they again diverge into ribbons and are partly obscured by vegetation immediately to the northeast of the quarry area. Although mapping of the individual channel sandbodies is incomplete, one ribbon sandbody has been traced for a linear distance of 7.2 km, and has a sinuosity ratio of 1.22 (see Plate I; Schumm, 1963).

Each channel sandbody is 8 to 25 m thick at its center and 50 m to probably as much as 100 m wide. Well-exposed sandbodies are of apron-channel geometry with symmetric channel cuts, thinning laterally into stringer sandstones that are interbedded with mudstones in the K-5 and K-6 areas.

The channel sandstones show large-scale trough cross-bedding and minor convolute bedding in their lower parts and grouped sets of planar cross-bedding above, with localized bioturbation at the top of the upper sets. Multiple preferentially cemented, linear tubelike arrays of small-scale foreset beds are often preserved in surface exposures of the upper parts of channel sandbodies. These structures parallel the trend of the ribbon sandbodies and can be followed for a distance greater than the width of the sandbody. Orientation of foreset beds indicates that the direction of flow was to the east (see Seeland, 1978b). Three of the channel sandbodies are subarkosic arenites, composed of quartz (37 to 46 percent) and feldspar (6 to 8 percent) grains and calcite cement (35 to 40 percent); the presence of feldspar grains suggests close proximity to the source area. A fourth channel sandbody contains a large proportion of intraformational material (16.5 percent, mostly sand-size mudclasts) in addition to quartz grains (64 percent) and calcite cement (5 percent) throughout most of its extent. A 1.0- to 2.5-m-thick matrix-supported intraformational conglomerate composed of subangular, friable cobbles of carbonaceous sandstone and mudstone and claystone pebbles crops out persistently at the base of this sandbody.

The integrity of each individual channel sandstone west of the quarry area suggests that streams were laterally fixed and that major stream movement in that area was by avulsion rather than by channel migration (Kraus and Middleton, 1987). Based on observed stratigraphic relations, movement was to the north, across a swath about 2 km wide toward the basin periphery during the interval of time represented. Movement by avulsion of the same stream system was much more restricted just to the north of the quarry area, as suggested by the multistoried/multilateral character of the channel sandstone complex.

K-5 area. Extensive exposures of the Lost Cabin Member lie north of the sandstone complex in the K-5 area. This area has been illustrated in Osborn (1929, Fig. 47) and Stucky (1988c, upper right, p. 16) and represents the stratigraphic sequence that typifies the Lost Cabin Member (Granger, 1910; Tourtelot, 1948; Stucky, 1984a, 1984b). The K-5 exposures are characterized throughout by variegated layers of red and drab silty mudstones. Individual bands rarely exceed 2 m in thickness. Many red mudstones show red/gray/green mottling near their tops, horizonation, slickenside surfaces between sediment blocks, calcareous glaebules, and bioturbation features (rhizoliths and burrows) that strongly suggest they are paleosols (Stucky, 1984a, 1984b; see Bown and Kraus, 1981a, 1981b, 1987). Gray mudstones often show primary depositional features (graded bedding and planar laminae) and sharp contacts with adjacent units. Minor occurrences of rhizoliths and calcareous glaebules in the gray units suggest that they were subaerially exposed for a short period of time prior to their burial by overlying units. The upper surfaces of both red and gray mudstones are flat, indicating little topographic relief in the area during the period of their deposition.

The clay fraction of most of the red and gray mudstones is dominated by montmorillonite, mainly smectite, with a minor amount of illite. One 2-m-thick red mudstone, however, shows a distinct change in clay composition from a dominance of smectite at the base, to smectite and illite in the middle, to kaolinite and smectite at the top. The upper third of this unit shows marked gray-green-yellow-red color mottling and rhizoliths and the lower third contains a high proportion of calcite. The lower two-thirds of this unit is dark red in color. This bed is the same as the "dark red stratum" of Granger (1910) and the "maroon shale" of Guthrie (1971; see Stucky, 1984a; hereafter referred to as the "dark red stratum"). The "dark red stratum" is a paleosol with unique spectral properties that allow it to be mapped on TIMS remote sensing images (Stucky and others, 1987). Based on field studies and image data, the "dark red stratum" crops out throughout the Buck Spring area and can be used for correlating exposures across either the channel sandstone network or vegetated areas. A 0.2- to 0.5-m-thick tabular greenish-gray sandstone lies above the "dark red stratum" in the K-5 exposures. Pockets of conglomerate, composed of abraded fossil bone and Precambrian pebble clasts, occur at the base of this sandstone in small 3- to 5-m-wide scours cut into the top of the "dark red stratum" to a depth of approximately 0.25 m. The base of the greenish-gray sandstone is thus an erosion surface on the "dark red stratum."

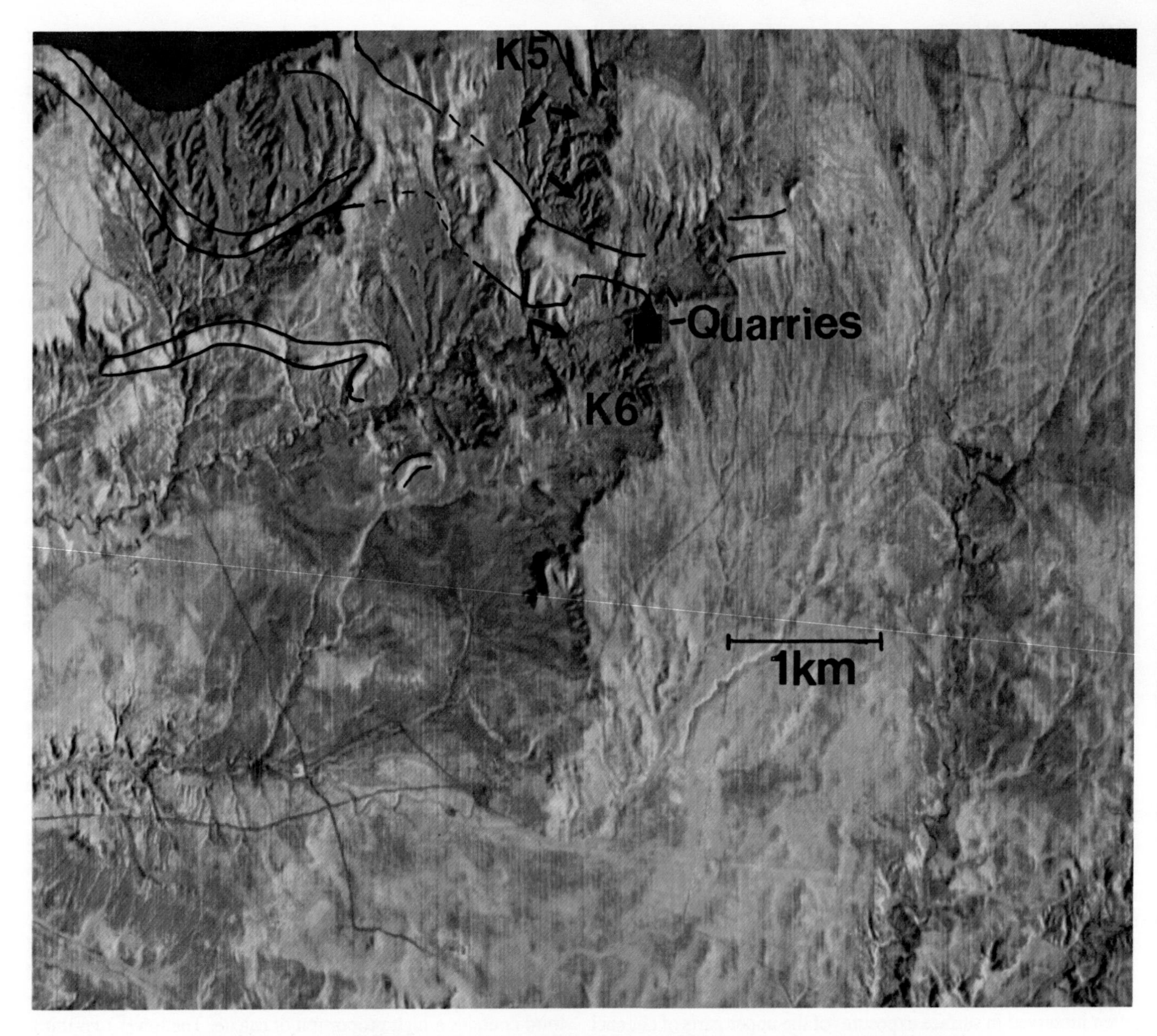

Plate I. Remote sensing image derived from Thermal Infrared Mulstispectral Scanner data (d-stretch false-color composite of bands 1, 3, and 5) of the Buck Spring area. Sandstones in the channel sandstone complex show up as yellow or yellow/lime green bands in the image. The "dark red stratum" at K-5 and K-6 is marked by a thin purple band on either side of the channel sandstone complex (arrows). Variegated and gray mudstones of the Lost Cabin Member of the Wind River Formation are red in color.

Outcrops of variegated mudstones above the "dark red stratum" extend from the north-central part of Section 15,T.38N., R.89W. into the SE¼ of Section 22, just to the north of the channel sandstone network. In the latter area, gray mudstones become more common in the lower part of the variegated sequence. Red mudstones above the "dark red stratum" cannot be distinguished from gray mudstones in the TIMS image (Plate I).

K-6 quarry area. The stratigraphic sequence in the immediate vicinity of the quarries at K-6 is dominated by gray mudstones, contrasting markedly with the variegated sequence in the K-5 area. At the base of the K-6 sequence is a 2-m-thick red mudstone. Although this unit is truncated by the channel sandstone complex, stratigraphic position, spectral properties (laboratory and remote sensing data), and lithology indicate that it is equivalent to the "dark red stratum" to the north. The greenish-gray sandstone that overlies this bed in the K-5 exposures is, however, not present above the "dark red stratum" in the quarry exposures; the sequence instead includes 15 m of thin tabular gray mudstones with occasional interbedded limestones, bluish-gray claystones, and fine-grained sandstones (Figs. 3 and 4). Although

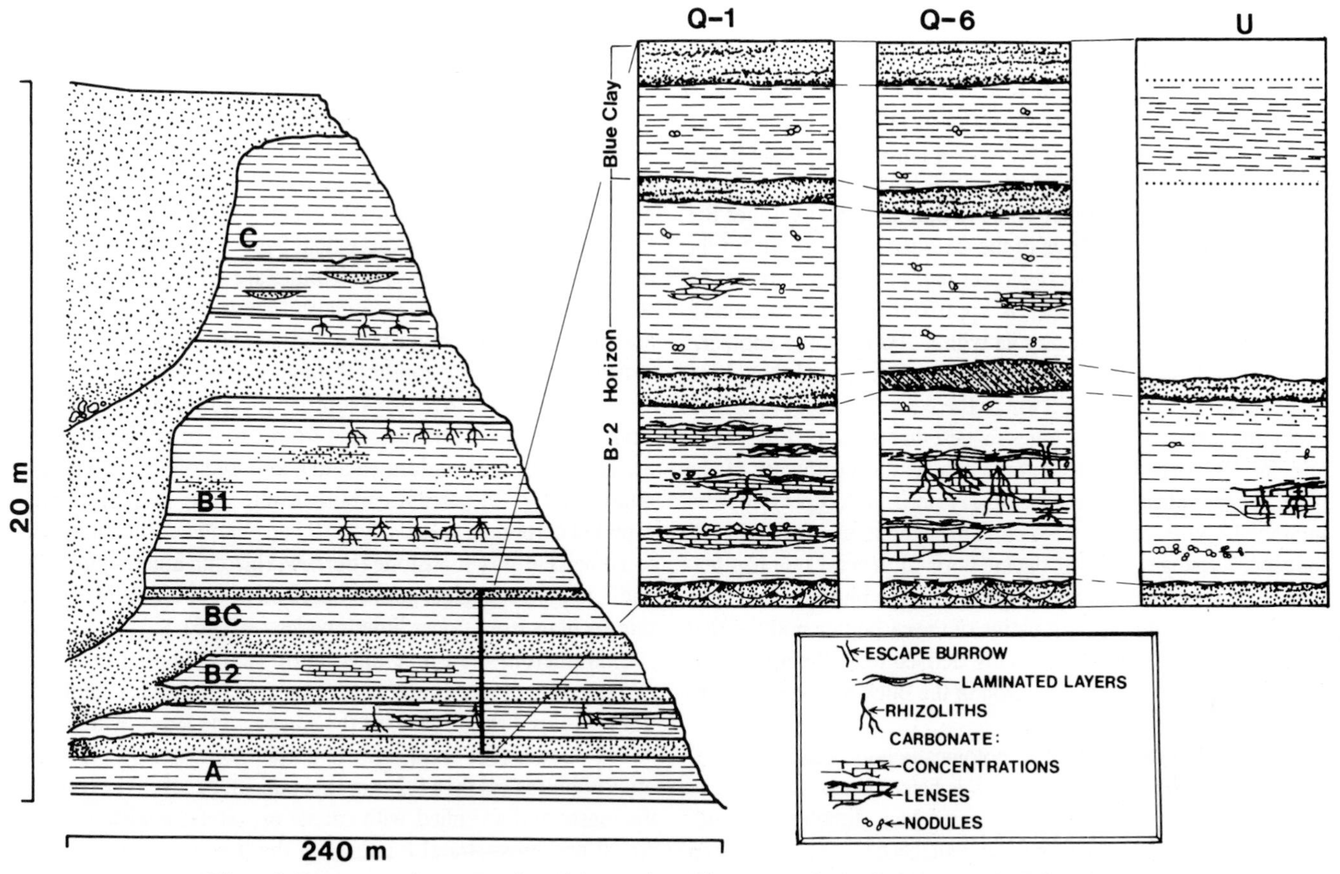

Figure 3. Diagrammatic cross section of the stratigraphic sequence in the K-6 area and the B-2 horizon. Three stratigraphic columns on right are composites based on the lithology, from left to right, of the two major quarry sites (Q-1 and Q-6), and a poorly fossiliferous part of the B-2 horizon (A). Vertebrate fossils are much more common in parts of the B-2 horizon, which have bioturbated limestone lenses or layered limestones.

interbedded limestones become much less common or absent at a distance of greater than 600 m away from the channel sandstone complex, the general character of the gray sequence is maintained throughout the exposures in areas presumed to be part of the distal flood plain, some 2 km to the south.

The gray sequence is designated the "B sequence" and includes the richly fossiliferous deposits of the Buck Spring Quarries. Tabular sandstone units within the "B sequence" vary from 0.15 to 1.2 m thick and are sandstone stringers tied to one apron-channel sandstone in the channel sandstone complex. The lithology of the fossil-producing units in the "B sequence" is discussed below.

Above the "B sequence" is a 10-m-thick interval of variegated mudstones (K-6, "C sequence") that are very similar in lithology to the variegated mudstones in the exposures above the "dark red stratum" to the north of the channel sandstone network. Except for the occurrence of the "dark red stratum," no direct lithological relationship can be established between the K-6 and K-5 exposures.

Figure 4. View of Quarry-6 in the gray sequence at K-6. The B-1, B-2, and "blue clay" (BC) horizons are indicated. The variegated sequence (VAR) above the B is shown in the foreground.

Lithology of the Buck Spring Quarries horizon

Of major importance at K-6 is the 6-m-thick series of beds within the "B sequence" that provides the abundant fossil vertebrates excavated from the quarries (Fig. 3). Several fossiliferous layers are present and include a lower B-2 horizon (2.5 to 3.0 m), a middle "blue clay" horizon (0.25 to 0.35 m), and an upper B-1 horizon (2.5 to 3.0 m). Five of the six quarries occur within the B-2 horizon; the sixth is in the lower part of the B-1 horizon. Of the two major quarries, Q-1 is approximately one-half meter above Q-6. All excavated sites lie within 200 m of one another.

The "B sequence" is laterally persistent and has been mapped for approximately 3 km to the west in surface exposures. Preliminary stratigraphic studies suggest that the B-1, "blue clay," and the B-2 may be traced for 4 km to the northwest of the quarry area, indicating that the quarry horizon occupied approximately 12 km^2 of surface area. Throughout its extent the "B sequence" changes little in thickness or in general outcrop features. The sequence can be recognized by the bluish color of the middle horizon and the dark patina of the carbonate nodules that litter the surface, most of which are derived from the upper B-1. The "blue clay" is the only horizon in the Buck Spring exposures that contains dolomite.

The B-2 horizon appears lithologically homogeneous in surface outcrop. However, the horizon is heterogeneous in test and quarry excavations and includes three basic lithologies: mudstone, sandstone, and limestone. Overall, clast size in the horizon increases toward the channel sandstone complex. Individual beds within the B-2 layer vary in thickness from laminations less than 1 mm to beds as much as 30 cm, all of which are separated by sharp contacts.

The sandstone units are persistent across the B-2 horizon, but are no more than 0.3 m thick. Laterally, these sandstones appear to correlate with stringers from one of the apron-channel sandstones in the network to the north. The units show planar bedding or fine-scale cross-laminations, both of which are disrupted only occasionally by vertical and horizontal (less common) root traces less than 5 mm in diameter, and vertical burrows 2.0 to 2.5 cm in diameter.

The mudstone units vary in thickness from 5 to 25 cm and are laterally impersistent, usually not extending more than 20 m in sampled areas. Some individual mudstone units (presumably single-event deposition) exhibit graded bedding with very fine sand or intraformational conglomerates (subrounded mudstone and claystone pebbles, calcareous nodules, plant and bone debris) at their base. The sand-size fraction includes intraclasts of mudstone and, occasionally, limestone. Small calcareous glaebules (approximately 1 cm in diameter) occur in mudstones that are not associated with limestone lenses. Where present, the glaebules occur in a 2-m-thick zone of the B-2 horizon about 0.3 m below the upper contact with the "blue clay." Bioturbation is more common in mudstones than in sandstones, but less so than in some limestones. In addition to lignified roots and vertical burrows, root casts (up to 3 cm in diameter) filled with calcite-cemented sandstone also occur. Vertical burrows in mudstones overlying limestones are infilled with limestone sediment from below.

Limestone units occur in random pockets within the B-2 horizon in an area between 250 and 600 m away from the channel. They are more common in the lower part of the horizon, and are much more restricted in occurrence than mudstones and sandstones, occupying no more than 5 percent of the total volume of the B-2 sediments. Fossil vertebrates are extremely abundant in the limestones as well as in subjacent mudstones. The limestones are of two types—bioturbated lenses and laminated layers—which grade into one another through the B-2 horizon.

Bioturbated lenses. Bioturbated limestone lenses vary from 5 to 20 cm in thickness and from 1 to 4 m in diameter (Fig. 5). The limestone is indurated and composed of 50 to 60 percent micrite, 20 to 25 percent carbonate silt, 10 to 15 percent well-rounded carbonate peloids, 5 to 10 percent quartz silt and sand, 5 to 10 percent silty mudstone (burrow, rhizolith and fracture fills), and 2 to 5 percent algal grains (based on 300 point counts on four thin sections from upper limestone lens at Q-6).

Numerous mudstone-filled burrows that are 0.5 to 1 cm in diameter are concentrated in the centers of the lenses. Bioturbation features decrease toward the edges of the lenses where alternating laminae of mudstone and limestone occur. Abundant microtubules, less than 0.2 mm in diameter, occur in the center of the lenses and are filled with calcite or, rarely, mudstone. Horizontal and cross-stratal fractures in the limestone are also filled with mudstone. These fractures are generally less than 1 mm wide, but some reach 5 mm. A thin veneer of iron oxide is present along some bedding planes and at the interface between the limestone and mudstone infilling of both burrows and fractures. The upper surfaces of the bioturbated lenses have either thin laminae of limestone or poorly sorted laminae consisting of sand-size rip-up clasts derived from the bioturbated limestones and mudstones.

Mudstones and sandstones adjacent to limestone lenses are also bioturbated. Silty mudstone units immediately underlying the lenses have a higher proportion of rhizoliths and burrows than do those lying lateral to or above the lenses. For example, at the base of the most fossiliferous limestone lens at Q-6, multiple horizontal, clay-filled root casts (2 to 3 mm in diameter) extend 7.5 cm below the center of the lens. One meter away from the lens center, these root casts are nearly absent, whereas lignified vertical roots become more common. In the mudstone immediately overlying the Q-6 limestone lens, all burrows are vertical and infilled with limestone. These burrows extend upward 10 cm to the top of the mudstone and are 3 to 4 cm in diameter; these may be invertebrate evacuation structures, suggesting that the overlying mudstone was deposited rapidly (H. Lang, oral communication, 1986).

Laminated layers. Layers of alternating mudstone and limestone laminae are much more restricted in occurrence than are bioturbated limestone lenses. These layers are no more than 5 cm thick and extend for only 1 or 2 m. Lithologically they are

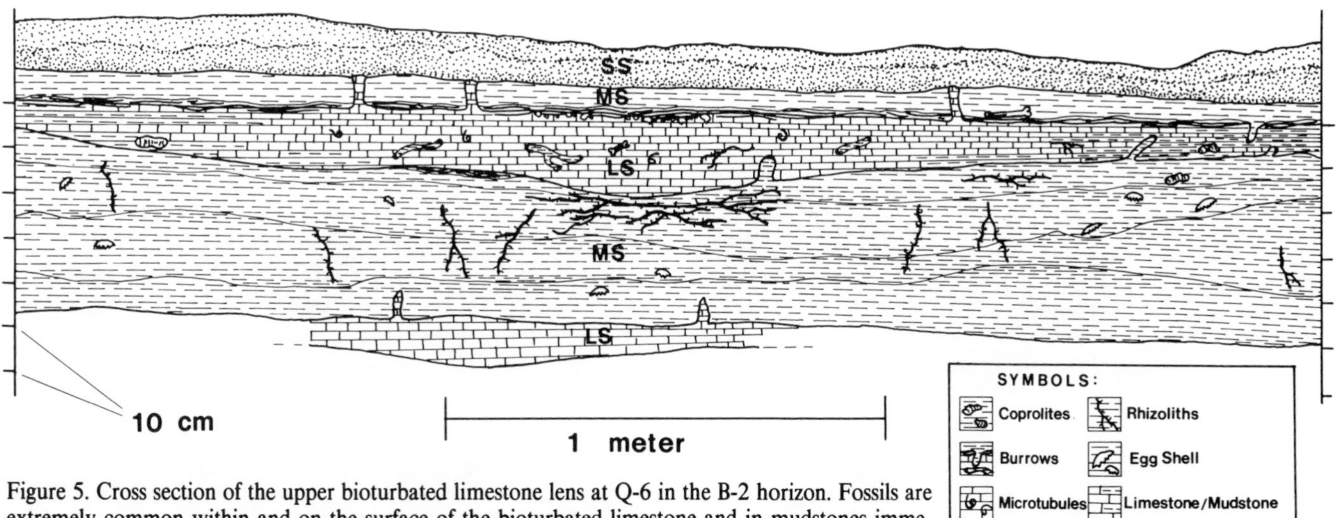

Figure 5. Cross section of the upper bioturbated limestone lens at Q-6 in the B-2 horizon. Fossils are extremely common within and on the surface of the bioturbated limestone and in mudstones immediately underlying it. Specimens of larger mammals, including skull and associated skeletal remains, are more common on the surface of the lens. Key to symbols: LS, Limestone; MS, Mudstone; SS, Sandstone.

much like the bioturbated lenses, except that they contain relatively equal amounts of mudstone and limestone, are poorly consolidated, and are less frequently disrupted by bioturbation. Limestone/mudstone laminar couplets are between 0.05 and 0.5 mm thick and occur at a frequency of between 20 and 60 couplets per cm. Abundant isolated fossil bones and coprolites occur within these limestone layers on laminar surfaces.

Occurrence and preservation of fossil vertebrates

The B-2 horizon is rich in fossil vertebrates, especially in the 6-hectare area of surface exposures that contain limestone units. Within this area, fossils are much more common in limestones and adjacent mudstones than in parts of the B-2 horizon that lack limestones. The fossil bones range in size from those of the smallest to the largest known vertebrates for the Lostcabinian (modern shrew- to tapir-size, that is *Uintasorex*- to *Coryphodon*-size). All quarry remains are exceptionally well preserved and include complete skulls and jaws, partial skeletons, skull, jaw and postcranial elements, and isolated teeth (Figs. 6, 7, 8, 9, 10).

Partially articulated or associated remains of single individuals occur in five lithologic contexts: (1) at the base of mudstones that overlie limestone lenses (articulated or associated; e.g., *Hyracotherium vasacciense, Lambdotherium popoagicum*), (2) within mudstones, often in intraformational conglomerates (articulated or associated: e.g., cf. *Armintodelphys blacki, Palaeictops multicuspis, Notharctus venticolis, Coryphodon* sp., *Hapalodectes* sp., *Shoshonius cooperi, Glyptosaurus donohoei, Xestops* sp.); (3) on bedding planes of laminated limestones (associated: *Xestops* sp., *Necrosaurus* sp., *Uintasorex* sp.); (4) in coprolites (associated: *Uintasorex* sp., cf. *Armintodelphys dawsoni, Nyctitherium* sp., small glyptosaurine anguids); and, rarely, (5) within bioturbated limestone lenses (associated or articulated: *Uintasorex* sp., a creodont, and a serpent).

Complete or nearly complete skull remains occur in the same contexts and include specimens of two amphibians, *Xestops,* cf. *Necrosaurus,* a sebecosuchian, cf. *Armintodelphys dawsoni, Scenopagus* sp., *Palaeictops multicupsis,* 2 species of *Didelphodus,* a chiropteran, *Microsyops* spp. (2 specimens), *Shoshonius cooperi* (5 specimens), *Notharctus venticolis* (3 specimens), *Microparamys,* several other small rodents, a creodont, *Didymictis, Lambdotherium popoagicum,* and *Hyracotherium vasacciense.* Disarticulated skulls of *Glyptosaurus donohoei, Xes-*

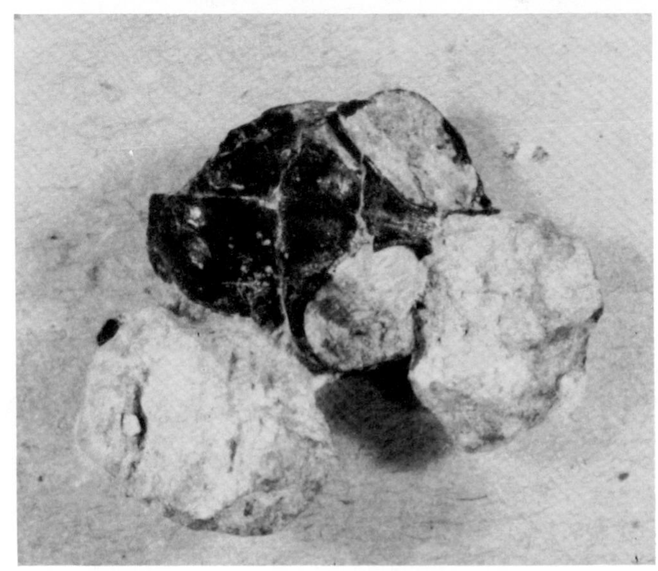

Figure 6. Dorsal view of a slightly crushed skull of *Shoshonius cooperi,* from 6 cm below the top of the bioturbated limestone lens at Q-6. Skull is approximately 2 cm in diameter.

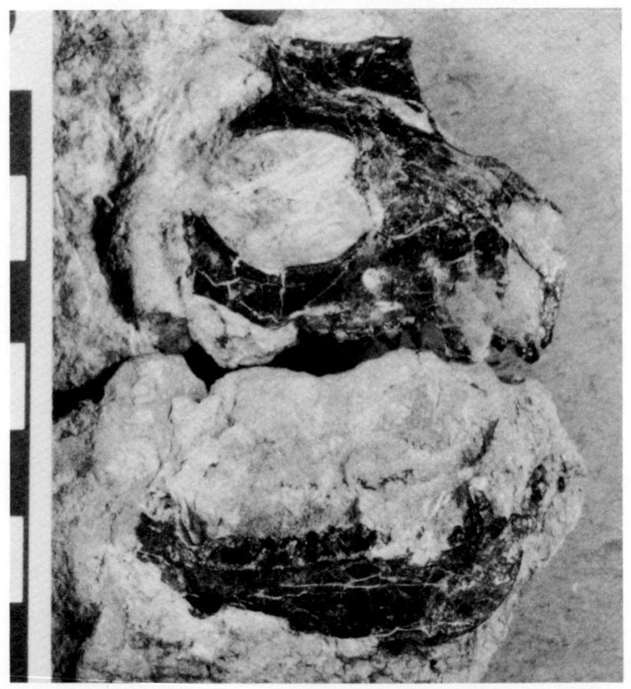

Figure 7. Lateral view of skull and lower jaws of *Notharctus venticolis.* The skull preserves the complete dentition, but the occipital region was removed or destroyed prior to burial. Specimen from the bioturbated limestone at Q-6.

Figure 8. Skull, lower jaws, and a portion of the forelimb of *Armintodelphys dawsoni,* preserved in a partly disaggregated coprolite. Specimen is from a mudstone unit at Q-1.

Figure 9. Lower jaw of *Scenopagus* sp. that preserves the complete dentition. Specimen was recovered from laminated limestone layer at Q-6.

Figure 10. Associated skull and lower jaws of a creodont from the center of the bioturbated limestone lens at Q-6 (see Fig. 4). Left and right maxillaries have been separated along the midline of the skull.

tops, and *Hapalodectes* are known from the mudstones. A skull of a multituberculate has been recovered from a calcareous glaebule on the surface of the B-1 horizon.

Isolated bones and teeth occur in all lithologies of the B-2 horizon. All teeth preserve the enamel intact and many of the isolated specimens either have coprolitic material adhering to their surfaces or are found in units that contain coprolites and/or fragments of disaggregated coprolites.

Smaller limb bones from the bioturbated limestones show more breakage than do those from mudstones and laminated limestones. Large and small cranial remains are often compressed along bedding planes to less than half of their original height.

However, some of the nearly complete skulls are crushed outward and disarticulated, suggesting that they may have been distorted either syndepositionally or prior to burial (*Hyracotherium;* a creodont). The occipital portion of the neurocranium was broken away from the skull prior to burial on some of the specimens (*Microsyops, Notharctus,* one specimen of *Shoshonius*). Limb bones of larger mammals rarely have impact or spiral fractures. Only a few of the larger bones show surface pitting, which may have resulted from either predepositional weathering or destruction of the surface by the penecontemporaneous infauna and plant roots.

In bioturbated limestones, fossil bones show random orientation, presumably as a result of infaunal activity. Vertical burrows that overlie limestone lenses occasionally preserve vertically oriented bone shafts and dentaries. Several coprolites perserving partially associated skeletons also lie parallel to bedding planes in the laminated limestones and mudstones.

Although cranial remains are more common than postcranial remains, the relative proportions of elements of the forelimb, hindlimb, foot, and axial skeleton are nearly equal (Fig. 11). This feature of the assemblage, in addition to the broad size range of the animals preserved, suggests that hydraulic sorting affected the accumulation only minimally (see Korth, 1979). None of the bones from the B-2 horizon have glaebule material adhering to their surfaces, which indicates that paleosol development was only incipient.

Recent weathering of the B-2 horizon has severely damaged many of the smaller bones, which are often fractured parallel and perpendicular to the long axis. The fragile nature of these smaller bones, coupled with erosion, undoubtedly affects the composition of the fauna recovered by surface prospecting.

Fragmentary fossil eggshells occur in all lithologies of the B-2 horizon. Nearly complete fossil eggs of large size (3 to 5 cm in diameter) occur within bioturbated limestones, and the few lizard eggshells occur in laminated limestone layers (see Hirsch and others, 1987).

Invertebrate and plant fossils

Invertebrate fossils are very rare in the quarry horizon. Three specimens of what appear to be abdominal segments of a fresh-water crustacean have been recovered from Q-1, and a single specimen of cf. *Biomphalaria* was recovered from Q-6. A few fragments of gastropod opercula have been found on the surface of the B-2 horizon. As mentioned above, algal skeletal remains compose from 2 to 5 percent of the material in bioturbated limestones. These specimens have not been identified.

Plant debris is relatively common in all excavated sites in the B-2 horizon. Most of the primary plant material consists of stem and root fragments replaced by calcium carbonate and occur as lag in mudstone units and within laminated limestones. The calcareous plant remains are fragmentary slivers and chunks of woody material; their rough surface texture suggests they were partly decomposed prior to burial. Carbonaceous or lignified roots in life position are also present in mudstone units. Root casts and molds are especially common at the base of the limestone unit at Q-6 as discussed above.

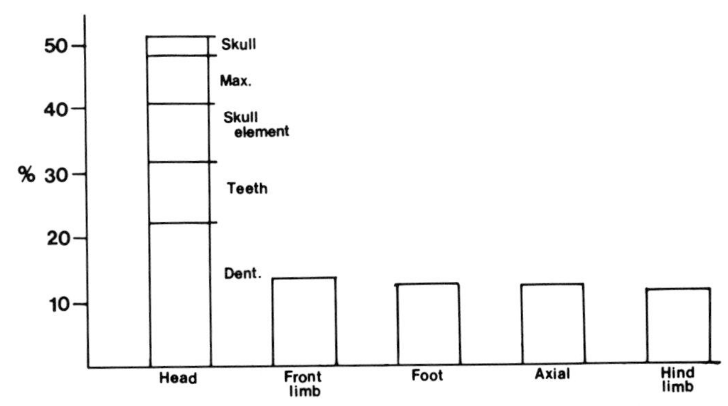

Figure 11. Percentage representation of parts of the head (Max., maxillaries; Dent., dentaries), foot, forelimb, hindlimb, and axial skeleton of all identifiable specimens excavated (n = 677) from Q-1 and Q-6 during 1984. Many cross sections of postcranial bones could not be specifically identified and are not included in the calculations.

Well-preserved leaf, stem, seed, and flower fossils do, however, occur in a carbonaceous shale that lies at the base of a ribbon sandbody, approximately 15 m above the "B sequence."

Origin of the fossil accumulation

Three types of processes appear to account for the exceptional preservation of the fossil vertebrates at the Buck Spring Quarries: natural death other than predation, predator accumulate, and hydraulic transport. The few articulated remains of smaller vertebrates, and perhaps some of the larger vertebrates as well, appear to have died by natural means other than predation, and were buried rapidly.

However, predators appear to be responsible for accumulation of the majority of the fossil vertebrate materials. Evidence that most of the smaller bones were derived from disaggregated mammalian coprolites includes: (1) many of the small bones have coprolitic material adhering to their surfaces; (2) complete coprolites with bones are common in mudstones and laminated limestones; and (3) isolated teeth always preserve the enamel (see Fisher, 1981). At least two different types of coprolites occur; smaller coprolites about 1 cm in diameter are more common and preserve complete bones of small vertebrates whereas larger coprolites about 2 cm in diameter preserve complete elements of smaller vertebrates as well as chips of bones from larger vertebrates. Specimens from smaller coprolites are either in partial articulation or in random position. Rare isolated and/or associated larger bones with impact and spiral fractures, and disarticulated but associated remains of larger mammals may represent predator scatters or kill sites.

Hydraulic transport of some of the smaller bones is indicated by (1) the high number of isolated bones and teeth, (2) their random distribution within limestone and mudstone units, (3) their occurrence in intraformational conglomerates at the base of mudstones, and (4) their association with disaggregated coprolitic materials. The absence of surface abrasion on the bones and the association of small bones with either intraformational conglomerates or mudstones that preserve intraclasts of sand-particle size suggests only minimal transport. These intraclasts would have been completely disaggregated over a short transport distance.

Paleoenvironment of the Buck Spring Quarries

Lost Cabin Member sediments of the Buck Spring area were deposited on flood plains and in channel cuts by presumably low-sinuosity, fixed-channel streams that flowed in an easterly direction. Evidence here is the ribbon-channel/apron-channel geometry of the sandstones in the channel sandstone network, the lateral relations of the tabular mudstone units in the K-6 and K-5 areas, and general trend and sedimentary features of the sand-bodies. Stream movement was primarily by avulsion, with apparent displacement of channels up to 1 km in some areas. This agrees with analyses of other Wind River sandstones (Seeland, 1978a, 1978b).

The thick accumulation of mudstones in overbank areas along the basin axis, their overall tabular form, and the nearly straight course of the channels suggest that downstream base-level controls (Smith and Smith, 1980) may have been responsible for the sediment accumulation in the Buck Spring area during Lost-cabinian time. Such control may have been present in the Casper Arch area near Arminto and Hell's Half Acre. Penecontemporaneous uplift is suggested by thinning of the Lost Cabin Member in this area and the unconformity between the Lysite and Lost Cabin Members along the eastern edge of the Wind River Basin (Keefer, 1965). The younger age of the Lost Cabin Member in this area (early Bridgerian) further suggests that uplift along the Casper Arch occurred penecontemporaneously with the deposition of Lost Cabin Member sediments in the Buck Spring area. However, the structure of subsurface and basement rock may have had some local influence as inferred from the convergence and confluence of the four channel sandstones into a sheet-like sandbody just to the north of the quarry area. This convergence suggests that the region was topographically low on the Lost-cabinian landscape. Subsurface control of present landform is also suggested by the geomorphology of the northeastern part of the basin (unpublished data).

Variegated mudstones in the K-5 and K-6 areas may represent depositional cycles that involved relatively rapid deposition (most gray units = 10^{-2} to 10^{2} years) followed by long-term periods of subaerial exposure and paleosol development (red paleosols = 10^{3} to 10^{4} years; see Bown, 1979; Bown and Kraus, 1981a, 1981b, 1987; Retallack, 1984; Kraus, 1987). This is suggested by the preservation of original depositional features in gray sediments and the well-developed paleosol characteristics in red mudstones that disrupt original features.

Thin, tabular, fine-grained sandstones that are interbedded with variegated and gray mudstone sequences preserve graded bedding and represent splay deposits from overbank flooding. Rapid deposition of some of these units is inferred from the vertical "evacuation" burrows filled with underlying sediment that reach the top of the units. Silty mudstones preserving primary depositional features also represent splay deposits as does the mud deposited as suspended load during overbank flooding.

Mapping of the quarry area indicates that the B-2 horizon is most fossiliferous between 250 and 600 m away from the main fluvial channel. Conditions for fossil preservation were unique as the limestone lithology of the horizon is not known from elsewhere in the Wind River Formation. Beaumont (1979), Demicco and others (1987), and Gingerich (1987) have described similar lithologies elsewhere. The preservation of vertebrate bone, egg shell, invertebrate remains, and carbonaceous and calcareous plant debris imply that Ph and Eh were relatively constant, between 6 and 8, and 0 and –200, respectively (Retallack, 1984). As in other areas, thin mudstone and fine-grained sandstones in the quarry horizon were deposited during overbank flooding.

The limestones, however, appear to have been derived from organic sources (skeletal grains, peloids, microtubules); they may have been precipitated and/or deposited in standing water and may represent the remnants of blue green algal mats (Beaumont, 1979; Demicco and others, 1987). The limited extent of the limestone units indicates that they were deposited in small ponds and/or well-drained swamps (Coleman, 1966). Periodic desiccation of bioturbated limestone lenses is inferred from syndepositional fractures infilled with mudstone, and is also supported by the presence of complete eggs of large and small size. Additionally, alternating couplets of limestone and mudstone as well as calcareous and carbonaceous plant roots suggest periodic, if not seasonal, changes in climatic and depositional conditions that allowed plant growth. The partly decomposed plant material and small glaebules in the B-2 horizon indicate that the area was episodically above ground-water level and subjected to some subaerial exposure and paleosol development. The paleosols of the B-2 horizon are classified as entisols—soil horizons preserving many of the original sedimentary features, but also preserving some evidence of plant-root disruption and eluviation (Bridges, 1978). The variable presence of carbonate-replaced and lignified root material may imply variable oxidation/reduction episodes as well. As such, the quarry deposits probably represent a well-drained swamp or ponded area adjacent to a stream. Judging from the size of the limestone units, small pools or ponds less than 10 m in diameter may have persisted for several years across this area.

Additional evidence for only temporary ponding is provided by the fossil vertebrates. Remains of exclusively aquatic vertebrates—fish, trionychid turtles, and crocodylids—are extremely rare and fragmentary (see below).

The majority of the fauna is composed of terrestrial lizards (see Hirsch and others, 1987) and terrestrial and arboreal mammals (Table 1; collections acquired through 1986, see below).

TABLE 1. RELATIVE ABUNDANCE OF FOSSIL MAMMALS FROM THE BUCK SPRING QUARRIES*

Species	Number of Specimens	Surface
Multituberculata		
cf. *Ectypodus* sp.	8	*
Marsupialia		
Peratherium innominatum	9	
Armintodelphys blacki	5	1
A. dawsoni	6	
Peratherium marsupium	5	1
Peradectes chesteri	3	
Insectivora		
Nyctitherium spp.	41	
Centetodon sp. A	28	
Scenopagus edenensis	27	
Talpavus spp.	26	
Scenopagus priscus	10	
Centetodon sp. B	9	
Macrocaranion sp.	6	1
"Scenopagus" sp.	1	
Chiroptera		
Chiroptera sp. A	2	1
Chiroptera sp. B	2	
"Proteutheria"		
Apatemys sp. A	9	1
Palaeictops multicuspis	4	1
"Palaeoryctidae" sp	3	
Apathemys sp. B	2	
Didelphodus sp. A	1	1
Didelphodus sp. B	1	
Amaramnus n. sp.	1	
Rodentia		
Knightomys sp.	22	
Pauromys sp.	7	
Microparamys sp	4	1
?*Mattimys* sp.	3	
Paramys copei	3	2
cf. *Thisbemys*		1
?Dermoptera		
Uintasorex sp.	38	1
Microsyops scottianus	5	10
Microsyops n. sp.	5	1
Primates		
Shoshonius cooperi	27	2
Phenacolemur sp. A	5	
Notharctus venticolis	5	2
Notharctus nunienus	2	1
Absarokius noctivagus	1	
Phenacolemur sp. B	1	
Creodonta		
Prolimnocyon antiquus	2	
Hyaenodontidae sp. A	1	1
Hyaenodontidae sp. B	1	
Oxyaena sp.	1	
Creodonta sp.	1	
Carnivora		
Viverravus lutosus	5	1
?*Miacis* sp.	1	
Didymictis altidens	1	1
Carnivora sp. indet.	1	1
Condylarthra		
Hyopsodus paulus	4	15
Hyopsodus wortmani	1	
Hapalodectes sp.	1	
Phenacodus vortmani	1	5
Hyopsodus walcottianus		3
Phenacodus primaevus		2
Artiodactyla		
Diacodexis secans	1	2
Diacodexis woltonensis		1
Bunophorus sinclairi		2
Perissodactyla		
Hyracotherium craspedotum	3	2
H. vasacciense	3	6
Lambdotherium popoagicum	3	8
Heptodon ventorum		1
Heptodon cf. *posticus*		1
Pantodonta		
Coryphodon sp.		1
Palaeanodonta		
Palaeanodonta sp. A	1	
Palaeanodonta sp. B		1

*An asterisk in the Surface column indicates recovery from the B-1 horizon only.

The articulated and scattered remains of larger mammals such as *Coryphodon, Hyracotherium,* and *Lambdotherium,* as well as the relatively abundant coprolitic remains of predators, suggest that most if not all of the fauna lived in the Buck Spring area. Associated and articulated remains of small arboreal and terrestrial mammals and lizards suggest that multistoried gallery forest was in close proximity, presumably in more upland areas. Evidence for large plants, save for rhizoliths 3 cm in diameter, is wanting from the immediate vicinity of the quarries.

Time represented by the quarry deposits

The sharp contacts between sandstone, mudstone, and limestone units as well as poorly developed paleosol features in the B-2 horizon suggest a relatively rapid period of accumulation and deposition. The sharp contacts of mudstones and sandstones and those of the varvelike laminar couplets of mudstone/limestone in the limestones may represent seasonal events or cycles of deposition. The occurrence of between 20 and 60 couplets per cm in laminated limestones suggests that each limestone unit encompasses between 100 and 600 depositional cycles. If these cycles represent yearly intervals, the several limestone units within a single vertical section of the B-2 fossil horizon account for no more than several thousand years of deposition. The limited temporal duration of the extremely rich assemblage from the B-2 horizon will allow study of evolutionary and ecological changes and patterns on a scale not often preserved in the Paleogene terrestrial record of vertebrates.

PALEOECOLOGY OF THE QUARRY ASSEMBLAGE

The excellent quality of preservation of the cranial and postcranial materials from the Buck Spring B-2 horizon promises to provide one of the most complete records of the adaptive complex and community structure of a single assemblage of Eocene vertebrates from North America. The quarry materials provide enough information to accurately determine individual body size, locomotor ability, food preference, food acquisition strategy, and population structure for many of the species. Our analyses are in a preliminary stage; nevertheless, some basic generalizations about the community structure of the B-2 assemblage are possible.

Species richness and faunal composition. The species richness of the B-2 assemblage is among the highest for Tertiary faunas that have been recovered from single fossiliferous beds (Stucky, 1989b). Minimally, the B-2 assemblage includes 65 species of mammals, 22 species of lizards, 3 snakes, 2 birds, 2 turtles, 2 crocodylians, 4 amphibians, and 3 fishes.

The mammalian assemblage consists of an abundance of marsupials, primitive placentals, insectivores, primates, rodents, and ungulates that typify the Lostcabinian Land-Mammal Subage (Table 1; see Stucky, 1984c). Importantly, many of the most abundant mammals from the quarry assemblage (especially the small insectivores, marsupials, rodents, and primates) are extremely rare or unknown from early Eocene localities in the Wind River Basin and elsewhere.

The carnivores and creodonts appear to be underrepresented. This is probably a sampling bias due to both the small number of specimens in the B-2 sample and the low original population size of each predator. Carnivores and creodonts present at other North American Lostcabinian or slightly younger localities but absent from the Buck Spring assemblage include *Prototomus, Arfia, Machaeroides, Uintacyon, Vulpavus, Vassocyon,* and several species of *Miacis.* Continued collecting of the quarry deposits should add some of these taxa to the assemblage.

Seven genera of mammals absent from the assemblage are either extremely rare in the Wind River Formation (*Stylinodon, Alveojunctus, Ignacius*) or occur at many localities but at very low frequencies (*Thryptacodon, Bathyopsis, Esthonyx, Paleosinopa*). These taxa may be habitat restricted. Three other missing genera (*Loveina, Meniscotherium,* and *Niptomomys*) occur at older Lostcabinian localities in the Wind River Basin and thus may have been extinct or extirpated by the time of deposition of the quarry sediments.

The lizard assemblage is the most taxonomically diverse for any known North American Paleogene locality (Hirsch and others, 1987), due to the abundance of small specimens. However, other Paleogene lower vertebrate assemblages are not well documented. This high diversity appears to reliably reflect diversity patterns for lizard faunas of Lostcabinian age, based on published faunal reports (Estes, 1983) and our unpublished studies of the excellent Carnegie Museum collections of Paleogene lizards. *Glyptosaurus donohoei* is the most abundant lizard from surface collections, and small anguids (*Xestops* and diploglossines), xantusiids, and iguanids dominate the excavated samples.

Aquatic or marginally aquatic lower vertebrates are rare in the B-2 horizon, and include only a few carapace fragments of turtles (1 trionychid, 1 undetermined), two partial associated skeletons of cf. *Allognathosuchus,* a partial skull of the first North American sebecosuchian (near *Sebecus*; this animal may have been terrestrial), and excavated materials of fishes (less than 5 scales of a lepisosteid, a dentary and vertebral centrum of *Amia,* and scales and bones of a small advanced teleost preserved in two coprolites). The amphibians (2 frogs and 2 salamanders) are also rare, represented by two skulls and several isolated ilia and vertebrae. In contrast, aquatic lower vertebrates (except for the small teleost and some of the amphibians) are common in most mammal-rich surface, excavated, and screen-washed assemblages from a diversity of depositional environments (e.g., red mudstones or paleosols, conglomerates including meandering stream lag and braidplain deposits; Stucky, 1984b). Aquatic vertebrates, including other turtles and crocodylids, dominate the "blue clay" horizon and the "dark red stratum" above and below the B-2 horizon, respectively.

The mammalian species composition and numbers of individuals from quarry excavations and surface prospecting of the B-2 horizon differ considerably (Table 1). Surface collections are biased toward larger mammals such as *Hyracotherium, Phenacodus, Lambdotherium, Microsyops,* and *Hyopsodus,* whereas excavated samples are dominated by small vertebrates. This bias is a

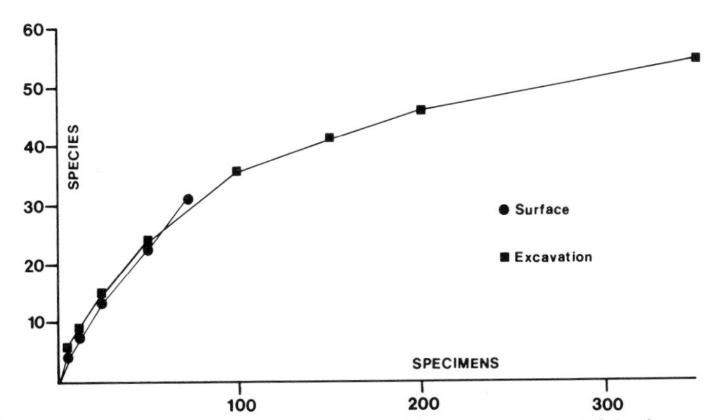

Figure 12. Rarefaction analysis of the assemblage from the B-2 horizon. Note the similarities in the rarefaction curves for species recovered from surface prospecting and from quarry excavations. The similarities in the two curves suggest that both sampling techniques predict a similar pattern of species diversity. Analysis based on collections made during 1984, 1985, and 1986.

product of sampling and preservation. Larger specimens are more readily spotted during surface collection and are also more resistant to weathering. Small specimens found in excavations show more in situ breakage from recent weathering and invasion by plant roots. As a result, small bones are usually destroyed to a depth of several centimeters in the subsurface prior to their exposure on the surface of the B-2 horizon. Excavated samples are assumed to be at least ordinally representative of the relative abundances of the species in the living community.

Despite the differences in preservation and representation between surface and quarry collections, a rarefaction analysis (Simberloff, 1978) of a maximum count of individuals (associated skeletal specimens counted as one individual, isolated jaws, skulls, and maxillaries counted as one individual each; isolated postcranials and teeth not counted) of the two samples reveals a similar pattern of diversity; the two rarefaction curves are nearly identical in slope and height (Fig. 12). This suggests that, despite the sampling biases, both surface and excavated samples yield reliable estimates of the relative alpha diversity or species richness of the assemblage for a given sample size.

Comparison of these rarefaction curves to those of other Eocene assemblages (recovered from surface and screen-washed samples and from a variety of depositional environments; McKenna, 1960; Delson, 1971; Robinson, 1966; West, 1973; Rose 1981a, 1981b; Storer, 1984; Eaton, 1984; Stucky, 1984a; Kihm, 1984; unpublished data) indicates that the quarry assemblage is among the most diverse known, not only for the Eocene, but also for the entire Tertiary of North America. The rarefaction curves for surface and excavated samples are extremely similar in slope and height to those of penecontemporaneous assemblages (Lostcabinian and Gardnerbuttean) and are steeper and higher than for those of faunas of older (Graybullian, early early Eocene) and younger age (Uintan, late middle Eocene and younger; Stucky, 1989b). These initial results—similarity between excavated and surface samples and similarity of these to penecontemporaneous assemblages—imply that estimates of species richness for Eocene assemblages from a broad spectrum of depositional environments may be compared directly, even if the samples are biased by different taphonomic and depositional factors.

The high species richness of the B-2 and other Lostcabinian assemblages is coupled with low species turnover (beta diversity) among known faunas that cross the early/middle Eocene boundary. These diversity patterns are correlated with a rise in diversity of the plants as well as a peak in tropicality in North America during the Eocene (see Stucky, 1989a).

Body size. Mammals from the B-2 assemblage include the complete spectrum of sizes, from largest to smallest, among known Lostcabinian taxa. Lower vertebrates of large body size (for example, large crocodylians, *Amia uintensis,* and the larger amphibians) are absent.

A subjective body-size classification of the mammals corresponding to the system of Nesbit Evans and others (1981) includes four categories: I, <1 kg; II, 1 to 10 kg; III, 10 to 45 kg; IV, >45 kg. Roughly, 55 percent of the species of mammals from the B-2 horizon are classified in size class I, 31 percent in size class II, 13 to 15 percent in size class III, and 3 to 5 percent in size class IV. This body-size distribution (and the species richness) is very similar to that recorded for modern African lowland and montane forest mammal communities (Nesbit Evans and others, 1981), except that categories III and IV are somewhat depauperate. This pattern is common to all North American mammalian faunas of early Paleogene age and is similar to that of nonvolant faunas from llanos and woodland habitats in Venezuela (Eisenberg and others, 1979). The dominance of small-bodied mammalian species is even more apparent when numbers of individuals in each size category are considered. The eleven most abundant taxa in the assemblage have an estimated body mass of less than 1 kg.

The abundance of small mammals from the B-2 horizon and the metabolic necessities of small size strongly imply that food and water were readily available during the Lostcabinian and not subject to seasonal shortages (Pennyciuck, 1979). This may apply to most mammalian assemblages from the Paleocene through middle Eocene as well. Assuming that the foraging areas of the smaller mammals were small, the consistent occurrence of the same suite of small-mammal species during the Lostcabinian—through approximately 500,000 years and across broad areas in the Rocky Mountain region (Stucky, 1984c, 1989a)—also implies that habitats either were not patchy over those large areas or were patchy only in the sense of a small checkerboard pattern or multistoried canopy woodland. This conclusion runs counter to the generalization that tropical systems promote species diversity through patchiness sufficient for population isolation (McCoy and Connor, 1980).

Feeding adaptations and trophic relationships. Of all paleobiological inferences, these are at present among the most subjective. Dentitions of Paleocene and Eocene mammals are difficult to classify into well-defined categories of feeding adaptation based solely on gross dental features; most Paleogene mammals are much more similar in overall morphology to one another than are modern species and are thus less distinctive in features that signify a particular feeding strategy. As such, inferences based on an analogy with modern mammals in the same clade (at the level of Order and Family) may obscure the breadth and limits of feeding adaptations among the Paleogene mammals.

Classification of species into such broad groups as carnivore, herbivore, insectivore, and omnivore may be misleading because of the general dental similarity among almost all Paleogene taxa. Mammals assigned to these broad categories will show a high degree of overlap in morphology, and as a result, the trophic groups will grade into one another. Ecomorphological analyses that associate and contrast taxa on the basis of phenetic similarity and functional morphology promise to provide more realistic models of feeding adaptations among Paleocene and Eocene mammals (Rickleffs and Travis, 1980; Stucky, 1989a).

Notwithstanding our own advice and caution, the species of mammals from the B-2 horizon are classified below into carnivores, insectivores, herbivores, and "omnivores" by major features of the dentitions. Inclusion in a category does not imply an obligatory feeding adaptation or food preference.

Vertebrate predators, identified by the possession of a carnassial or shearing dentition, probably included the creodonts, carnivores, and *Hapalodectes,* and account for ten species or approximately 15 percent of the taxa. Some of the smaller species, especially in the Order Carnivora, probably also ate insects and fruit, as do their modern counterparts.

Small taxa with high-cusped piercing dentitions, including the insectivores, chiropterans, marsupials, and a palaeoryctid, are considered insectivorous and make up 23 percent of the assemblage. Some of the "larger" mammals in this category (e.g., *Armintodelphys blacki, Peratherium marsupium*) may have utilized other high-energy resources such as seeds, fruits, and vertebrate flesh.

Facultative herbivores include the perissodactyls, phenacodontids, hyopsodontids, artiodactyls, *Coryphydon,* incipiently lophodont rodents (sciuravids and some of the larger ischyromyids), notharctine primates, and perhaps *Microsyops* (18 taxa, 28 percent of species). These herbivores have low-crowned dentitions that emphasize lateral shear and crushing, indicative of folivory, frugivory, and/or seed predation (see Kay and Hylander, 1978). Among the herbivores, only *Coryphodon* and the species of *Heptodon* have lophodont dentitions that suggest obligate folivory. Some of the smaller species with more bunodont dentitions, such as *Diacodexis secans, D. woltonensis, Hyopsodus paulus, Notharctus,* and the rodents, probably relied more on fruit and seeds than leafy vegetation.

Remaining taxa (33 percent of species) have generalized tritubercular dentitions ("proteutheres", small primates, and rodents, *Uintasorex*) or unique dentitions (multituberculates, palaeanodonts) and are difficult to assign to the carnivore, herbivore, or insectivore categories. This group may include food specialists (such as the two palaeanodonts) but most were probably food generalists or "omnivores." Because of their small body size, they probably relied on high-energy food resources such as seeds, fruits, plant gum, and insects.

In terms of numbers of individuals, carnivores, insectivores, herbivores, and "omnivores" represent 4, 45, 12, and 39 percent, respectively for quarry samples, and 5, 37, 24, and 34 percent for combined quarry and surface samples.

A greatly enlarged or hypertrophied tooth at the front of the lower jaw occurs in all rodents, *Ectypodus, Apatemys, Phenacolemur,* and the microsyopids and suggests a specialized food acquisition and ingestion strategy. The anterior incisors of the rodents are evergrowing and show well-developed distal wear facets from gnawing activity. The other taxa with long and sharply pointed anterior teeth, probably practiced small object manipulation or hard-substrate piercing (Kay and Hylander, 1978; also see Happel, 1988).

Locomotor ability. The evidence at hand indicates that all of the larger mammals in body size classes III and IV were terrestrial, but none were highly specialized for high-speed cursorial locomotion (Rose, 1982, 1985; unpublished data). This conclusion probably also applies to most of the mammals in size categories I and II. However, the primates and microsyopids were arboreal (14 percent of the species), and the chiropterans volant (Jepsen, 1970). The anterior limbs of palaeanodonts are highly modified for digging (Rose, 1987). Some of the marsupials, "proteutheres," multituberculates, insectivores, rodents, and carnivores were probably scansorial. The abundance of small arboreal mammals suggests a forested habitat with an upper canopy. More detailed studies of locomotor function and size, may determine, in part, the structure and spacing within the understory (see Dubost, 1979).

ACKNOWLEDGMENTS

We wish to thank H. R. Lang, J. D. Swarts, S. Rose, N. Obermiller, and D. Kron for their discussions on the paleontology and geology of the quarry sites and for their assistance in the field. We also express our gratitude to all who have lent a hand in the excavations over the past four field seasons. Our work in Wyoming has greatly benefitted from the help and hospitality of Jim, Mary Helen and Rob Hendry, Pat and Bill Spratt, Zane and Ginger Fross, and John and Helen Lumley.

Thanks are also due to Tom Bown and Ken Rose for extending their invitation to us to participate in the symposium and to Mary Kraus and Robert Hunt who improved the manuscript with their comments. C. Inouye provided the x-ray diffractions and spectral analyses of field specimens, H. Lang provided the TIMS image, and J. Bennett wrote the microcomputer program for the rarefaction analysis. This research was supported by NSF grant BSR-8402051, NASA grants NAGW-0949, and NAGW-1803, and grants from the M. Graham Netting research fund of The Carnegie Museum of Natural History (to L. K. and R. S.).

REFERENCES CITED

Beaumont, E. A., 1979, Depositional environments of Fort Union sediments (Tertiary, northwest Colorado) and their relation to coal: American Association of Petroleum Geologists Bulletin, v. 63, p. 194–217.

Bown, T. M., 1979, Geology and mammalian paleontology of the Sand Creek facies, Lower Willwood Formation (lower Eocene), Washakie County, Wyoming: Geological Survey of Wyoming Memoir 2, p. 1–151.

Bown, T. M., and Kraus, M. J., 1981a, Lower Eocene alluvial paleosols (Willwood Formation, northwest Wyoming, U.S.A.) and their significance for paleoecology, paleoclimatology, and basin analysis: Palaeogeography, Palaeoclimatology, Palaeoecology, v. 34, p. 1–30.

—— , 1981b, Vertebrate fossil-bearing paleosol units (Willwood Formation, northwest Wyoming, U.S.A.); Implications for taphonomy, biostratigraphy, and assemblage analysis: Palaeogeography, Palaeoclimatology, Palaeocology, v. 34, p. 31–56.

—— , 1987, Integration of channel and floodplain suites; 1, Developmental sequence and lateral relations of alluvial paleosols: Journal of Sedimentary Petrology, v. 57, p. 587–601.

Bridges, E. M., 1978, World soils: Cambridge University Press, 2nd edition, 128 p.

Coleman, J. M., 1966, Ecological changes in a massive freshwater clay sequence: Transactions of the Gulf Coast Association of Geological Societies, v. 16, p. 159–174.

Dawson, M. R., Stucky, R., Krishtalka, L., and Black, C. C., 1986, *Machaeroides simpsoni*, new species, oldest known sabertooth creodont (Mammalia) of the Lost Cabin Eocene: University of Wyoming Contributions to Geology Special Paper 3, p. 177–182.

Delson, E., 1971, Fossil mammals of the Wasatchian Powder River local fauna, Eocene of northeast Wyoming: Bulletin of the American Museum of Natural History, v. 146, p. 309–364.

Demicco, R. V., Bridge, J. S., and Cloyd, K. C., 1987, A unique freshwater carbonate from the upper Devonian Catskill Magnafacies of New York State: Journal of Sedimentary Petrology, v. 57, p. 327–334.

Dubost, G., 1979, The size of African forest artiodactyls as determined by the vegetation structure: African Journal of Ecology, v. 17, p. 1–17.

Eaton, J. G., 1982, Paleontology and correlation of Eocene volcanic rocks in the Carter Mountain area, Park County, southeastern Absaroka Range, Wyoming: University of Wyoming Contributions to Geology, v. 21, p. 153–194.

Eisenberg, J. F., O'Connell, M. A., and August, P. V., 1979, Density, productivity, and distribution of mammals in two Venezuelan habitats, *in* Eisenberg, J. F., ed., Vertebrate ecology in the northern neotropics: Washington, D.C., Smithsonian Institution Press, p. 187–207.

Estes, R., 1983, Encyclopedia of paleoherpetology; Part 10A; Sauria Terrestria Amphisbaenia, p. 1–245.

Fisher, D. C., 1981, Crocodilian scatology, microvertebrate concentrations, and enamel-less teeth: Paleobiology, v. 7, p. 262–275.

Gingerich, P. D., 1987, Early Eocene bats (Mammalia, Chiroptera) and other vertebrates in freshwater limestones of the Willwood Formation, Clark's Fork Basin, Wyoming: Ann Arbor, University of Michigan Museum of Paleontology Contribution 27, p. 275–320.

Granger, W., 1910, Tertiary faunal horizons in the Wind River Basin, Wyoming, with description of new Eocene mammals: Bulletin of the American Museum of Natural History, v. 28, p. 235–251.

Guthrie, D. A., 1971, The mammalian fauna of the Lost Cabin Member, Wind River Formation (lower Eocene) of Wyoming: Annals of the Carnegie Museum, v. 43, p. 47–113.

Happel, R., 1988, Seed-eating by west African cercopithecines, with reference to the possible evolution of bilophodont molars: American Journal of Physical Anthropology, v. 75, p. 303–328.

Hirsch, K., Krishtalka, L., and Stucky, R. K., 1987, Revision of trhe Wind River faunas, early Eocene of central Wyoming; Part 8, First fossil lizard egg (?Gekkonidae) and list of associated lizards: Annals of Carnegie Museum, v. 56, p. 223–230.

Jepsen, G. L., 1970, Bat origins and evolution: Biology of Bats, v. 1, p. 1–64.

Kay, R. F., and Hylander, W. L., 1978, The dental structure of mammalian folivores with special reference to Primates and Phalangeroidea (Marsupialia), *in* Montgomery, G. G., ed., The ecology of arboreal folivores: Washington, D.C., Smithsonian Institution Press, p. 173–196.

Keefer, W. R., 1965, Stratigraphy and geologic history of the uppermost Cretaceous, Paleocene, and lower Eocene rocks in the Wind River Basin, Wyoming: U.S. Geological Survey Professional Paper 495–A, p. A1–A76.

—— , 1970, Structural geology of the Wind River Basin, Wyoming: U.S. Geological Survey Professional Paper 495–D, p. D1–D35.

Kihm, A. J., 1984, Early Eocene mammalian faunas of the Piceance Creek Basin, northwestern Colorado [Ph.D. thesis]: Boulder, University of Colorado, 381 p.

Korth, W. W., 1979, Taphonomy of microvertebrate fossil assemblages: Annals of Carnegie Museum, v. 48, p. 235–285.

—— , 1982, Revision of the Wind River faunas, early Eocene of central Wyoming; Part 2, Geologic setting: Annals of Carnegie Museum, v. 51, p. 57–78.

Kraus, M. J., 1987, Integration of channel and floodplain suites; 2, Vertical relations of alluvial paleosols: Journal of Sedimentary Petrology, v. 57, p. 602–612.

Kraus, M. J., and Middleton, L. J., 1987, Contrasting architecture of two alluvial suites in different structural settings, *in* Ethridge, F. G., Flores, R. M., and Harvey, M. D., eds., Recent developments in fluvial sedimentology: Society of Economic Paleontologists and Mineralogists Special Publication 39, p. 253–262.

Krishtalka, L., and Stucky, R. K., 1983, Revision of the Wind River faunas, early Eocene of central Wyoming; Part 3, Marsupialia: Annals of Carnegie Museum, v. 52, p. 205–228.

—— , 1985, Revision of the Wind River faunas, early Eocene of central Wyoming; Part 7, Revision of *Diacodexis* (Mammalia, Artiodactyla): Annals of Carnegie Museum, v. 54, p. 413–486.

Krishtalka, L., Black, C. C., and Riedel, D. W., 1975, Paleontology and geology of the Badwater Creek area, central Wyoming; Part 10, A late Paleocene mammal fauna from the Shotgun Member of the Fort Union Formation: Annals of Carnegie Museum, v. 45, p. 179–212.

Krishtalka, L., and 9 others, 1987, Eocene (Wasatchian through Duchesnean) chronology of North America, *in* Woodburne, M. E., ed., Cenozoic biochronology of North America: Berkeley, University of California Press, p. 77–117.

Love, J. D., 1978, Cenozoic thrust and normal faulting, and tectonic history of the Badwater area, northeastern margin of the Wind River Basin, Wyoming: Wyoming Geological Association 30th Annual Field Conference Guidebook, p. 235–238.

McCoy, E. D., and Connor, E. F., 1980, Latitudinal gradients in the species diversity of North American mammals: Evolution, v. 34, p. 193–203.

McKenna, M. C., 1960, Fossil Mammalia from the early Wasatchian Four Mile local fauna, Eocene of northwest Colorado: University of California Publications in Geological Sciences, v. 37, p. 1–130.

Nesbit Evans, E. M., Van Couvering, J.A.H., and Andrews, P., 1981, Palaeoecology of Miocene sites in western Kenya: Journal of Human Evolution, v. 10, p. 99–116.

Osborn, H. F., 1929, The titanotheres of ancient Wyoming, Dakota, and Nebraska: U.S. Geological Survey Monograph 55, 953 p.

Pennycuick, C. J., 1979, Energy costs of locomotion and the concept of "foraging radius," *in* Sinclair, A.R.E., and Norton-Griffiths, M., eds., Serengeti; Dynamics of an ecosystem: University of Chicago, p. 164–184.

Retallack, G. J., 1984, Completeness of the rock and fossil record; Estimates using fossil soils: Paleobiology, v. 10, p. 59–78.

Rickleffs, R. E., and Travis, J., 1980, A morphological approach to the study of avian community organization: Auk, v. 97, p. 321–338.

Robinson, P., 1966, Fossil Mammalia of the Huerfano Formation, Eocene of

Colorado: New Haven, Connecticut, Yale University Bulletin of Peabody Museum of Natural History, v. 21, p. 1–95.

Rose, K. D., 1981a, The Clarkforkian Land-Mammal Age and mammalian faunal composition across the Paleocene–Eocene boundary: University of Michigan Papers in Paleontology, v. 26, p. 1–197.

——, 1981b, Composition and species diversity in Paleocene and Eocene mammal assemblages; An empirical study: Journal of Vertebrate Paleontology, v. 1, p. 367–388.

——, 1982, Skeleton of *Diacodexis,* oldest known artiodactyl: Science, v. 216, p. 621–623.

——, 1985, Comparative osteology of North American dichobunid artiodactyls: Journal of Paleontology, v. 59, p. 1203–1226.

——, 1987, New skeletal remains of Eocene palaeanodonts [abs.]: Journal of Vertebrate Paleontology, v. 7, supplement to n. 3, p. 24A.

Schumm, S. A., 1963, Sinuosity of alluvial rivers of the Great Plains: Geological Society of America Bulletin, v. 74, p. 1089–1100.

Seeland, D., 1978a, Sedimentology and stratigraphy of the lower Eocene Wind River Formation, central Wyoming: Wyoming Geological Association 13th Annual Field Conference Guidebook, p. 181–191.

——, 1978b, Eocene fluvial drainage patterns and their implications for uranium and hydrocarbon exploration in the Wind River Basin, Wyoming: U.S. Geological Survey Bulletin 1446, 21 p.

Setoguchi, T., 1978, Paleontology and geology of the Badwater Creek area, central Wyoming; Part 16, The Cedar Ridge local fauna (late Oligocene): Bulletin of Carnegie Museum of Natural History, no. 9, p. 1–61.

Simberloff, D., 1978, Use of rarefaction and related methods in ecology, *in* Dickson, K. L., Cairns, J., Jr., and Livingston, R. J., eds., Biological data in water pollution assessment; Quantitative and statistical analyses: American Society for Testing and Materials, p. 150–165.

Sinclair, W. J., and Granger, W., 1911, Eocene and Oligocene of the Wind River and Bighorn Basins: Bulletin of the American Museum of Natural History, v. 30, p. 85–117.

Smith, D. G., and Smith, N. D., 1980, Sedimentation in anastomosed river systems; Examples from alluvial valleys near Banff, Alberta: Journal of Sedimentary Petrology, v. 50, p. 157–164.

Storer, J. E., 1984, Mammals of the Swift Current Creek local fauna (Eocene: Uintan, Saskatchewan: Saskatchewan Culture and Recreation, Museum of Natural History, Natural History Contribution 7, 158 p.

Stucky, R. K., 1984a, Revision of the Wind River faunas, early Eocene of central Wyoming; Part 5, Geology and biostratigraphy of the upper part of the Wind River Formation, northeastern Wind River Basin: Annals of Carnegie Museum, v. 53, p. 231–294.

——, 1984b, Revision of the Wind River faunas, early Eocene of central Wyoming; Part 6, Stratigraphic sections and locality descriptions, upper part of the Wind River Formation, northeastern Wind River Basin: Annals of Carnegie Museum, v. 53, p. 295–323.

——, 1984c, The Wasatchian–Bridgerian Land Mammal Age boundary (early to middle Eocene) in western North America: Annals of Carnegie Museum, v. 53, p. 347–382.

——, 1988, Geology from afar; Satellite and aircraft remote sensing of ancient fossil deposits: Carnegie Magazine, v. 59, p. 12–17.

——, 1989a, The anatomy of continental diversity; Mammalian faunal dynamics in the North American Eocene, *in* Stenseth, N. C., ed., Coevolution in ecosystems; The Red Queen hypothesis: Cambridge University Press (in press).

——, 1989b, The evolution of land mammal diversity in North America during the Cenozoic: Current Mammalogy, v. 2, p. 375–432.

Stucky, R. K., and Krishtalka, L., 1982, Revision of the Wind River faunas, early Eocene of central Wyoming; Part 1, Introduction and Multituberculata: Annals of Carnegie Museum, v. 51, p. 39–56.

——, 1983, Revision of the Wind River faunas, early Eocene of central Wyoming; Part 4, The Tillodontia: Annals of Carnegie Museum, v. 52, p. 375–391.

Stucky, R. K., Lang, H. R., Krishtalka, L., and Redline, A. D., 1987, Analysis of Paleocene/Eocene depositional environments; Preliminary TM and TIMS results, Wind River Basin, Wyoming: Ann Arbor, Michigan, Proceedings, International Geoscience and Remote Sensing Symposium, 1987, p. 1163–1168.

Tourtelot, H. A., 1948, Tertiary rocks in the northeastern part of the Wind River Basin, Wyoming: Wyoming Geological Association Third Annual Field Conference Guidebook, p. 53–67.

West, R. M., 1973, Geology and mammalian paleontology of the New Fork–Big Sandy area, Sublette County, Wyoming: Fieldiana Geology, v. 29, p. 1–193.

Wood, H. E., Jr., and 6 others, 1941, Nomenclature and correlation of the North American continental Tertiary: Geological Society of America Bulletin, v. 42, p. 1–48.

MANUSCRIPT ACCEPTED BY THE SOCIETY JUNE 12, 1989

Printed in U.S.A.

Geological Society of America
Special Paper 243
1990

Mammals of the Bridgerian (middle Eocene) Elderberry Canyon Local Fauna of eastern Nevada

Robert J. Emry
Department of Paleobiology, National Museum of Natural History, Smithsonian Institution, Washington, D.C. 20560

ABSTRACT

The first Eocene vertebrate assemblage known from the Great Basin, the Elderberry Canyon Local Fauna, occurs in rocks referred to the Sheep Pass Formation near Ely, Nevada. Approximately 40 taxa are now known, including small anuran amphibians, small reptiles, birds, and mammals. The mammalian component consists of: the insectivorans *Apatemys bellus, Pantolestes longicaudus,* a tiny apternodont, at least one nyctitheriid, and at least four other taxa representing dormaalid and/or erinaceid erinaceomorphs; an epoicotheriid palaeanodont, cf. *Tetrapassalus mckennai;* the primates *Notharctus tenebrosus, Trogolemur myodes,* and two species of uintasoricines; the rodents *Reithroparamys delicatissimus, R.* cf. *R. huerfanensis, Sciuravus* sp., *Microparamys* sp., *Pauromys* sp., *Mattimys* sp., and two new genera; the hyaenodont *Sinopa minor;* two viverravid carnivores including *Viverravus;* the condylarth *Hyopsodus paulus;* the perissodactyls *Hyrachyus modestus, Hyrachyus affinis, Helaletes nanus, Isectolophus latidens,* and a new genus and species; and the artiodactyl *Antiacodon pygmaeus.*

Greatest faunal similarity is with the Black's Fork Member (or lower), Bridger Formation, and other early Bridgerian localities such as Powder Wash in the Douglas Creek Member of the Green River Formation in northeastern Utah. The age of the Elderberry Canyon Local Fauna can confidently be called early Bridgerian.

The Elderberry Canyon Fauna is preserved in carbonate rocks believed to have been deposited in a shallow, warm, heavily vegetated, permanent, hardwater lake. The mammals lived on marshy wetland terrain adjacent to the lake, although some faunal elements may have been transported in from more distant habitats.

INTRODUCTION

This chapter is a preliminary account of the first Eocene vertebrate fauna known from the Great Basin. Except for the rather limited assemblage of Chadronian mammals from Titus Canyon (Stock, 1949), near Death Valley, California, no other Paleogene vertebrate faunas are known from this province. The fossil bone concentration that has produced this new Eocene assemblage, herein called the Elderberry Canyon Local Fauna, was discovered in the summer of 1975 by Dr. Thomas Fouch, his daughter Melissa Fouch, and Forrest G. Poole of the U.S. Geological Survey, who collected several blocks, amounting to about one-tenth cubic meter (about three cubic feet), of bone-bearing limestone from a locality near Ely, Nevada (Fig. 1). When Fouch submitted the material to Dr. G. Edward Lewis, also of the Lakewood, Colorado, survey office, for identification, Lewis suggested he send the material to the National Museum in Washington for identification by Dr. C. Lewis Gazin (at that time Paleogene mammals collected by Survey geologists from the Denver center were routinely sent to Gazin for identification, while Neogene mammals were identified by Lewis and retained in Lakewood). When the blocks of bone-bearing limestone arrived in Washington, Gazin, who was then in his sixth year of retirement though still maintaining an office in the museum, brought the material to

Emry, R. J., 1990, Mammals of the Bridgerian (middle Eocene) Elderberry Canyon Local Fauna of eastern Nevada, *in* Bown, T. M., and Rose, K. D., eds., Dawn of the Age of Mammals in the northern part of the Rocky Mountain Interior, North America: Boulder, Colorado, Geological Society of America, Special Paper 243.

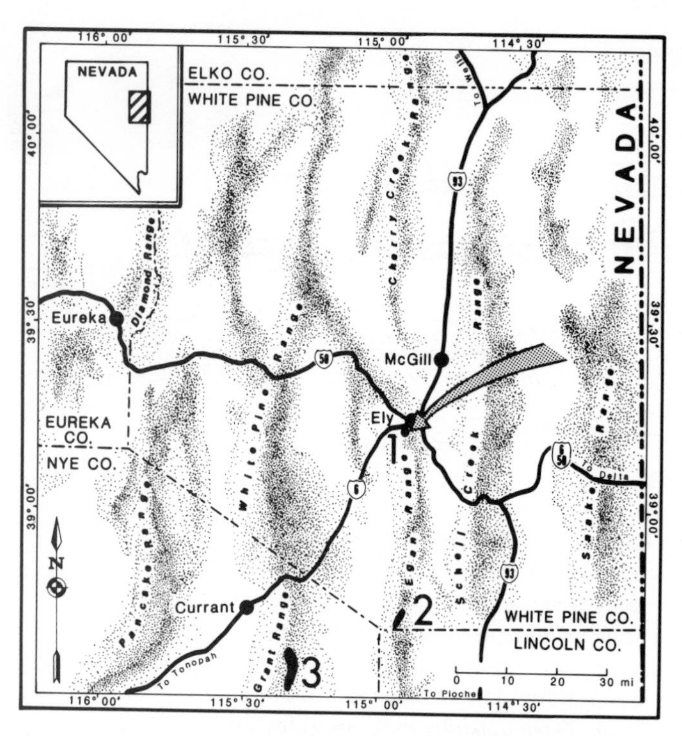

Figure 1. Map indicating approximate location of Elderberry Canyon Quarry (1), the type area of the Sheep Pass Formation in Sheep Pass Canyon (2), and an additional area of referred Sheep Pass Formation (3) from which Fouch (1979) reported the occurrence of *Nyctitherium* cf. *N. velox.*

my attention and suggested that our laboratory prepare enough of the material to provide an age determination for Fouch. When preparation revealed that the bones were those of the Eocene ceratomorph *Hyrachyus,* I contacted Fouch, who generously offered to show me the locality in Elderberry Canyon, and others in eastern Nevada, the following summer.

At the Elderberry Canyon locality the fossil bones are concentrated in a bed of limestone about 25 to 30 cm thick at the base of a ledge of limestone 2 m, and more, in thickness (Fig. 2). The bone-bearing units immediately overlie a poorly lithified calcareous mudstone, which in the natural outcrop provided an erosional undercut beneath the bone-bearing unit. Additional blocks of the limestone could be collected by undercutting them and prying them out. Most of the bones are concentrated in beds below a minor erosional disconformity within the larger limestone unit (Fig. 3), although bones occur rarely in limestone above this erosional surface.

Additional blocks of limestone were collected in 1976, and when it was determined through preparation that a variety of mammals were represented, more bone-bearing rock was collected in 1977, 1979, and 1980. By 1980 it was no longer possible to extract more of the fossiliferous unit without removing the overlying rock. So in 1983, with the assistance of Dan Chaney and Arnold Lewis of the U.S. National Museum's Vertebrate Paleontology preparation laboratory, and a gasoline-powered jackhammer, the rock was removed from an area approximately 2 by 3 m, and the productive layer was collected from most of this area. This rock is still being processed.

Laboratory preparation of the material has been a tedious, time-consuming process, most of which has been accomplished by Frederick Grady and Dan Chaney. Only a millimeter or two of rock can be removed in a dilute formic acid bath before the rock must be washed and dried so that all of the exposed bone can be systematically hardened before the next acid cycle can proceed. Preservation of teeth is usually quite good, and even the jaws of smaller mammals are often very well preserved. The skulls and larger bones are usually very much compressed.

The collection continues to grow, and additional taxa are regularly added to the faunal list. The number of vertebrate taxa represented is now approximately 40. The number is uncertain because a variety of small anuran amphibians and small reptiles (lizards and possibly snakes) are not yet sorted and identified. At least two different birds are present, but the avian material recovered to date is not sufficiently well preserved to afford positive identifications. At least 30 kinds of mammals have been recognized, but some identifications are still uncertain.

Invertebrates reported from the site by Fouch (1979) include bivalve and gastropod molluscs, and unidentified ostracodes. Plant root casts and other impressions also occur in the limestone.

Abbreviations used

Acronym prefixes denoting institutional collections are: AMNH = American Museum of Natural History, New York; CMNH = The Carnegie Museum of Natural History, Pittsburgh; USNM = National Museum of Natural History (U.S. National Museum), Washington, D.C.; YPM = Yale Peabody Museum, New Haven. When used with skeletal elements or in dental notations, L = left and R = right. With measurements, AP = anteroposterior, and TR = transverse. L.F. = local fauna.

GEOLOGIC SETTING

The Elderberry Canyon Quarry and its fauna occur in a sequence of conglomerate and carbonate beds on the east flank of the Egan Range just south of Ely, Nevada. The bone bed is in limestone overlying conglomerate about 50 m above the base of the section, which is locally more than 200 m thick (Fouch, 1979). This sequence has been referred to the Sheep Pass Formation, a name applied by Winfrey (1958) to a 980-m-thick sequence of nonmarine carbonate and terrigenous units well exposed in Sheep Pass Canyon, which is also in the Egan Range about 50 km (30 mi) south of Ely (locality no. 2 of Fig. 1). Kellogg (1964) measured 1,014 m of section in Sheep Pass Canyon, described the formation in more detail, and mapped its distribution in the southern Egan Range.

The Paleogene rocks of east-central Nevada are character-

Figure 2. View of Elderberry Canyon Quarry as seen looking northward across head of Elderberry Canyon. The bone-bearing horizon is between the pointers, partially obscured by the spoil pile. Schell Creek Range is seen in right distance.

ized generically as "prevolcanic" and "synvolcanic" (Fouch, 1979), with units formed prior to the Oligocene generally free of volcanic material, and those of Oligocene and younger age containing abundant volcanic detritus. In its type area, the Sheep Pass Formation is nonvolcanic, and characterizes the "prevolcanic" phase of sedimentation. The sequence in Elderberry Canyon, however, is nonvolcanic in the lower part, but grades upward into carbonate units that contain tuffaceous material (Fouch, 1979). The Elderberry Canyon Local Fauna occurs in the lower, nonvolcanic part. Fouch believes that the sequence at Elderberry Canyon may be younger than any part of the Sheep Pass Formation at its type area in Sheep Pass Canyon (locality no. 2, Fig. 1).

Fouch (written communication, 1987) suggests that the Sheep Pass Formation at its type area is probably equivalent to most of the Maastrichtian and Paleocene North Horn Formation of central Utah, the Paleocene and early Eocene Flagstaff Member, and the early Eocene lower part of the main body of the Green River Formation of northeastern Utah, whereas the sequence at Elderberry Canyon may be somewhat younger, probably equivalent to Eocene parts of the Elko Formation of northeastern Nevada and to the Mahogany oil-shale bed and younger parts of the Green River Formation in northeastern Utah and northwestern Colorado. The mammalian assemblage indicates an age no younger than early Bridgerian for the fossiliferous part of the Sheep Pass Formation exposed in the lower part of the Elderberry Canyon section. See Krishtalka and others (1987) for further discussion and correlation of Bridgerian faunas.

Fouch (1979; written communication, 1987) interprets coarser fractions of the Elderberry Canyon section as having been deposited on alluvial fans and fan deltas that extended into the margins of a lake, and the fine-grained sediments in a warm, heavily vegetated, shallow, permanent lake, with slow-moving currents. The prevolcanic Paleogene rocks of eastern Nevada, as a whole, have such abrupt changes in lithofacies, and such limited exposures, that Fouch (1979) found it difficult to determine the extent of individual lakes or lake phases, but thought the evidence most compatible with deposition in a series of smaller separated lakes, perhaps temporarily connected in both space and time. The depositional environment of the Elderberry Canyon Local Fauna may not have been very different from that of the Bridger fauna. The preponderance of carbonates in east-central Nevada may tell less about the depositional environment than it does about the composition of the sediment source.

Figure 3. Elderberry Canyon Quarry showing floor and face exposed during 1983 excavation. Bones are concentrated in beds above the interface indicated at the points of the triangles and below the erosional disconformity at the points of the arrows. A few bones were encountered above this erosional break.

SYSTEMATIC PALEONTOLOGY

As mentioned above, the Elderberry Canyon Quarry has root casts and other plant impressions and several kinds of invertebrate animals that will not be further discussed here. Among the vertebrates, small anuran amphibians and small reptiles are represented by many specimens that are not yet sorted and identified. Birds are represented by several limb elements too poorly preserved to allow identification, though size alone confirms the presence of at least two taxa. The systematic paleontology for purposes of this report will be limited to systematic paleomammalogy.

Class Mammalia
Superorder Insectivora

The Insectivorans, as the inclusive group recognized by Novacek (1986), are the most abundant in terms of number of specimens, and the most diverse in terms of species represented, of the mammals of the Elderberry Canyon assemblage. Included are at least: a very small apternodontid, the apatemyid *Apatemys,* the pantolestid *Pantolestes,* one and more likely two nyctitheriids, a possible leptictid, and at least four genera of erinaceid and/or dormaalid erinacemorphs. The apatemyid and the pantolestid are confidently identified and systematically treated below. The other insectivorans are either not yet firmly identified, or represent new taxa, and will be the subject of other studies.

Family Apatemyidae Matthew, 1909

Apatemys Marsh, 1872

The specimens of *Apatemys* from Elderberry Canyon are identified in the context of the analysis of Bown and Schankler (1982), which seems less counterintuitive than West's conclusions (1973b), although the species assignment here would also probably agree with West's systematics.

Apatemys bellus Marsh, 1872
Figure 4A, B, C

Referred specimens. All numbers with USNM prefix. 417406, left ramal fragments with M_1; 417444, left ramal fragment with part of canine and P_4; 417420, LP^4; 417421, LM^1; 417422, LM^1; 417423, LM^1; 417424, RM^1; 417425, RM^2.

Discussion. The sample from Elderberry Canyon is small, but nevertheless some specimens do have characters with systematic value. USNM 417406 and 417444 are believed to be parts of the

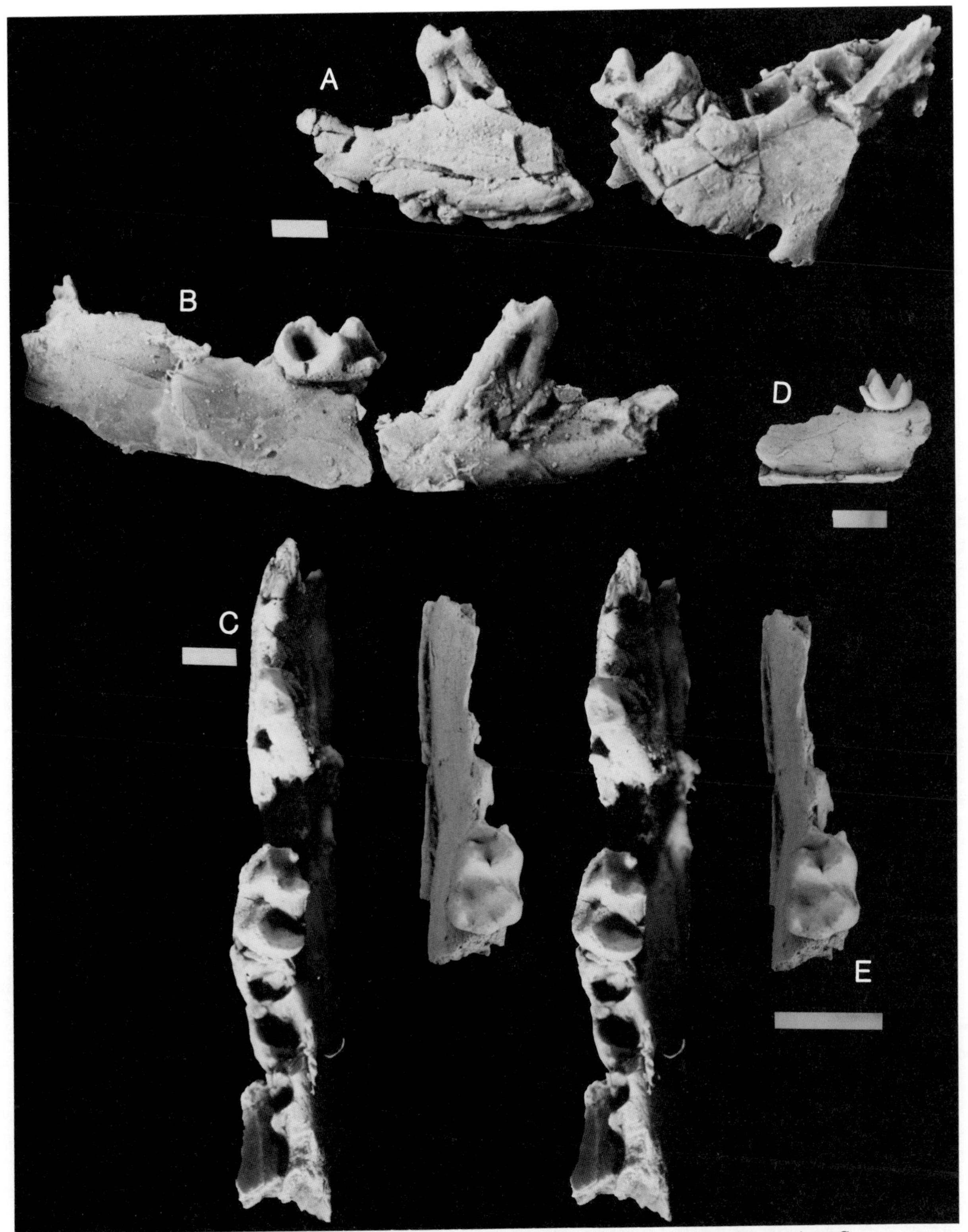

Figure 4. *Apatemys bellus* and *Pantolestes longicaudus* from Elderberry Canyon Quarry. A to C, *Apatemys bellus,* USNM 417444 (anterior section) and USNM 417406 (posterior section), in (A) buccal, (B) lingual, and (C) stereo occlusal views. D to E, *Pantolestes longicaudus,* USNM 417361, in (D) buccal and (E) stereo occlusal views. A to C approximately ×10, bar = 1 mm. D approximately ×2, E approximately ×4, bars = 5 mm.

same jaw, although the two parts have no definite contact; hence the two parts bear separate catalog numbers. Both jaw parts appear to be apatemyid; they were found near each other on one block of limestone, and *Apatemys* is not otherwise common in the assemblage. It seems improbable that complementary parts of two individuals of a rare taxon would have been preserved so close together, and without additional parts of either individual. The combination of the two jaw parts does, however, make up an unusual, somewhat enigmatic specimen. The posterior part with M_1 is clearly and comfortably assigned to *A. bellus.* However, the anterior part has a two-rooted P_4 (which occurs in some *A. bellus*), but also has two alveoli between the enlarged canine and P_4 (which does not seem to be the case in any known apatemyid). In previously known apatemyids, the procumbent bladelike premolar anterior to P_4 has but a single root at its posterior end. Of the two alveoli present here in USNM 417406, the more posterior is much the smaller. If the two jaw fragments are correctly associated, then the specimen is apparently unique in having either a double-rooted P_3 (also sometimes termed P_2 in the literature) or, less likely, two single-rooted premolars. Given the lack of certain association between the two parts, it seems best to refer it to *A. bellus,* where it fits best in terms of M_1 morphology and size, and which is also known to have a double-rooted P_4 in some specimens (McKenna, 1963; West, 1973). Due to breakage, the number of mental foramina cannot be determined for the Elderberry Canyon material, although the posterior section of the jaw has one foramen beneath M_1.

Measurements in mm. USNM 417406: P_4, AP = 1.1, TR = 0.7; M_1, AP = 2.1, TR = 1.4. USNM 417420: P^4, AP = 1.69, TR = 1.51. USNM 417421: M^1, AP = 1.77, TR = 2.26. USNM 417422: M^1, AP = 1.84, TR = 2.09. USNM 417423: M^1, AP = 1.47, TR = 2.12. USNM 417424: M^1, AP = 1.65, TR = 1.87. USNM 417425: M^2, AP = 1.42, TR = 1.65.

Family Pantolestidae Cope, 1884

***Pantolestes* Cope, 1872b**
***Pantolestes longicaudus* Cope, 1872b**
Figure 4D, E

Referred specimens. USNM 417361, ramal fragment with M_1; USNM 417417, RP^4; USNM 417418, RP^3; USNM 417419, LM^1.

Discussion. The few teeth assigned here are clearly *Pantolestes,* and are referred to *P. longicaudus* on the basis of size. They compare well with specimens from the lower Bridger Formation assigned to *P. longicaudus.*

Measurements in mm. USNM 417361: M_1, AP = 5.0, TR = 3.6. USNM 417417: P^4, AP = 4.08, TR = 4.09. USNM 417418: P^3, AP = 3.84, TR = 2.24. USNM 417419: M^1, too damaged for useful measurements.

Order Chiroptera
Suborder Microchiroptera, incertae sedis.
Figure 5

Material. USNM 417350, skull, mandible, associated partial postcranial skeleton; USNM 417351, left mandible with P_4–M_3. Several uncataloged isolated teeth.

Discussion. At least two species of bat are present in the Elderberry Canyon assemblage, each represented by material sufficiently informative to allow identification or description. This necessarily will be taken up in future studies, when it is possible to assemble the critical comparative material.

? Order Pholidota
Suborder Palaeanodonta Matthew, 1918
Family Epoicotheriidae Simpson, 1927

***Tetrapassalus* Simpson, 1959**
? *Tetrapassalus,* cf., *T. mckennai* Simpson, 1959

Referred material. USNM 417422, right maxilla with canine, three postcanine teeth, and an alveolus for a fourth postcanine; USNM 417353, both mandibular rami, each with one incisor, canine, three postcanines and alveoli for two additional postcanines, and associated right humerus and right femur.

Discussion. The Elderberry Canyon epoicothere is represented by material that would certainly be adequate to identify it with certainty, were it not for inadequacies in the type material of taxa with which it must be compared. The presence of five postcanine tooth loci in each dentary, whereas the type and only specimen of *T. mckennai* appears to have but four, would seem to put this referral out of the question. I nevertheless refer it provisionally to *T. mckennai* because of its similar size, and because I believe it is not certain that *T. mckennai* had only four dentary postcanine teeth. Bridgerian epoicothere taxa (*T. mckennai* Simpson, 1959, and *T. proius* West, 1973a) are known from only a single specimen each. Consequently we know nothing of intraspecific variation in these small palaeanodonts, and they have little value in biostratigraphic correlation. For purposes of this report, I simply record the presence of an epoicotheriid palaeanodont in the Elderberry Canyon Local Fauna, with provisional taxonomic allocation. A more elaborate description and discussion will be the subject of another study.

Measurements in mm. USNM 417353, depth of jaw, lingually, below last postcanine tooth is approximately 2.4, both right and left, versus about 2.6 in *T. mckennai.*

Order Primates

At least three primate genera occur in the Elderberry Canyon Quarry: the adapid *Notharctus,* and the omomyids

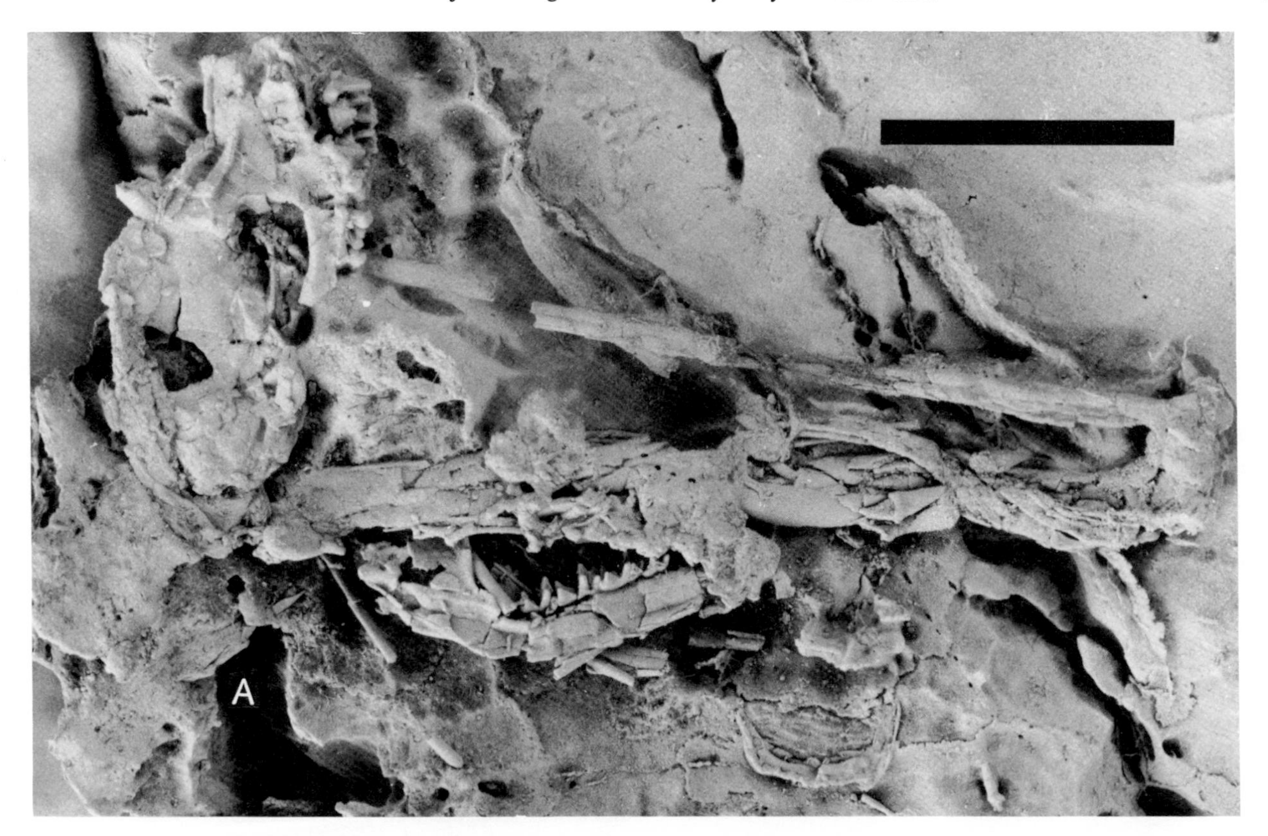

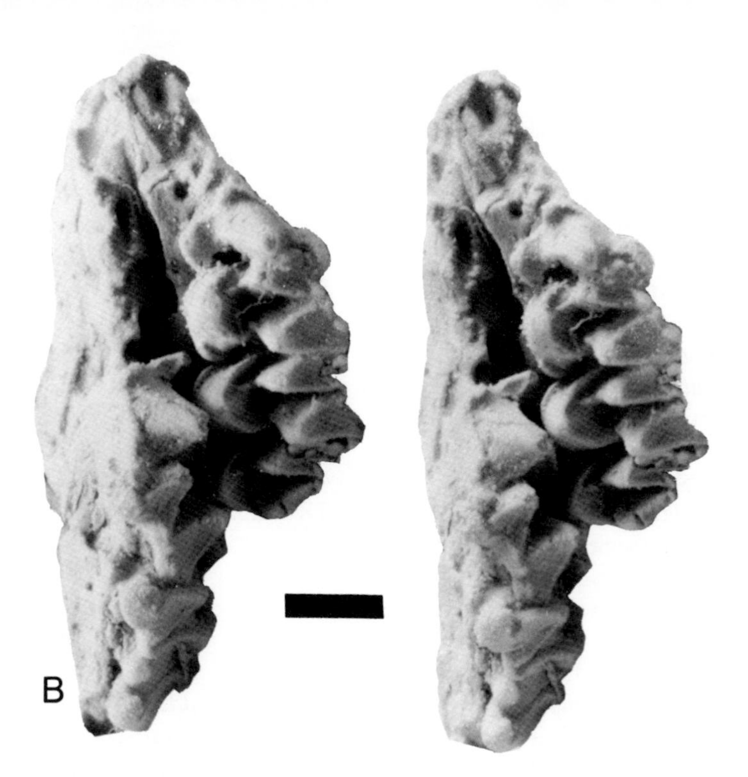

Figure 5. Microchropteran bat (USNM 417350) from Elderberry Canyon Quarry. A, overall view showing skull (upper left), left mandibular ramus, and associated postcranial elements, approximately ×5. B, enlarged stereo view of palate, approximately ×14.5, scale bar = 1 mm.

Trogolemur and *Uintasorex*. There is considerable uncertainty in the literature as to whether *Uintasorex* is an omomyid or not, a microsyopid or not, or whether it is a primate or an insectivore. Having no further insight than previous authors into this question, I place it in Omomydae following Krishtalka (1978) and Krishtalka and Schwartz (1978).

Family Adapidae Trouessart, 1879

Notharctus Leidy, 1870b
Notharctus tenebrosus Leidy, 1870b

Referred specimen. USNM 417415, RM^2.
Discussion. Notharctus is represented by a single tooth, but there can be little doubt about its identification. It can be matched almost exactly, in morphology and size, by numerous specimens of *Notharctus tenebrosus* from the Black's Fork Member of the Bridger Formation.
Measurements in mm. USNM 417415, AP = 5.8, TR = 7.1 at the widest point. The latter measurement should be considered a maximum, because the tooth is slightly displaced along an anteroposterior fracture.

Family Omomyidae Trouessart, 1879

Trogolemur Matthew, 1909
Trogolemur myodes Matthew, 1909
Figure 6

Referred material. USNM 417355, left mandibular ramus with P_3–M_3 and root of I_1; USNM 417356, right mandibular ramus with I_1, P_3–M_3; USNM 417390, LM^2; USNM 417396, RM^2; USNM 417391, RM^3; USNM 417401, buccal portion of M^1 or M^2; USNM 417402, buccal portion of RM^1 or M^2; USNM 417389; LM_1; USNM 417392, LM_2?.
Discussion. The small size, relatively deep mandible, greatly enlarged I_1 and elongate M_3 talonid confirm the identification of this small primate as *Trogolemur,* at present the best-represented primate in the Elderberry Canyon assemblage. Included in the referred material is a specimen directly comparable to the type (AMNH 12559, a right mandible with P_2–M_3), allowing confident identification at the species level as *T. myodes.* In the Elderberry Canyon material, the enamel is somewhat more wrinkled in the talonid basins of M_3 than it appears to be in the type, but other than this, the dental morphology seems to be nearly identical in all comparable details with that of the type, which it also matches quite closely in size. USNM 417356 includes the complete I_1, not previously reported in *Trogolemur.* As expected from the large, deep root previously known, the I_1 is greatly enlarged (Fig. 6), slightly flattened transversely, and sharply pointed. The tip of this incisor appears to be recurved in Figure 6C, but this is due to displacement at a fracture; before breakage, the tip would have continued along the same curve as the rest of the incisor.

Maxillary dentition has not previously been reported for *Trogolemur.* The upper teeth referred here are not articulated nor directly associated with undoubted *Trogolemur* lower teeth, but there would seem to be little doubt that they are correctly allocated. The upper teeth are of a comparably small primate, no other lower dentition in the fauna can be associated with them; and no other upper dentitions can be associated with the *Trogolemur.* Although the greatly enlarged lower incisor is somewhat out of character for an omomyid, the upper molars do not seem to be particularly unusual, though they do seem to have some advanced characters. They resemble the molars of *Strigorhysis* (Bown, 1979) in some respects, including having well-developed conules, in having the strong enamel expansion of the base of the protocones, and in the relatively elevated posterior margin of the trigon, with confluent postprotocone crista and postcrista (Fig. 6). The teeth differ from those of *Strigorhysis* in their smaller size, lack of enamel crenulations, and in their more sharply pointed, labially leaning paracones and metacones.

Continued processing of limestone from the Elderberry Canyon site should produce additional specimens of this small primate, increasing the chances of finding more definite associations of upper and lower teeth. For now the maxillary dentition is known just from these tentative allocations.

Trogolemur was long known from but a single specimen, the type, described by Matthew in 1909. A second Bridger specimen was recognized in the Yale Peabody Museum collections and reported by Gazin (1958), and Robinson (1968) referred two isolated lower teeth from Badwater Creek Locality 5a. McKenna (1980) listed cf. *Trogolemur myodes* in the assemblage from bone bed A of the type section of the Tepee Trail Formation in northwestern Wyoming, but he did not indicate the nature of the material. If this identification is correct, the taxon may range from early Bridgerian to early Uintan, suggesting that its limited occurrences may be related more to ecologic than to temporal factors.
Measurements in mm. USNM 417356: M_{1-3} = 6.3; P_1–M_3 = 8.45; P_3, AP = 1.0, TR = 0.9; P_4, AP = 1.1, TR = 1.25; M_1, AP = 1.5, TR = 1.5; M_2, AP = 1.4, TR = 1.6; M_3, AP = 2.1, TR = 1.4. USNM 417355: M_{1-3} = 5.3; P_1–M_3 = 7.0; P_3, AP = 1.1, TR = 0.8; P_4, AP = 1.2, TR = 1.2; M_1, AP = 1.7, TR = 1.5; M_2, AP = 1.4, TR = 1.6; M_3, AP = 2.0, TR = 1.3. USNM 417389: M_1, AP = 1.7, TR = 1.5. USNM 417390: M^2, AP = 1.6, TR = 2.6. USNM 417391: M_3, AP = 1.3, TR = 2.2. USNM 417396: M^2, AP = 1.6, TR = 2.3. USNM 417401: ?M^1, AP = 1.6. 417402: ?M^1, AP = 1.6.

Uintasorex Matthew, 1909
Uintasorex species

Discussion. At least two taxa—if not two genera, then two species of *Uintasorex*—occur in the Elderberry Canyon assemblage. Their study is too tentative to include taxonomic conclusions here. One taxon is represented by two mandibular rami and isolated upper and lower teeth, and appears to be smaller than previously described species. The second taxon is represented by mandible and maxilla fragments and isolated teeth, is slightly

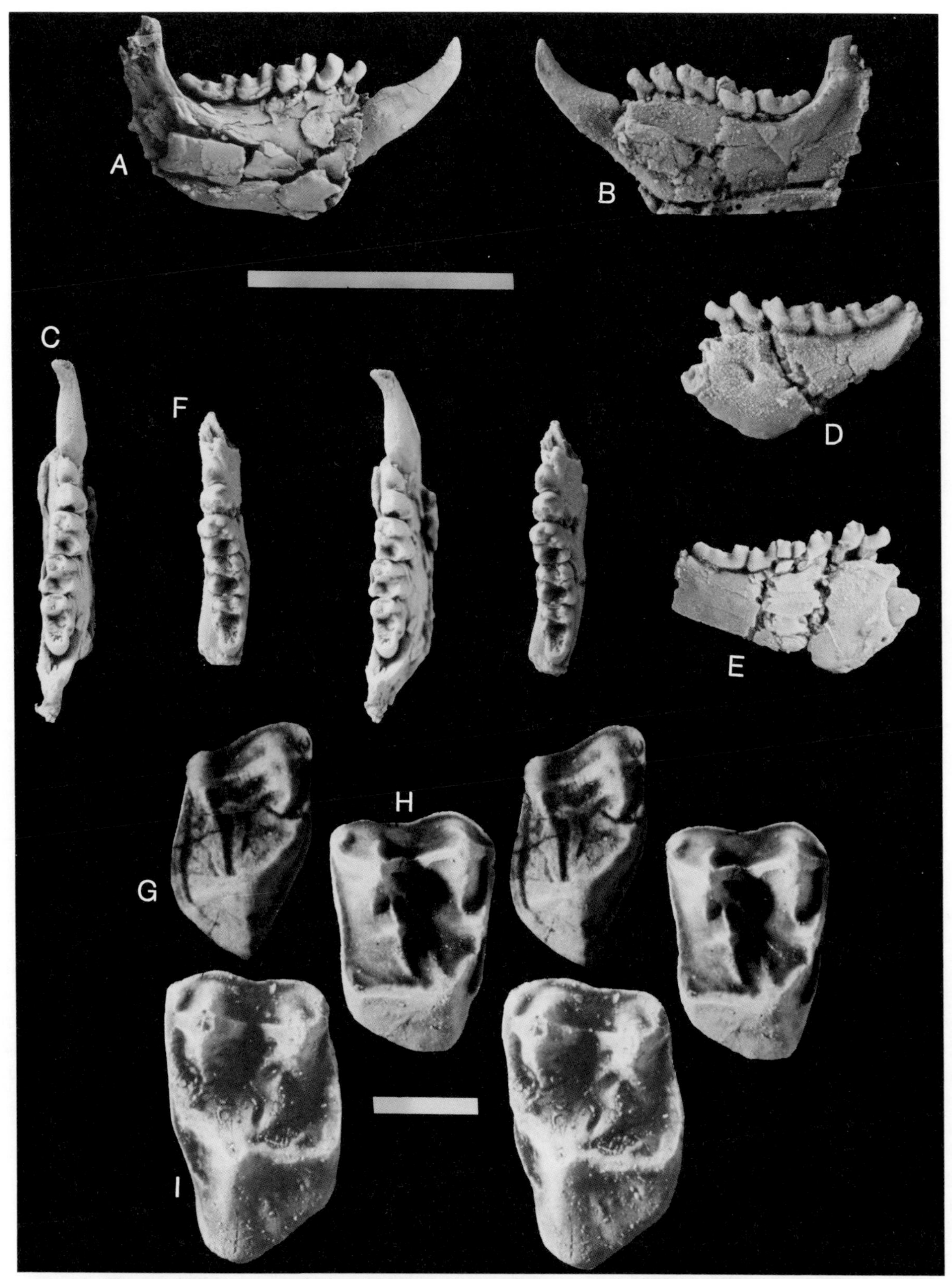

Figure 6. *Trogolemur myodes* from Elderberry Canyon Quarry. A to C, USNM 417356, right mandibular ramus in (A) buccal, (B) lingual, and (C) stereo occlusal views. D to F, USNM 417355, left mandibular ramus in (D) buccal, (E) lingual, and (F) stereo occlusal views, G., USNM 417391, right M3/, occlusal stereogram. H, USNM 417396, right M2/, occlusal stereogram. I, USNM 417390, left M2/, occlusal stereogram. A to F approximately ×5, bar = 1 cm. G to I, approximately ×20, bar = 1 mm.

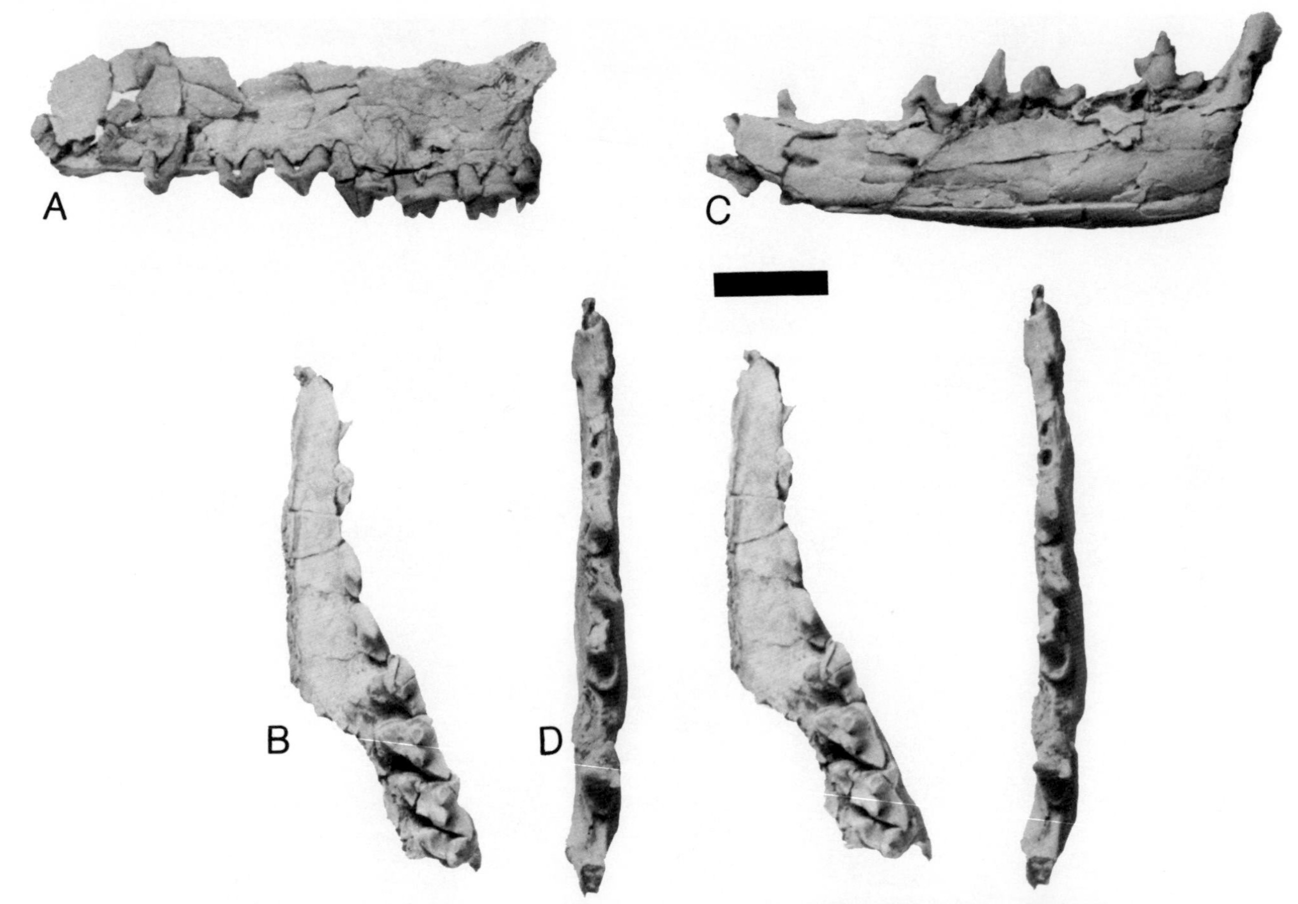

Figure 7. *"Sinopa" minor* from Elderberry Canyon Quarry. A to B, USNM 417345, left maxilla in (A) buccal and (B) stereo occlusal views, C to D, USNM 417346, left mandibular ramus in (C) buccal and (D) stereo occlusal views. All approximately ×1.5, bar = 1 cm.

larger, and may be referable to *U. parvulus* Matthew (1909), although this identification should be considered tentative. More detailed information will be presented in a separate study.

Order Rodentia

Discussion. The rodents of the Elderberry Canyon assemblage are the subject of an unfinished study, to be published separately with Dr. William Korth. A summary is given here of the present, somewhat tentative systematic conclusions of that study. At least eight taxa of rodents are present in the fauna, representing four families. Most of the material consists of isolated teeth, with only three specimens having two or more teeth definitely associated. Approximately 100 isolated rodent teeth have now been cataloged. It is perplexing, and perhaps ecologically or taphonomically significant, that better rodent material has not been recovered. Insectivorans and primates, as small, and surely as delicate, as most of the rodents represented, are known from fewer but better specimens.

The rodents presently identified are: *Reithroparamys delicatissimus* (Leidy, 1871b); *Reithroparamys* cf. *R. huerfanensis* Wood, 1962; *Sciuravus* sp.; *Microparamys* new species; a possible new genus cf. *Microparamys; Pauromys* new species; a new genus comparable to *Simimys;* and *Mattimys* sp.

Reithroparamys delicatissimus is previously known only from Bridgerian faunas. The undetermined *Sciuravus* is larger than *S. wilsoni,* the only known Wasatchian species, and is similar to *S. eucrastidens* and *S. nitidus,* both Bridgerian species. *Reithroparamys huerfanensis* is known heretofore only from the early Bridgerian upper Huerfano Formation of Colorado. The other rodents are either new, inconclusively identified, or not well enough known to afford an age assessment.

Order Creodonta
Family Hyaenodontidae Leidy, 1869

Sinopa Leidy, 1871a
Sinopa minor Wortman, 1902
Figure 7

Referred specimens. USNM 417345, left maxilla with P^1–M^3; USNM 417344 right maxilla with P^1–M^3; USNM 417346 left mandible with P_1, P_{3-4}, M_1, M_3.
Discussion. Comprehensive study of Eocene hyaenodonts, now

underway by others, will surely result in reassignment of this species, but it is not certain that it belongs in *Proviverra* Rütimeyer, 1862, where other species of *"Sinopa"* have been assigned. For purposes of this report it is more important to recognize that the Elderberry Canyon material is closely comparable to material from the lower Bridger Formation. Whatever its eventual higher taxonomic placement, it surely is conspecific with specimens from Bridger "B" in the USNM collections, and there seems no reason to doubt that this material represents the species described by Wortman (1902) as *Sinopa minor.* Within the Bridger Formation this small hyaenodont is not known to occur above the Blacks Fork Member, or lower Bridger.
Measurements in mm. USNM 417344: P^1–M^3 = 34.8; P^{1-4} = 21.7; M^{1-3} = 13.2; P^1, AP = 3.7, TR = 1.3; P^2, AP = 4.4, TR = 1.9; P^3, AP = 6.0, TR = 2.2; P^4, AP = 6.7, TR = 4.7; M^1, AP = 6.1, TR = 5.1; M^2, AP = 5.7, TR = 6.5; M^3, AP = 3.1, TR = 6.2. USNM 417345: P^1–M^3 = 36.7; P^{1-4} = 22.9; M^{1-3} = 13.6; P^1, AP = 3.8, TR = 1.3; P^2, AP = 4.7. TR = 1.8; P^3, AP = 5.3, TR = 2.3; P^4, AP = 5.8, TR = 4.9; M^1, AP = 5.9, TR = 5.5; M^2, AP = 5.7, TR = 5.8; M^3, AP = 3.3, TR = 5.4. USNM 417346: P_1–M_3 = 39.1; P_1, AP = 1.9, TR = 1.1; P_3, AP = 5.6, TR = 2.1; M_1, AP = 6.1, TR = 3.2; M_3, AP = 6.6, TR = 3.6.

Order Carnivora
Family Viverravidae Wortman and Matthew, 1899

I use the Family Viverravidae in the sense of Gingerich and Winkler (1985), who essentially perpetuate Matthew's (1909, 1915) scheme of dividing the Miacidae into two groups, those that lack M^3/M_3 while retaining functional M^2/M_2 (= Viverravinae), and those that retain M^3/M_3 (= Miacinae). Gingerich and Winkler recognize these groups at family level, the Viverravidae and Miacidae, respectively, and include in each family a number of genera unknown to Matthew.

Two "miacid" taxa occur in the Elderberry Canyon assemblage. One is comfortably included in *Viverravus.* The other is clearly a viverravid, as the family is recognized by Gingerich and Winkler. However, this second taxon has a combination of characters that would seem to leave it an orphan under the scheme proposed by Flynn and Galiano (1982), where it would be excluded from their Viverravidae, as they restrict the family to only *Viverravus* and *Simpsonictis,* and would also be excluded by some characters from their Didymictidae, which includes other miacids with and without M^3/M_3.

***Viverravus* Marsh, 1872**
***Viverravus minutus* Wortman, 1901**
Figure 8D, E

Referred specimen. USNM 417349, left mandibular ramus with P_2, P_4–M_2.
Discussion. This specimen has at least some of the characters that Flynn and Galiano (1982) considered important in recognizing *Viverravus.* The metaconid of M_1 is directly lateral to the protoconid, and if not lower than the paraconid (as Flynn and Galiano note is true for *Viverravus*), at least does not appear to be higher. The specimen lacks P_3, so its size relative to P_4 cannot be determined; the P_3 alveoli suggest that in its anteroposterior dimension it was not larger than P_4, but this would not necessarily mean the P_3 was smaller overall. The cristid obliqua of M_{1-2} is aligned anteroposteriorly rather than obliquely. When compared with *Viverravus* in the USNM collections from the Bridger Formation, USNM 417349 is most similar to specimens identified as *V. gracilis* and *V. minutus,* which seem to differ from each other in no important characters other than size. USNM 417349 is smaller than specimens assigned to *V. gracilis* and slightly larger than those assigned to *V. minutus.* I refer it to the latter species on the basis of its size. It might be noted that this particular specimen differs from all Bridger *Viverravus* that I was able to observe, in its extreme reduction of the metaconid and paraconid of M_2.
Measurements in mm. USNM 417349: P_2, AP = 2.4, TR = 1.0; P_4, AP = 4.3, TR = 2.0; M_1, AP = 5.5, TR = 2.9; M_2, AP = 3.1, TR = 1.8.

Viverravidae, incertae sedis
Figure 8A, B, C

Material. USNM 417347, skull, very much flattened dorsoventrally, with most postcanine dentition; USNM 417348, both mandibular rami, each with C–M_2.
Discussion. Though the skull and mandible are cataloged individually because they were not articulated nor indisputably associated, I am confident that they represent not only the same taxon, but the same individual. The skull and mandible were found near each other, and they complement each other in size and occlusal morphology; this, coupled with the extreme rarity of carnivore material otherwise in the Elderberry Canyon collection, would seem to argue strongly for association. The specimens are comparable in size to *Viverravus minutus,* though perhaps slightly smaller. The maxillary dentition is considerably disoriented and disaligned, though some individual teeth are essentially intact. As in *Viverravus,* M^3/M_3 is absent. The dental morphology of the upper teeth resembles that of *Viverravus* in some respects, but the M^1 is more reminiscent of *Protictis* in its large, curving parastyle cusp and its lack of a hypocone. M^{1-2} lack conules, resembling *Viverravus* in this character. The lower dentition also both agrees and disagrees with that of typical *Viverravus.* For example, Flynn and Galiano (1982) point out that the anterior accessory cusp of P_4 in Viverravidae is well ahead of the main cusp and far lingual to the anteroposterior axis of the tooth. USNM 417348 agrees, except that the cusp is not inflected lingually. The cristid obliqua of M_{1-2} is parallel to the anteroposterior axis of the tooth, as is the case in all *Viverravus,* according to Flynn and Galiano. An important character of *Viverravus,* considered unusual and derived by Flynn and Galiano, is its enlarged P_3, which is larger overall than P_4 (although they mention that the anteroposterior dimension of P_3 may not exceed that of P_4, they state that the main cusp is always higher in *Viverravus*). This is not the case in

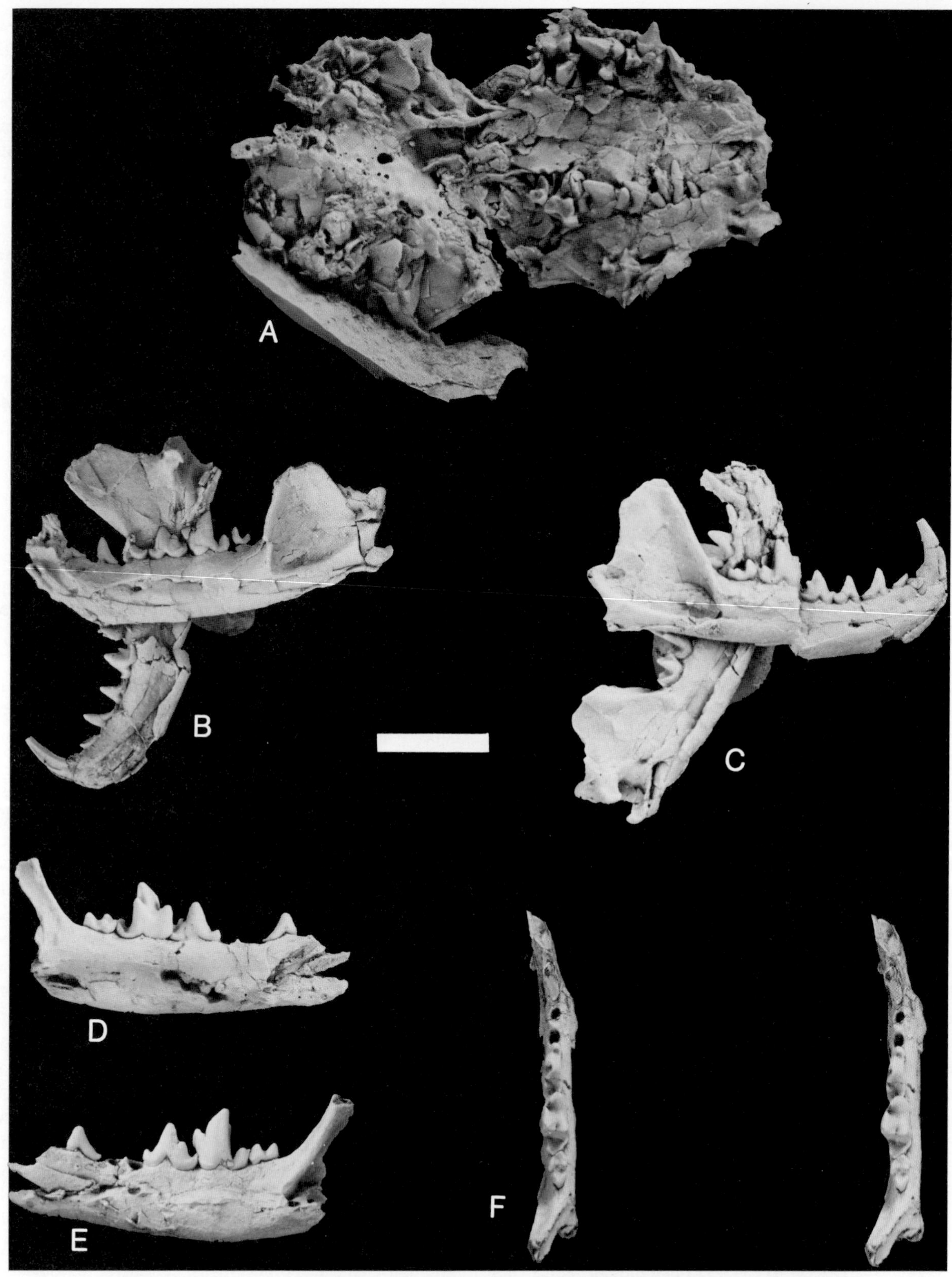

Figure 8. Carnivores from Elderberry Canyon Quarry. A, USNM 417347, Viverravidae, skull in ventral view. B to C, USNM 417348, Viverravidae, pair of mandibles in (B) left lateral and (C) right lateral views. D to E, USNM 417349, *Viverravus minutus*, left mandibular ramus in (D) lingual, (E) buccal, and (F) occlusal stereo views. All approximately ×2, bar = 1 cm.

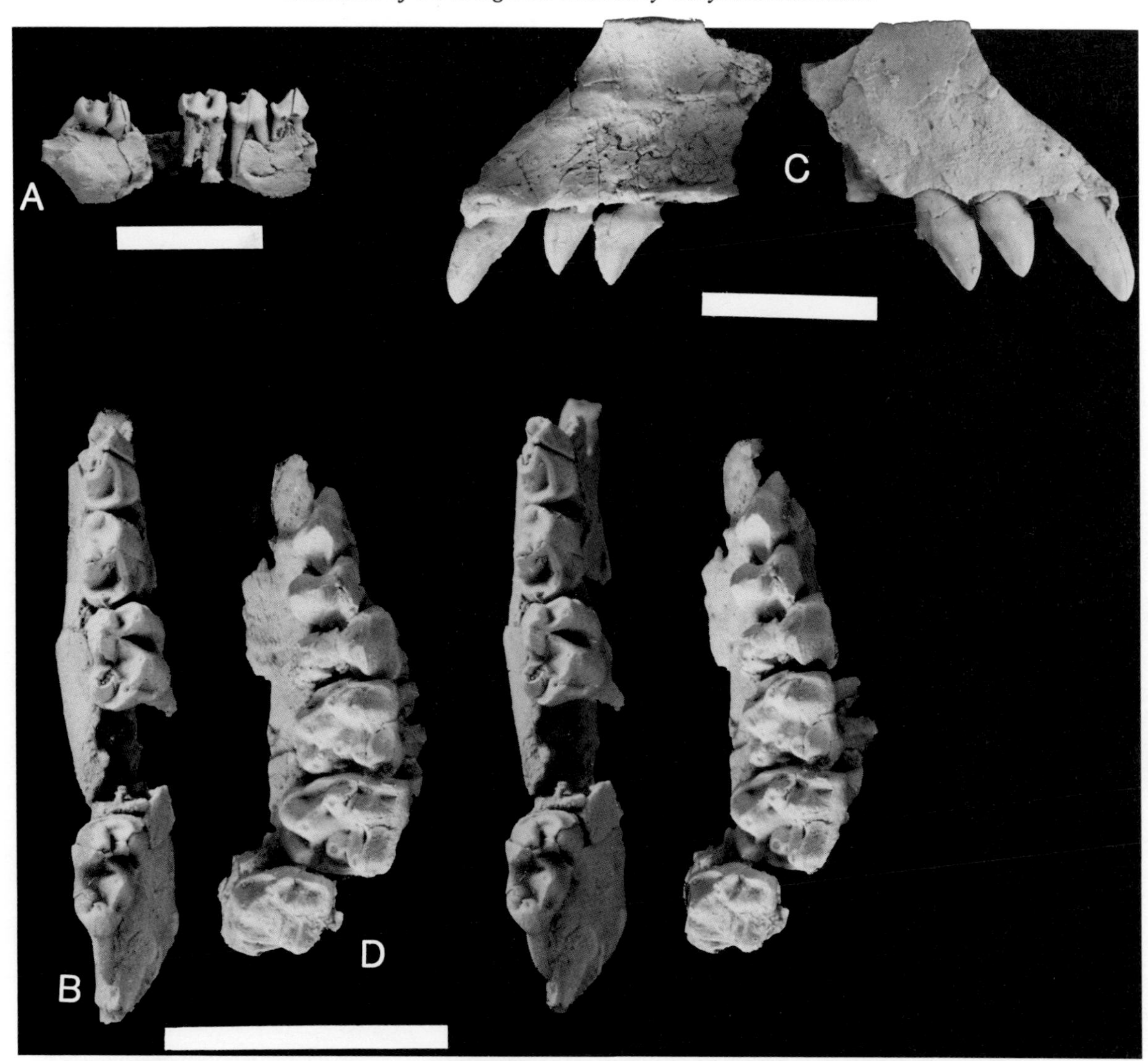

Figure 9. *Hyopsodus paulus* from Elderberry Canyon Quarry. A to B, USNM 417371, right P/3-M/1, M/3, in (A) buccal and (B) stereo occlusal views. C, USNM 417370, right premaxilla in medial and lateral views. D, USNM 417358, left maxilla in stereo occlusal view. A approximately ×2, bar = 1 cm. B and D approximately ×4, bar = 1 cm. C approximately ×5, bar = 5 mm.

USNM 417348; P_3 is smaller than P_4 in all dimensions, and moreover, is shorter than P_2.

These Elderberry Canyon specimens have characters that exclude them from *Viverravus* as that genus is defined by Flynn and Galiano (1982), but they would also be excluded by one character or another from both miacid families, the Viverravidae and Didymictidae, as these were characterized by Flynn and Galiano. For now, I prefer to simply assign the material to Viverravidae in the sense of Gingerich and Winkler (1985), and leave its generic assignment unsettled. The material may represent a new genus, but it might also be preferable to place it in *Viverravus* by defining that genus less restrictively than Flynn and Galiano did.

Measurements in mm. USNM 417347: M^1, AP = 3.3, TR = 4.2; M^2, AP = 1.7, TR = 3.0. USNM 417348: overall length of right mandible = 32.6; RP_1, AP = 1.5, TR = 0.9; RP_2, AP = 2.3, TR = 1.1; RP_3, AP = 2.6, TR = 1.1; RP_4, AP = 3.4, TR = 1.5; RM_1, AP = 5.0; RM_2, AP = 2.9; LP_1, AP = 1.75; LP_2, AP = 2.4, TR = 1.1; LP_3, AP = 2.3; LP_4, AP = 2.6; LM_1, AP = 4.6, TR = 2.9; LM_2, AP = 2.95, TR = 1.8.

Order Condylarthra
Family Hyopsodontidae Lydekker, 1889

Hyopsodus Leidy, 1870a
Hyopsodus paulus Leidy, 1870a
Figure 9

Referred specimens. All numbers with USNM prefix. 417358, left maxilla with P^2–M^3; 417370, right premaxilla with I^{1-3}; 417371, right mandible with P_3–M_1, M_3; 417428, LP^{1-3}; 417429, RM^2; 417430, LP^{2-3}; 417431, RP_4; 417432, RM^1;

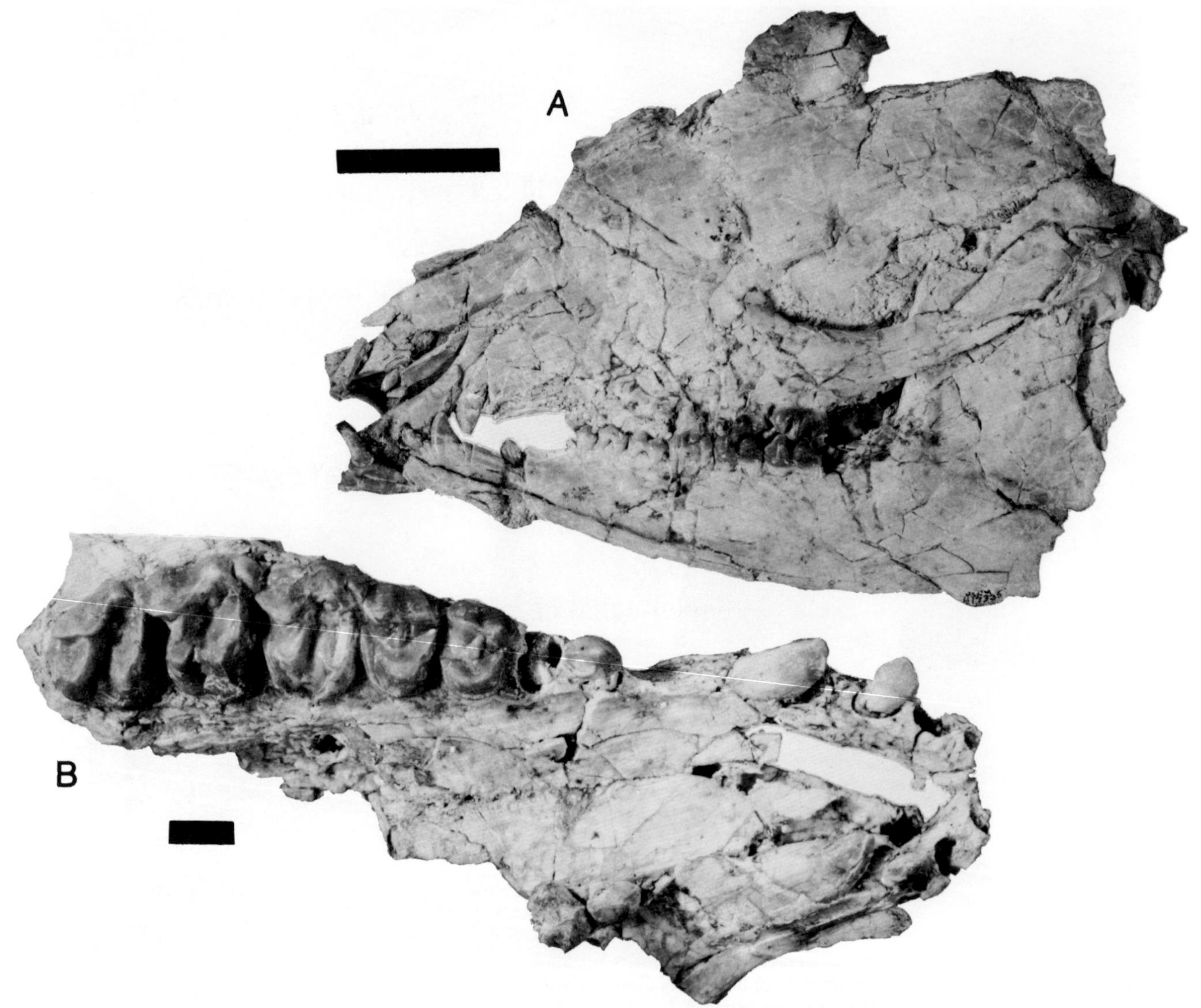

Figure 10. *Hyrachyus modestus* from Elderberry Canyon Quarry. A, USNM 417335, juvenile skull and mandibles in lateral view. B, USNM 417327, partial skull in occlusal view. A approximately ×0.5, bar = 5 cm. B approximately ×1, bar = 1 cm.

417433, LM^3; 417434, LP^4; 417435, LM^3; 417436, RM^1; 417437, LM_1; 417438, RM_2; 417439, RM^1 or M^2; 417440, LP^4–M^1; 417441, LP^4.

Discussion. The specimens of *Hyopsodus* recovered to date indicate that no more than one species is present in the Elderberry Canyon assemblage. In size and dental characteristics the specimens correspond most closely to material identified as *H. paulus* in the USNM collections from the Bridger Formation. The type of *H. paulus* is from the lower Bridger; Gazin (1976) found that the USNM collections have 868 identifiable specimens of *H. paulus* from the lower Bridger and none from the upper Bridger. This would seem to indicate that *H. paulus* is restricted to the lower Bridger. However, West (1979) placed *Hyopsodus marshi* and *H. despiciens* (both of which Gazin considered valid species) in the synonymy of *H. paulus,* thereby extending the range of *H. paulus* through the upper Bridger. The Elderberry Canyon material would be referred to *H. paulus* under either Gazin's or West's taxonomy, although under West's it would be a considerably less precise temporal indicator. If West's taxonomy is correct though, and *H. marshi* and *H. despiciens* are not distinguishable from *H. paulus* without knowing their provenance within the Bridger Formation, then any specimen not from the Bridger would be a less precise temporal indicator at any rate.

Measurements in mm. USNM 417358: P^2, AP = 2.2, TR = 1.8; P^3, AP = 2.4, TR = 2.9; P^4, AP = 2.5, TR = 4.1; M^1, AP = 3.4, TR = 4.3; M^2, AP = 3.4, TR = 4.9; M^3, AP = 3.0, TR = 3.9. USNM 417371: P_3, AP = 2.95, TR = 1.6; P_4, AP = 3.3, TR = 2.2; M_1, AP = 3.5, TR = 2.9; M_3, AP = 4.1, TR = 3.7. USNM 417428: P^1, AP = 1.91, TR = 1.33; P^2, AP = 2.45, TR = 2.20; P^3, AP = 2.67, TR = 2.80. USNM 417429: M^2, AP = 3.76, TR = 4.91. USNM 417430: P^2, AP = 2.73, TR = 1.88; P^3, TR = 2.97. USNM 417431: P_4, AP = 3.09, TR = 3.89. USNM 417432: M^1,

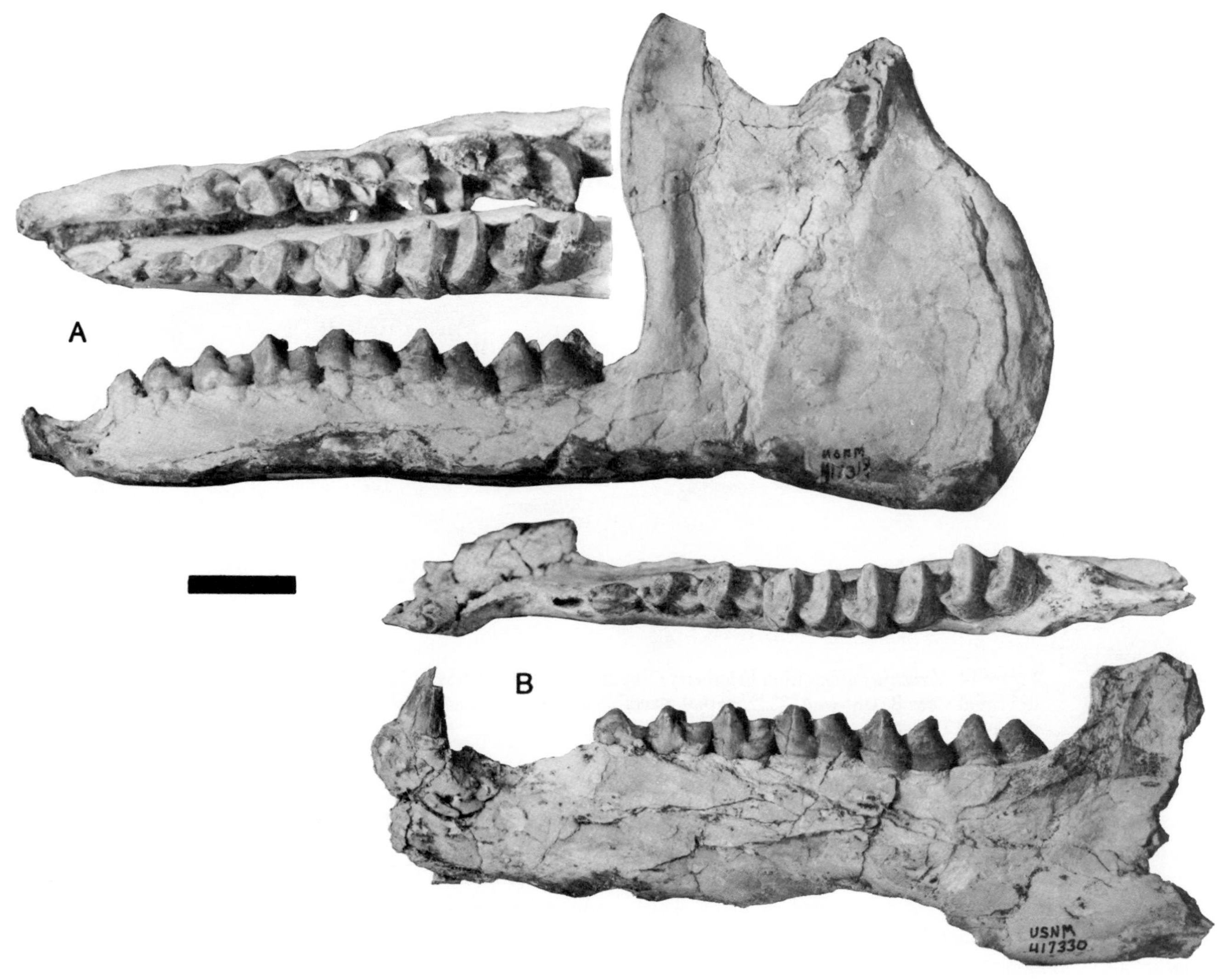

Figure 11. *Hyrachyus modestus* from Elderberry Canyon Quarry. A, USNM 417319, mandibles in lateral and occlusal views. B, USNM 417330, left mandibular ramus in lateral and occlusal views. All approximately ×1, bar = 2 cm.

AP = 3.15, TR = 3.52. USNM 417433: M^3, AP = 2.83, TR = 3.92. USNM 417434, P^4, AP = 2.19, TR = 3.23. USNM 417435, M^3, AP = 2.78, TR = 4.17. USNM 417436: M^1, AP = 3.57, TR = 4.37. USNM 417437, M^1, AP = 3.41, TR = 2.33. USNM 417438: M_2, AP = 3.45, TR = 2.21. USNM 417440: M^1, AP 3.6, TR = 4.07.

Order Perissodactyla
Superfamily Rhinocerotoidea Gill, 1872

Hyrachyidae Wood, 1927

***Hyrachyus* Leidy, 1871c**
***Hyrachyus modestus* (Leidy), 1870a**
Figures 10, 11

Referred specimens. All numbers with USNM prefix. 417319, right and left mandibles, each with P_1–M_3; 417320, RM^{1-3}; 417321, partial skull with RP^2–M^3 and LC, P^3–M^3; 417322, LM^{1-3}; 417323, palate with RC, P^1–M^3, LC, P^3–M^2, and mandible with C, P_2–M_3; 417324, anterior part of skull with RC, P^{1-2} and LP^1; 417325, LM^3; 417326, skull fragment with LM^3; 417327, partial skull with RI^3, C, P^1, P^3–M^3 and LI^3, C, P^{1-2}; 417328, mandibles with RI_{1-3}, C, P_1–M_3 and LI_{1-3}; 417329, skull and mandibles; 417330, mandibles with right and left C, P_2–M_3; 417331, RM^2; 417335, skull and mandibles articulated with deciduous dentition plus M^{1-2}; 417336, skull fragment with LM^3 and mandible fragment with LM_{2-3}; 417337, right maxilla with P^3–M^1; 417338, skull and mandible fragments with LM_3; 417407, Right and left mandibles each with dP_{3-4}, M_{1-2}; 471408, right and left mandibles with dentition badly fractured; 417409, right and left mandibles not completely prepared, 417410, right and left mandibles each with P_2–M_3; 417411, left mandible with P_3–M_3; 417412, right mandible with P_4–M_3; 417414, incomplete M_{2-3}.

Figure 12. *Hyrachyus affinis* from Elderberry Canyon Quarry. A, USNM 417333, anterior part of skull in ventral view. B, USNM 417332, partial mandible in right lateral and occlusal views. All approximately ×0.75, bar = 5 cm.

Discussion. *Hyrachyus modestus,* the largest mammal yet recognized in the Elderberry Canyon assemblage, is the most abundantly represented larger mammal. As the list of referred specimens indicates, the species is represented by many skulls, jaws, and other dental remains, but uncataloged postcranial elements are also common.

The dentition is very well preserved in a number of specimens, and can be matched very closely by specimens assigned to *H. modestus* in the USNM collections from the lower Bridger Formation, and agree with *H. modestus* in the sense of Radinsky (1967). In both size and details of the dental morphology, no distinctions can be made between Bridger specimens and those from Elderberry Canyon.

Measurements in cm. Measurements of *H. modestus* are all included in Table 1.

Hyrachyus affinis (Marsh), 1871
Figure 12

Referred specimens. All with USNM prefix. 417332, right mandibular ramus with P_4–M_3 and left mandibular ramus with P_1–M_1; 417333, partial skull with RI^3, P^1–M^1 and LI^1, P^1–M^1; 417334, right mandibular ramus with dP_{2-4} and left mandibular ramus with dP_1, dP_3–M_1.

Discussion. A few specimens of *Hyrachyus* are distinctly smaller than the others in all dimensions, and closely match specimens in the USNM collections identified as *Hyrachyus affinis.* The Elderberry Canyon specimens accord well with the description and illustrations of *H. affinis* by Wood (1934). The type of *H. affinis* (YPM 12530) is from Bridger "Horizon B." The Elderberry Canyon material is closer in size to *H. affinis* from the lower Bridger than it is to the slightly smaller but otherwise similar specimens from the upper Bridger variously termed *H. gracilis* or *H. affinis gracilis* in the literature. Radinsky (1967) placed *H. affinis* in the synonymy of *H. modestus,* essentially because it could not be separated statistically on size, even though a slight bimodal distribution was seen in some measurements. Without attempting to revise the taxonomy of *Hyrachyus* here, it seems equally defensible to refer to the smaller forms as *H. affinis* as to refer to them as the smaller specimens of *H. modestus.*

Measurements in cm. All measurements of the specimens referred to *H. affinis* are given in Table 2.

? Hyrachyidae
Undescribed genus and species

Material. USNM 417339, left mandibular ramus with P_2–M_3 and alveolus for P_1; USNM 417340, right and left maxillaries each with P^1–M^3; USNM 417341, right mandible fragment with M_3; USNM 417342, right mandible fragment with P_{3-4} and part of P_2; unnumbered mandible fragment with part of symphysis and lower border of right ramus.

Discussion. The familial referral of the new taxon is somewhat surprising, inasmuch as *Hyrachyus* is one of the largest Bridgerian

Figure 13. *Helaletes nanus* from Elderberry Canyon Quarry. USNM 417359, right mandibular ramus in lateral and occlusal views. Approximately ×0.75, bar = 5 cm.

ceratomorphs, and the new taxon is the smallest ceratomorph yet recorded. It is about 10 percent smaller than *Selenaletes scopaeus* Radinsky, 1966, from which it differs in the defining characters of *Selenaletes,* and may approach the lower size limit for perissodactyls in general. Known only from the Elderberry Canyon assemblage, the new taxon is not, of course, useful in biostratigraphic correlation. It does increase the diversity of perissodactyls at Elderberry Canyon to at least five species.

Superfamily Tapiroidea Gill, 1872
Family Helaletidae Osborn, in Osborn and Wortman, 1892

***Helaletes* Marsh, 1872**
***Helaletes nanus* (Marsh), 1871**
Figure 13

Referred specimens. USNM 417359, left mandibular ramus with P_2–M_3; USNM 417413, right maxillary fragment with incomplete P^2, complete P^{3-4}, incomplete M^1, and an associated M_2.
Discussion. The size, absence of P_1, and the presence of a discrete though small hypoconulid on M_3 confirm the identification of the jaw as *Helatetes.* While the ramus has all the cheek teeth in place, a fracture running the length of the tooth row obscures the dental details. It is referred to *H. nanus* principally on the basis of size, and the morphologic details that can be discerned are not inconsistent with this specific allocation. The P^3 and P^4 of USNM 417413 appear to be slightly lower crowned than those of most *H. nanus* from the Bridger Formation, but otherwise the details of cusp and loph arrangement agree quite well.
Measurements in mm. USNM 417413: P^3, AP = 6.8; P^4, AP = 7.2, TR = 9.6; M^1, AP = 9.3; M_1, AP = 8.6, TR = 6.1. USNM 417359: P_2–M_3 = 51.8; M_3, AP = 11.0, TR = 8.5; other teeth are too damaged to obtain useful measurements.

Family Isectolophidae Peterson, 1919

***Isectolophus* Scott and Osborn, 1887**
***Isectolophus latidens* (Osborn and others), 1878**

Referred specimen. USNM 417416, LM^1.
Discussion. This taxon is represented by a single tooth intermediate in size between *Hyrachyus* and *Helaletes.* The tooth is slightly larger than homologous teeth identified as *I. latidens* (and/or *"Parisectolophus" latidens*) in the USNM collections from the Bridger Formation. It is morphologically similar though in having the protoloph shorter than the metaloph and raked obliquely backward lingually so that it is not parallel to the metaloph. The metacone is large, contributing to the high, prominent ectoloph typical of *Isectolophus.* Radinsky (1963) placed the genus *Parisectolophus,* which Peterson (1919) had created for the species *I. latidens,* into the synonymy of *Isectolophus,* and assigned all known Bridgerian *Isectolophus* to *I. latidens.* For want of a better alternative, I assign USNM 417416 to this same species, although the tooth cannot really do more than establish the presence of an isectolophid tapiroid in the Elderberry Canyon Local Fauna.
Measurement in mm. USNM 417416, LM^1, maximum width at metaloph = 14.2; no other useful measurements can be made.

Order Artiodactyla
Family Dichobunidae Gill, 1872

Genus *Antiacodon* Marsh, 1872
***Antiacodon pygmaeus* (Cope) 1872a**
Figure 14

Referred specimens. An uncataloged astragalus, and USNM 417343, a left mandibular ramus with P_{1-4} and parts of M_{1-3}.

TABLE 1. MEASUREMENTS IN CENTIMETERS OF THE DENTITIONS OF *HYRACHYUS MODESTUS* FROM THE ELDERBERRY CANYON LOCAL FAUNA AND THE BRIDGER"B"

USNM #	P1/		P2/		P3/		P4/		M1/		M2/		M3/		P1-4/	M1-3/	P1/-M3/
	L	W	L	W	L	W	L	W	L	W	L	W	L	W	L	L	L
Elderberry Canyon Local Fauna Sample																	
417320 r	....	....	....	....	....	....	....	....	....	....	1.82	1.95	....	....	....	....	....
417321 r	....	....	0.84	1.31*	1.14	1.65*	1.35	1.75	....	....	1.79	1.89*	1.65	2.09	....	....	....
417321 l	....	....	....	....	1.15	1.55	1.52	1.74	....	....	1.78	....	1.60	2.01	....	....	....
417322 l	....	....	....	....	....	....	....	....	1.71	1.75*	1.85	2.05	1.65	1.92	....	....	....
417323 r	0.90	0.69	0.90	1.13	1.25	1.54	1.20	....	....	....	....	....	....	....	4.15	....	....
417323 l	....	....	1.01	....	1.11	....	1.33	....	1.55*	....	1.90*	....	....	....	....	....	....
417324 r	0.83	....	1.01	....	....	....	....	....	....	....	....	....	....	....	....	....	....
417325 l	....	....	....	....	....	....	....	....	....	....	....	....	1.56	1.77	....	....	....
417326 r	....	....	....	....	....	....	....	....	....	....	....	....	....	1.75	....	....	....
417327 r	0.95	0.69	....	....	1.24	1.52	1.35	1.74	1.76	2.05	1.99	2.31	1.59	2.00	4.22	5.04	9.10
417327 l	0.92	0.68	....	1.13	....	....	....	....	....	....	....	....	....	....	....	....	....
417329 r	0.83	0.57	0.90	1.00	1.24	1.35	1.12	1.50	1.68	1.80	1.97*	2.00	1.84	2.12	4.10	5.16	9.00
417329 l	0.85	....	0.91	....	1.11	....	1.54	1.67	....	....	1.90	1.98	1.75	1.98	4.22	4.85	8.95
417331 r	....	....	....	....	....	....	....	....	....	....	1.94	2.08	....	....	....	....	....
Bridger "B" Sample																	
13412 r	0.84	0.57	....	....	1.16	1.46	1.36	1.68	1.68	1.96	1.92	2.16	1.81	2.01	4.01	5.19	9.20
13412 l	0.83	....	0.95	....	1.18	....	1.35	....	1.63	....	1.95	....	1.83	....	4.29	5.20	9.28
23659 r	....	....	1.06	1.52	1.18	1.54	1.33	1.71	1.76	1.79	1.96	1.95	1.80	1.90	....	....	....
Statistics																	
Elderberry Canyon Local Fauna																	
Number	6	4	6	3	7	4	7	5	4	3	9	7	7	8	4	3	3
Maximum	0.95	0.96	1.01	1.13	1.25	1.55	1.54	1.75	1.76	2.05	1.99	2.31	1.84	2.12	4.22	5.16	9.10
Minimum	0.83	0.57	0.84	1.00	1.11	1.35	1.12	1.50	1.55	1.75	1.78	1.89	1.56	1.75	4.10	4.85	8.95
Bridger "B"																	
Number	2	1	2	1	3	2	3	2	3	2	3	2	3	2	2	2	2
Maximum	0.84	0.57	1.06	1.52	1.18	1.54	1.36	1.71	1.68	1.96	1.96	2.16	1.82	2.01	4.29	5.20	9.28
Minimum	0.83		0.95		1.16	1.46	1.33	1.68	1.63	1.79	1.92	1.95	1.80	1.90	....	....	....

TABLE 1. MEASUREMENTS IN CENTIMETERS OF THE DENTITIONS OF *HYRACHYUS MODESTUS* FROM THE ELDERBERRY CANYON LOCAL FAUNA AND THE BRIDGER"B" (continued)

USNM #	P/1		P/2		P/3		P/4		M/1		M/2		M/3		P1-/4	M1-/3	P/1-M/3
	L	W	L	W	L	W	L	W	L	W	L	W	L	W	L	L	L
Elderberry Canyon Local Fauna Sample																	
417319 r	0.72	0.44	0.95	0.62	1.15	0.85	1.33	1.00	1.88*	1.22*	1.94*	1.27*	1.96*	1.15*	4.03	5.37	9.48
417319 l	0.65	0.44	0.91	0.59	1.16	0.89	1.30	0.96	1.60	1.20	1.75	1.34	2.05	1.28	4.01	5.62	9.21
417323 l	....	....	0.95	....	1.18	....	1.10	....	1.55	....	1.80	....	1.80	....	3.75	5.35	9.10
417328 r	0.80	0.52	0.93	0.63	1.16	1.28*	1.32*	0.98	1.65	1.20	1.84	1.32	1.89	1.31	4.00	5.51	9.49
417329 r	....	....	....	....	....	....	....	....	....	....	....	....	....	....	....	4.90	....
417329 l	....	....	....	....	....	....	....	....	1.46	1.06	1.80	1.15	1.85	1.21	....	4.90	....
417330 r	....	....	....	....	....	....	....	....	....	....	....	....	....	....	4.30	4.91	9.35
417330 l	....	....	0.95	0.57	1.20	0.71	1.32	0.95	1.51	1.05*	1.80	1.28	1.83	1.23	4.22	5.21	9.26
417411 l	....	....	....	....	....	....	1.35	1.10	1.68	1.25*	1.95	1.33	1.95	1.35	....	5.41	....
417412 r	....	....	....	....	....	....	1.18	....	1.36	1.17	1.66	....	1.99	1.22	....	....	....
Bridger "B" Sample																	
12739 l	....	....	....	....	....	....	....	....	1.72	1.16	1.82	1.32	1.97	1.22*	....	5.41	....
13421 r	....	....	....	....	1.32	0.80	1.35	0.94	1.67	1.11	1.75	1.21	1.90	1.22	....	5.19	....
13421 l	....	....	....	....	....	....	....	....	1.65	1.12	1.75	1.20	1.89	1.20	....	5.21	....
23651 l	....	....	....	....	1.25	0.84	1.35	0.93	1.71	1.11	1.83	1.22	1.96	1.28	....	5.51	....
Statistics																	
Elderberry Canyon Local Fauna																	
Number	3	3	5	4	5	3	5	4	5	5	6	5	6	5	4	7	5
Maximum	0.80	0.52	0.95	0.63	1.20	0.89	1.33	1.00	1.65	1.22	1.94	1.32	2.05	1.31	4.22	5.62	9.49
Minimum	0.65	0.44	0.91	0.57	1.15	0.71	1.10	0.95	1.46	1.05	1.75	1.15	1.80	1.15	4.00	4.90	9.10
Bridger "B"																	
Number	....	....	....	....	2	2	2	2	4	4	4	4	4	4	....	4	....
Maximum	....	....	....	....	1.32	0.84	1.35	0.94	1.72	1.16	1.83	1.32	1.97	1.28	....	5.51	....
Minimum	....	....	....	....	1.25	0.80	1.35	0.93	1.65	1.11	1.75	1.20	1.89	1.20	....	5.19	....

*estimated values.

TABLE 2. MEASUREMENTS IN CENTIMETERS OF THE DENTITIONS OF *HYRACHYUS AFFINIS* FROM THE ELDERBERRY CANYON LOCAL FAUNA AND THE *HYRACHYUS AFFINIS* FROM THE BRIDGER "B"

USNM #	P1/		P2/		P3/		P4/		M1/		M2/		M3/		P1-4/	M1-3/	P1/-M3/
	L	W	L	W	L	W	L	W	L	W	L	W	L	W	L	L	L
Elderberry Canyon Local Fauna Sample																	
417333 r	0.76	0.55	0.79	0.94	0.91	1.28	1.15	1.42	....	....	....	....	....	....	....	....	....
417333 l	0.73	0.56	0.79	1.04	0.93	1.32	1.12	1.55	....	....	....	....	....	....	....	....	....
Bridger "B" Sample																	
23656 r	....	....	0.85	0.94	0.99	1.28	1.11	1.50	1.49	1.65	1.68	1.80	1.51	1.71	....	4.52	....
23656 l	0.78	....	0.83	0.93	1.04	1.30	1.16	1.45	1.40	1.68	1.66	1.80	1.46	1.71	3.82	4.57	8.22
23661 r	....	....	....	....	....	....	....	....	1.38	1.50	1.55	1.65	1.40	1.60	....	4.29	....
23661 l	....	....	....	....	0.96	1.13	1.05	1.42	1.26	1.51	1.53	1.61	1.42	1.61	....	4.23	....
USNM #	P/1		P/2		P/3		P/4		M/1		M/2		M/3		P/1-4	M/1-3	P/1-M/3
L	W		L		W		L		W		L		W		L	W	L
W	L		W		L		L		L								
Elderberry Canyon Local Fauna Sample																	
417332 r	....	....	0.79	....	0.89	....	0.98*	....	1.32	....	1.62	....	1.60	....	3.49	4.88	8.40
417332 l	0.55	0.38*	0.85	0.55*	0.95	0.69	1.05	0.80*	1.49	1.12*	....	....	....	....	3.43	....	....
Bridger "B" Sample																	
23653 l	....	....	....	....	1.10	0.66*	1.12	0.82	1.50	0.95	1.62	1.05	1.68	1.07	....	4.80	....
23654 l	....	....	0.88	0.54	1.09	0.80	1.22	0.88	1.36	1.09	1.71	....	1.85	1.25	3.80	4.96	8.72

*estimated values.

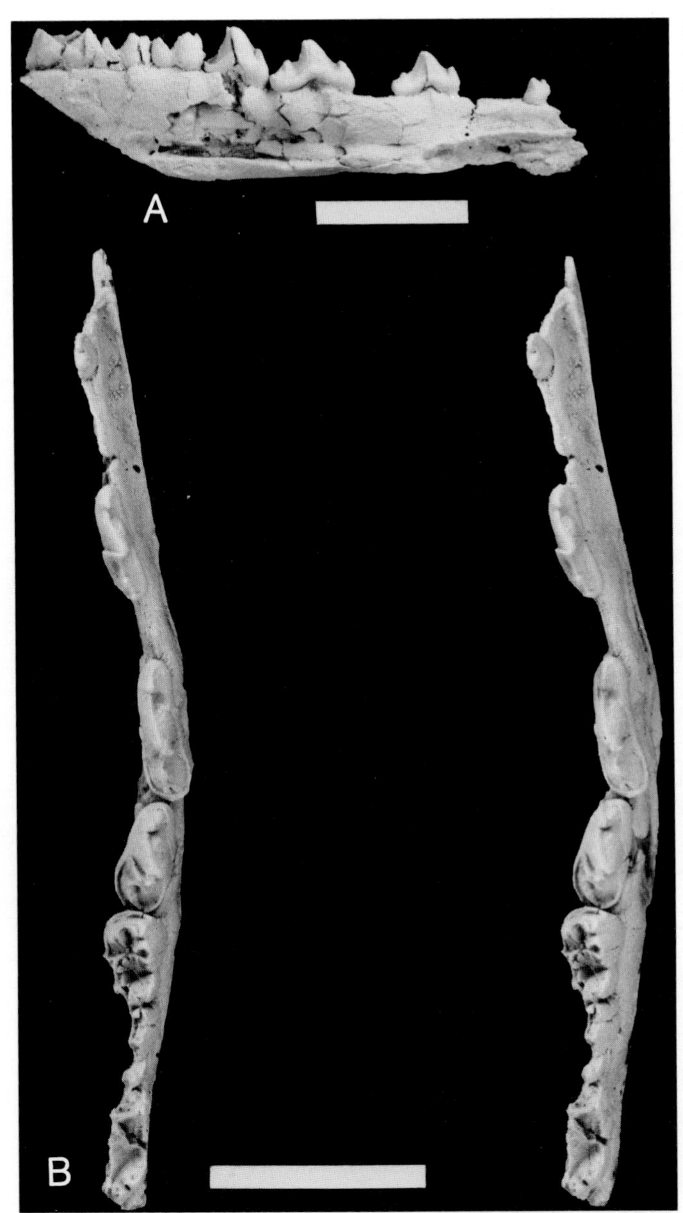

Figure 14. *Antiacodon pygmaeus* from Elderberry Canyon Quarry. USNM 417343, left mandibular ramus in (A) buccal and (B) stereo occlusal views. A approximately ×2, B approximately ×3, bars = 1 cm.

Discussion. The astragalus, though clearly of a small artiodactyl, is poorly preserved and otherwise indeterminate. It is referred to this taxon only because it is the only artiodactyl represented by dental material in the Elderberry Canyon assemblage. The mandibular ramus is assigned to *Antiacodon* because of the spacing of its premolars (diastemata between P_1 and P_2, and between P_2 and P_3), and although the buccal portions of the molars are missing, the lingual parts that are preserved show that the molars retained distinct paraconids, a characteristic feature of *Antiacodon.* A few specimens of *Antiacodon* from the Bridger Formation have enough of the anterior part of the jaw preserved to show that P_2 was separated by diastemata from both P_1 and P_3. The specimen (CMNH 10930) described by Burke (1969) from the Powder Wash locality in the Green River Formation of Utah also shows this premolar spacing. In the Elderberry Canyon specimen the ramus appears to be even more anteriorly elongated, and the interdental spaces greater, than in the Bridger and Green River specimens, although the measurements show that the teeth themselves are similarly sized. Burke (1969) noted the differences in diastema length, but suggested that the length might vary with ontogenetic age, the specimens with longer diastemata having the teeth considerably worn. The Elderberry Canyon specimen, with even longer diastemata, is only slightly worn, seemingly negating Burke's suggestion. Much larger samples will be needed to assess the taxonomic importance of this variation.

The parts of the molars preserved in the Elderberry Canyon specimen compare closely with those of *Antiacodon pygmaeus* from the Brider Formation. The P_4, however, lacks the distinct metaconid seen in the homologous tooth of most Bridger specimens (West, 1984). In this respect it appears to be more like the Powder Wash specimen described by Burke (1969), who noted that the very small metaconid of his specimen was not typical of *A. pygmaeus.* While acknowledging the differences, it seems best, for the present, to assign USNM 417343 to *Antiacodon pygmaeus* on the basis of measurements of the teeth, and to leave the taxonomic significance of the recognized differences to be resolved eventually by larger samples.

Measurements in mm. USNM 417343: P_1, AP = 2.1, TR = 1.7; P_2, AP = 5.0, TR = 1.4; P_3, AP = 7.1, TR = 2.0; P_4, AP = 4.7, TR = 2.4; M_1, AP = 3.1, TR = 2.8; M_2, AP = 4.6; M_3, AP = 4.8; P_{1-4} = 24.9; M_{1-3} = 12.5; P_1–M_3 = 37.45; diastema P_{1-2} = 4.5; diastema P_{2-3} = 3.0.

DISCUSSION

The Elderberry Canyon assemblage, as presently constituted, shares the greatest faunal similarity with the lower part of the Bridger Formation, although the Elderberry Canyon Local Fauna, as might be expected with a sample from a single site, is less diverse than the lower Bridger fauna as a whole. The Elderberry Canyon Local Fauna has some conspicuous absences. For example, several kinds of turtles are common in the Bridger Formation, but not a single piece of turtle bone has been discovered to date at Elderberry Canyon. The large mammals, brontotheres, for example—characteristic, common, and conspicuous in the Bridger Formation—are not yet recognized in the Elderberry Canyon Local Fauna, where the ceratomorph *Hyrachyus* is the largest mammal identified. No marsupials have been seen in the Elderberry Canyon assemblage, although comparably small eutherians are abundantly represented by well-preserved specimens. Several small rodents, including *Microparamys,* known from the Bridger Formation, are represented at Elderberry Canyon, but the larger and more common Bridger rodent *Paramys,* which elsewhere occurs almost invariably with *Microparamys,* has not been found at Elderberry Canyon.

The presence, or predominance, of hedgehog-like insectivorans, and the concurring presence of uintasoricines and *Trogolemur* suggest similarity with the fauna of bone bed A of the type section of the Tepee Trail Formation in northwestern Wyoming (McKenna, 1980). The Tepee Trail assemblage is clearly younger than the Elderberry Canyon assemblage. McKenna (1980) believes the Tepee Trail assemblage is younger than known Bridgerian faunas, and would best be classified as early Uintan. McKenna cautioned that the age assessment afforded by the Tepee Trail assemblage is complicated by peculiarities of the fauna that he believed might be due to sampling a paleoecology not previously sampled, perhaps at higher paleoaltitude and near volcanic vents. The similarities seen in the Elderberry Canyon assemblage can hardly be ascribed to these same ecological conditions, which further complicates the interpretation, but nevertheless, some paleoecologic control seems likely.

Figure 15 lists the mammalian elements of the Elderberry Canyon Local Fauna, as they are presently identified, and indicates occurrences of the taxa elsewhere. It becomes readily apparent that the greatest similarity, much of it at the species level, is with the lower part, or Black's Fork Member, of the Bridger Formation. Several taxa in the Nevada assemblage were previously known only in the fauna of the lower Bridger.

The Elderberry Canyon assemblage is geologically older than the numerous Eocene mammalian local faunas from southern California (see Krishtalka and others, 1987) summarized by Golz and Lillegraven (1977), which are Uintan and younger. Though Bridgerian faunas are not represented in southern California (see Krishtalka and others, 1987), there is significant faunal similarity, some even at the species level, between southern California and the Rocky Mountain region in earliest Uintan faunas, after which time endemism becomes progressively greater through the end of the Eocene. Lillegraven (1979) noted this and postulated that in Bridgerian time faunal interchange was relatively unrestricted between the Rocky Mountain region and southern California, by way of a corridor of tropical to subtropical lowlands that included southern Nevada. The Elderberry Canyon assemblage occurs along what would have been the northern margin of Lillegraven's "gangplank of lowlands." Its faunal composition surely demonstrates that interchange was relatively free, at least between Nevada and the Rocky Mountain region in Bridgerian time, and establishes a fairly typical "Bridger fauna" much closer to southern California than previously known. The composition of the Elderberry Canyon fauna thus fits the expectations projected by analysis of the Eocene assemblages of the Pacific Coast and Rocky Mountain regions.

ELDERBERRY CANYON LOCAL FAUNA TAXON	WASATCHIAN	BRIDGERIAN				UINTAN
		Powder Wash	Lower Bridger	Upper Bridger	Tabernacale Butte	
Insectivora						
Apatemys bellus	X	X	X	X	X	X
Pantolestes longicaudus			X	X		
Apternodontidae						
Nyctitheriidae						
Dormaaliidea						
Erinaceidae						
Chiroptera (two forms)						
Palaeanodonta						
Tetrapassalus cf. ***T. mckennai***					X	
Primates						
Notharctus tenebrosus			X			
Trogolemur myodes			X			?X
Uintasoricinae						
Rodentia						
Reithroparamys delicatissimus			X		X	
R. cf. ***R. huerfanensis***	X		X			
Microparamys new sp.						
cf. ***Microparamys*** new gen. et sp.						
Sciuravus sp.	X	X	X	X	X	X
Pauromys new sp.						
cf. ***Simimys*** new gen. et sp.						
Mattimys sp.	X					
Creodonta						
Sinopa minor			X			
Carnivora						
Viverravus minutus			X			
Viverravus incertae sedis						
Condylarthra						
Hyopsodus paulus			X			
Perissodactyla						
Hyrachyus modestus	X		X	X		
Hyrachyus affinis			X	X		
Hyrachyine ***incertae sedis***						
Helaletes nanus			X	X		
Isectolophus latidens			X	X		
Artiodactyla						
Antiacodon pygmaeus	X	X	X			

Figure 15. List of mammalian taxa presently known from Elderberry Canyon Local Fauna. "X" indicates occurrence of these taxa elsewhere in Bridgerian faunas, and in pre-Bridgerian (Wasatchian) and post-Bridgerian (Uintan) time. Occurrences of some problematical taxa are explained in the text.

ACKNOWLEDGMENTS

The fundamental contribution, without which this report could not have been written, nor even contemplated, was of course the discovery of the fossil concentration in Elderberry Canyon by Thomas D. Fouch, who then gave generously of his time to guide me to the site, assisted in collecting more of the material, and also showed me most of the other Tertiary deposits and fossil localities in east-central Nevada.

Able assistance in the increasingly difficult job of collecting the rock in the field has been provided, at one time or another, by Dan Chaney, John Flynn, Frederick Grady, Arnold Lewis, and Michael Pechacek. Several tons of limestone from Elderberry Canyon have literally gone down the drain in the course of the tedious preparation process, administered for more than a decade now, first by Frederick Grady and then by Dan Chaney, both of the Vertebrate Paleontology Preparation Laboratory at the USNM. Figure 1 was rendered by Mary Parrish. Most of the specimen photography was by Victor Krantz and most of the photographic printing was by Dan Chaney, who also assisted in many other ways. Many useful comments resulted from the careful critical reviews by Dr. William Turnbull of the Field Museum of Natural History and Dr. Robert M. West of the Cranbrook Institute of Science. The manuscript was typed, through its several drafts, by Diane Cloyd. My sincere thanks to all who helped complete this report.

REFERENCES CITED

Bown, T. M., 1979, New Omomyid primates (Haplorhini, Tarsiiformes) from middle Eocene rocks of west-central Hot Springs County, Wyoming: Folia Primatologica, v. 31, p. 48–73.

Bown, T. M., and Schankler, D., 1982, A review of the Proteutheria and Insectivora of the Willwood Formation (lower Eocene), Bighorn Basin, Wyoming: U.S. Geological Survey Bulletin 1523, p. 1–79.

Burke, J. J., 1969, An Antiacodont from the Green River Eocene of Utah: Kirtlandia, no. 5, p. 1–7.

Cope, E. D., 1872a, Description of some New Vertebrata from the Bridger Group of the Eocene: Proceedings of the American Philosphical Society, v. 12, p. 460–465.

——, 1872b, Second account of New Vertebrata from the Bridger Eocene: Proceedings of the American Philosophical Society, v. 12, p. 466–468.

——, 1884, The Vertebrata of the Tertiary formations of the West, Book 1, *in* Hayden, F. V., ed., Report of the United States Geological Survey of the Territories: U.S. Geological Survey, v. 3, 1009 p.

Flynn, J. J., and Galiano, H., 1982, Phylogeny of early Tertiary Carnivora, with a description of a new species of *Protictis* from the middle Eocene of northwestern Wyoming: American Museum Novitates no. 2725, p. 1–64.

Fouch, T. D., 1979, Character and paleogeographic distribution of Upper Cretaceous (?) and Paleogene nomarine sedimentary rocks in east-central Nevada, *in* Armentrout, J. M., Cole, M. R., and TerBest, H., eds., Cenozoic paleogeography of the western United States; Pacific Coast Paleogeography Symposium 3: Pacific Section, Society of Economic Paleontologists and Mineralogists, p. 97–111.

Gazin, C. L., 1958, A review of the middle and upper Eocene Primates of North America: Smithsonian Miscellaneous Collections, v. 136, p. 1–112.

——, 1976, Mammalian faunal zones of the Bridger middle Eocene: Smithsonian Contributions to Paleobiology, no. 26, p. 1–25.

Gill, T., 1872, Arrangement of the families of mammals and synoptic tables of characters of the subdivisions of mammals: Smithsonian Miscellaneous Collections, v. 11, p. 1–98.

Gingerich, P. D., and Winkler, D. A., 1985, Systematics of Paleocene Viverravidae (Mammalia, Carnivora) in the Bighorn Basin and Clark's Fork Basin, Wyoming: Ann Arbor, University of Michigan Contributions from the Museum of Paleontology, v. 27, p. 87–128.

Golz, D. J., and Lillegraven, J. A., 1977, Summary of known occurrences of terrestrial vertebrates from Eocene strata of southern California: University of Wyoming Contributions to Geology, v. 15, p. 43–65.

Kellogg, H. E., 1964, Cenozoic stratigraphy and structure of the southern Egan Range, Nevada: Geological Society of America Bulletin, v. 74, p. 685–708.

Krishtalka, L., 1978, Paleontology and geology of the Badwater Creek Area, central Wyoming; Part 15, Review of the late Eocene primates from Wyoming and Utah, and the Plesitarsiiformes: Annals of the Carnegie Museum, v. 47, p. 335–360.

Krishtalka, L., and Schwartz, J. H., 1978, Phylogenetic relationships of Plesiadapiform-tarsiiform primates: Annals of the Carnegie Museum, v. 47, p. 515–540.

Krishtalka, L., and 10 others, 1987, Eocene (Wasatchian through Duchesnean) biochronology of North American, *in* Woodburne, M. O., ed., Cenozoic mammals of North America; Geochronology and biostratigraphy: Berkeley, University of California Press, p. 77–117.

Leidy, J., 1869, The extinct mammalian fauna of Dakota and Nebraska, including an account of some allied forms from other localities, together with a synopsis of the mammalian remains of North America: Journal of the Academy of Natural Sciences of Philadelphia, series 2, v. 7, p. 1–472.

——, 1870a, Remarks on a collection of fossils from the western Territories: Proceedings of the Academy of Natural Sciences of Philadelphia, 1870, p. 109–110.

——1870b, [Descriptions of *Palaeosyops paludosus, Microsus cuspidatus,* and *Notharctus tenebrosus*]: Proceedings of the Academy of Natural Sciences of Philadelphia, 1870, p. 113–114.

——, 1871a, Remains of extinct mammals from Wyoming: Proceedings of the Academy of Natural Sciences of Philadelphia, 1871, p. 113–116.

——, 1871b, Notice of some extinct rodents: Proceedings of the Academy of Natural Sciences of Philadelphia, 1871, p. 230–232.

——, 1871c, Report on the vertebrate fossils of the Tertiary formations of the West, *in* Hayden, F. V., ed., Preliminary report of the United States Geological Survey of Wyoming and portions of contiguous territories (2nd annual report), Part 4: U.S. Geological Survey, p. 340–370.

Lillegraven, J. A., 1979, A biogeographical problem involving comparisons of later Eocene terrestrial vertebrate faunas of western North America, *in* Gray, J., and Boucot, A. J., eds., Historical biogeography, plate tectonics, and the changing environment: Corvallis, Oregon State University Press, p. 333–347.

Lydekker, R., 1889, Untitled, *in* Nicholson, H. A., and Lydekker, R., eds., A manual of paleontology for use of students with a general introduction on the principles of paleontology, 3rd ed.: Edinburgh and London, Wm. Blackwood and Sons, v. 2, p. 889–1474.

Marsh, O. C., 1871, Notice of some new fossil mammals from the Tertiary Formation: American Journal of Science and the Arts, v. 2, p. 35–44.

——, 1872, Preliminary description of new Tertiary mammals, Parts 1–4: American Journal of Science, series 3, v. 4, p. 122–128, 202–224, 504.

Matthew, W. D., 1909, The Carnivora and Insectivora of the Bridger Basin, middle Eocene: Memoirs of the American Museum of Natural History, v. 9, part 6, p. 289–567.

——, 1915, A revision of the lower Eocene Wasatch and Wind River faunas; Part 1, Order Ferae (Carnivora), Suborder Creodonta: Bulletin of the American Museum of Natural History, v. 34, p. 4–103.

——, 1918, Edentata, *in* Matthew, W. D., and Granger, W., eds., A revision of the lower Eocene Wasatch and Wind River faunas; Part 5, Insectivora (continued), Glires, Edentata: Bulletin of the American Museum of Natural History, v. 38, p. 565–657.

McKenna, M. C., 1963, Primitive Paleocene and Eocene Apatemyidae (Mammalia, Insectivora) and the primate-insectivore boundary: American Museum Novitates, no. 2160, p. 1–39.

——, 1980, Late Cretaceous and early Tertiary vertebrate paleontological reconnaissance, Togwotee Pass area, northwestern Wyoming, *in* Jacobs, L. L., ed., Aspects of vertebrate history; Essays in honor of Edwin Harris Colbert: Museum of Northern Arizona Press, p. 321–343.

Novacek, M. J., 1986, The skull of leptictid insectivorans and the higher-level classification of eutherian mammals: Bulletin of the American Museum of Natural History, v. 183, p. 1–112.

Osborn, H. F., and Wortman, J. L., 1892, Fossil mammals of the Wasatch and Wind River beds; Collection of 1891: Bulletin of the America Museum of Natural History, v. 4, p. 81–147.

Osborn, H. F., Scott, W. B., and Speir, F., 1878, Paleontological report of the Princeton Scientific Expedition of 1877: Princeton College Contributions of the Museum of Geology and Archaeology, no. 1, p. 49–53, 135.

Peterson, O. A., 1919, Report upon the material discovered in the upper Eocene of the Uinta Basin by Earl Douglass in the years 1908–1909, and by O. A. Peterson in 1912: Annals of the Carnegie Museum, v. 12, p. 127–130.

Radinsky, L. B., 1963, Origin and early evolution of North American Tapiroidea: New Haven, Connecticut, Yale University Bulletin of the Peabody Museum of Natural History, no. 17, p. 1–106.

——, 1966, A new genus of early Eocene tapiroid (Mammalia, Perissodactyla): Journal of Paleontology, v. 40, p. 740–742.

——, 1967, *Hyrachyus, Chasmotherium,* and the early evolution of helaletid tapiroids: American Museum Novitates, no. 2313, p. 1–23.

Robinson, P., 1968, The paleontology and geology of the Badwater Creek area, central Wyoming; Part 4, Late Eocene primates from Badwater, Wyoming, with a discussion of material from Utah: Annals of the Carnegie Museum, v. 39, p. 307–326.

Rütimeyer, L., 1862, Eocaene Saugethiere aus dem Gebiet des Schweizerischen

Jura: Allgemeine Schweizerische Gesellschaft, neue Denkschritte, v. 19, p. 1–98.

Scott, W. B., and Osborn, H. F., 1887, Preliminary report on the vertebrate fossils of the Uinta Formation: Proceedings of the American Philosphical Society, v. 24, p. 255–264.

Simpson, G. G., 1927, A North American Oligocene edentate: Annals of the Carnegie Museum, v. 17, p. 283–298.

——, 1959, A new middle Eocene edentate from Wyoming: American Museum Novitates, no. 1959, p. 1–8.

Stock, C., 1949, Mammalian fauna from the Titus Canyon Formation, California; Contributions to paleontology 8: Carnegie Institute of Washington Publication 584, p. 229–244.

Trouessart, E. L., 1879, Catalogue des mammiferes vivants et fossiles: Revue Magazine Zoologie, v. 7, p. 219–285.

West, R. M., 1973a, An early middle Eocene epoicotheriid (Mammalia) from southwestern Wyoming: Journal of Paleontology, v. 47, p. 929–931.

——, 1973b, Review of the North American Eocene and Oligocene Apatemyidae (Mammalia: Insectivora): Lubbock, Texas Tech University Special Publications of the Museum, no. 3, p. 1–42.

——, 1979, Paleontology and geology of the Bridger Formation, southern Green River Basin, southwestern Wyoming; Part 3, Notes on *Hyopsodus:* Milwaukee, Wisconsin, Milwaukee Public Museum Contributions in Biology and Geology, no. 25, p. 1–52.

——, 1984, Paleontology and geology of the Bridger Formation, southern Green River Basin, southwestern Wyoming; Part 7, Survey of Bridgerian Artiodactyla, including description of a skull and partial skeleton of *Antiacodon pygmaeus:* Milwaukee, Wisconsin, Milwaukee Public Museum Contributions in Biology and Geology, v. 56, p. 1–47.

Winfrey, W. M., Jr., 1958, Stratigraphy, correlation, and oil potential of the Sheep Pass Formation, east-central Nevada: Rocky Mountain Section, American Association of Petroleum Geologists, p. 77–82.

Wood, A. E., 1962, The early Tertiary rodents of the Family Paramyidae: Transactions of the American Philosophical Society, v. 52, part 1, p. 1–261.

Wood, H. E., 2nd, 1927, Some early Tertiary rhinoceroses and hyracodonts: Bulletin of American Paleontology, v. 13, p. 161–269.

——, 1934, Revision of the Hyrachyidae: Bulletin of the American Museum of Natural History, v. 65, p. 181–295.

Wortman, J. L., 1901–1902, Studies of Eocene mammalia in the Marsh Collection, Peabody Museum; Part 1: Carnivora: American Journal of Science, v. 11, p. 333–348, 437–450; v. 12, p. 143–154, 193–206, 281–296, 377–382, 421–432; v. 13, p. 39–46, 115–128, 197–206, 433–448; v. 14, p. 17–23.

Wortman, J. L., and Matthew, W. D., 1899, The ancestry of certain members of the Canidae, Viverridae, and Procyonidae: Bulletin of the American Museum of Natural History, v. 12, p. 109–138.

Manuscript Accepted by the Society June 12, 1989

Printed in U.S.A.

Geological Society of America
Special Paper 243
1990

Plagiomenids (Mammalia: ?Dermoptera) from the Oligocene of Oregon, Montana, and South Dakota, and middle Eocene of northwestern Wyoming

Malcolm C. McKenna
Frick Curator, Department of Vertebrate Paleontology, American Museum of Natural History, New York, New York 10024

ABSTRACT

Two new genera and species of plagiomenids (Mammalia, ?Dermoptera, Plagiomenidae) are described from the North American Uintan (middle Eocene) and Chadronian (early Oligocene). A third genus and species, *Ekgmowechashala philotau,* from the early and middle Arikareean (late Oligocene) of the northern United States, is removed from the primate family Omomyidae and placed in the Plagiomenidae. All three newly recognized plagiomenids are placed in the Ekgmowechashalinae, sister subfamily to the subfamily Plagiomeninae (new rank). Ekgmowechashaline plagiomenids are somewhat primate-like, as is the plagiomenine genus *Worlandia,* but the Plagiomenidae are usually considered to be allied to the living colugos of southeast Asia, order Dermoptera. Analysis of that relationship is placed outside the scope of this paper.

***Tarka stylifera,* the earliest known ekgmowechashaline, occurs in the type section of the Tepee Trail Formation, early Uintan (Shoshonian: late medial Eocene) of northwestern Wyoming. This locality falls in paleomagnetic Chron C20R, interpreted to be close to 47.5 Ma in age. A second, more primitive but later-occurring ekgmowechashaline genus and species, *Tarkadectes montanensis,* is from a nominally early Oligocene level (Chadronian) in the Kishenehn Formation of northern Montana. *Ekgmowechashala* is known from lower dentitions from the early Arikareean Sharps Formation of South Dakota and probably from an upper dentition reported from middle Arikareean rocks in the John Day Formation of Oregon. *Ekgmowechashala* is placed with the other two genera because of lower cheek-tooth morphology, but it lacks the enlarged incisor of *Tarka.* Ekgmowechashalines are hypothesized here to be primarily frugivores, folivores, and nectar- and exudate-feeders.**

Until now, known undoubted plagiomenids were restricted to the Paleocene and early Eocene (Wasatchian). The newly recognized post-Wasatchian occurrences are all in the northern part of the United States and are in keeping with previously known plagiomenid geographic distribution, which ranged from northern Wyoming to the Canadian arctic and possibly beyond.

INTRODUCTION

This paper places on record *Tarka* and *Tarkadectes,* two new genera of an early Tertiary family of mammals, Plagiomenidae Matthew, 1918. A previously known genus, *Ekgmowechashala,* already awarded its own subfamily in the primate family Omomyidae because of morphological incongruence there, is here transferred to the Plagiomenidae in part because what were anomalies in a primate rubric are compatible with plagiomenid affinities. Additional evidence supports this action. Within the Plagiomenidae, all three genera are nevertheless maintained in the available subfamily Ekgmowechashalinae Szalay, 1976, whereas

McKenna, M. C., 1990, Plagiomenids (Mammalia: ?Dermoptera) from the Oligocene of Oregon, Montana, and South Dakota, and middle Eocene of northwestern Wyoming, *in* Bown, T. M., and Rose, K. D., eds., Dawn of the Age of Mammals in the northern part of the Rocky Mountain Interior, North America: Boulder, Colorado, Geological Society of America, Special Paper 243.

advanced members of the Plagiomenidae as previously understood are placed in the subfamily Plagiomeninae Matthew, 1918 (new rank). A previously named plagiomenid subfamily, Worlandiinae Bown and Rose, 1979, is reduced to tribal rank within the subfamily Plagiomeninae. These changes modify both the known diversity and known stratigraphic range of the Plagiomenidae, but do not help to solve the long-standing problem of the bearing of plagiomenids on the origin of the mammalian order Dermoptera.

Historical summary of plagiomenid classification

Our knowledge about plagiomenids has grown slowly. In his revision of the early Eocene Wasatchian faunas of the United States, Matthew (1918, p. 598–602) described the mammalian genus *Plagiomene,* basing it on *Plagiomene multicuspis* from the Willwood Formation of the Bighorn Basin, northwestern Wyoming. Because of the phenetic distinctiveness of the new genus, Matthew placed *Plagiomene* in a new family, Plagiomenidae, allocating it to the order Insectivora, which he regarded as something of a scrapbasket. Although Matthew discussed dental similarities of *Plagiomene* to desman moles, he was more impressed with similarities to the living colugos (order Dermoptera, family Galeopithecidae = Cynocephalidae, "flying lemurs") of southeast Asia, suggesting provisional phyletic affinity with them even though he refrained from placing the Plagiomenidae in the Dermoptera instead of the Insectivora.

In the same pages (Matthew, 1918, p. 600) Matthew mentioned, as possibly representative of plagiomenid ancestry, an undescribed Paleocene genus and species from the Paskapoo beds of Canada. This animal was later named *Elpidophorus elegans* by Simpson (1927), who thought it to be questionably an oxyclaenid. Matthew also remarked in passing (Matthew, 1918, p. 600, 607) that certain trituberculate teeth from the Cretaceous Lance Formation were similar to those of *Elpidophorus;* however, I do not know which particular Lance trituberculate teeth Matthew had in mind. Possibly, the reference was to lower teeth of the then-unrecognized marsupial genus *Glasbius,* but the matter cannot be decided (see Clemens, 1966, p. 24–34).

Next, Simpson (1928, 1929a, b) placed a new genus and species from the Clarkforkian of Bear Creek Coal Mine of Montana, *Planetetherium mirabile,* in the Plagiomenidae. He regarded the plagiomenids as members of the Insectivora. A few years later, Simpson (1935) described (with some reservations) as mixodectids the Torrejonian genus and species *Eudaemonema cuspidata.* Then, a year later, he named a new species of *Elpidophorus, E. patratus,* from Scarritt Quarry in the Tiffanian of the Crazy Mountain Field, Montana (Simpson, 1936). He regarded the mixodectids and microsyopids as a single family, Mixodectidae, not even divisible into two subfamilies as Matthew (1915, p. 467) had proposed. Simpson placed the Mixodectidae in the Insectivora (sensu lato) and suggested that the mixodectids and plagiomenids may have had a common ancestry. A year later, Simpson (1937) considered the matter in greater detail and concluded that both *Eudaemonema* and *Elpidophorus* were mixodectids (including microsyopids) and that there was a possible relationship between mixodectids (sensu lato) and plagiomenids. However, he doubted that there was convincing evidence to link the mixodectids (and plagiomenids) to the Dermoptera. Still later, Simpson (1945) became convinced, on the basis of then-unpublished material of plagiomenids in the Princeton Collection (now at Yale University), that the plagiomenids of the Paleocene and early Eocene of North America are indeed early dermopterans related to the modern colugos of southeast Asia, but he did not also transfer the mixodectids. He regarded the plagiomenids and colugos as deserving of ordinal rank, coordinate with the Insectivora and Chiroptera. Romer (1966, 1968), however, suggested that the colugos, including the plagiomenids, might as well be placed within the order Insectivora, remarking that if flying squirrels were only to be accorded subfamilial rank among the rodent family Sciuridae, why should the colugos be treated any differently among the insectivorans? Nevertheless, he continued to use the term Dermoptera. He also believed that the Mixodectidae of the Paleocene are possibly related to the Dermoptera. Similar views were expressed by Van Valen (1967), whose ideas were made known to Romer before publication. Sloan (1969, Fig. 6) placed *Elpidophorus, Planetetherium,* and *Plagiomene* in an ascending phyletic sequence that arose from mixodectids. The latter were depicted as stem-dermopterans. However, Butler (1972) placed the plagiomenids in Romer's (1966) paraphyletic scrapbasket Proteutheria, a philosophical step backward because the "proteutherians" are impossible to define except by their lack of apomorphies that characterize excluded descendants.

Russell and others (1973) and Rose (1973, 1975) continued to think of plagiomenids as dermopterans, as have nearly all authors since then. Schwartz and Krishtalka (1976, p. 4, 5, Fig. 4) republished a drawing published by Rose (1973) of a jaw of *Plagiomene multicuspis* and reinterpreted its dental formula. They believed *Plagiomene* to be a dermopteran and to have 5 premolars but no canine. Krishtalka (1976), West and others (1977), Rose and Simons (1977), West and Dawson (1978), Van Valen (1979), and Szalay and Drawhorn (1980) also regarded plagiomenids as dermopterans. Szalay and Drawhorn identified some isolated foot bones from northwestern Colorado and southern Wyoming as belonging to dermopterans, but this does not necessarily mean that the bones are those of plagiomenids; no plagiomenid dentitions have been found there, and thus the foot bones are more likely to belong to other mammal taxa occurring at the same sites. Novacek (1980) also accepted arguments for inclusion of the Plagiomenidae in the Dermoptera, but pointed out that the evidence is virtually restricted to dental features. In their exemplary study of the dermopteran auditory region based on living dermopterans, Hunt and Korth (1980) were noncommittal about relationships of the early Tertiary plagiomenids. However, Rose (1982) once more regarded the Plagiomenidae as dermopterans, mentioning a content of six unspecified Paleocene and Eocene genera from North America and one (*Placentidens*) from the early Eocene of Europe.

As can be seen from this brief historical summary, the family Plagiomenidae generally has been relegated, on the basis of dental anatomy, to the otherwise Recent mammalian order Dermoptera, whose living representatives are confined to southeastern Asia and adjacent islands. Plagiomenid cranial and postcranial anatomy has not been described. However, this may be rectified soon, because important new cranial as well as associated postcranial material of previously known kinds of mixodectids, microsyopids, and plagiomenids has been collected since the last published reviews of these groups. The new material is under study variously by F. S. Szalay, S. G. Lucas, M. J. Novacek, R.D.E. MacPhee, M. Cartmill, M. R. Dawson, and K. D. Rose.

New plagiomenid diversity

The new plagiomenid diversity established in this paper falls within the already-named subfamily Ekgmowechashalinae Szalay, 1976, previously considered to be omomyid primates. The earliest-occurring ekgmowechashaline is the new genus and species *Tarka stylifera* from the early Uintan (Shoshonian: late medial Eocene; paleomagnetic Chron C20R, approximately 47.5 Ma) of northwestern Wyoming (Berggren and others, 1985; Flynn, 1986). A second new genus and species, *Tarkadectes montanensis,* is a more primitive but later-occurring animal from the Chadronian (nominally early Oligocene) of northern Montana. The third genus and species, *Ekgmowechashala philotau,* occurs in latest Oligocene (early Arikareean) deposits of South Dakota. An unassociated fragmentary maxilla thought to belong to *Ekgmowechashala* (although not necessarily to *E. philotau*) has been reported from the middle Arikareean of Oregon. Hitherto, *Ekgmowechashala* has resided uneasily in the omomyid Primates. *Tarka, Tarkadectes,* and *Ekgmowechashala* are unusual, long-surviving elements in the Eocene and Oligocene mammalian fauna. They are hypothesized here to be somewhat primate-like plagiomenids primarily adapted to feeding on leaves, fruits, insects, gums, and saps. Nevertheless, phase I chewing facets (Kay and Hiiemae, 1974; Kay and Hylander, 1978) of *Tarkadectes* suggest a relatively greater proportion of fibrous food in the diet of *Tarkadectes. Ekgmowechashala* does not possess an enlarged incisor. However, although the anterior dentition is not yet known in *Tarkadectes, Tarka* does possess an enlarged incisor and has reduced teeth between the enlarged incisor and the molars. Enlarged incisors might have been useful to animals engaged in feeding on exudates, but of course, other uses can also be envisioned. In *Tarka* the posterior cheek-teeth are widened and are abundantly supplied with additional cusps and crenulations, useful to folivores, frugivores, and exudate-feeders alike.

Thus far, *Tarkadectes montanensis* is known only from p4–m2 of the fragmentary type specimen, a jaw from a nominally early Oligocene level in the Kishenehn Formation near the U.S.–Canada border. *Tarka stylifera* is represented by seven specimens, mostly lower jaws collected over a period of 23 years, solely from TTA (Tepee Trail, bone bed A) Quarry in the type section of the Tepee Trail Formation, northeast of the town of Dubois, Wyoming. *Ekgmowechashala* is known from seven lower dentitions from the late Oligocene (early Arikareean) Sharps Formation of South Dakota. An unassociated partial maxilla from middle Arikarrean beds in the John Day Formation in Oregon has been referred to the genus as well (Rose and Rensberger, 1983). Ekgmowechashalines have not been reported from other localities.

All three ekgmowechashaline genera appear in the stratigraphic record without apparent phylogenetic roots recorded from immediately older rocks of the same area. They thus apparently represent reintroductions either from distant areas or from environments not normally sampled by paleontological collecting. *Ekgmowechashala* cannot be descended from *Tarka* unless an unlikely reversal of incisor enlargement is invoked. Further, the more derived *Tarka* occurs earlier than the generally more plesiomorphous *Tarkadectes.* Moreover, at the site producing *Tarka,* at least one other long-surviving element occurs, *Protictis (Protictoides) aprophatos* Flynn and Galiano, 1982. Faunal composition and relative abundances of taxa are unusual at this site, suggestive of an environment not often sampled.

Associated fauna

The fauna collected from TTA Quarry (Chron C20R, approximately 47.5 Ma) includes 37 species of vertebrates, but only a few of them have been mentioned or described adequately in the literature (Love, 1939; Lewis, 1973; Novacek, 1977; Rose, 1978; McKenna, 1980; MacFadden, 1980; Flynn and Galiano, 1982). The known fauna from TTA Quarry is as follows:

Salamander
Scincidae, new genus and species, cf. *Contogenys* and *Paracontogenys*
Iguanidae, new genus and species
Neoplagiaulacid multituberculate, genus and species indeterminate
Herpetotherium sp.
Peradectes sp.
Palaeictops, new species
Erinaceid insectivoran, cf. *Ocajila* sp.
Oligoryctes sp.
Microchiropteran bat, genus and species indeterminate
Epoicothere, cf. *Tetrapassalus* sp.
Apatemys sp.
Uintasorex parvulus
Second uintasoricine species
Nyctitherium sp.
Tarka stylifera (this paper)
Trogolemur, cf. *Trogolemur myodes*
Phenacolemur or *Ignacius* sp.
Omomyid, genus and species indeterminate
?*Reithroparamys* sp.
Microparamys, cf. *M. dubius*
Sciuravus, new species

?*Protadjidaumo typus*
Small eomyid rodent
Rodent: family, genus and species indeterminate
Second rodent: family, genus and species indeterminate
Hyopsodus, new species, cf. *H. uintensis*
Protictis (Protictoides) aprophatos
Epihippus uintensis
Dilophodon minusculus
Amynodon advenus
Titanothere, genus and species indeterminate
Chalicothere, genus and species undescribed
Achaenodon sp.
Dichobunid sp.
?Homacodont, genus and species indeterminate
Low-crowned selenodont artiodactyl, genus and species indeterminate

None of these Shoshonian (late medial Eocene) taxa is so unusual an animal as *Tarka stylifera* but, like *Tarka* and *Ekgmowechashala, Protictis (Protictoides) aprophatos* has phylogenetic roots in the Paleocene yet lacks known temporally intermediate links with that ancestry. Conspicuously absent from the collections thus far made from TTA Quarry are fishes, turtles, crocodiles, and notharctid primates. Unexpectedly, the commonest mammal is the *Ocajila*-like erinaceid. Carnivores are poorly represented and creodonts are lacking. Uintatheres are absent as well at TTA Quarry, but they occur nearby in the Tepee Trail Formation at a slightly lower stratigraphic level. Flynn (1986) has published on the biostratigraphy and magnetostratigraphy of this unusual site.

Institutional abbreviations

AMNH: American Museum of Natural History, New York, New York.
CMNH: Carnegie Museum of Natural History, Pittsburgh, Pennsylvania.
LACM: Los Angeles County Museum, Los Angeles, California.
UCMP: Museum of Paleontology, University of California, Berkeley, California.
UM: University of Michigan Museum of Paleontology, Ann Arbor, Michigan.
USGS: United States Geological Survey.

Other abbreviations

C20R: paleomagnetically reversed part of paleomagnetic Chron C20.
I, C, P, M: upper incisor, canine, premolar, molar, respectively.
i, di, c, p, dp, m: lower incisor, deciduous (milk) incisor, canine, premolar, deciduous (milk) premolar, molar, respectively.
Ma: million years before present (MYBP).
TTA: Tepee Trail Formation, bone bed A (Quarry).

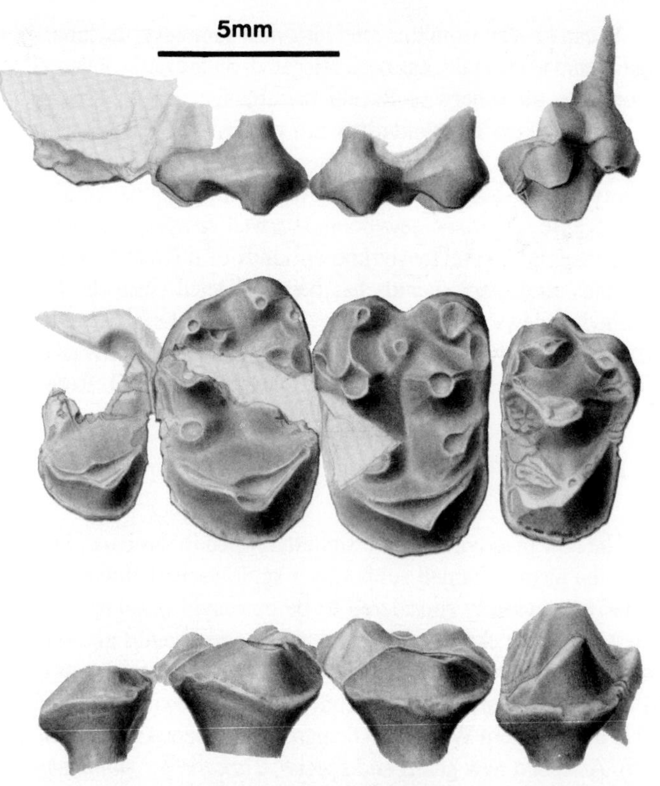

Figure 1. *Tarka stylifera.* Composite labial (top), occlusal (middle) and lingual (bottom) drawings of P4, M1, M2, M3, based upon AMNH 95713, damaged right M1, M2, M3, and 88294, left P4 and lingual half of M1 (reversed). See 5-mm bar for scale. See Figure 2 for stereoscopic photographs of these specimens.

SYSTEMATICS

?Order DERMOPTERA
Family Plagiomenidae Matthew, 1918, p. 598
Subfamily Ekgmowechashalinae (Szalay, 1976, p. 349)

Diagnosis. This subfamily differs from plagiomenine plagiomenids in that (as interpreted) P4 loses the metacone, the heel of p4 reduces, the stylar cusps of P4–M3 hypertrophy (condition unknown in *Tarkadectes*), the molar metastylids enlarge, and p4–m3 gain a stylar cusp on the cingulum, posterolabial to the protoconid. In contrast to plagiomenines, ekgmowechashalines lack the large and nearly completely molarized P3/p3, probably do not equip P2 with a metacone (inferred from reduction of anterior lower premolars), and do not have P4/p4 so large as or larger than M1/m1 (see Fig. 12).

Tarka, new genus

Type. Tarka stylifera McKenna, new species.
Locality, associated fauna, age, and distribution. As for the type species.

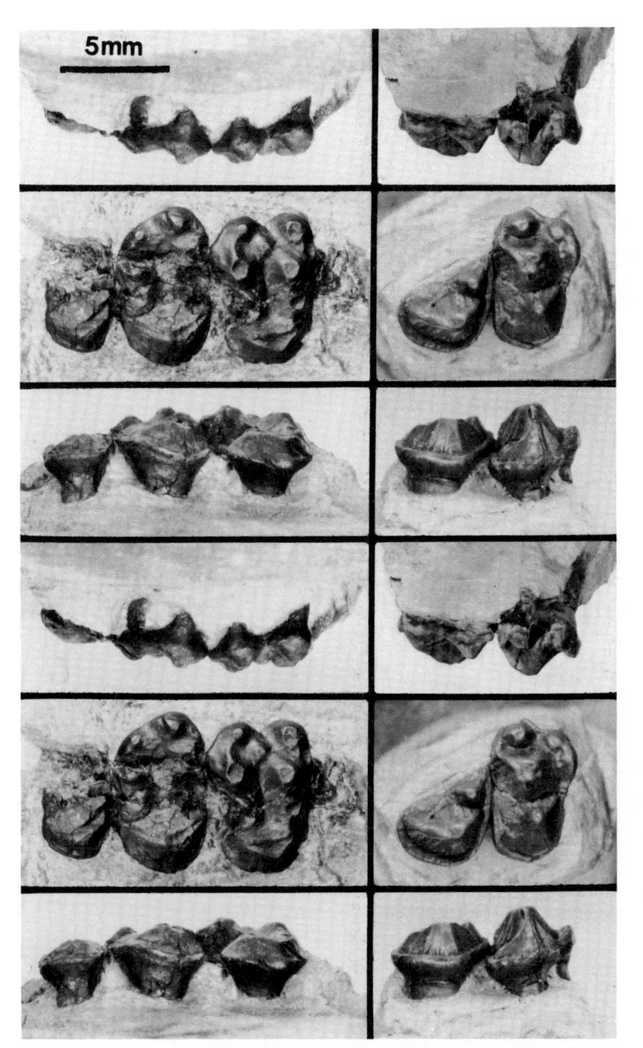

Figure 2. *Tarka stylifera.* Stereoscopic photographs of AMNH 95713, damaged right M1, M2, M3, and AMNH 88294, left P4 and lingual half of M1 (reversed for ease of viewing), labial, occlusal, and lingual views. See 5-mm bar for scale. See Figure 1 for composite drawing.

Etymology. Named for Chester Tarka, scientific illustrator, teacher, and philosopher.
Diagnosis. As for the type species.

Tarka stylifera, new species
(Figs. 1 to 8)

Type. AMNH 113133, left lower jaw (Figs. 3, 4) of an old individual with teeth identified here as enlarged ?il (or, less probably, i2, di1, or di2), ?c, ?p1 (or dp1), p3, p4, m1, m2, and roots of m3 (conventional notation).
Hypodigm. AMNH 113133 (type); 99594, left lower jaw (Figs. 5, 6) of a young individual with part of an enlarged alveolus for an enlarged incisor, posterior wall of an alveolus for ?c, alveolus of ?p1 (or dp1), and four cheek-teeth—p3, p4, m2, and damaged m1 (entoconid and posterolabial stylar cuspid); 95733, left lower jaw (Figs. 5, 6) of a young individual with part of enlarged incisor alveolus, roots of ?p1 (or dp1) and m3, p3, damaged p4 and m1 (stylar cuspids), and m2; 97260, left lower jaw (Figs. 7, 8) of an old individual with m2, m3, and anterolabially damaged m1; 113871, right m3 with damaged trigonid of a young individual (Figs. 7, 8); 88294, left P4 and lingual half of M1 in maxillary fragment (Figs. 1, 2); 114009, worn ?right M1; and 95713, damaged right M1, M2, and M3 in maxillary fragment (Figs. 1, 2). All known specimens are from a single producing layer, bone bed A, unit 24, at least 170 m above the local base of the type section of the Tepee Trail Formation in "tuff, hard, fine-grained, olive drab, andesitic, ledge-forming" (Love, 1939, p. 75; Flynn, 1986), TTA Quarry, NW¼,NE¼,NE¼, Sect. 4,T.43N,R.104W, USGS East Fork Basin Quadrangle, 7.5 minute Series (topographic), 1967, Fremont County, northwestern Wyoming.
Age. Shoshonian Subage, early Uintan Land Mammal Age, Chron C20R of paleomagnetic stratigraphy, late medial Eocene, approximately 47.5 Ma (Golz, 1976; Berggren and others, 1978; Berggren and others, 1985; Flynn, 1986).
Etymology. The trivial name refers to the numerous styles and stylids borne by the cheek-teeth of this animal.
Diagnosis. Short-faced plagiomenids with eight teeth in dentary: ?i1; two single-rooted, small teeth behind ?i1; and, following a brief diastema, a small p3, a p4 with well-developed trigonid but reduced, broad heel tucked under the trigonid of m1; molars decreasing in size from M1/m1 to M3/m3; p4 and m2 with one large labial stylar cusp, m1 with two; lower molars with extra cusps between protoconid and metaconid; paraconid lacking on m2 and m3; lower molars with ledge running labiad from metastylid on posterior face of trigonid; enamel highly crenulated when unworn; P4 non-molariform; upper cheek-teeth with large stylar cusps and "V"-shaped trough running labiad from labial side of protocone (see also the cladogram depicted in Fig. 12 for differential diagnoses within the Ekgmowechashalinae).
Description and discussion. The lower jaw is somewhat shortened and deep, especially in comparison with *Plagiomene* or *Ceutholestes* and other lipotyphlans possibly closely related to Dermoptera. The two rami were joined but not fused at a large oval symphysis (see Fig. 3). A large anterior mental foramen lies beneath the root of the tooth identified here as ?p1 (or dp1), in the base of a depression in the labial side of the ramus, bordered in front by the swelling around the root of ?i1 and in the rear by another swelling around the anterior root of p3. A second, smaller mental foramen lies under p3 in AMNH 99594 and 113133 and just behind that point in AMNH 95733. Both mental foramina have somewhat raised borders. At the rear of the cheek-tooth row, the coronoid process rises steeply as is known to be the case in *Plagiomene* and *Elpidophorus elegans,* but as in most known mixodectid and plagiomenid jaws, it is badly broken in all known jaws of *Tarka.*

The enamel on all teeth of *Tarka stylifera* is somewhat crenulated, as for example in certain petauristine sciurids and *Thisbemys,* and is covered with small, irregular bumps. For this

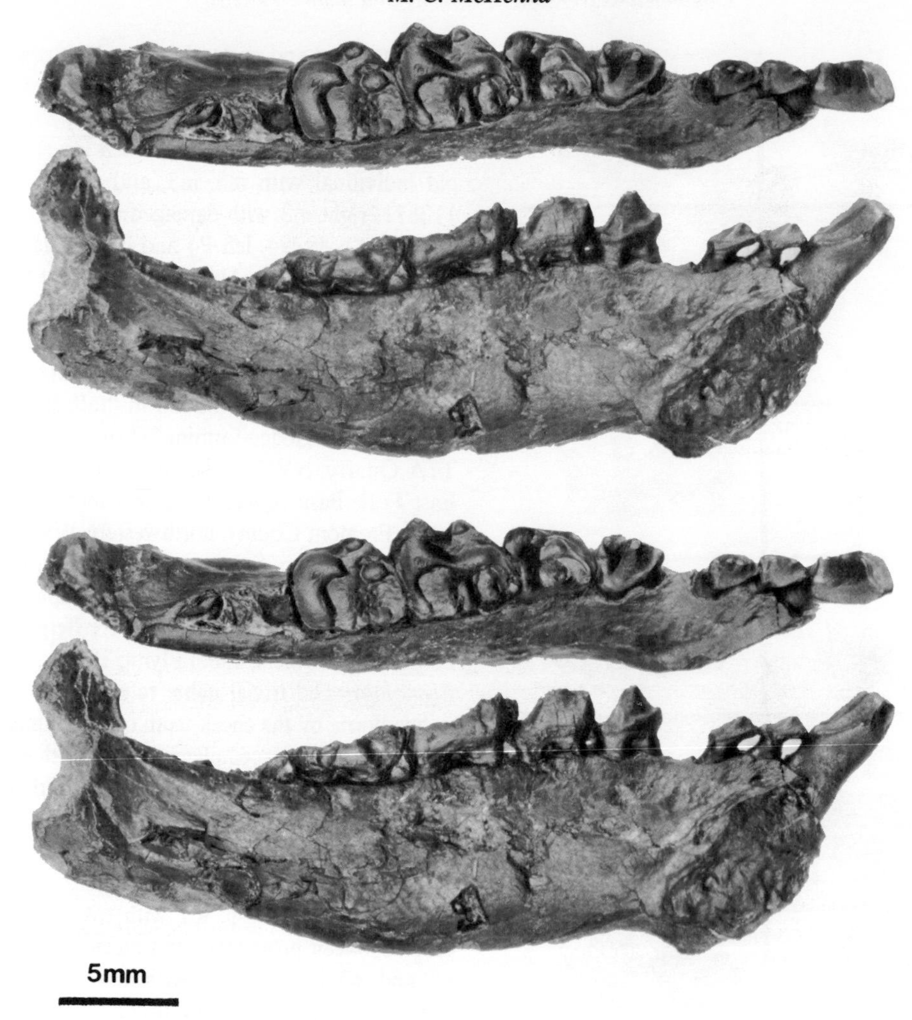

Figure 3. *Tarka stylifera.* Stereoscopic photographs of AMNH 113133, type specimen, left lower jaw with il, ?c, ?pl (or dpl), p3, p4, m1, m2, and roots of m3, occlusal and lingual views. See 5-mm bar for scale. See Figure 4 for labial view.

reason, some structures, particularly in the molars, are difficult to identify.

Several uncertainties affect the identification of some teeth assigned to *Tarka stylifera.* First, no field association was found between the upper and lower teeth here assigned to a single taxon. However, they do occur in the same quarry. They are assigned to one taxon because of size, occlusion, and morphology. Second, P4 is assigned because of size and the morphological similarity of the accompanying half of M1 (AMNH 88294) to that of AMNH 95713. That P4 and M1 of AMNH 88294 might actually be P3 and P4 has been considered and rejected because this would imply that p4 would be completely molariform, known not to be the case. Rather, P4/p4 appear to have been "demolarized" by loss of a former P4 metacone and reduction of the true heel of p4 (compensated by the addition of a new posterolabial stylar cuspid). Third, a relatively large, isolated m3 (AMNH 113871) is assigned on the basis of morphology and the assumption that m3 of AMNH 97260 is at the small end of variation in size. This last assignment is the most troublesome, but the only known m3 in a jaw is also small relative to m2 of the same individual; moreover, morphology is usually much stronger evidence for relationship than size, especially in an m3.

The enlarged anterior lower incisor is here identified as ?i1 by comparison with *Mixodectes* (Szalay, 1969), *Worlandia* (Rose, 1982), and *Plagiomene* (Matthew, 1918; Rose, 1973, 1982). However, this identification is not certain because homology has not yet been demonstrated adequately. The crown projected high above the level of the succeeding antemolar teeth. In AMNH 113133, an old individual, the tip is worn below the level where one can determine with certainty whether it was single-cusped or double-cusped as in *Worlandia, Plagiomene,* or in a lower incisor that Rose and I believe probably belongs to *Planetetherium* (Simpson, 1928, Fig. 12B; Rose, 1973). Certainly, the crown in AMNH 113133 is not tricuspate, nor is it multicuspate as in *Litolestes* (Schwartz and Krishtalka, 1976), *Amphidozotherium* (Sigé, 1976), or *Ceutholestes* (Rose and Gingerich, 1987). No trace of a crease is present on the anterior face of the crown of AMNH 113133 and none exists posteriorly. If the unworn crown

Figure 4. *Tarka stylifera.* Stereoscopic photograph of AMNH 113133, type specimen, left lower jaw with il, ?c, ?pl (or dpl), p3, p4, m1, m2, and roots of m3, labial view. See 5-mm bar for scale. See Figure 3 for occlusal and lingual views.

was apically divided, the division must have begun above the level of wear in the existing specimen and the division would have been minor, in contrast to the other genera in which the structure of the anterior teeth is known. Inasmuch as the tooth is also somewhat narrower from side to side than in *Plagiomene* and *Worlandia,* I doubt that its apex was divided significantly. This conclusion is supported by the fact that the crown is parallel-sided rather than splayed anteriorly as in the other genera. Posterolabially, posteriorly, and medially, three small crests descend from the apex. The posterolabial crest terminates in a small but distinct posterolabial heel cusp just above the base of the enamel; the two other crests terminate in a somewhat larger basal heel cusp that is at the posterolingual base of the crown. The lingual crest has the character of a cingulum; the central crest is a strengthening ridge; and the labial crest is a cutting edge, now dulled by wear in its surviving lower part. If the labial crest once possessed a small cuspule, no trace survives. Together, the anteriormost incisors of the two lower jaws form a powerful mechanism for dealing with insects, bark, fruits, nuts, or other hard food items. Such a structure is compatible with a diet including exudates, suggested by the cheek-tooth morphology. Wear was heavy before the death of AMNH 113133, both apically and along the crests running down from the apex. The root of ?i1 is set deeply in the bone of the mandibular symphysis, which, while massive, is unfused. The tooth is set a little more steeply than in *Worlandia* and much more steeply than in *Plagiomene.* The jaw is correspondingly deeper than in *Worlandia* and much deeper than in *Plagiomene.*

Behind ?i1 and anterior to p3 there are two single-rooted teeth, followed by a short diastema. Because the diastema can be seen fairly well in three specimens at different stages of dental wear, I think it unlikely that a tooth was present at this site in young individuals of *Tarka stylifera.* The identification of the two surviving single-rooted teeth is uncertain but, if one indulges in an analogy with *Worlandia,* the pair must be drawn from some combination of i2, c, p1 (or dp1), and p2. The tooth identified by Rose (1982) as i2 of *Worlandia* would seem to be a prime candidate for loss because of its occupancy of part of the alveolus of what is identified as i1 (Rose, 1982; not Bown and Rose, 1979), so I speculate that the same process happened independently in the ancestry of *Tarka stylifera.* What is called i3 presumably went the same route in both genera (unless what is identified as i1 is actually i2, and "i2" of *Worlandia* is actually i3). This would leave c, p1 (or dp1), and p2 to choose from. If the diastema of *Tarka stylifera* represents a phylogenetic rather than an ontogenetic loss, then the two teeth immediately behind ?i1 are

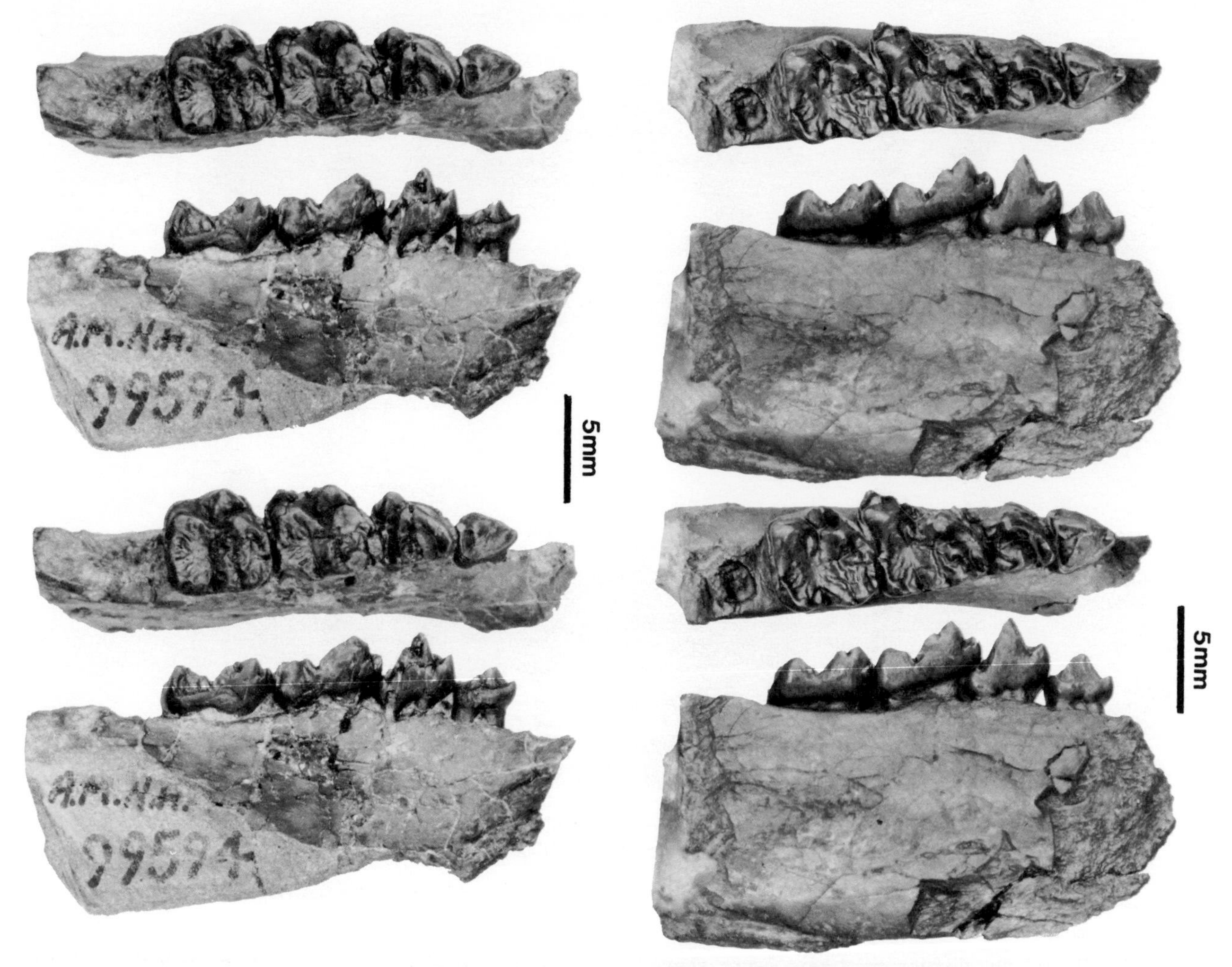

Figure 5. *Tarka stylifera.* Left: stereoscopic photographs of AMNH 99594, left lower jaw fragment with p3, p4, m1, and m2, occlusal and lingual views. Right: stereoscopic photographs of AMNH 95733, left lower jaw fragment with p3, p4, m1, and m2, occlusal and lingual views. See 5-mm bars for scale. See Figure 6 for labial views.

probably c and p1 (or dp1), respectively, p2 having been suppressed along with i2 and i3 sometime in the ancestry of the genus.

If one does not accept the analogy of *Worlandia* in identifying these teeth, then all three anteriormost teeth of *Tarka stylifera* could be some sequence derived from within i1, i2, i3, c, p1 (or dp1), and p2. However, the teeth will be treated as ?i1, ?c, and ?p1 (or dp1) here, based primarily on the *Worlandia* model. Unless a reversal in evolution occurred (see Fig. 12, characters 15–37), *Worlandia* did not give rise to *Tarka stylifera,* so the matter is not quite secure.

The tooth I identify as probably the lower canine, c (called p1 by Schwartz and Krishtalka, 1976), has a blade-like crown that projects far forward beyond its single root, as in *Litolestes* and *Plagiomene* (in which c has two roots). The tooth (and its more posterior companion) is quite small, but this is not because of crowding by the large root of ?i1. There was plenty of room in the horizontal ramus of the jaw for the development of both ?c and ?p1 (or dp1). Rather, ?c seems to have been undergoing reduction due to some other cause and would presumably have been lost eventually in some even shorter faced descendant of *Tarka stylifera,* were such a line to exist. The bladelike crown consists of a large protoconid, equipped with a sharp, anteroventrolingually directed paralophid. The paralophid does not terminate in a distinct paraconid; rather, it appears to join the anterior end of a broad, lingual cingular shelf. Posteriorly, ?c possesses a wide heel, continuous with the aforementioned shelf. A notch above the root, at the lingual side of the heel, may be caused by wear. At the posterolabial corner of the tooth a distinct heel cusp appears to be present, but, if so, it is heavily worn in the single example available. Labially, on the posterior base of the protoconid, the faintest trace of a tiny stylar cusp, a mere roughening of the enamel, can be seen under appropriate lighting. The root of the ?canine is large in its alveolus but sharply decreases in diame-

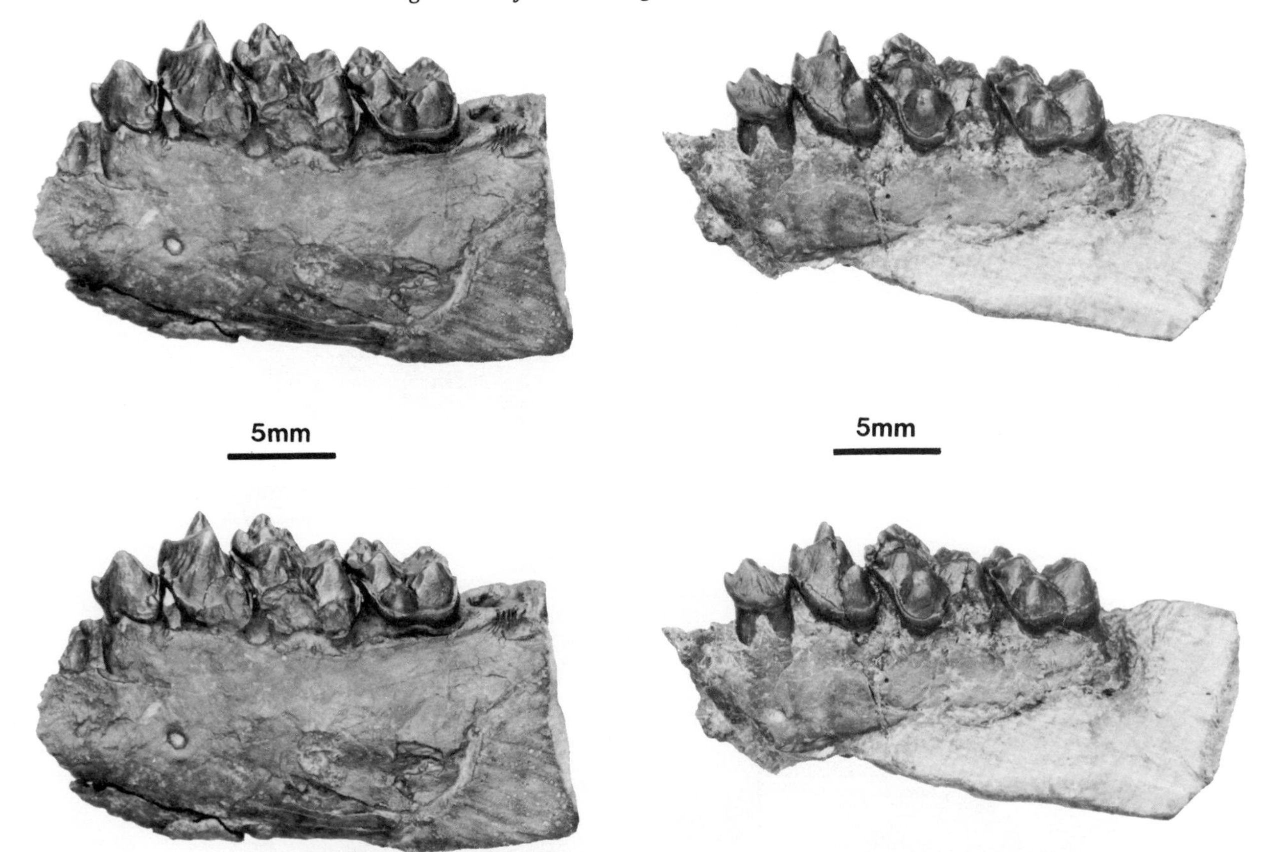

Figure 6. *Tarka stylifera.* Right: stereoscopic photograph of AMNH 99594, left lower jaw fragment with p3, p4, m1, and m2, labial view. See 5-mm bar for scale. See left part of Figure 5 for occlusal and lingual views. Left: stereoscopic photograph of AMNH 95733, left lower jaw fragment with p3, p4, m1, and m2, labial view. See 5-mm bar for scale. See right part of Figure 5 for occlusal and lingual views.

ter just beneath the crown. In some respects, ?c of *Tarka stylifera* resembles the hatchet-like tooth in the anterior lower dentition of apatemyids. The tooth also resembles the lower canine of *Litolestes,* identified as p1 by Schwartz and Krishtalka (1976).

The tooth identified questionably as the first lower premolar (or dp1) differs from ?c, although it is about the same size, in that its crown does not project so far forward above the root. It is less blade-like, not so high anteriorly, and appears to have possessed a tiny paraconid, now mostly worn away in the available example. Like ?c, it possesses a lingual cingular shelf. The heel is shorter than that of the ?canine, but a tiny posterolabial heel cusp seems to have been present. As with the ?canine, in proper light a trace of a stylar cusp or roughened area can be seen on the labial side of the protoconid. The root of ?p1 (or dp1) is not so constricted at the base of the crown as that of the ?canine.

Following the diastema that presumably represents the phylogenetic loss of p2, is a low, non-molariform p3 with two roots. Its triangular crown is mostly a trigonid, with a high central protoconid, a low but sharply distinct paraconid at the anterior end, and a very weak metaconid on two of the three available examples. On p3 of the type specimen, a metaconid is not even weakly expressed; possibly its traces have been worn away. The paraconid lies somewhat lingual to the base of the anterior crest of the protoconid and serves as the anterior terminus of a strong lingual cingulum. The cingulum acts as a sharply defined shelf that runs to the rear, turns to become the low heel, and finally ends at the posterolabial base of the protoconid. Along its course, the cingulum is joined by a weak vertical crest beneath the site of the metaconid. It also gives rise to a low cusp midway across the heel. Before it ends at the base of the protoconid after crossing the heel, the cingulum bears a small labial stylar cuspule.

The last lower premolar, p4, is about twice the size of p3 and is double rooted, but it is not molariform, although its well-developed trigonid is functionally part of the molar series. Unlike p3, p4 possesses a curved, high, and strong paralophid (paracristid) that in heavy wear incorporates the paraconid rather than leads to it as on p3. However, in early wear the paralophid (paracristid) may be slightly notched as in AMNH 99594. A very high metaconid lies lingual to the protoconid, overtopping it and separated from the whole paralophid by a curved trough. The majority of the metaconid is conical, but the posterolabial surface forms a steep cliff, lingual to the middle of which is a sharp, vertical ridge of enamel that divides the face into broad labial and narrow lingual grooves. A shear facet is present on this ridge and is continuous beneath the posterior trigonid notch with the shear facet of the posterolingual corner of the protoconid. A large,

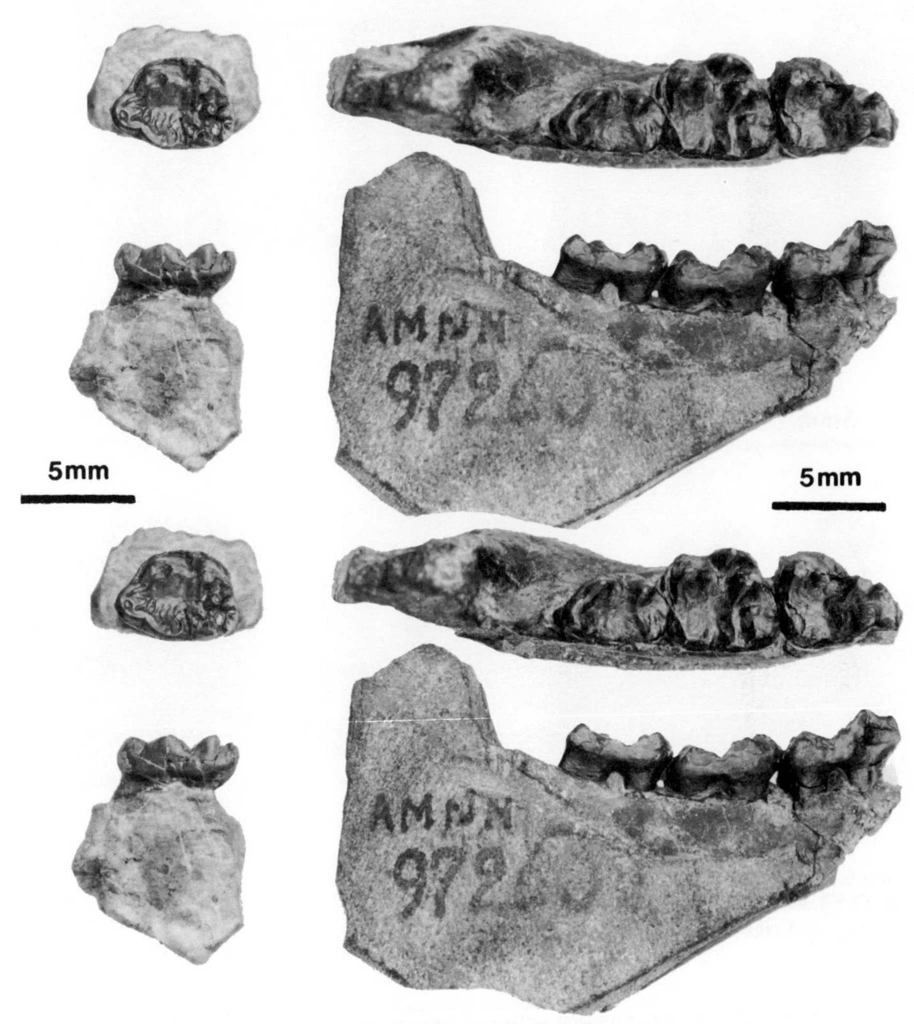

Figure 7. *Tarka stylifera.* Left: stereoscopic photographs of AMNH 113871, unworn right m3 with slight damage to trigonid (reversed for ease of comparison). Right: stereoscopic photographs of AMNH 97260, fragmentary left lower jaw with damaged and worn m1, m2, and m3, occlusal and lingual views. See 5-mm bars for scale. The isolated m3 shown reversed here is larger than that of AMNH 97260, but clearly belongs to no other taxon. See Figure 8 for labial views.

conical stylar cusp lies at the posterolabial base of the protoconid. In contrast to p3, there is no lingual cingulum, but a well-defined, though uneven, labial cingulum runs from the anterior base of the paraconid around the labial base of the crown, going completely around the large stylar cusp to join or become the narrow heel crest. The latter terminates in a low cuspule at the posterolingual corner of the tooth. In the middle of the posterior cingulum or heel crest is another low cuspule. The entire heel of p4 is hidden under the closely appressed trigonid of m1 and is evidently suppressed.

The largest and most complex molar of *Tarka stylifera* is m1. It possesses a bewildering array of extra cusps and crenulations, but displays a weak paraconid, closely appressed to the base of the metaconid. In its basic structure, a trigonid and a broad, basined talonid wider than the trigonid can be discerned, occupying the lingual two-thirds of the tooth. However, the labial third consists of two large stylar cusps. One or another of the stylar cusps is broken away on all specimens except the worn AMNH 113133. In contrast to the marsupial genus *Glasbius,* which is superficially similar, the more anterior of these two stylar cusps is always larger than the posterior one, rising nearly as high as the protoconid, at whose labial (slightly posterolabial) base it lies. The posterior labial stylar cusp, also conical, is low and does not rise much above the labial cingulum from its origin near the base of the hypoconid. Another extra cusp lies on the trigonid between the protoconid and the metaconid, closer to the latter than to the former. Together with the metaconid, protoconid, and anterior labial stylar cusp, it contributes to an unusual row of four major cusps, marching across the wide trigonid. Posterior to the metaconid is a metastylid that is rather more elaborate than that of *Elpidophorus elegans.* This forms the lingual apex of a ledge that crosses the rear wall of the trigonid toward a posterolingually directed crest from the protoconid, which it nearly meets at a vertical gully at about the middle of the posterior trigonid wall.

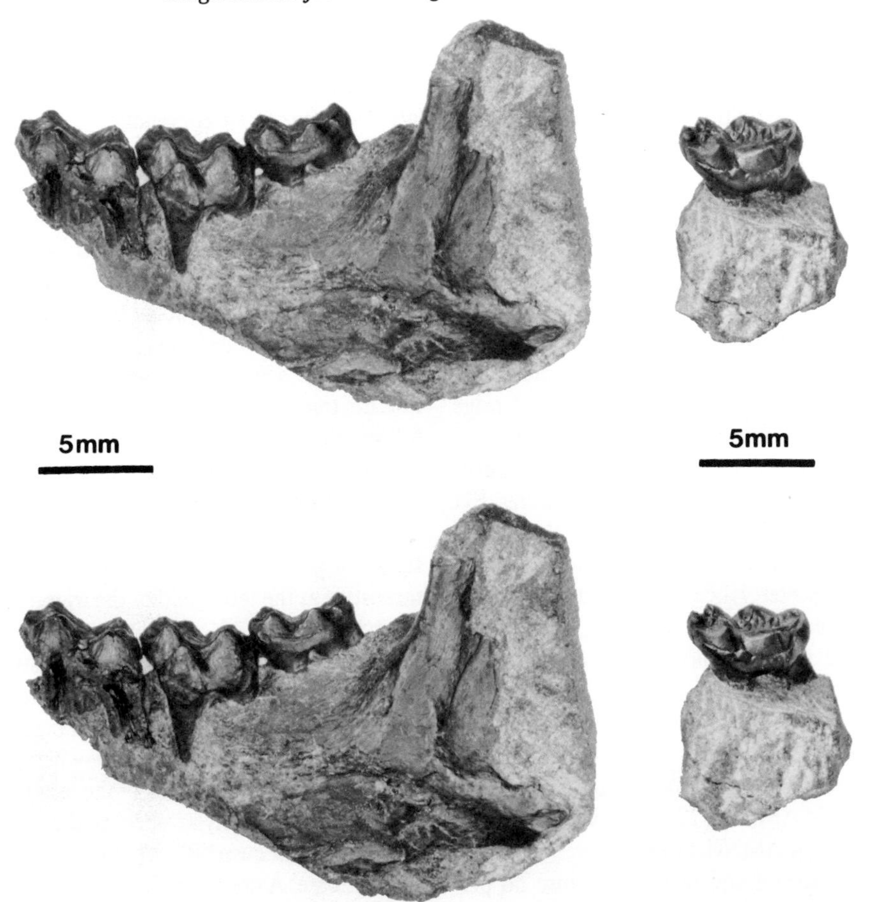

Figure 8. *Tarka stylifera.* Left: stereoscopic photograph of AMNH 97260, fragmentary left lower jaw with damaged and worn m1, m2, and m3, labial view. Right: stereoscopic photograph of AMNH 113871, unworn right m3 with slight damage to trigonid (reversed from original for ease of comparison). See 5-mm bars for scale. The "left" m3 shown reversed here is larger than that of AMNH 97260, but clearly belongs to no other taxon. See Figure 7 for occlusal and lingual views.

This ledge is represented by a line of cusps in *Tarkadectes.* The talonid is dominated labially by a strong hypoconid with nearly horizontal wear facets leading into the talonid basin and by a highly crenulated entoconid (and anteriorly placed "entostylid") at the posterolingual corner of the tooth. A hypoconulid (as such) is not present. What is taken here to be the posterior part of a cristid obliqua is present labially, but from its origin at the anterior of the hypoconid it does not cross the tooth beyond the posterolingual base of the protoconid. If present, the mesoconid is merely one of several anterior crenulations on the cristid obliqua. A cingulum surrounds three sides of the tooth but is absent on the lingual side.

The second lower molar has most of the features of m1 but is smaller, narrower across the talonid than the trigonid, lacks the paraconid, possesses a larger metastylid, and lacks the posterior labial stylar cusp that is so prominent on m1. The ledge running labiad from the metastylid is stronger than on m1. When unworn, this ledge is seen to comprise a series of tiny coalesced cuspules that cross the tooth and rise to join the protoconid apex. The ledge plus protoconid and the weak paralophid (paracristid) surround the central part of the trigonid, which consists of a smaller metaconid than that of m1 plus a proliferation of cuspules that seem homologous with the extra trigonid cusp of m1 that occurs between the protoconid and metaconid of that tooth. When unworn, the talonid of m2 is a sea of crenulations, but heavy wear removes most of them, as in AMNH 113133. An "entostylid", anterior to the entoconid, and a transverse groove and ridge anterior to the hypoconid are fairly constant features of the talonid. No true hypoconulid is present, but the cingulum does swerve dorsad just behind the hypoconid, although it does not swell or bear a cusp. A nearly horizontal wear facet on the hypoconid indicates that chewing motions were primarily from side to side.

The third lower molar is known from two examples, one in the jaw with m1 and m2 (AMNH 97260) and the other a large, isolated tooth (AMNH 113871). The m3 of AMNH 97260 is reduced compared to m1 and m2 of the same specimen, but the tooth is also heavily worn and none too well preserved. To make matters worse, AMNH 97260 was recovered from especially hard matrix and could not be prepared fully. The large, isolated m3, on the other hand, is well preserved except for the loss of the anterior part of its metaconid and of a lingual cuspule within the central chaos of the trigonid apex. Like m2, m3 lacks a paraconid, that structure having been subsumed in a cuspidate paralophid (paracristid) as on m2. A central, anterior cuspule on the paralophid (paracristid) resembles a paraconid but is apparently not homologous with a true paraconid. On AMNH 97260, details of the trigonid are vague, but at least one extra trigonid cusp exists between the protoconid and metaconid. On AMNH 113871, two main cuspules lie between the protoconid and metaconid. The prominent posterior ledge of the trigonid seen on m1 and m2 is only weakly expressed on m3 and is worn away on AMNH 97260. On AMNH 113871, only a very weak metastylid is indicated at the base of the metaconid. A groove and ridge anterior to the hypoconid is preserved on both specimens and resembles those of m2. A large, high entoconid and a rearward-projecting "third lobe" complete the talonid's main features. Six labially directed ridgelets run from the entoconid and "third lobe" toward a straight, anteroposteriorly oriented, central talonid gutter. A cingulum follows the posterolabial, labial, and anterolabial base of the crown, continuously from the "third lobe" to the anterior part of the trigonid. On this structure in AMNH 113871 a small stylar cusp appears on the cingulum at the posterolabial base of the protoconid, but no stylar cusps are present on the cingulum of AMNH 97260. The large m3 of AMNH 113871 thus resembles the other molars of *Tarka stylifera* more than does the less well-preserved m3 of AMNH 97260, known actually to belong to *Tarka stylifera.* That AMNH 113871 represents a primate instead has been considered and rejected because no primate of nearly comparable size is otherwise known from the collection from TTA Quarry.

P4 is known from a single example, AMNH 88294. The tooth is accompanied by the lingual half of an M1. These are identified as P4 and M1, rather than as a P3 and molariform P4, respectively, because of the non-molariform p3 and posteriorly compressed p4 of lower jaws believed to represent the same taxon. P4 is a transversely oriented, rectangular tooth dominated by a high, rather centrally set paracone and a low, anterolingually placed protocone. No protoconule (paraconule) or metaconule is present. A transverse, crest-like parastyle is followed posterolabially by another stylar cusp at the anterolabial base of the paracone. Next, a large, posterolabially damaged stylar cusp lies posterolabial to the paracone. Together, these stylar cusps make the labial cingulum area highly distinctive. The large stylar cusp possesses anterior and posterior crests. From the paracone apex an anterior crest runs to the parastyle, and a posterior crest runs from the apex to the rear of the tooth, where it turns abruptly labiad to join the posterior crest of the large stylar cusp. There is no metacone. A dull crest connects the paracone apex lingually to a similar crest from the protocone. An anterodorsolabially directed crest from the protocone apex runs to join the anterior cingulum lingual to its elevation as the parastyle. Also from the protocone's apex, a crest runs posterodorsolingually to the somewhat swollen posterolingual corner of the crown; however, no hypocone exists (in the sense of a projecting cusp). P4 is completely surrounded with a cingulum except where stylar cusps are constructed upon it and, possibly, at the posterolabial corner, where the morphology is hidden under the overlapping M1. The whole crown of P4 is crenulated, especially so on the posterior slopes of the paracone and protocone and on the lingual slope of the latter.

M1 is larger than M2 by a small margin. It is significantly larger than M3 or P4. Salient characteristics are a straight, transverse, "V"-shaped trough from the labial side of the protocone to a point between the paracone and metacone, well-developed protoconule (paraconule) and (probably) metaconule, four main stylar cusps on a wide stylar shelf, a posteriorly crested protocone but no trace of a hypocone, and a prominent cingulum that follows the anterior, lingual, and labial borders of the crown and doubtless once served as the precursor of the stylar shelf, now hypertrophied. The stylar cusps are conveniently (but perhaps not accurately) identified as, from anterior to posterior: a large labial stylar cusp "B" in the position of a stylocone; a small conical mesostyle at the labial end of the trigon basin, lingual to the level of stylar cusps "B" and "D"; another large labial stylar cusp ("D"); and a small metastyle. The parastyle is either absent due to the hypertrophy of stylar cusp "B" or, more likely, is represented by a crest curving from the anterior base of stylar cusp "B" to the anterolabial base of the paracone, where it is met by an anteriorly directed crest from the apex of the latter. Dominating the anterolabial part of the stylar shelf is the comparatively very large, lobe-like, stylar cusp "B", which preserves a bit of cingulum at its anterior base. A crest connects the anterior part of stylar cusp "B" to the parastyle. Inboard and rearward from stylar cusp "B" is the smaller, conical mesostyle, which blocks the labial end of the transverse trough that crosses the crown. Together, the mesostyle, stylar cusp "B," and the paracone form an equilateral triangle similar to that seen in *Worlandia, Planetetherium,* and *Plagiomene.* However, in those genera the mesostyle is more closely associated with the paracone and does not block the labial exit of the trigon basin. In *Planetetherium, Worlandia,* and *Plagiomene accola* (Rose, 1981) an additional, small stylar cuspule forms between stylar cusp "B" and the mesostyle. A lobe-like stylar cusp "D" follows, set labially, as is the more anterior stylar cusp "B". A crest from stylar cusp "D" connects to the rear of the labial face of the metacone. The low metastyle completes the impressive stylar array; it is placed close to the metacone and posterolabial to it, but far inboard of the two major stylar lobes of the labial side of the tooth. A crest runs from the posterior side of the apex of the metacone to the metastyle. The enamel of M1 is crenulated in unworn condition (AMNH 88294), but the crenulations disappear with wear (AMNH 95713). The three roots of the tooth are small for a tooth of this crown size and give the impression that the wide crown is weakly supported on stilts. A similar condition is to be seen in undescribed arctic paromomyids (McKenna, in preparation).

In the only available example, M2 is damaged by a crack. The mesostyle (if present) and parts of the paracone, metacone, and protoconule (paraconule) are now missing. From what can be seen now, M2 is very similar to M1, but the more posterior of

TABLE 1. MEASUREMENTS OF *TARKA STYLIFERA* AND, WHERE NOTED, *TARKADECTES MONTANENSIS*

	(in mm)
?il-m2 length (AMNH 113133, present degree of wear)	25.3
p3-m2 Length (AMNH 113133)	16.0
(AMNH 95733)	17.2
(AMNH 99594)	17.8
m1-m3 length (AMNH 97260)	16.0
(AMNH 113133, estimated)	16.0
p4-m2 length (CMNH 40818, *Tarkadectes montanensis*)	10.2
(AMNH 113133)	13.1
(AMNH 95733)	14.1
(AMNH 99594)	14.8
Depth (labial) of jaw beneath trigonid of m1	
(CMNH 40818, *Tarkadectes montanensis*)	6.7
(AMNH 113133, estimated)	more than 6.1
(AMNH 95733)	less than 10.2
Depth (lingual) of jaw beneath trigonid of m1	
(CMNH 40818, *Tarkadectes montanensis*)	6.8
(AMNH 113133)	more than 7.8
(AMNH 95733)	less than 11.6
?i1 maximum diameter, base of crown at heel, normal to root	
(AMNH 113133)	2.6
Lower ?canine a-p crown length (AMNH 113133)	2.1
Lower ?canine crown width (AMNH 113133)	1.4
?p1 (or dp1) a-p crown length (AMNH 113133)	2.1
?p1 (or dp1) crown width (AMNH 113133)	1.5
Diastema ?p1 (or dp1) -p3 at crown (AMNH 113133)	1.8
Length p3 (AMNH 113133, parallel to tooth row)	2.9
(AMNH 95733)	3.5
(AMNH 99594)	3.0
Width p3 (AMNH 113133, normal to length)	2.7
(AMNH 95733)	3.2
(AMNH 99594)	2.6
Length p4 (CMNH 40818, *T. montanensis*, parallel to tooth row)	2.6
(AMNH 113133)	3.9
(AMNH 95733, estimate)	4.7
(AMNH 99594)	4.6
Width p4 (CMNH 40818, *T. montanensis*, normal to length)	2.4
(AMNH 113133)	3.7+
(AMNH 95733, estimate)	3.9
(AMNH 99594)	3.9
Length m1 (CMNH 40818, parallel to tooth row)	4.0
(AMNH 97260)	5.3
(AMNH 113133)	5.8
(AMNH 95733)	6.2
(AMNH 99594)	6.3
Width trigonid m1 (CMNH 40818, *T. montanensis*, normal to length, at cingulum, parallel to occlusal plane)	2.9
(AMNH 113133)	5.1
(AMNH 99594)	5.8
Width talonid m1 (CMNH 48108, *T. montanensis*, normal to length, estimated)	3.1
(AMNH 113133)	5.5+
(AMNH 95733, estimated)	5.8
(AMNH 99594, estimated)	5.5
Length m2 (CMNH 40818, parallel to tooth row)	4.0
(AMNH 97260)	5.2
(AMNH 113133)	5.0+
(AMNH 95733)	5.2
(AMNH 99594)	5.4
Width trigonid m2 (CMNH 40818, *T. montanensis*, normal to length)	3.0
(AMNH 97260)	5.0+
(AMNH 113133)	5.3+
(AMNH 95733)	5.4
(AMNH 99594)	6.1
Width talonid m2 (CMNH 48108, *T. montanensis*, normal to length)	3.1
(AMNH 97260)	4.3+
(AMNH 113133)	4.2+
(AMNH 95733)	5.0
(AMNH 99594, estimate)	5.3
Length m3 (AMNH 97260, parallel to tooth row)	4.8
(AMNH 113871)	5.5
Width trigonid m3 (AMNH 97260, normal to length)	3.0
(AMNH 113871)	3.7
Width talonid m3 (AMNH 97260, normal to length)	2.9
(AMNH 113871)	4.2
Length P4 (AMNH 88294, at paracone)	3.9
(AMNH 88294, at protocone)	2.9
Width P4 (AMNH 88294)	6.3
Length M1-M3 (AMNH 95713, at level of conules, estimated)	12.1
Width M1 (AMNH 95713, anterior half of tooth)	6.9
Width M2 (AMNH 95713, anterior half of tooth)	7.1

its two major stylar lobes is smaller than on M1 and is not connected directly to the apex of the metacone. The protoconule (paraconule) and metaconule are large. The presence of a large metaconule reinforces the probability of occurrence of a similar cupsule on the damaged M1. The metastyle is either minute or absent.

Because of damage, little can be said of the only example of M3. However, it is substantially smaller than either M1 or M2. Its labial half is now missing except for fragments of the paracone and protoconule (paraconule). There is a large protocone, a tiny metaconule, and a posterodorsolabially sweeping crest from the protocone apex. There is no hypocone. The small size of M3 of AMNH 95713 correlates well with the reduced m3 of AMNH 97260 but not with AMNH 113871, unless the latter accompanied still larger anterior molars not yet represented in our collections.

Measurements. See Table 1.

Tarkadectes, new genus

Type. Tarkadectes montanensis, new species.
Locality, age, and distribution. As for the type species.
Etymology. Allusion to *Tarka* and *Mixodectes.*
Diagnosis. As for the type species (see also Fig. 12).

Tarkadectes montanensis, new species
(Fig. 9)

Type. CMNH 40818, left lower jaw fragment with p4–m2 and part of alveolus of m3.
Hypodigm. Type specimen only.
Locality and age. Paola Siding Locality, Middle Fork of the Flathead River, Flathead County, Montana. Upper sequence of the Coal Creek Member of the Kishenehn Formation, about 300 m stratigraphically above beds that yield a fission-track date of 33.2 ± 1.5 Ma (Constenius and Dyni, 1983). On paleontological grounds, the nominally early Oligocene level in the Canadian Kishenehn (Russell, 1954) is currently regarded as medial Chadronian in age (Ostrander, 1985), but how that level fits with the Chadronian part of the Kishenehn on the Montana side of the international boundary is not accurately known.
Etymology. For the state of Montana.
Diagnosis. Lower cheek-teeth not so wide as those of *Tarka* nor equipped with huge, labial, stylar cusps as in that genus. Weak entostylid on anterior slope of entoconid of m2 (condition unknown on m1 because of breakage, but presumably it was stronger). Phase I chewing facets of main cusps enhanced. Line of cusps across anterior of talonid, rather than a shelf; these cusps also with phase I facets. Size smaller than *Tarka,* larger than the Kishenehn species of the lipotyphlan genus *Thylacaelurus* (McKenna in Van Valen, 1965, p. 394; Szalay, 1969, p. 242; sometimes thought to be possibly a dermopteran, e.g., Van Valen, 1967, pp. 261, 271; Rose and Simons, 1977; Krishtalka and Setoguchi, 1977; Storer, 1984). See also Figure 12 for differential diagnoses within the Ekgmowechashalinae.
Description. The jaw of *Tarkadectes montanensis* is relatively deeper with respect to lower cheek-tooth size than that of *Tarka stylifera.* The fourth lower premolar is double-rooted and consists of a well-developed trigonid and a low cingulum that serves as a heel. The posterior part of the cingulum is overlapped by the trigonid of m1. The paralophid (paracristid) is straight and leads from the protoconid anterolinguad and downward to join the labial side of the paraconid. The anterolabial side of the paralophid is bevelled near its top by a wear facet (wear surface 2 of Kay and Hiiemae, 1974), which faces anterolabiodorsad. This facet, involved in phase I chewing, is part of a family of wear surfaces seen on m1 and m2 (Kay and Hylander, 1978). The paraconid is low and conical, and has undergone apical wear. The angular protoconid is the highest and most massive cusp. It possesses strongly wrinkled enamel on its anterolabial and posterolabial faces and has two long, descending crests on its posterior face in addition to a short crest that goes toward the metaconid. Lingual to the protoconid, the high metaconid is a triangular cusp. Leading from its apex, sharp crests descend labiad, anterolinguad, and posterolinguad. The descending crests of the protoconid, together with weaker ones on the concave posterior face of the metaconid, give the posterior wall of the trigonid a striated appearance. A strong cingulum runs posterolabiad from the anterior base of the paraconid to a point above the labial side of the anterior root, where it turns posteriad to skirt the labial base of the protoconid. At the posterolabial corner of the tooth the cingulum overhangs the root and turns linguad to form the heel. The cingulum ends lingually at a small cuspule just behind the lingual base of the posterior wall of the trigonid. The entire cingulum is cuspidate, with small elevations at the anterolabial corner of the tooth, at the posterolabial corner, and part way across the heel between the bases of the descending crests of the protoconid. The small cingulum cusp at the posterolabial corner of the tooth is evidently homologous with the larger cusp at the same position on p4 of *Tarka.* The enamel of p4 is wrinkled.

The first lower molar possesses the plan of the main cusps of the same tooth in *Tarka,* but lacks large labial cingulum cusps. Instead, the strong, beaded labial cingulum bears only tiny cuspules at the comparable sites. All unworn enamel surfaces of m1 are wrinkled. The trigonid opens lingually at a sharp notch, but is nonetheless compressed. The paraconid is lingually placed, lying slightly posterior to the lingual termination of the paralophid. The latter has a wear facet (number 2, for phase I chewing; Rose and Simons, 1977) that faces anterodorsolabiad, as on p4. From the hypoconid a line of cuspules marches anterolinguad across the front of the talonid basin, ending in a strong metastylid that merges with the metaconid. From the hypoconid to the metastylid, this line of cuspules is bevelled by anterodorsolabially facing wear facets parallel to that of the paralophid. These facets are part of the phase I chewing facet system (Kay and Hiiemae, 1974). The entoconid is broken away but would have been large when whole and would doubtless have had a prominent phase I wear surface. Because of breakage, the morphology of the entostylid, presumed to be present, is unknown.

The second lower molar differs from m1 in that the paraconid is more centrally placed and the paralophid is consequently shorter. There is no cingulum cusp at the posterolabial base of the protoconid, in strong contrast to the large cusp of *Tarka.* The row of cuspules from the hypoconid to the metastylid is weaker than on m1 but is not so weak nor disorganized as in *Tarka.* Phase I chewing facets face anterodorsolabiad on the crest from the hypoconid to the next cusp of the row. A beaded cingulum runs from the paraconid base around the labial side of the tooth, becomes obscure at the base of the hypoconid, and then continues past a weak hypoconulid to end at the entoconid. The entoconid is small, bears an anterior crest with a basal entostylid, and has a prominent anterolabial phase I chewing facet that faces the talonid basin. The unworn enamel surfaces of m2 are wrinkled, although they are less wrinkled than in *Tarka.*

The third lower molar is broken away with the posterior part of the jaw, but an interdental wear facet on m2 and enough of the m3 alveolus are present to confirm its occurrence, in proba-

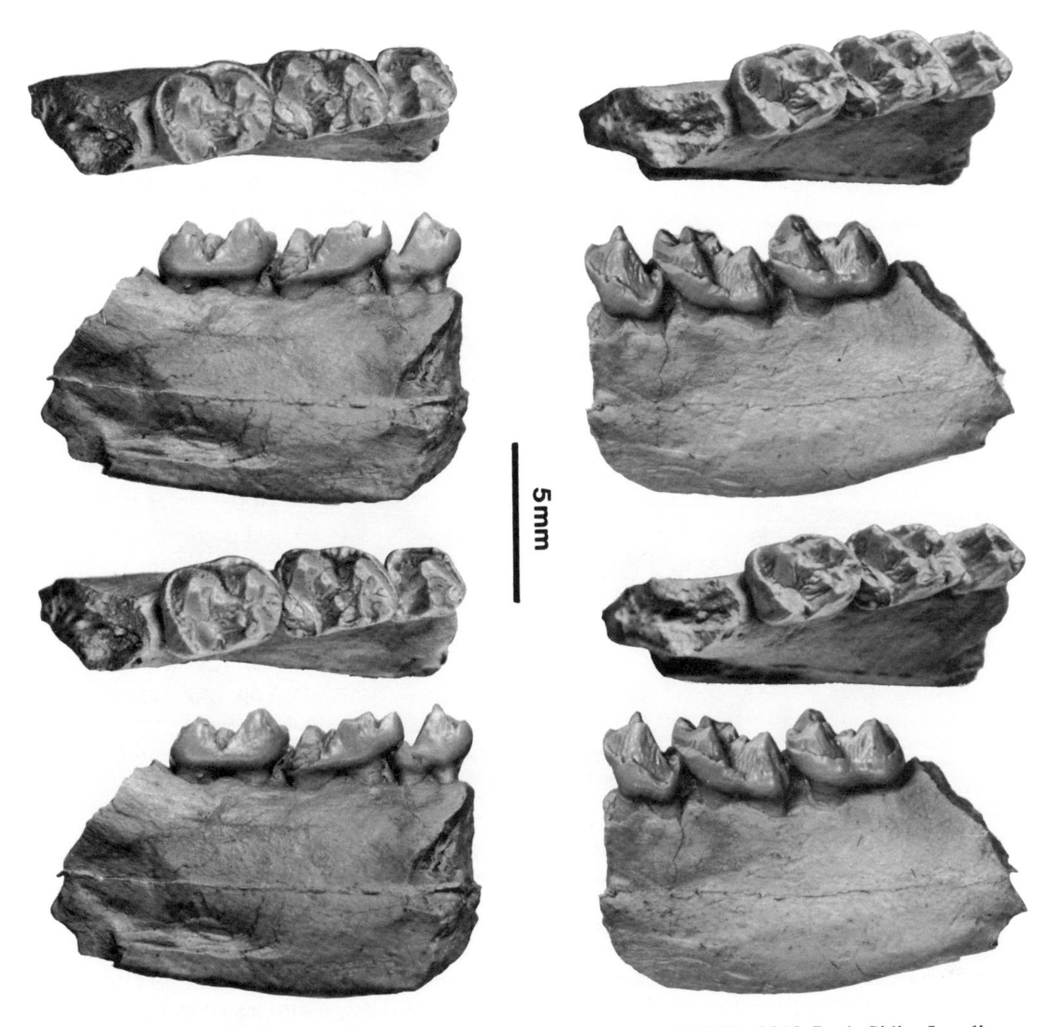

Figure 9. *Tarkadectes montanensis.* Epoxy cast of type specimen, CMNH 40818, Paola Siding Locality, upper sequence of Coal Creek Member of the Kishenehn Formation, Middle Fork of the Flathead River, Flathead County, Montana. A fission track date of 33.2 ± 1.5 ma has been obtained about 300 m stratigraphically below the site. Left top: occlusal view of lower jaw fragment with p4–m2 and part of alveolus of m3; left bottom, lingual view of same. Right top: view looking anteroventrolabiad, parallel with phase I wear surfaces. Right bottom: labial view. All views stereoscopic. See 5-mm bar for scale.

ble contrast to *Thylacaelurus,* which appears to have lost M3 and therefore probably m3 as well (but see Storer, 1984).

Discussion. *Tarkadectes montanensis* lacks the extreme enamel crenulation of *Tarka stylifera* and also possesses only a weak adumbration of the massive labial stylar cusps that so effectively widen p4–m2 in *Tarka,* increasing the area of enamel surface of *Tarka's* cheek-teeth and therefore presumably adapting it for a diet involving a substantial proportion of exudates or fruit. In *Tarkadectes,* m1 and m2 possess a row of cuspules across the anterior end of the talonid basin that in *Tarka* exists as a complexly wrinkled shelf. These cuspules in *Tarkadectes* are worn by phase I chewing action. *Tarkadectes* also lacks the new cusp between the protoconid and metaconid seen in *Tarka.* Moreover, *Tarkadectes* is significantly smaller than *Tarka.* Clearly, *Tarkadectes* was less modified than *Tarka* with regard to complexity of the enamel wrinkling, but was more involved in phase I power stroke chewing during feeding that processed fibrous materials instead of or in addition to insects, fruits, gums, sap, and other soft food.

Tarkadectes montanensis and *Thylacaelurus montanus* are both known from the Kishenehn Formation, so the question arises: might they be the same taxon? However, Kishenehn *Thylacaelurus montanus* is significantly smaller than *Tarkadectes* and has apparently lost M3 as well as developed a peculiarly specialized P4 (Szalay, 1969, Fig. 8). Whatever the affinities of *Thylacaelurus montanus* (e.g., Van Valen, 1965, 1967), it is clearly not the same animal as *Tarkadectes montanensis.*

Measurements. See Table 1.

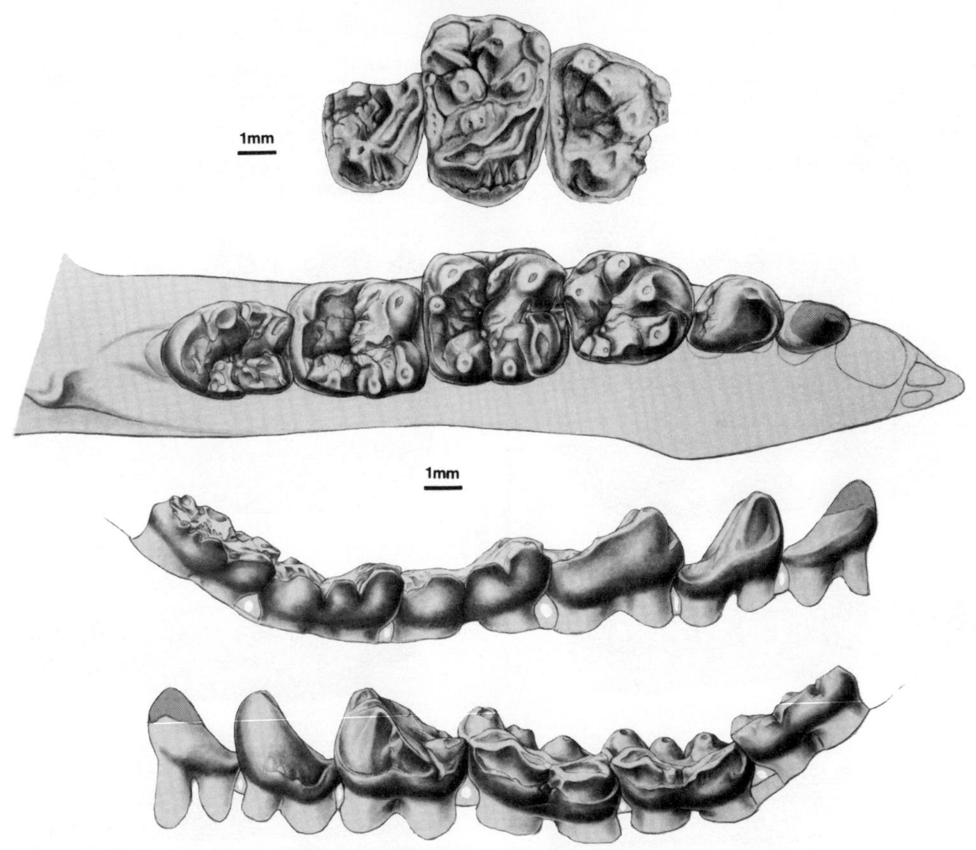

Figure 10. Below: composite reconstruction of occlusal, lingual, and labial views of the lower dentition of *Ekgmowechashala philotau* from the late Oligocene (early Arikareean) of South Dakota, reprinted from Szalay (1976, Fig. 133). Note that the spacing between anterior teeth is depicted as somewhat more crowded than indicated by photographs (Szalay, 1976, Figs. 129, 130). Above: Occlusal view of left P4–M2 (labial part of M2 missing), UCMP 128231, from the middle Arikareean of the John Day Formation, Oregon, reprinted (but reversed for ease of comparison) from Rose and Rensberger (1983, Fig. 2). This specimen was referred to *Ekgmowechashala* sp. (or a closely related taxon) by Rose and Rensberger. See 1-mm bars for scale.

Affinities of *Ekgmowechashala* Macdonald, 1963 (Fig. 10)

Macdonald (1963, 1970) and Szalay (1976) described the type and various referred specimens of this enigmatic late Oligocene mammal from South Dakota, placing it in the primate family Omomyidae but voicing doubts about the allocation. Szalay worked only with the lower jaw and dentition, all that was then available (but see his Table 27). He believed that *Ekmowechashala* was possibly derived from something like the omomyid genus *Rooneyia* Wilson, 1966. Nevertheless, Szalay awarded *Ekgmowechashala* its own subfamily. Rose and Rensberger (1983), who referred an unassociated upper dentition from Oregon to *Ekgmowechashala,* concurred. Yet, the genus does not make a very good bedfellow among undoubted omomyids (Szalay and Delson, 1979). See Figure 10 (top) for an occlusal view of this specimen, reversed for ease of comparison. On the basis of new information that was not available to previous authors, I suggest that *Ekgmowechashala* need not be compared exclusively with primates; instead, I believe it is a primitive plagiomenid allied with *Tarka* and *Tarkadectes.*

The lower jaw and the dentition of *Ekgmowechashala philotau* have been described and figured extensively by Szalay (1976, p. 349, 354–362; see Fig. 10, bottom three views) on the basis of seven specimens. Unfortunately, associated maxillae are not known. As Szalay noted (1976, p. 354), "the mandibular symphysis of [*Ekgmowechashala*], in lacking any traces of distinguishable inferior or superior transverse tori, is significantly different from the symphysis of any other known omomyid." The symphysis of *Tarka stylifera,* however, also lacks tori (Fig. 3). The anterior lower teeth of *Ekmowechashala* are poorly known or merely indicated to be present. Two small pits that are apparently alveoli for small lower incisors lie anterior to what has been interpreted to be an enlarged lower canine tooth in LACM 9207 (Szalay, 1976, Fig. 129, stereophoto). The alveoli identified by Szalay (1976, p. 349) are interpreted by him to hold i1 and i2. There is thus no enlarged tooth at the anterior end of the lower dentition. A short diastema separates the tooth identified as the lower canine from a double-rooted tooth called p2 on LACM 9207, but this information appears to have been discounted in Szalay's reconstructions (Szalay, 1976, Figs. 133–134, composite drawings), which suggest a somewhat more primate-like crowding of these teeth. A double-rooted p2 exists in *Plagiomene,* but not in *Worlandia* or *Tarka.* Until an undamaged specimen is found, the number of incisors and the nature of the canine tooth will remain speculative in *Ekgmowechashala,* but at least we know that i1 was not enormously enlarged. Loss of a lower incisor and p1 (?dp1), coupled with enlargement of the lower canine, would be expected if *Ekgmowechashala* were derived from an unspecialized animal near the ancestry of *Plagiomene,* but *Tarka* and *Worlandia* had specialized their anterior lower dentitions in a different way from such a base and are therefore not available as ancestors for *Ekgmowechashala* without invoking reversals (see Fig. 12). Thus the anterior lower dentition of *Ekgmowechashala* only weakly supports my hypothesis, if at all. However, the enlarged lower canine, short diastema, and double-rooted p2 are hardly features of omomyids, which generally crowd these anterior lower teeth. If *Ekgmowechashala* were really a primate, its lower canine and p2 would be weak evidence for relationship with higher primates, not omomyids.

Behind the tooth identified as p2 there is a small, double-rooted p3 as in *Elpidophorus* and *Tarka.* This tooth is not so molariform as p3 of *Plagiomene* nor, especially, that of *Worlandia.* The heel of p3 is short and there is neither paraconid nor metaconid. Thus p3 is similar to that of *Mixodectes, Elpidophorus,* and *Eudaemonema.* It does resemble some omomyid teeth, but the crown is not blade-like.

The fourth lower premolar is a strong, semimolariform tooth in *Ekgmowechashala,* but is unlike any omomyid p4, even that of *Microchoerus.* It has been interpreted to lack a paraconid, so that the anterolingual cusp is called the metaconid (Szalay, 1976). The rest of the trigonid of p4 is made up of a large protoconid set labially but not at the tooth's edge, somewhat posterior to the level of the metaconid. On the posterior slope of the trigonid a very prominent metastylid occurs lingually. A large hypoconid is present posterolabially, well in from the labial border of the tooth, which is supplied with a broad posterolabial cingulum. Between the protoconid and the hypoconid, labial to the level of their apices, is a huge stylar cusp like that of *Tarka.* The heel of p4 is slightly tucked under the trigonid of m1.

The lower molars resemble those of *Tarka* and *Worlandia* generally, but not in all details. As in *Tarka, Plagiomene, Planetetherium,* and *Worlandia,* the molars progressively decrease in size significantly from m1 to m3. The paraconid of these teeth is completely suppressed, as in rodents and in the posterior molars of *Tarka.* That this suppression is convergent is suggested by the paraconids of m1 and m2 of *Tarkadectes.* In *Tarkadectes* the paraconid of m1 is still strong, but that of m2 is reduced and shifted toward the midline of the tooth.

The lower molars of *Ekgmowechashala* are rectangular in outline, especially m1 and m2, and lack the extra trigonid cusp displayed by *Tarka* between the protoconid and metaconid. In this regard, *Ekgmowechashala* is merely plesiomorphous, as are *Tarkadectes* and the plagiomenines. A stylar cusp is present behind the protoconid on all the lower molars, but is not large. As in *Tarka* and *Tarkadectes,* an enlarged metastylid has become a major cusp on the lingual side of the tooth, similar to the metastylid (= mesostylid) of certain rodents but not homologous with the somewhat similarly placed entostylid of *Plagiomene.* Among omomyids, *Shoshonius* and *Washakius* have a small metastylid, but in them it is closely appressed to the posterolingual base of the metaconid and is not involved in a system of small cuspules and crenulations on the anterior side of the talonid basin. Other omomyids do not even have these weak similarities to *Ekgmowechashala.* On m1 the talonid cusps are recognizable as such, but the distance from the hypoconid apex to that of the entoconid is only slightly greater than the distance from the protoconid apex to the metaconid apex, in contrast to omomyids and *Plagiomene.* The hypoconulid is present as a distinct, labially placed cuspule, not as a mere wrinkle in the posterior cingulum near the tooth's midline as in omomyids.

The second lower molar is offset lingually from m1. Its talonid basin is a confused mass of wrinkles, but the metastylid, hypoconid, and entoconid are discernible as such. The hypoconulid, while still adumbrated at its labial position seen in m1, is now a minor eminence on the posterior cingulum adjoining the hypoconid.

The third lower molar is also offset lingually somewhat. Homologies of its talonid cusps are doubtful, but the wide, rounded posterior wall contrasts with the pinched, narrow projection of most omomyids.

Rose and Rensberger (1983) described UCMP 128231 from middle Arikareean beds in the John Day Formation of Oregon (Fig. 10, top). They referred it to *Ekgmowechashala* or a closely related taxon (Rose and Rensberger, 1983, p. 108), refraining from assigning it to the type species, *E. philotau.* The specimen is reasonably (although not certainly) assigned to an ekgmowechashaline, despite not occurring in company with an ekgmowechashaline lower dentition, and has relatively low-crowned upper cheek-teeth with bulbous cusps, poorly developed crests, and crenulated enamel. These features apply as well to *Tarka, Planetetherium,* and *Worlandia* among other mammals with similar features (upper teeth are not yet known for *Tarkadectes* nor for *Ekgmowechashala philotau* from South Dakota). Rose and Rensberger also stated that P4 is submolariform, by which they meant that a metacone is present. However, this cusp can also be interpreted as a stylar cusp. The low cingulum girdling P4 is highly unusual if the animal is an omomyid, but would be in keeping with P4 of primitive plagiomenids. M1 was described by Rose and Rensberger as though its largest stylar cusp (as interpreted here) were the paracone, and as though the stylar cusp at the posterolabial corner of the tooth were the metacone. In the absence of knowledge about the peculiar stylar cusps of *Tarka,* such interpretations are natural and conservative. However, the cusps identified by Rose and Rensberger as the paraconule (protoconule) and metaconule seem to me to be plausibly interpretable in a more novel way as the true paracone and metacone, with the parastyle and large, central, labial, stylar cusp forming the usual plagiomenid "V"-pattern with the true paracone at its lingual (acute) angle. Rose and Rensberger did not mention a small swelling on the preprotocrista that might be a true paraconule, from which a short crest projects posterolabiad, nor a worn cusp between the hypocone and true metacone (their metaconule), that seems more likely to me to be a true metaconule. The hypocone is not a swollen "cingulum hypocone" as in many omomyids; rather, it is an expansion of the postprotocrista. M2 is significantly smaller than M1, a contrast with omomyids and primitive plagiomenids, and a similarity to advanced plagiomenids. The paracone and metacone of M2, if I have reidentified them correctly, are broken away on their labial sides. The hypocone is quite large, as in *Rooneyia,* but it seems to have formed in the same manner as that of M1, whereas that of *Rooneyia* is clearly cingulum-derived from something like the condition seen in *Washakius.* As in M1, the enamel is wrinkled and a cingulum surrounds the preserved part of the tooth except for a short distance at the posterior base of the hypocone. A disorganized area of enamel wrinkles occupies the region that I suggest is homologous with a true metaconule. Rose and Rensberger did not state whether an interdental wear facet exists as a result of prolonged contact with an M3, but M3 was presumably present and was smaller than M2 if the lower dentition is a guide. Whatever its taxonomic assignment, UCMP 128231 from the John Day of Oregon, like the lower dentition, can be interpreted to be less omomyid-like (and more plagiomenid-like) than Rose and Rensberger believed.

Thus, primarily on the basis of the lower dentition but secondarily because of a plausible reinterpretation of a referred upper dentition, a case can be made for removal of *Ekgmowechashala* from the Omomyidae and for association of *Ekgmowechashala* with a mammalian group to which it has not been compared previously: Plagiomenidae (sensu lato). Important similarities to *Plagiomene, Planetetherium, Tarkadectes, Tarka,* and *Worlandia* lead me to transfer *Ekgmowechashala* and the subfamily based on it, Ekgmowechashalinae, to the Plagiomenidae (Fig. 12; see also this figure for differential diagnoses within the Ekgmowechashalinae). It is no longer necessary to postulate cryptic omomyid ancestry for an odd late Oligocene genus from South Dakota and Oregon that still occurred after other primates were rare or extinct in the United States. However, if *Ekgmowechashala* is seen as plagiomenid, these occurrences in the northern half of the United States are in keeping with the occurrence of *Tarka* in upland Eocene deposits of northwestern Wyoming, of *Tarkadectes* in Chadronian deposits at the U.S.–Canadian border, of *Elpidophorus* in Montana and Canada, and of plagiomenids that in the Eocene ranged from northern Wyoming to the high Canadian arctic and possibly beyond.

FEEDING ADAPTATIONS

Because of the extreme hardness of the volcaniclastic matrix in which vertebrates are found at TTA Quarry, most specimens, including those of *Tarka stylifera,* have had to be carved from the matrix with steel needles or airbrasive techniques. Hard rocks are necessarily broken apart at the quarry to expose cross sections of contained specimens. At the moment of discovery, weak parts of fossils can be lost. For these reasons, the surfaces of teeth of prepared specimens from TTA Quarry are not often suitable for study of wear scratches incurred during life. No such studies are attempted here because wear scratches are surely frosted by airbrasive action or overlain by further damage that inevitably accompanied preparation with needles. Nevertheless, several things can be said about the mode of chewing in *Tarka stylifera.* If the symphysis of the lower jaw is oriented in its natural position in the vertical plane, the labial wear on the lower teeth is seen to be substantial, including even the labial stylar cusps, resulting in an occlusal plane that is ventrolabially to dorsolingually oriented rather than horizontal. The palate, therefore, was probably somewhat arched, so that the upper cheek-teeth faced partly inward as well as ventrad. Chewing was in part from side to side, as demonstrated by the wear surfaces on the hypoconids, but was also in part a pounding action, as shown by the apical wear on most of the other cusps of both the upper and lower molars. The apparent verticality of the ascending ramus is in harmony with this, because the bite force would be concentrated over the molars. The tooth questionably identified here as the first lower incisor was also heavily affected by apical wear, whatever its duties were. Little evidence of phase I wear facets occurs in known specimens of *Tarka stylifera* but AMNH 95733 has traces of them and they are easily seen on the type specimen of *Tarkadectes montanensis.*

The proliferation of cusps and crenulation of the cheek-teeth of ekgmowechashaline plagiomenids suggest a diet involving a substantial fraction of fruit (with contained insects; see Redford and others, 1984), nectar, gums, or sap. Possible analogues are to be seen in various rodents and in certain members of the early Cenozoic primate families Picrodontidae (Williams, 1985) and Omomyidae (Macdonald, 1963, 1970; Rose and Rensberger, 1983; Roth, 1985) or in the marsupial groups Glasbiinae, Caroloameghiniinae, and Polydolopidae (Clemens, 1966; Marshall, 1982a, 1982b). However, it must be admitted that dietary interpretation has a large speculative element in even the most information-rich paleontological investigations.

ENVIRONMENT IN WHICH EKGMOWECHASHALINE PLAGIOMENIDS LIVED

In the Bighorn Basin, *Plagiomene* is generally rare but locally common; *Worlandia* and *Planetetherium* are each abundant at one quarry only, but are otherwise exceedingly rare (Rose, 1981, and personal communication, 1988). All known specimens of *Tarka stylifera* are from a single site, TTA Quarry in the type section of the Tepee Trail Formation. These facts suggest (but do not prove) that these animals were adapted to some sort of special habitat or habitats.

The environmental conditions occupied by the ekgmowechashaline plagiomenids are a matter of some interest in view of the extraordinary dental features of these animals, but little is known. Late Oligocene *Ekgmowechashala* could have inhabited patches of forest from Oregon to what are now the western High Plains, but its occurrence in the early Arikareean of South Dakota is difficult to explain. So far as I am aware, the flora of the Sharps Formation of South Dakota, in which the type species occurs, is almost unknown, but the conditions of deposition are generally regarded on the basis of fossil soils as temperate and even arid and steppe-like, with trees restricted to stream-side areas (Retallack, 1983). The lack of enlarged incisors in *Ekgmowechashala* may mean that its diet was more general than one based in large part on the consumption of gums and saps, because enlarged incisors in animals with such diets are often used to break open tree bark. If *Ekgmowechashala* were merely frugivorous, a steady supply of fruit would have been necessary, but that requirement is difficult to reconcile with prevailing ideas of the environment of the High Plains in the latest Oligocene. On the other hand, Chadronian *Tarkadectes* is known from an intermontane basin, with extensive forests presumably nearby. The conditions of deposition of the upper sequence of the Coal Creek Member of the Kishenehn Formation, in which the single known specimen of *Tarkadectes montanensis* was found, have not been determined but are under study by K. N. Constenius.

Potentially, when paleobotanical work in the Tepee Trail Formation is studied in more detail, somewhat more can be said about the environment in which *Tarka stylifera* lived. The flora of the Tepee Trail Formation is not adequately known as yet, but both pollen and leaves occur. While making a comparison with the "lowland, humid, warm-temperature conditions" indicated by the flora of the "Early Basic Breccia" and Tatman formations of northwestern Wyoming, Dorf (1953) briefly mentioned that "a florule from the type locality of the Late Eocene Tepee Trail formation of the southern Absarokas shows similarity to the more temperate flora of the Green River formation of Colorado and Utah, confirming the suspicion that the latter, which differs from the type Green River flora, may also be of Late Eocene age." I assume that the florule came from the type section as well as from the type locality, but the exact level is not known to me. However, an unpublished pollen study by E. B. Leopold based on materials from the type section of the Tepee Trail Formation was mentioned by Leopold and MacGinitie (1972, p. 159). The level of these specimens is known. Twelve forms of pollen were recovered from this site (Fig. 11, H), about 4.6 m (15 ft) above the base of unit 19 of the type section of the Tepee Trail Formation (J. D. Love, personal communication, 1979). This pollen bearing site thus occurs close beneath TTA Quarry, from which all specimens of *Tarka stylifera* have been recovered. Unfortunately, no detailed study has been made of the florule represented. No other work seems to have been done on the fossil plants of the Tepee Trail Formation.

The nearby Kisinger Lakes flora (Rohrer, 1966) was originally thought to occur in the Tepee Trail Formation. For that reason it might be thought to yield approximate information about the climatic conditions at the level of TTA Quarry (provided that those conditions were essentially unchanged in whatever temporal gap separates them). However, later work by Rohrer (in MacGinitie, 1974) has shown that the Kisinger Lakes flora and the more basinward Tipperary flora (Berry, 1930) occur instead in Bridgerian rocks (paleomagnetic Chron 21) of the underlying Aycross Formation. The pertinence of both of these well-known floras is thus reduced. At best, it can be said that during at least part of Aycross time "tropical or near-tropical [conditions prevailed], with a pronounced winter dry season. The annual precipitation was between 35 and 55 inches. The average annual temperature was between 19 and 23 degrees C. The average temperature of the coldest month was not below 15 degrees C. The climate was either frostless, or, possibly, with rare, light frosts" (MacGinitie, 1974, p. 1).

The sediments of bone bed A, in which TTA Quarry was developed, are a hard, light blue, tuffaceous mudstone that weathers to a greenish-brown color at generally cliff-forming outcrops. The mudstone is without apparent internal stratification and was deposited at the top of an extensive cut-and-fill sequence in the type Tepee Trail section (Fig. 11, a). Immediately above bone bed A occurs a boulder conglomerate with individual boulders of up to more than a meter in diameter. These rocks were laid down in an upland setting, close to the scene of long-continued, violent volcanic activity. The cause of the concentration of vertebrate remains has not been fully determined, but the absence of gars, turtles, and crocodilians suggests that the site was not a substantial pond or oxbow of any long duration. Possibly the tuffaceous mudstone arrived as a slurry resulting from local sheet flooding, carrying with it the predominantly disarticulated and fragmented bones of vertebrates from nearby ash-covered slopes.

CONCLUSIONS

Two new genera and species of plagiomenids (Mammalia, ?Dermoptera) are described from North American middle Eocene and nominally lower Oligocene sediments. A third genus and species, *Ekgmowechashala philotau,* from upper Oligocene deposits of the northern United States, is transferred from the primate family Omomyidae to the Plagiomenidae. Previously, known undoubted plagiomenids were restricted to the Paleocene and early Eocene. All three newly recognized plagiomenids are placed in the Ekgmowechashalinae, the sister subfamily to the subfamily Plagiomeninae (new rank). The Plagiomenidae are usually considered to be allied to the living dermopterans of southeast Asia; however, analysis of that relationship is placed outside the scope of this chapter.

If correctly placed, these three newly recognized plagiomenids expand the known diversity of the Plagiomenidae (Fig. 12). The earliest of these probable folivores, frugivores, and nectar- and exudate-feeders, *Tarka stylifera,* occurs in the type section of the Tepee Trail Formation, early Uintan (Shoshonian: late medial Eocene, paleomagnetic Chron C20R, approximately

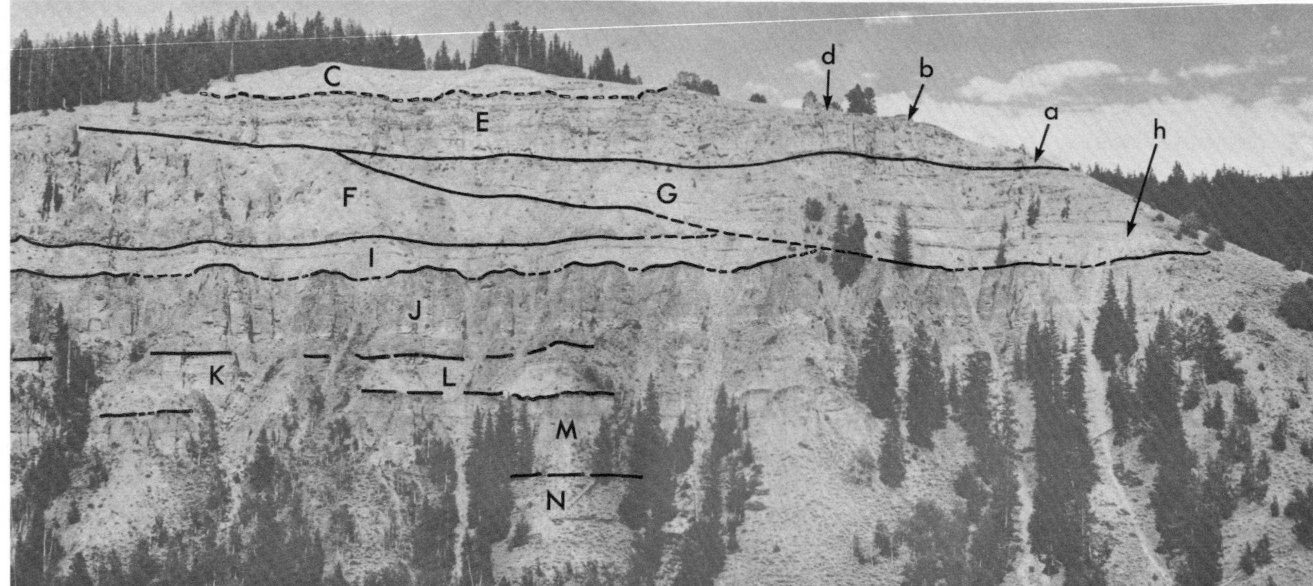

Figure 11. Top: panoramic view east across East Fork Canyon at part of type section of the Tepee Trail Formation. Bottom: interpretation. Site of vertebrate fossil localities in this view is 305 m (1,000 ft) west of east line, 91.4 m (300 ft) south of north line of Sect. 4,T.43N.,R.104W., East Fork Basin Quadrangle, Fremont County, Wyoming. Height of visible part of face is about 168 m (550 ft). a, TTA Quarry in bone bed A = unit 24 of type section (Love, 1939, p. 75); b, bone bed B in unit 27 of type section (Love, 1939); C, soft green and gray tuff; d, bone bed D (Love, 1939), 5.4 m (17.7 ft) above bone bed B, site of specimen of *Achaenodon* sp. identified as *Parahyus vagus* by Lewis (1973); E, hard green cliff-forming tuff; F, debris-flow conglomerate lens; G, carbonaceous shale and tuff; h, horizon of uranium-bearing (maximum uranium content 0.022 percent U) carbonaceous shale, about 4.6 m (15 ft) above base of unit 19 of type section (Love, 1939, p. 75) (spectrographic analysis indicates unusually high amounts of barium, copper, molybdenum, and vanadium; 12 forms of pollen also occur [J. D. Love, E. B. Leopold, personal communications, 1979]); I, soft, even-bedded sequence; J, hard, cliff-forming bedded tuff; K, light yellow soft tuff; L, light yellow tuff; M, hard, cliff-forming tuff; N, soft tuff. Composite of two photographs by J. D. Love, August 20, 1971.

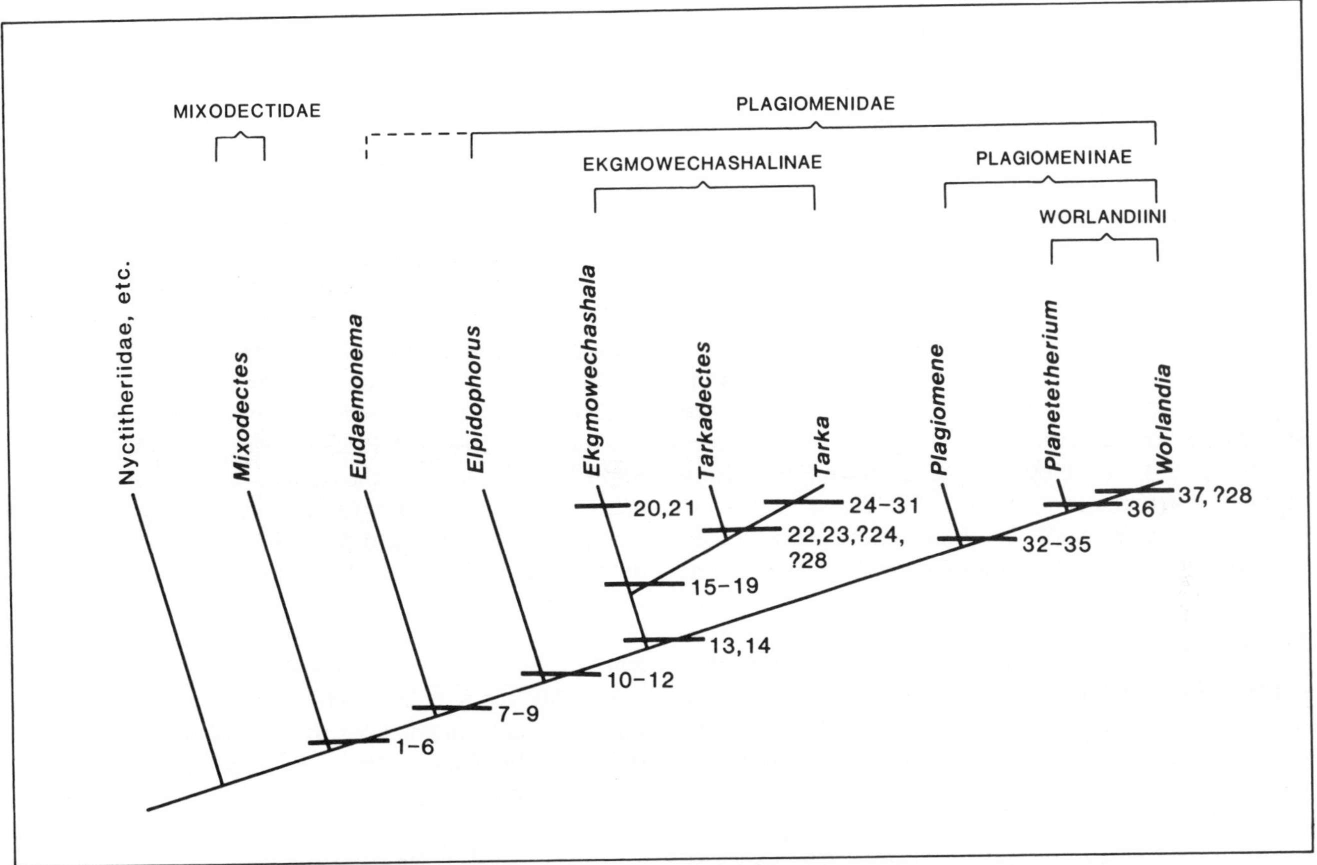

Figure 12. Cladogram depicting hypotheses of relationships of nine genera of plagiomenid and plagiomenid-like mammals from the Paleogene of North America. Living dermopterans, *Thylacaelurus, Placentidens,* picrodontids, microsyopids, bats, and primates are omitted deliberately, but the lipotyphlan family Nyctitheriidae is added as an unsupported outgroup (character polarities at the most plesiomorphous node of cladograms are always ambiguous in any case). The broader affinities of mixodectids, microsyopids, and plagiomenids are currently under study by others, based upon cranial and skeletal material. Hypothetically synapomorphous (shared-derived) characters at the nodes are as follows: 1, I1–2/il–2 large (one of which is later lost); 2, lower molars with low trigonids; 3, lower molars with strong anterolabial cingulum; 4, hypoconids of m1–m3 labially extended (later modified in *Ekgmowechashala*); 5, upper molars with 5 stylar cusps; 6, M1–M2 with hypocone (later to be lost; possibly, the hypocone was added earlier); 7, P3–P4 with metacone (later reversed in Ekgmowechashalinae); 8, P4/p4 with all molar cusps; 9, m3 talonid narrower; 10, hypocone reduced; 11, metaconid and entoconid of lower molars elevated with respect to protoconid and hypoconid; 12, paralophid (paracristid) of lower molars appressed to front of protolophid (protocristid), restricting trigonid basin; 13, lower molars with entostylid (entoconulid), crown pattern wrinkled; 14, molars decrease in size from M1/m1–M3/m3; 15, P4 loses metacone; 16, heel of p4 reduces; 17, hypertrophy of P4–M3 stylar cusps (unknown in *Tarkadectes*); 18, molar metastylids enlarge; 19, p4–m3 with small stylar cusp on cingulum, posterolabial to protoconid; 20 p4–m3 metastylids well separated from metaconids; 21, hypertrophy of p4 stylar cusp posterolabial to protoconid; 22, cusp row forms (apparently by modification of the anterior end of the cristid obliqua) from hypoconid to metastylid (lower molars); 23, heel of p4 tucked under trigonid of m1; 24, hypocone submerges in posterior protocone crest, loses identity; 25, p4–m2 stylar cusps posterolabial to protoconid hypertrophy; 26, m1 with stylar cusp at labial base of hypoconid; 27, size increases; 28, il enlarges at expense of other incisors; 29, extra cusp (or cusps) form between protoconid and metaconid on lower molars; 30, phase I chewing facets deemphasized; 31, cusp row from hypoconid to metastylid becomes shelf; 32, P3/p3 large; 33, P3/p3 nearly completely molarized; 34, P2 with metacone; 35, P4/p4 as large as or larger than M1/m1; 36, size of animal smaller; 37, P3/p3 completely molarized, with metacone well separated from paracone.

47.5 Ma) of northwestern Wyoming. A second, more primitive but later-occurring genus and species, *Tarkadectes montanensis,* is from a nominally early Oligocene level (Chadronian) in the Kishenehn Formation of northern Montana. All three ekgmowechashaline plagiomenid genera are somewhat primate-like but, in company with other plagiomenids, possibly have other affinities. Although related to the plagiomenid genera *Worlandia, Planetetherium, Plagiomene, Elpidophorus,* and *Eudaemonema,* the three known ekgmowechashalines lack the molarization of p3 and p4 seen in those genera. Presumably, the posterior premolars of *Ekgmowechashala, Tarka,* and *Tarkadectes* had become "demolarized" as the jaw was shortened and the posterior upper and lower cheek-teeeth were widened by the addition of strong labial stylar shelves and stylar cusps. The enamel of unworn cheek-teeth is wrinkled in *Tarkadectes* and *Ekgmowechashala* and is heavily crenulated in *Tarka.* Two lower incisors and a small premolar, believed to be p2, are absent in *Tarka,* whose remaining lower incisor is enlarged. The diet of *Tarka* and *Tarkadectes* probably emphasized fruits, insects, gums, and sap, but the lack of enlarged lower incisors in *Ekgmowechashala* may indicate that exudate feeding was not emphased in that genus. Judged from the prominent phase I chewing facets of its known cheek-teeth, *Tarkadectes* probably consumed a higher proportion of fibrous food than *Tarka.* Among plagiomenines, *Worlandia* is also somewhat primate-like in habitus. Both *Worlandia* and *Tarka* possess somewhat shortened jaws, suggesting a shoretened face, and both have an enlarged lower incisor. However, the most recent common ancestor of *Worlandia* and other plagiomenines with *Ekgmowechashala, Tarka,* and *Tarkadectes* must have lived in the Paleocene. Such an ancestor would have resembled nyctitheriids, *Mixodectes, Eudaemonema,* and, especially, *Elpidophorus.*

ACKNOWLEDGMENTS

J. D. Love introduced me to the Tepee Trail Formation and its unusual fauna more than 35 years ago. He has been of constant help ever since. He also took the photograph and provided the photographic negative and the geological interpretation from which Figure 11 was prepared. M. R. Dawson kindly permitted the description here of *Tarkadectes montanensis,* collected by K. N. Constenius. J. Shumsky prepared and cast it. O. Simonis performed the difficult work of preparation of specimens of *Tarka stylifera* from refractory matrix from TTA Quarry and C. Tarka made the illustrations of both *Tarka* and *Tarkadectes.* I also thank L. Meeker, C. Tarka, E. Heck, and J. M. Winsch for help with drafting and for photographic enhancement of the view of the type section of the Tepee Trail Formation depicted in Figure 11. F. S. Szalay and K. D. Rose provided photographic prints of their illustrations of *Ekgmowechashala.* T. M. Bown, M. R. Dawson, E. R. Dumont, J. G. Eaton, J. J., Flynn, H. Galiano, P. D. Gingerich, L. Krishtalka, R.D.E. MacPhee, E. Manning, M. J. Novacek, K. D. Rose, P. N. Shive, K. A. Sundell, F. S. Szalay, and A. R. Wyss offered helpful comments, although one or two of them grumbled about transferring *Ekgmowechashala* from the Omomyidae to the Plagiomenidae. Numerous graduate students, American Museum of Natural History employees, members of my family, and volunteers over the years have participated in the collection of many of the specimens described here. I warmly thank all these generous people.

REFERENCES CITED

Berggren, W. A., Kent, D. V., and Flynn, J. J., 1985, Paleogene geochronology and chronostratigraphy, *in* Snelling, N. J., ed., The chronology of the geological record: Geological Society of London Memoir 10, p. 141–195.

Berggren, W. A., McKenna, M. C., Hardenbol, J., and Obradovich, J. D., 1978, Revised Paleogene polarity time scale: Journal of Geology, v. 86, p. 67–81.

Berry, E. W., 1930, A flora of Green River age in the Wind River Basin of Wyoming: U.S. Geological Survey Professional Paper 165, p. 55–81.

Bown, T. M., and Rose, K. D., 1979, *Mimoperadectes,* a new marsupial, and *Worlandia,* a new dermopteran, from the lower part of the Willwood Formation (early Eocene), Bighorn Basin, Wyoming: Ann Arbor, University of Michigan, Contributions of the Museum of Paleontology, v. 25, no. 4, p. 89–104.

Butler, P. M., 1972, The problem of insectivore classification, *in* Joysey, K. A., and Kemp, T. S., eds., Studies in vertebrate evolution: New York, Winchester Press, p. 253–265.

Clemens, W. A., Jr., 1966, Fossil mammals of the type Lance Formation, Wyoming; Part 2, Marsupialia: University of California Publications in Geological Sciences, v. 62, p. 1–122.

Constenius, K. N., and Dyni, J. R., 1983, Lacustrine oil shales and stratigraphy of part of the Kishenehn Basin, northwestern Montana: Golden, Colorado School of Mines Mineral and Energy Resources, v. 26, no. 4, p. 1–16.

Dorf, E., 1953, Succession of Eocene floras in northwestern Wyoming [abs.]: Geological Society of America Bulletin, v. 64, no. 12, pt. 2, p. 1413.

Flynn, J. J., 1986, Correlation and geochronology of middle Eocene strata from the western United States: Palaeogeography, Palaeoclimatology, Palaeoecology , v. 55, p. 335–406.

Flynn, J. J., and Galiano, H., 1982, Phylogeny of early Tertiary Carnivora, with a description of a new series of *Protictis* from the middle Eocene of northwestern Wyoming: American Museum Novitates, no. 2725, p. 1–64.

Golz, D. J., 1976, Eocene Artiodactyla of southern California: Natural History Museum of Los Angeles County Science Bulletin, no. 26, p. 1–85.

Hunt, R. M., Jr., and Korth, W. W., 1980, The auditory region of Dermoptera; Morphology and function relative to other living mammals: Journal of Morphology, v. 164, p. 167–211.

Kay, R. F., and Hiiemae, K. M., 1974, Jaw movement and tooth use in Recent and fossil primates: American Journal of Physical Anthropology, v. 40, no. 2, p. 227–256.

Kay, R. F., and Hylander, W. L., 1978, The dental structure of mammalian folivores with special reference to Primates and Phalangeroidea (Marsupialia), *in* Montgomery, G. G., ed., The ecology of arboreal folivores: Washington, D.C., Smithsonian Institution Press, p. 173–196.

Krishtalka, L., 1976, North American Nyctitheriidae (Mammalia, Insectivora): Carnegie Museum Annals, v. 46, art. 2, p. 7–28.

Krishtalka, L., and Setoguchi, T., 1977, Paleontology and geology of the Badwater Creek area, central Wyoming; Part 13, The late Eocene Insectivora and Dermoptera: Carnegie Museum Annals, v. 46, art. 7, p. 71–99.

Leopold, E. B., and MacGinitie, H. D., 1972, Development and affinities of Tertiary floras in the Rocky Mountains, *in* Graham, A., ed., Floristics and paleofloristics of Asia and eastern North America: Amsterdam, Elsevier, p. 147–200.

Lewis, G. E., 1973, A second specimen of *Parahyus vagus* Marsh, 1876: U.S. Geological Survey Journal of Research, v. 1, no. 2, p. 147–149.

Love, J. D., 1939, Geology along the southern margin of the Absaroka Range, Wyoming: Geological Society of America Special Paper 20, 134 p.

Macdonald, J. R., 1963, The Miocene faunas from the Wounded Knee area of western South Dakota: American Museum of Natural History Bulletin, v. 125, p. 139–238.

——, 1970, Review of the Miocene Wounded Knee faunas of southwestern South Dakota: Los Angeles County Museum of Natural History Science Bulletin, v. 8, p. 1–82.

MacFadden, B. J., 1980, Eocene perissodactyls from the type section of the Tepee Trail Formation of northwestern Wyoming: University of Wyoming Contributions to Geology, v. 18, no. 2, p. 135–143.

MacGinitie, H. D., 1974, An early middle Eocene flora from the Yellowstone-Absaroka volcanic province, northwestern Wind River Basin, Wyoming: University of California Publications in Geological Sciences, v. 108, p. 1–103, with chapters by E. B. Leopold and W. L. Rohrer.

Marshall, L. G., 1982a, A new genus of Caroloameghiniinae (Marsupialia: Didelphoidea: Didelphidae) from the Paleocene of Brazil: Journal of Mammalogy, v. 63, no. 4, p. 709–716.

——, 1982b, Systematics of the extinct South American marsupial family Polydolopidae: Fieldiana Geology, new series, no. 12, p. 1–109.

Matthew, W. D., 1915, Part 4, Entelonychia, Primates, Insectivora (part), *in* Matthew, W. D., and Granger, W., A revision of the lower Eocene Wasatch and Wind River faunas: American Museum of Natural History Bulletin, v. 34, art. 14, p. 429–483.

——, 1918, Part 5, Insectivora (continued), Glires, Edentata, *in* Matthew, W. D., and Granger, W., eds., A revision of the lower Eocene Wasatch and Wind River faunas: American Museum of Natural History Bulletin, v. 38, p. 565–657.

McKenna, M. C., 1980, Late Cretaceous and early Tertiary vertebrate paleontological reconnaissance, Togwotee Pass area, northwestern Wyoming, *in* Jacobs, L. L., ed., Aspects of vertebrate history: Flagstaff, Museum of Northern Arizona Press, p. 321–343.

Novacek, M. J., 1977, A review of Paleocene and Eocene Leptictidae (Eutheria: Mammalia) from North America: PaleoBios, no. 24, p. 1–42.

——, 1980, Cranioskeletal features in tupaiids and selected Eutheria as phylogenetic evidence, *in* Luckett, W. P., ed., Comparative biology and evolutionary relationships of tree shrews: New York, Plenum Press, v. 2, p. 35–93.

Ostrander, G. E., 1985, Correlation of the early Oligocene (Chadronian) in northwestern Nebraska, *in* Martin, J. E., ed., Fossiliferous Cenozoic deposits of western South Dakota and northwestern Nebraska: Rapid City, South Dakota School of Mines and Technology, Dakoterra, v. 2, part 2, p. 205–231.

Redford, K. H., da Fonseca, G.A.B., and Lacher, T. E., Jr., 1984, The relationship between frugivory and insectivory in Primates: Primates, v. 25, no. 4, p. 433–440.

Retallack, G. J., 1983, Late Eocene and Oligocene paleosols from Badlands National Park, South Dakota: Geological Society of America Special Paper 193, 82 p.

Rohrer, W. L., 1966, Geologic quadrangle map, Kisinger Lakes Quadrangle, Fremont County, Wyoming: U.S. Geological Survey Geologic Quadrangle Map GQ-724, scale 1:24,000.

Romer, A. S., 1966, Vertebrate paleontology: Chicago, Illinois, University of Chicago Press, 3rd ed., 468 p.

——, 1968, Notes and comments on vertebrate paleontology: Chicago, Illinois, University of Chicago Press, 304 p.

Rose, K. D., 1973, The mandibular dentition of *Plagiomene* (Dermoptera, Plagiomenidae): Breviora, no. 411, p. 1–17.

——, 1975, *Elpidophorus,* the earliest dermopteran (Dermoptera, Plagiomenidae): Journal of Mammalogy, v. 56, no. 3, p. 676–679.

——, 1978, A new Paleocene epoicotheriid (Mammalia), with comments on the Palaeanodonta: Journal of Paleontology, v. 52, no. 3, p. 658–674.

——, 1981, The Clarkforkian Land-Mammal Age and mammalian composition across the Paleocene–Eocene Boundary: Ann Arbor, University of Michigan Papers on Paleontology, no. 26, p. 1–197.

——, 1982, Anterior dentition of the early Eocene plagiomenid dermopteran *Worlandia:* Journal of Mammalogy, v. 63, no. 1, p. 179–183.

Rose, K. D., and Gingerich, P. D., 1987, A new insectivore from the Clarkforkian (earliest Eocene) of Wyoming: Journal of Mammalogy, v. 68, no. 1, p. 17–27.

Rose, K. D., and Rensberger, J. M., 1983, Upper dentition of *Ekgmowechashala* (omomyid primate) from the John Day Formation, Oligo–Miocene of Oregon: Folia Primatologica, v. 41, p. 102–111.

Rose, K. D., and Simons, E. L., 1977, Dental function in the Plagiomenidae; Origin and relationships of the mammalian order Dermoptera: Ann Arbor, University of Michigan Contributions of the Museum of Paleontology, v. 24, no. 20, p. 221–236.

Roth, C., 1985, Kauzyklus und Usurfacetten von *Microchoerus* Wood, 1844 (Omomyiformes, Primates): Mainzer geowissenschaftliche Mitteilungen, v. 14, p. 287–306.

Russell, D. E., Louis, P., and Savage, D. E., 1973, Chiroptera and Dermoptera of the French early Eocene: University of California Publications in Geological Sciences, v. 95, p. 1–57.

Russell, L. S., 1954, Mammalian fauna of the Kishenehn Formation, southeastern British Columbia: Annual Report of the National Museum of Canada for 1952–1953, Bulletin no. 132, p. 92–111.

Schwartz, J. H., and Krishtalka, L., 1976, The lower antemolar teeth of *Litolestes ignotus,* a late Paleocene erinaceid (Mammalia, Insectivora): Annals of the Carnegie Museum, v. 46, art. 1, p. 1–6.

Sigé, B., 1976, Insectivores primitifs de l'Éocène supérieur et Oligocène inférieur d'Europe occidentale. Nyctithériidés: Mémoires du Muséum National d'Histoire Naturelle, Nouvelle Série, Série C, Sciences de la Terre, v. 34, no. 1, p. 1–140.

Simpson, G. G., 1927, Mammalian fauna and correlation of the Paskapoo Formation of Alberta: American Museum Novitates, no. 268, p. 1–10.

——, 1928, A new mammalian fauna from the Fort Union of southern Montana: American Museum Novitates, no. 297, p. 1–15.

——, 1929a, A collection of Paleocene mammals from Bear Creek, Montana: Annals of the Carnegie Museum, v. 19, no. 2, p. 115–122.

——, 1929b, Third contribution to the Fort Union fauna at Bear Creek, Montana: American Museum Novitates, no. 345, p. 1–12.

——, 1935, New Paleocene mammals from the Fort Union of Montana: Proceedings of the U.S. National Museum, v. 83, no. 2981, p. 221–224.

——, 1936, A new fauna from the Fort Union of Montana: American Museum Novitates, no. 873, p. 1–27.

——, 1937, The Fort Union of the Crazy Mountain Field, Montana, and its mammalian faunas: U.S. National Museum Bulletin, v. 169, p. 1–287.

——, 1945, The principles of classification and a classificaiton of mammals: American Museum of Natural History Bulletin, v. 85, p. 1–350.

Sloan, R. E., 1969, Cretaceous and Paleocene terrestrial communities of western North America: North American Paleontological Convention Proceedings, Sept., 1969, pt. E, p. 427–453.

Storer, J. E., 1984, Mammals of the Swift Current Creek Local Fauna (Eocene: Uintan), Saskatchewan: Saskatchewan Culture and Recreation Department, Museum of Natural History Natural History Contributions, no. 7, p. 1–158.

Szalay, F. S., 1969, Mixodectidae, Microsyopidae, and the insectivore-primate transition: American Museum of Natural History Bulletin, v. 140, art. 4, p. 193–330.

——, 1976, Systematics of the Omomyidae (Tarsiiformes, Primates), taxonomy, phylogeny, and adaptations: American Museum of Natural History Bulletin, v. 156, art. 3, p. 157–450.

Szalay, F. S., and Delson, E., 1979, Evolutionary history of the Primates: New York, Academic Press, 580 p.

Szalay, F. S., and Drawhorn, G., 1980, Evolution and diversification of the Archonta in an arboreal milieu, *in* Luckett, W. P., ed., Comparative biology and evolutionary relationships of tree shrews: New York, Plenum Press, p. 35–93.

Van Valen, L., 1965, Paroxyclaenidae, an extinct family of Eurasian mammals: Journal of Mammalogy, v. 46, no. 3, p. 388–397.

——, 1967, New Paleocene insectivores and insectivore classification: American Museum of Natural History Bulletin, v. 135, art. 5, p. 217–284.

——, 1979, The evolution of bats: Evolutionary Theory, v. 4, p. 103–121.

West, R. M., and Dawson, M. R., 1978, Vertebrate paleontology and the Cenozoic history of the North Atlantic region: Polarforschung, v. 48, no. 1/2, p. 103–119.

West, R. M., Dawson, M. R., and Hutchison, J. H., 1977, Fossils from the Paleogene Eureka Sound Formation, N.W.T., Canada; Occurrence, climatic, and paleogeographic implications, *in* West, R. M., ed., Paleontology and plate tectonics: Milwaukee, Wisconsin, Milwaukee Public Museum Special Publications in Biology and Geology, no. 2, p. 77–93.

Williams, J. A., 1985, Morphology and variation in the posterior dentition of *Picrodus silberlingi* (Picrodontidae): Folia Primatologica, v. 45, p. 48–58.

Wilson, J. A., 1966, A new primate from the Oligocene, west Texas, preliminary report: Folia Primatologica, v. 4, p. 227–248.

MANUSCRIPT ACCEPTED BY THE SOCIETY JUNE 12, 1989

Printed in U.S.A.

Index

[Italic page numbers indicate major references]

Typeset by WESType Publishing Services, Inc., Boulder, Colorado
Printed in U.S.A. by JL Printing, Loveland, Colorado